Marine Science: An Ecological Approach

Marine Science: An Ecological Approach

Editor: Simon Oakenfold

RCALLISTO
REFERENCE

www.callistoreference.com

Callisto Reference,
118-35 Queens Blvd., Suite 400,
Forest Hills, NY 11375, USA

Visit us on the World Wide Web at:
www.callistoreference.com

ISBN: 978-1-63239-996-0 (Hardback)

Cataloging-in-Publication Data

Marine science : an ecological approach / edited by Simon Oakenfold.
   p. cm.
Includes bibliographical references and index.
ISBN 978-1-63239-996-0
1. Marine ecology. 2. Marine sciences. I. Oakenfold, Simon.
QH541.5.S3 M37 2018
577.7--dc23

Table of Contents

Permissions

List of Contributors

Index

Preface

This book aims to highlight the current researches and provides a platform to further the scope of innovations in this area. This book is a product of the combined efforts of many researchers and scientists, after going through thorough studies and analysis from different parts of the world. The objective of this book is to provide the readers with the latest information of the field.

Marine science is the study of the flora and fauna existing in the oceans. It also studies the physical as well as chemical properties of the ocean. It is an inter disciplinary field and branches out into sub-fields such as marine biology, physical oceanography, marine geology and many others. The objective of this book is to give a general view of the different areas of marine science, and its applications. It is a vital tool for all researching or studying this field as it gives incredible insights into emerging trends and concepts.

I would like to express my sincere thanks to the authors for their dedicated efforts in the completion of this book. I acknowledge the efforts of the publisher for providing constant support. Lastly, I would like to thank my family for their support in all academic endeavors.

Editor

Ecosystem Scale Acoustic Sensing Reveals Humpback Whale Behavior Synchronous with Herring Spawning Processes and Re-Evaluation Finds No Effect of Sonar on Humpback Song Occurrence in the Gulf of Maine in Fall 2006

Zheng Gong[1¤], **Ankita D. Jain**[2], **Duong Tran**[1], **Dong Hoon Yi**[2], **Fan Wu**[1], **Alexander Zorn**[1], **Purnima Ratilal**[1], **Nicholas C. Makris**[2*]

1 Department of Electrical and Computer Engineering, Northeastern University, Boston, Massachusetts, United States of America, **2** Department of Mechanical Engineering, Massachusetts Institute of Technology, Cambridge, Massachusetts, United States of America

Abstract

We show that humpback-whale vocalization behavior is synchronous with peak annual Atlantic herring spawning processes in the Gulf of Maine. With a passive, wide-aperture, densely-sampled, coherent hydrophone array towed north of Georges Bank in a Fall 2006 Ocean Acoustic Waveguide Remote Sensing (OAWRS) experiment, vocalizing whales could be instantaneously detected and localized over most of the Gulf of Maine ecosystem in a roughly 400-km diameter area by introducing array gain, of 18 dB, orders of magnitude higher than previously available in acoustic whale sensing. With humpback-whale vocalizations consistently recorded at roughly 2000/day, we show that vocalizing humpbacks (i) were overwhelmingly distributed along the northern flank of Georges Bank, coinciding with the peak spawning time and location of Atlantic herring, and (ii) their overall vocalization behavior was strongly diurnal, synchronous with the formation of large nocturnal herring shoals, with a call rate roughly ten-times higher at night than during the day. Humpback-whale vocalizations were comprised of (1) highly diurnal non-song calls, suited to hunting and feeding behavior, and (2) songs, which had constant occurrence rate over a diurnal cycle, invariant to diurnal herring shoaling. Before and during OAWRS survey transmissions: (a) no vocalizing whales were found at Stellwagen Bank, which had negligible herring populations, and (b) a constant humpback-whale song occurrence rate indicates the transmissions had no effect on humpback song. These measurements contradict the conclusions of Risch et al. Our analysis indicates that (a) the song occurrence variation reported in Risch et al. is consistent with natural causes other than sonar, (b) the reducing change in song reported in Risch et al. occurred days before the sonar survey began, and (c) the Risch et al. method lacks the statistical significance to draw the conclusions of Risch et al. because it has a 98–100% false-positive rate and lacks any true-positive confirmation.

Editor: Z. Daniel Deng, Pacific Northwest National Laboratory, United States of America

Funding: This research was supported by the National Oceanographic Partnership Program, the Census of Marine Life, the Office of Naval Research, the Alfred P. Sloan Foundation, the National Science Foundation, the Presidential Early Career Award for Scientists and Engineers, Northeastern University, and Massachusetts Institute of Technology. The authors thank David Reed for providing technical assistance. The funders had no role in study design, data collection and analysis, decision to publish, or preparation of the manuscript.

Competing Interests: Two of the co-authors of this manuscript are inventors of the patent US20060280030 ('Continuous, continental shelf-scale monitoring of fish populations and behavior') which is owned by MIT and was discovered under US Government Research Sponsorship, giving the US Government certain rights with regard to this patent. This patent involves ocean acoustic waveguide remote sensing of fish populations.

* Email: makris@mit.edu

¤ Current address: Department of Mechanical Engineering, Massachusetts Institute of Technology, Cambridge, Massachusetts, United States of America

Introduction

Passive acoustic survey methods employing hydrophones at fixed locations [1–15] or mobile platforms [16,17] have been widely used to detect, localize, track and study the behavior [1–9,13–15] and abundance [4,10–12] of whales. With our array situated on the northern flank of Georges Bank from September 19 to October 6, 2006 [18,19], we could detect and localize vocalizing whales over most of the Gulf of Maine, a roughly 400-km diameter area, including Georges and Stellwagen Banks, and

so monitor vocalization behavior over an ecosystem scale. This was possible because we used a large-aperture, densely-sampled, coherent hydrophone array with orders of magnitude higher array gain [20–25] than previously available in acoustic whale sensing. We detected roughly 2000 humpback whale vocalizations per day and used these to determine the corresponding whale locations over time by introducing a synthetic aperture tracking technique [26–29] and the array invariant method [30] to the whale sensing problem.

We find that the distribution of the vast majority of vocalizing humpback whales coincided with the primary time and location of Atlantic herring during their peak annual spawning period. During daylight hours, herring were found to be dispersed on the seafloor in deeper waters over wide areas of Georges Bank's northern flank [18]. At sunset, they would then rise and converge to form dense and massive evening shoals, which migrated to the shallow waters of Georges Bank for spawning, following a regular diurnal pattern [18]. We find the humpback whale vocalization behavior followed a similarly strong diurnal pattern, temporally and spatially synchronous with the herring shoal formation process, with vocalization rates roughly ten times higher at night than during daylight hours. At night, most humpback whale vocalizations originated from concentrated regions with dense evening herring shoals, while during daytime, their origins were more widely distributed over areas with significant but diffuse pre-shoal herring populations. These vocalizations are comprised of: (i) non-song calls, dominated by repetitive downsweep "meows" (approximately 1.44 second duration, 452 Hz center frequency, 170 Hz bandwidth, and 31 second repetition rate) which apparently have not been previously observed; and (ii) songs [2]. The repetitive non-song calls were highly diurnal and synchronous with the herring shoal formation process, consistent with hunting and feeding behavior. In contrast, songs occurred at a constant rate with no diurnal variation, and are apparently unrelated to feeding and the highly diurnal herring spawning activities.

Before and during Ocean Acoustic Waveguide Remote Sensing (OAWRS) survey transmissions [18,19], we measured constant humpback whale song occurrence, indicating these transmissions had no effect on humpback whale song. In addition, our data shows no humpback whale vocal activity originating from Stellwagen Bank, which had negligible herring populations [31,32], but vocalizing humpbacks located near Georges Bank, which had dense and decadally high herring populations [31], could be heard at Stellwagen Bank. These results are consistent with previous observations of humpback whale feeding activity in the Gulf of Maine and Stellwagen Bank which show humpback whales leave Stellwagen Bank for other regions plentiful in herring for feeding during the herring spawning season [33]. These results, however, contradict the conclusions of Risch et al. [34]. To investigate this contradiction, the Risch et al. statistical test [34] is applied to the annual humpback whale song occurrence time series reported from single sensor detections at Stellwagen Bank in time dependent ambient noise published by Vu et al. [35] and shown to false-positively find that humpback whales react to sonar 98–100% of the time over a yearly period when no sonars are present. A simple explanation for this severe statistical bias [36,37] is found upon inspection of the Vu et al. [35] multi-annual humpback whale song occurrence time series. The reported time series [35] have (i) inconsistencies in trend, (ii) large differences in song occurrence, and (iii) random correlation between years when no sonar is present. This shows that 98–100% of the time, the approach used in Risch et al. [34] mistakes natural variations in song occurrence for changes caused by sonar when no sonar is present. When the Risch et al. statistical test [34] is applied to the same humpback whale song occurrence data reported in Risch et al. [34] for 2008 and 2009, it false-positively finds humpback whales respond to sonar 100% of the time when no sonar is present. With the 98–100% false positive rate and the lack of any true positive confirmation for the Risch et al. statistical approach [34], the analysis of Risch et al. [34] lacks the statistical significance to draw the conclusions found in Risch et al. [34]. The fact that the reported reducing change in humpback whale song occurrence, to zero [34,35], occurred while the OAWRS

vessels were docked on the other side of Cape Cod from Stellwagen Bank, at the Woods Hole Oceanographic Institution, due to severe winds, days before OAWRS transmissions for active surveying began on September 26, 2006, yet no other explanation for this reduction than sonar is provided in Risch et al. [34], is consistent with a violation of temporal causality in the Risch et al. [34] study. Our data analysis indicates that the change in humpback whale song occurrence Risch et al. [34] reported is consistent with wind-dependent noise [20,23,38,39] limiting the single-hydrophone measurements of Risch et al. [34] to a small wind-speed-dependent fraction of the singing humpback whales and songs detected by our densely sampled, large aperture, coherent array. These findings are all consistent with the constant humpback whale song occurrence rates before and during OAWRS survey transmissions found with our wide-area towed array measurements.

Results and Discussion

2.1 Humpback whale behavior is synchronous with herring spawning processes during the peak annual Atlantic herring spawning period in the Gulf of Maine

Vocalizing humpback whales and spawning herring populations [18,19] were simultaneously localized and imaged over thousands of square kilometers during the peak annual spawning period of Atlantic herring in the Gulf of Maine by instantaneous passive and active OAWRS [18,40,41] techniques respectively in the Fall of 2006. We find humpback whale behavior in the Gulf of Maine to be highly coupled to peak herring spawning activities, which last for roughly one week but whose inception can vary [42,43] by many weeks from year to year. This coupled humpback whale and herring behavior occurs over too short a period to be accurately resolved by available seasonal, yearly or decadal averages [44,45], but can be well resolved by OAWRS methods. The high array gain [20–25] of the densely sampled large aperture coherent OAWRS passive receiver array used here enables detection of whale vocalizations either two orders of magnitude more distant in range or lower in signal-to-noise ratio (SNR) than a single hydrophone (Sections 3.1 and 3.5), which has no array gain. The array used here has 160 hydrophones with 4 nested 64-hydrophone subapertures. We determined whale bearings by beamforming and ranges by applying the instantaneous array invariant method [30] and synthetic aperture tracking techniques [26–29,46] to the whale sensing problem, leading to the spatial distribution of humpback whale call rate density shown in Figures 1, 2 and 3 over the period from September 22 to October 6, 2006, which coincided exactly with the peak annual herring spawning period [42]. Humpbacks are identified based on presence of song, as well as appropriate frequency content, duration, signature and repetition rate of calls.

We find that the vast majority of vocalizing humpback whales were spatially distributed in regions coinciding with the primary aggregations of spawning herring during the peak annual herring spawning period [18,42] in the Gulf of Maine (Figure 1). During this period, spawning herring populations instantaneously imaged by the active OAWRS system were found to regularly form massive dense shoals during evening hours along the northern flank of Georges Bank between water depths of 50 m and 200 m, which constituted the favorable shoal formation areas [18,19] (Figure 1). Water depths of 160 to 200 m were favored by spawning herring to form dense and massive evening shoals (> 0.20 fish/m^2), before migration to shallower water ($\approx$50 m) spawning grounds on Georges Bank [18]. The more diffusely scattered herring populations with lower areal population density

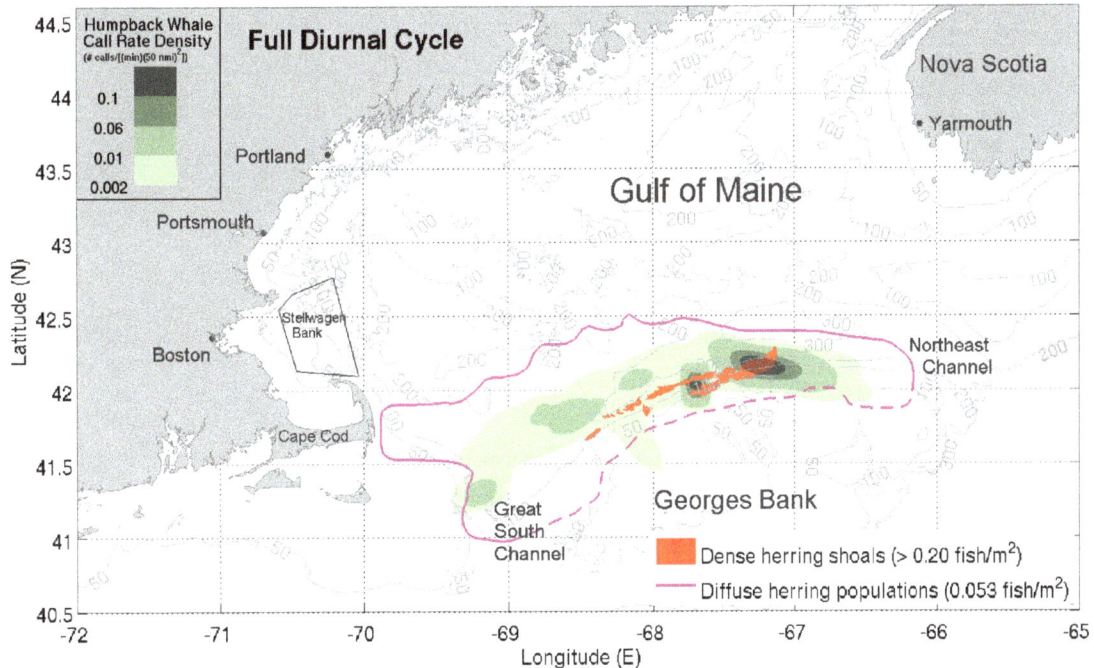

Figure 1. Distributions of vocalizing humpback whales and spawning herring populations in Fall 2006. Spatial distribution of vocalizing humpback whales coincides with the time and location of spawning Atlantic herring distributions in Fall 2006. Humpback whale vocalizations are found to be distributed along the northern flank of Georges Bank, coinciding with dense herring shoals (>0.20 fish/m², red shaded areas) imaged using active OAWRS system [18] and diffuse herring populations (≈0.053 fish/m², bounded by magenta line) obtained from conventional fish finding sonar (CFFS) line-transect data from NEFSC Annual Fall Herring Surveys [18,63]. The green shaded areas indicate the overall humpback whale call rate densities (number of calls/[(min) (50 nmi)²]) measured with our large aperture array. All data represent means between September 22 and October 6, 2006. The dashed magenta line represents the southern bound of the NEFSC survey tracks [18,63]. The black trapezoid indicates Stellwagen Bank [158].

(≈0.053 fish/m²) were found to be widely distributed between water depths of 50 m and 300 m, which include dense shoal formation areas [18], by concurrent Northeast Fisheries Science Center (NEFSC) line-transect ultrasound and trawl surveys [47], as shown in Figures 1, 2, and 3. At night, most vocalizing humpback whales were also found to be concentrated within water depths of 50 m to 300 m, in close proximity to the dense evening herring shoals (Figure 3). During daytime, vocalizing humpbacks were widely distributed within regions containing the more diffuse pre-shoal herring populations on the northern flank of Georges Bank and the Great South Channel (Figure 2). The observed high spatial correlation between the distribution of vocalizing humpback whales and the primary spawning herring populations in the Gulf of Maine is consistent with a mass feeding of humpback whales on herring that is synchronized with the peak herring spawning processes.

We find humpback whale vocalization behavior follows a strong diurnal pattern that is temporally synchronous with the regular herring shoal formation process [18]. The diurnal pattern is quantified by vocalization rates roughly ten times higher at night than during daylight hours (Figure 4(A)). The synchronization is quantified by a high correlation (0.82 at 0–15 minute time lag in Figure 4(B)) between time series of spawning herring shoal population density and humpback whale call rate (Figure 4(A)).

The mechanisms behind the observed synchronized diurnal pattern between humpback whales and spawning herring can be understood by examining the shoal formation process. In daytime, the herring are more widely distributed within thin layers roughly 5 m from the seafloor (on average 0.053 fish/m²) in deeper waters on the northern flank of Georges Bank (Figure 2A of Ref. [18]).

Near sunset local convergences of population density reach a critical threshold of 0.2 fish/m² after which coherent shoal formation waves appear (Figures 1 to 3 of Ref. [18]) and areal population density drastically increases at a rate of roughly 5 fish/m² per hour (Figure 3 of Ref. [18]) to form dense and massive shoals. Shoal formation in deeper waters after dusk allows herring spawning activities to proceed under the cover of darkness with reduced risk of predator attack [48,49]. The resulting roughly 50-fold increase in the areal population density of herring shoals, triggered by reduction in light levels, is closely followed (within 15 minutes) by a sudden order of magnitude increase in humpback whale call rate, as shown in Figure 4. The corresponding spatial focusing of vocalizing humpback whales from regions containing the overall dispersed herring populations in the day to those with dense shoals at night has been shown in Figures 2 to 3. Evening humpback whale vocalization rates remain high during the subsequent migration of herring shoals toward shallower spawning grounds on Georges Bank [18], and throughout the night until herring shoals dissipate as light levels increase at sunrise [18] (Figure 4). These findings are consistent with a feeding-behavior cause for the elevated humpback whale nocturnal vocalization rates and spatial focusing on dense shoals. The findings of vocal humpback whales exclusively in the vicinity of large spawning herring aggregates during the peak annual herring spawning period, and diurnal vocalization rates synchronized with diurnal herring spawning processes, also provide substantial evidence in favor of the theory that humpback whales leave areas with negligible herring populations, and migrate to primary herring spawning grounds in the Gulf of Maine where large

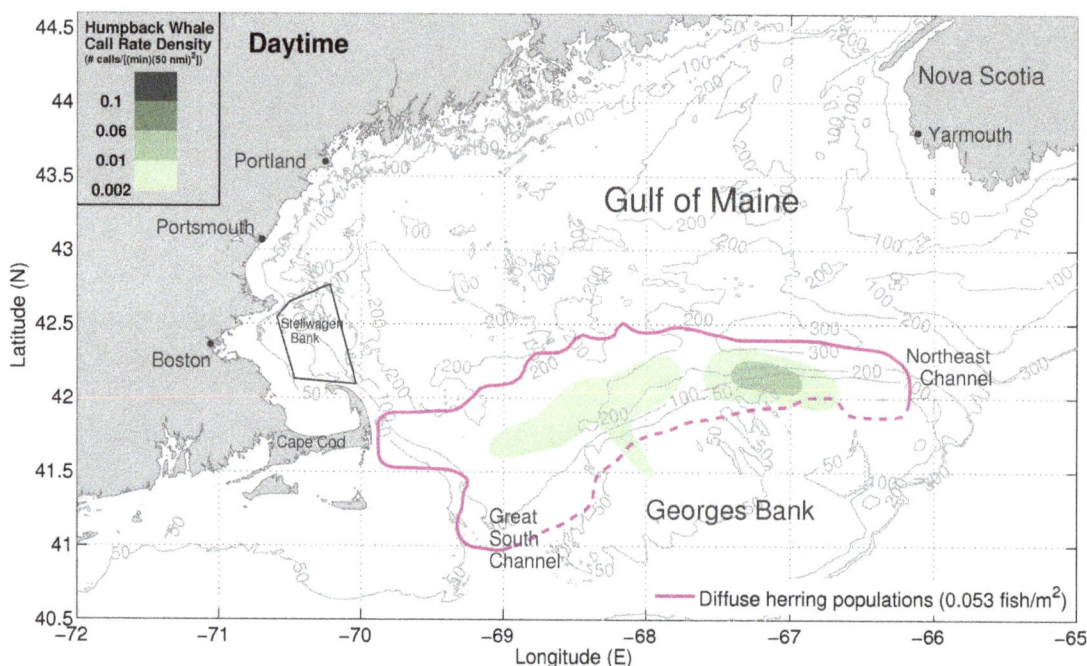

Figure 2. Daytime distributions of vocalizing humpback whales and diffuse herring populations. Spatial distribution of vocalizing humpback whales coincides with the locations of diffuse herring populations during daytime hours. In daylight, the vast majority of the humpback whale vocalizations originate within areas containing diffuse herring populations ($\approx$0.053 fish/m^2, bounded by magenta line) [63]. The green shaded areas indicate the daytime humpback whale call rate densities (number of calls/[(min) (50 nmi)^2]) measured with our large aperture array. All data represent daytime means between September 22 and October 6, 2006. The dashed magenta line represents the southern bound of the NEFSC survey tracks [18,63]. The daytime hours are between sunrise and sunset (06:00:01 to 18:00:00 EDT). The black trapezoid indicates Stellwagen Bank [158].

herring populations make hunting and feeding far more efficient [33].

The diurnal nature of observed humpback whale vocalizations (Figure 4) is comprised of a three-fold occurrence rate increase of repetitive non-song calls at night (Figure 5), which is consistent with communication [50–52] or prey echolocation [50,51,53] during feeding activities. "Meows" are the most frequently recorded non-song calls at night, followed by "bow-shaped" calls and "feeding cries". Repetitive "meows" are primarily uttered in series at night, in spatial and temporal synchronization with the formation of large spawning herring shoals. They are characterized by roughly 1.44 second duration, frequency modulated (537 Hz to 367 Hz) downsweep signals repeated at roughly 31 second intervals (Figures 6(A) and 7). Apparently, they have not been previously observed. These "meows" have significantly different spectral-temporal structure from "Megapclicks" [54], which are of much higher frequency, higher repetition rate, and lower source level, and have been previously associated with evening foraging activities. It has been suggested in Ref. [54] that "Megapclicks" could be "useful for some form of rough acoustic detection such as identifying the seafloor or other large target." Apart from communication, another possible function of "meows" could be to detect large targets, in particular large prey aggregations. Moreover, the range resolution for acoustic sensing using the finite time duration "meow" calls is $cT/2 \approx 1$ km [18,21,26,27,40,55–62], without matched filter pulse compression, where c is the sound speed and T is the time duration of the "meows," and so is consistent with echolocation of large herring shoals that typically exceed 1 km in horizontal extent [18,19,63–65]. Previously observed humpback whale "cries" [66] of roughly 0.4–8.2 second duration occur in a frequency band overlapping with that of "meows," but are characterized by shorter, frequency

modulated introductory and ending sections, separated by a relatively longer middle section with less frequency modulation, making them significantly different from the observed "meows." Individually uttered "meows", which only occurred intermittently with no pattern, were observed over the full diurnal cycle, and were far less numerous than repetitive "meows" uttered in series. The "bow-shaped" calls are the second most abundant humpback whale non-song vocalizations observed at night. Similar to the repetitive "meows," they are also primarily uttered in series at night. The "bow-shaped" calls are characterized by a repetition interval of roughly 58 seconds, a roughly 2.36 second duration, a frequency modulated (511 to 367 Hz) main downsweep section followed by a short upsweep coda (Figure 6(B)), and a repetition interval roughly 2 times longer than that of the repetitive "meows". The humpback whale "feeding cries" we observed are characterized by a roughly 3.18 second duration, frequency oscillating main pulse followed by a short highly frequency modulated coda (Figure 6(C)). They occurred only at night but far less frequently than the repetitive "meows" and "bow-shaped" calls, with a repetition interval of roughly 11 minutes. The "feeding cries" we observed are similar in frequency band and duration to individual "cries" previously observed in Alaskan humpback whale cooperative feeding [66], which is consistent with the calls we observed being related to cooperative humpback whale feeding on spawning herring.

Humpback whale songs (Figure 8) were found to lack diurnal variation across our observations during the peak annual herring spawning period (Figure 5), which is consistent with an invariance of singing behavior to diurnal feeding activities. Months before the herring spawning season and far from prime herring spawning grounds, absence of diurnal variation was previously observed in humpback whales singing north of the Great South Channel,

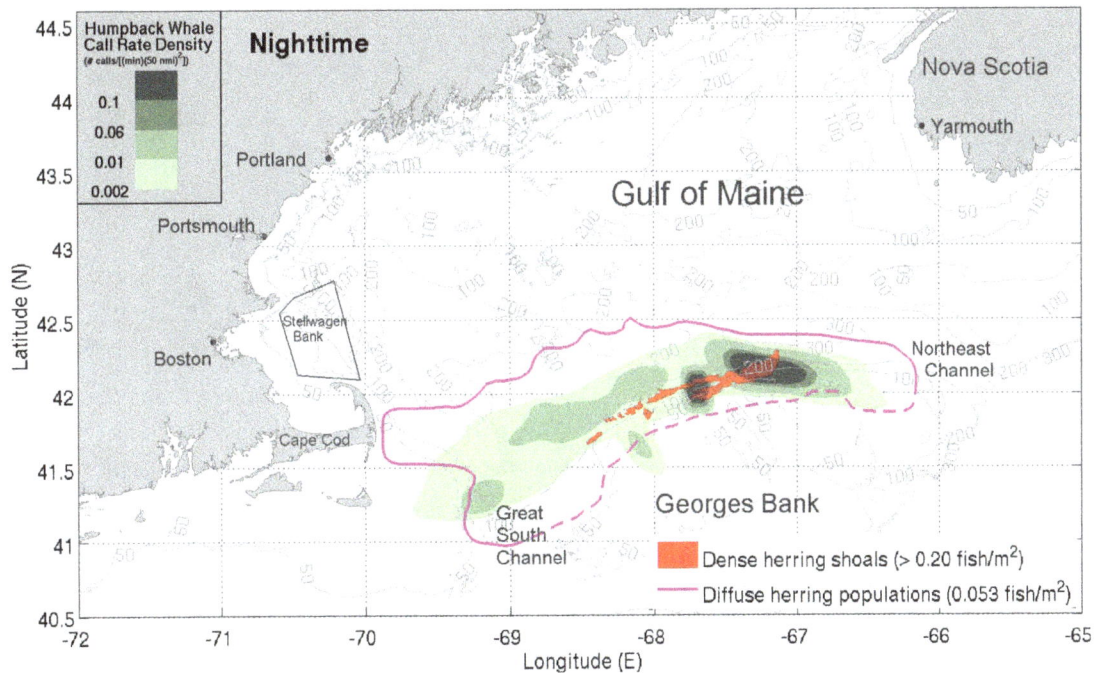

Figure 3. Nighttime distributions of vocalizing humpback whales and dense herring shoals. Spatial distribution of vocalizing humpback whales coincides with the locations of dense evening herring shoals during nighttime hours. At night, vocalizing humpback whales become concentrated at and near dense evening herring shoals (>0.20 fish/m^2, red shaded areas) that form along the northern flank of Georges Bank and call rates increase dramatically [18]. The green shaded areas indicate the nighttime humpback whale call rate densities (number of calls/[(min) (50 nmi)^2]) measured with our large aperture array. All data represent nighttime means between September 22 and October 6, 2006. The magenta line bounds the areas with diffused herring populations (≈ 0.053 fish/m^2). The dashed magenta line represents the southern bound of the NEFSC survey tracks [18,63]. The data shown are for nighttime hours between sunset and sunrise the next day (18:00:01 to 06:00:00 EDT). The black trapezoid indicates Stellwagen Bank [158].

which was thought to be potentially related to aseasonal mating [67]. In contrast, a diurnal pattern in acoustic energy was detected off of Western Maui, Hawaii, during the humpback whale breeding season with a single omni-directional hydrophone [5]. The increased acoustic energy at night was in the humpback whale vocalization band and attributed to humpback whale song choruses in breeding activities. The fact that songs occurred far less frequently than non-song calls in our observations by a factor of 4 (Figure 5), is consistent with humpback whale vocalization behavior that is closely related to primary seasonal activities.

2.2 Re-evaluation finds no effect of sonar on humpback whale song occurrence

Before and during OAWRS survey transmissions [18,19], we measured a constant humpback whale song occurrence rate, as shown in Figure 9, indicating no change of humpback song related to these transmissions over the entire survey area in the Gulf of Maine, a roughly 400-km diameter area, including Georges and Stellwagen Banks. Additionally, we find that the humpback whale song occurrence rate from Stellwagen Bank was constant before and during OAWRS survey transmissions, indicating no change of humpback song at Stellwagen Bank related to these transmissions. These direct measurements contradict the conclusions of Risch et al. [34].

To investigate this contradiction, we first follow the standard practice of checking for the bias [36,37] of a statistical test by applying the test to control data where no stimulus is present to determine the false positive outcome rate [68–70]. Since the bias of Risch et al. statistical test [34] was not checked in Risch et al.

[34], we do so here (Section 3.4) with the available annual humpback whale song occurrence data [35] from the same set of single sensors Risch et al. [34] used at Stellwagen Bank. We show that their statistical test false-positively finds whales react to sonar 98–100% of the time over a yearly period when no sonars are present. For example, when their statistical test is applied to annual humpback whale song occurrence data published in Ref. [35], with 2006 as the test year and 2008 as the control year, it false-positively finds whales react to sonar: (1) 100% of the time over the year before the "during" period; and (2) 98% of the time over the year when the "during" period is excluded from the test, as described in Section 3.4 and Table 1. Here the "during" period is defined as the 11-day period from September 26 to October 6 with active OAWRS survey transmissions, the "before" period is the 11-day period before the "during" period, and the "after" period is the 11-day period after the "during" period following the usage in Risch et al. [34]. When applied to the same humpback whale song occurrence data reported in Risch et al. [34] over the 33-day period from September 15 to October 17 for 2008 and 2009, with either of these two years as the test year and the other as the control year, the statistical test false-positively finds humpback whales respond to sonar 100% of the time when no sonar is present, as described in Section 3.4 and Table 2, indicating a self-contradiction in the Risch et al. [34] approach. No meaningful conclusions can be drawn from a statistical test with such high bias.

An explanation for the severe bias in the statistical test of Risch et al. [34] becomes evident upon inspection of the annual humpback whale song occurrence time series published in Ref. [35]. Very large natural variations within and across years are

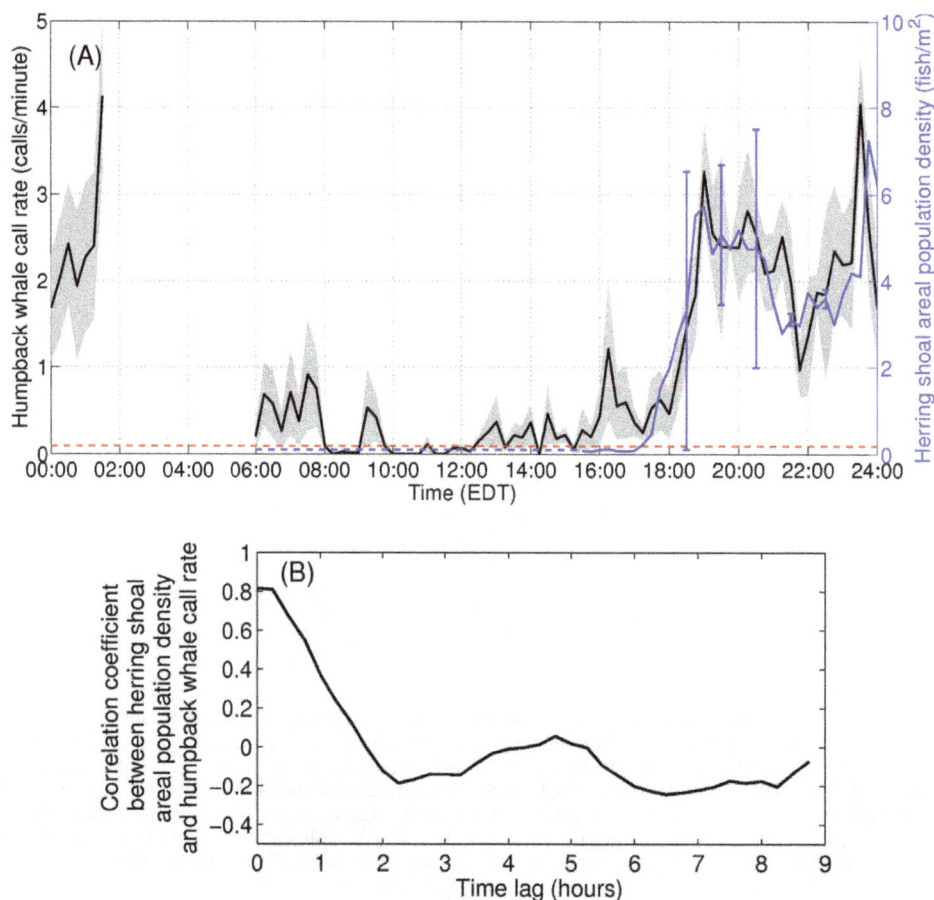

Figure 4. Humpback whale call-rate is synchronized with Atlantic herring shoal population density over a diurnal cycle. (A) Mean humpback whale call rate (black line within gray standard deviation over 15 minute bins) over a diurnal cycle and mean herring shoal areal population density (blue line with standard deviation indicated by the blue error bars) from September 28 to October 3. When the areal population density of the diffuse daytime herring populations reaches a critical threshold of approximately 0.2 fish/m^2 (red dashed line) near sunset, the herring population density drastically increases at a rate of roughly 5 fish/m^2 per hour [18] to form evening shoals. (B) Diurnal humpback whale call rate follows a synchronous pattern with 0.82 correlation coefficient and 0–15 minute time lag between the two time series in (A). The period from roughly 2–6 EDT contains a data gap.

common in the humpback whale song occurrence time series when no sonars are present, as can be seen in Figure 10. There are many periods lasting roughly weeks where high song occurrence episodes are found in one year but not in another, when no sonars are present (Figure 10). For the majority of the time, greater than 57%, the difference in the song occurrence across years when no sonars are present exceeds that of the "during" period (Figure 11), indicating that there is nothing unusual about such differences, which rather than "alterations" [34] are actually the norm. The statistical test used by Risch et al. [34] is overwhelmingly biased because it mistakes natural variations in humpback whale song occurrence 98–100% of the time for changes caused by sonar when no sonar is present, lacks any true positive confirmation and so lacks the statistical significance to draw the conclusions of Risch et al. [34].

Since the reported reducing change in humpback whale song occurrence, to zero [34,35], occurred in the "before" period (Figure 10) while the OAWRS vessels were inactive and docked on the other side of Cape Cod from Stellwagen Bank at the Woods Hole Oceanographic Institution due to severe winds days before OAWRS transmissions for active surveying began on September 26, 2006, the Risch et al. analysis [34] severely violates temporal

causality. Moreover, the annual humpback whale song occurrence time series are uncorrelated over 11-day periods across years, and the correlation coefficient obeys a random distribution peaking at zero correlation about which it is symmetric (Figure 12), showing that correlation in trend between years is random and quantitatively expected to be zero with roughly as many negative correlations as positive ones. In fact, the correlation coefficient between the humpback whale song occurrence across years smoothly transitions from negative values in the "before" period, showing no similarity or relation in trend between years just before the 2006 OAWRS survey transmission period, to some of the highest positive correlations obtained between years in the "during" period (Figure 12). This demonstrates high similarity and relation in trend between years during the 2006 OAWRS active survey transmission period, which contradicts the results of the Risch et al. [34] study. These causality violations are also discussed in the context of the measured temporal coherence of humpback whale song occurrence in Section 3.6.

It is well known that wind speed variation can lead to severe detection range limitations in passive sensors, especially a single sensor that has zero array gain [20,23,25,71]. Risch et al. [34] did not investigate the effect of wind dependent ambient noise on the

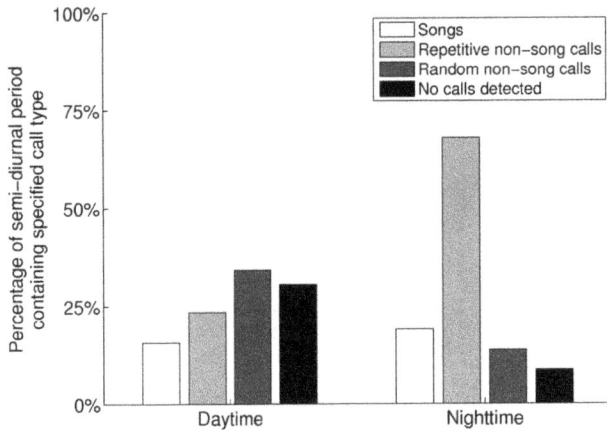

Figure 5. Percentage of semi-diurnal period containing different classes of humpback whale vocalizations for day and night. A roughly three-fold percentage increase is found at night for repetitive non-song calls, which are primarily responsible for the overall diurnal dependence of observed humpback whale vocalizations. Humpback whale songs showed negligible mean variation compared to standard deviations for day (15.7%±18%) versus night (19.1%±15%). Percentages were calculated using the approaches discussed in Section 3.2. The total percentage, the sum of all four categories, exceeds 100% because different call types could occur within overlapping time windows. The "No calls detected", however, is mutually exclusive with the other categories. Here the daytime hours are between sunrise and sunset (06:00:01 to 18:00:00 EDT) and nighttime hours are between sunset and sunrise the next day (18:00:01 to 06:00:00 EDT).

detection range of their single hydrophones located in the Stellwagen Bank (Figure 13). They did report that "Ambient noise levels over the whole analysis bandwidth (10–1000 Hz) and in the frequency band with most humpback whale song energy (70–300 Hz) did not vary dramatically within or between years." Wind speeds varied, however, from calm to near-gale conditions within a period of a few hours or days, many times over the 33-day period examined by Risch et al. [34], as is common for Fall in Stellwagen Bank [72]. These natural wind speed variations must have significantly changed the local wind-dependent noise level according to known physics [20,73]. Since noise "can have a tremendous, if not a dominating, influence on the detection range of any sonar system" [39], the dramatic changes in wind speed at Stellwagen Bank must have led to dramatic changes in the detection range of single sensors deployed there. The range at which signals, in this case humpback whale songs, can no longer be detected because they become indistinguishable from ambient noise is the detection range from the sensor. Since ambient noise is wind speed dependent, so is the detection range (Figure 13), and so is humpback whale song occurrence measured at that sensor if variations in wind speed cause the detection range to pass through the range of the singing humpback whales (Figure 14). In this case even if a whale sang at a constant rate, song occurrence measured at the sensor (Figure 15) would vary with local wind noise (Figure 14). Moreover, the annual humpback whale song occurrence reported in Ref. [35] had a standard deviation of 3.54 dB in the 33-day period examined by Risch et al. [34], which was less than the 3.8 dB standard deviation in ambient noise level reported by Risch et al. [34], and so local ambient noise variation could have caused all the variations in humpback whale song occurrence reported over that period.

Using the measured wind speeds at Stellwagen Bank [72], and the measured spatial distribution and constant rates of singing

humpback whales determined by our large aperture array, we determine the song occurrence detectable by a single hydrophone at Stellwagen Bank, as shown in Figure 15. We find it to match the song occurrence reported by Risch et al. [34] in the "before" and "during" periods with high accuracy, within ±18% of the reported means, which is much less than the standard deviation of the humpback whale song occurrence reported by Risch et al. [34]. This match shows that the variation in reported song occurrence from the "before" to "during" period is due to detection range limitations of the single sensor at Stellwagen Bank from wind-dependent ambient noise, and is not due to the song production rate, which we show to be constant. The constant song production and occurrence rates in the "before" and "during" periods measured by our large aperture array are unaffected by wind noise because the array gain was sufficiently high to make the detection range well beyond the range of the vocalizing whales for all wind conditions (Figure 13). Our data shows no humpback whale vocal activity originating from Stellwagen Bank in either the "before" or "during" periods, but vocalizing humpback whales located near Georges Bank could be heard at Stellwagen Bank during low wind noise conditions (Figure 13). In high wind noise, the single sensor mean detection range at Stellwagen Bank is too short to include the regions with measured singing humpback whales, but in low wind noise, it is large enough to include the regions with measured singing humpback whales as shown in Figure 13, making the mean song detection rate at Stellwagen Bank higher in lower wind noise. Noise from near gale force winds in the last 3 days of the "before" period, for example, caused a significant drop in the detection range of the single sensor and the corresponding significant drop in the song occurrence rate at Stellwagen Bank [35] while the OAWRS vessels were inactive and docked at the Woods Hole Oceanographic Institution. Since the OAWRS experiment was conducted only up to October 6, 2006, the vocalizing humpback whale distribution in the "after" period was not measured and we do not investigate the song occurrence for that period.

It has been previously shown that due to collapse of the herring stock at Stellwagen Bank, humpback whale populations drastically decline at Stellwagen Bank during the herring spawning period and correspondingly increase at other locations where spawning populations are large [33]. Moreover, in the Fall of 2006, herring populations were negligible in the Massachusetts Bay and Cape Cod area, including Stellwagen Bank [32], but in contrast were decadally high in the Georges Bank region [31], consistent with the theory that humpback whales migrate to locations with large spawning herring aggregations [33]. This phenomenon was not mentioned or investigated in Risch et al. [34], but it is highly relevant because the time period Risch et al. [34] focused on is centered exactly on the peak annual herring spawning period of the Gulf of Maine for 2006. Indeed, it has been previously shown by OAWRS in Ref. [18] and by annual NEFSC acoustic echosounding and trawl surveys in Refs. [63] and [43] that this peak annual herring spawning period occurred from the last week of September to the first week of October 2006 on Georges Bank. Based on the results of Ref. [33], it should then be expected that the Stellwagen Bank humpback whale population would be low at this time and the population at Georges Bank would be high, as has been confirmed in Section 2.1 for vocalizing humpback whales.

The levels of the various anthropogenic noises at Stellwagen Bank were not discussed in Risch et al. [34], but only OAWRS levels were selected for analysis and discussion without this context. It is recommended by the National Academy of Sciences (NAS), however, that "A comprehensive noise impact assessment

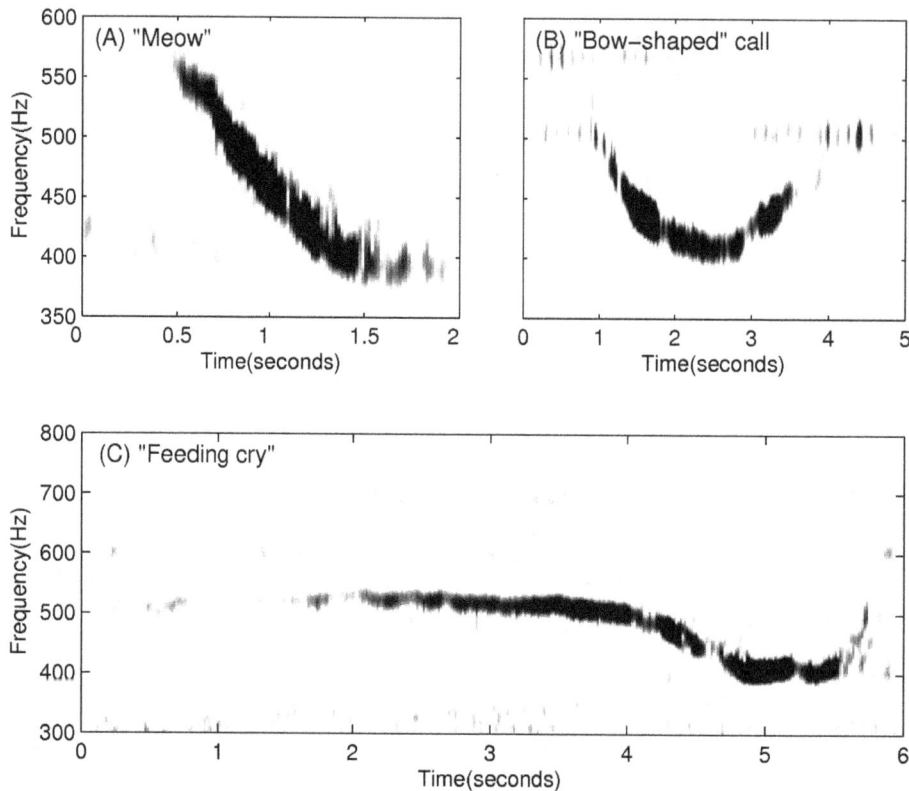

Figure 6. Spectrograms of a typical "meow", "bow-shaped" call and "feeding cry" observed during OAWRS 2006 experiment. (A) "Meow" is a roughly 1.4 second duration, frequency modulated downsweep signal (570 to 380 Hz) with a center frequency of roughly 475 Hz. (B) "Bow-shaped" call has a roughly 2.4 second duration, downsweep frequency modulated section (510 to 395 Hz) followed by a short upsweep coda with a center frequency of roughly 440 Hz. (C) "Feeding cry" consists of (1) a main section that lasts approximately 3.5 seconds with frequency oscillations between 500 Hz and 540 Hz and (2) a 2 second long frequency-modulated ending section.

would include additional specific data regarding both sound levels and sources throughout the area for which impacts are being assessed [74]." Such an impact assessment should include "all aspects of the acoustic environment" [75] to avoid the problem another impact assessment had of being evaluated as "misrepresentative of the existing soundscape [74]." Here the soundscape of anthropogenic noise sources at Stellwagen Bank, from highest to lowest intensity or loudest to most quiet is delineated in Tables 3 and 4, following these NAS recommendations, where it is seen that the reported OAWRS transmissions fell at the quietest end of the noise spectrum when audible. Shipping traffic, on the other hand, contributes most to the anthropogenic component of mean acoustic intensity at Stellwagen Bank by many orders of magnitude. Most anthropogenic sources of underwater noise listed in Tables 3 and 4 continuously operate [76,77] over a wide range of frequencies audible to whales, i.e. tens to hundreds of Hertz [20,39,77,78], and result in received levels that may exceed the currently recommended NOAA guideline of 120 dB re 1 μPa received level [79–83] in water for continuous noise [84] for a range of whale distances (Table 3). Even the maximum OAWRS received sound pressure level reported by Risch et al. [34] is orders of magnitude lower than the current 160 dB NOAA guideline for short duration signals such as the OAWRS 1–2 seconds duration pulse, and significantly lower than the 120 dB guideline for even continuous sources [84] which OAWRS is not. The maximum received acoustic intensities of OAWRS signals at Stellwagen Bank reported by Risch et al. [34] are the same as those of a quiet wooded forest or a quiet room with no conversation [85], whereas

the acoustic intensities received at Stellwagen Bank from shipping traffic are often the same as those of a busy roadway or a busy airport runway [26,85]. Risch et al. [34] reported that visual inspections of humpback whales in Stellwagen Bank were made during the OAWRS experiment, suggesting that humpback whales were within visible range of research vessels. Research vessels close enough to whales to sight them can easily have engine noise levels at the whales greatly exceeding the reported OAWRS levels over broader frequency bands and much greater time duration (Table 3).

Before and during OAWRS survey transmissions, we measured constant humpback whale song occurrence and production rates over our entire survey area roughly 400-km in diameter covering most of the Gulf of Maine, including Stellwagen Bank, indicating the transmissions had no effect on humpback whale song production rate. Using annual humpback whale song occurrence reported from single sensor detections at Stellwagen Bank [35] in time dependent ambient noise, we show the statistical test used by Risch et al. [34] for assessing the response of humpback whales to sonar transmission false positively finds humpback whales respond to sonar 98–100% of the time when no sonars are present. With this and the lack of any true positive confirmation for the Risch et al. [34] statistical approach, the analysis of Risch et al. [34] lacks the statistical significance to draw the conclusions of Risch et al. [34]. The fact that the Risch et al. [34] analysis only allows sonar causes for the reducing change reported in Risch et al. [34], yet the change occurred days before the sonar survey began, is consistent with a violation of temporal causality in the Risch et al. [34] study.

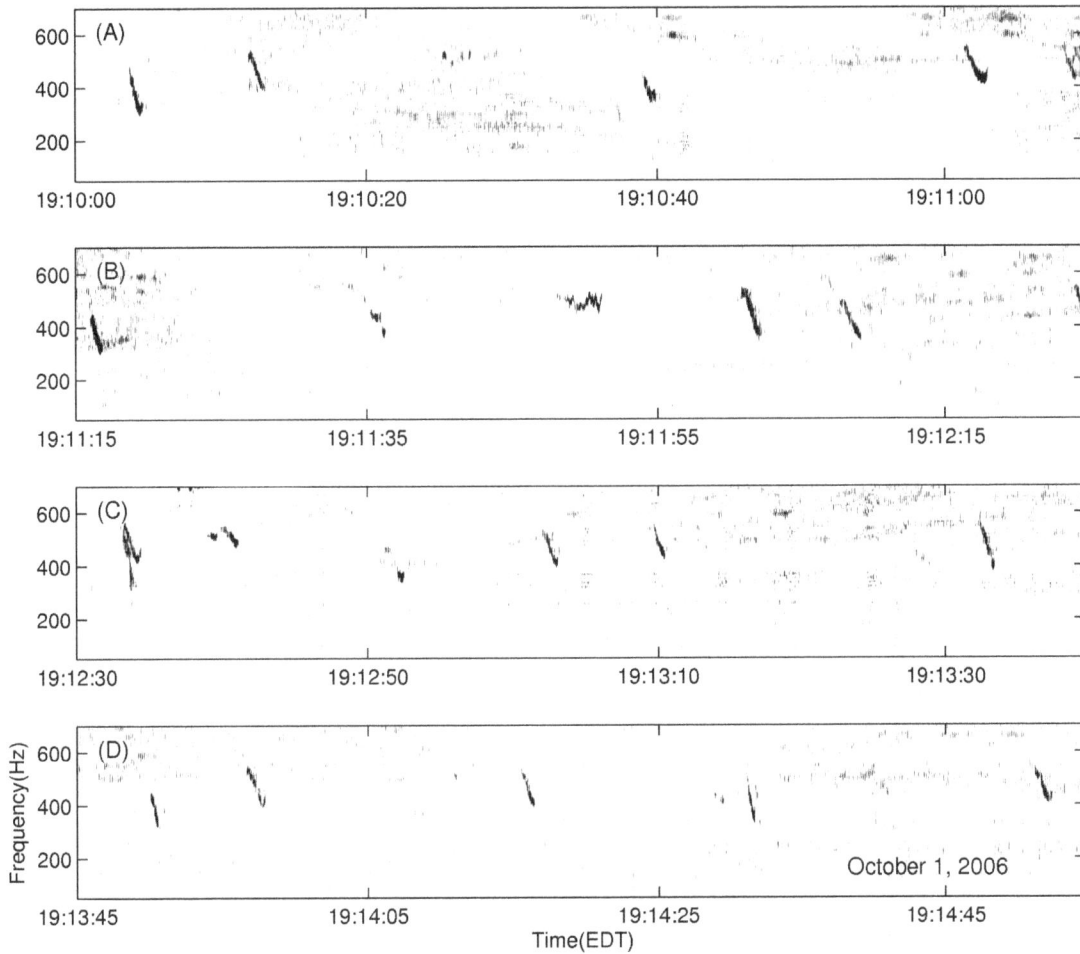

Figure 7. Spectrograms of typical repetitive "meows" observed during OAWRS 2006 experiment in the Gulf of Maine. Four 70-s time series containing repetitive meows are shown in (A) – (D) recorded 5-s apart, on October 1, 2006 between 19:10:00 EDT and 19:14:55 EDT.

The Risch et al. statistical test [34] mistakes natural variations in whale song reception, from such factors as natural variations in whale distributions [44], singing behavior [1,2], and ambient noise, for changes caused by sonar 98–100% of the time when no sonar is present. Before and during OAWRS survey transmissions, we find that the variations in song occurrence at Stellwagen Bank reported by Risch et al. [34] are consistent with the natural phenomena of detection range fluctuations caused by wind-dependent ambient noise, through well established physical processes [20,73]. Misinterpretation of natural phenomenon from flawed analytic methods such as biased testing and neglect of physical laws can have seriously negative consequences [86–90].

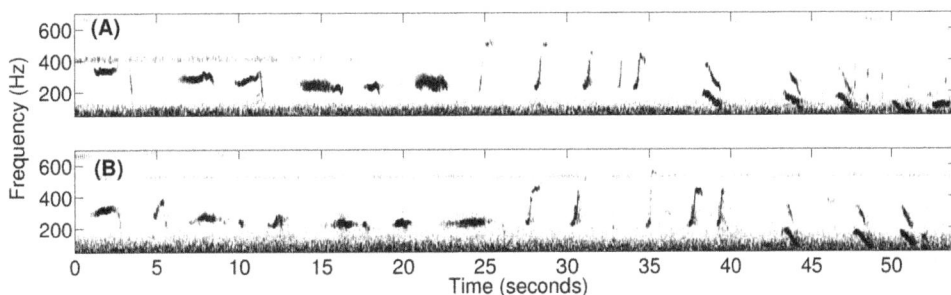

Figure 8. Spectrograms of a typical repeated humpback whale song theme observed during OAWRS 2006 experiment. A repeated humpback whale song theme, starting at (A) 23:17:44 EDT and (B) 23:49:01 EDT and each lasting roughly 1 minute, was recorded on October 2, 2006 from a singing humpback whale in the northern flank of Georges Bank.

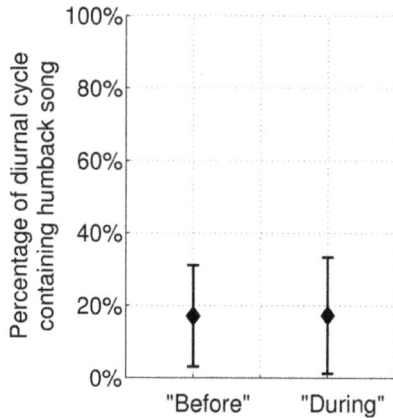

Figure 9. Humpback song occurrence rate is constant in the periods "before" and "during" OAWRS survey transmissions. The mean percentage of a diurnal cycle containing humpback whale song in the periods "before" and "during" OAWRS survey transmissions, as defined in Section 2.2, remains constant, indicating the transmissions had no effect on humpback whale song over the entire passive 400-km diameter survey area of the Gulf of Maine including Stellwagen Bank.

Materials and Methods

3.1 The passive receiver array

Acoustic recordings of whale vocalizations were acquired using a horizontal passive receiver line-array, the ONR five-octave research array [91], towed by Research Vessel *Oceanus* along designated tracks just north of Georges Bank [18,19], as shown in Figure 13. The multiple nested sub-apertures of the array contain a total of 160 hydrophones spanning a frequency range from below 50 to 3750 Hz for spatially unaliased sensing. A fixed sampling frequency of 8000 Hz [19] was used so that acoustic signals with frequency contents up to 4000 Hz were recorded without temporal aliasing. Two linear apertures of the array, the low-frequency (LF) aperture and the mid-frequency (MF) aperture, both of which consist of 64 equally spaced hydrophones with respective inter element spacing of 1.5 m and 0.75 m, were used to analyze humpback whale calls with fundamental frequency content below 1000 Hz. For humpback whale calls with frequency content below 500 Hz, the LF aperture was used, while for humpback whale calls with frequency content extending beyond 500 Hz up to 1 kHz, the MF aperture was used. The angular resolution $\beta(\phi, f_c)$ of the horizontal receiver array is $\beta(\phi, f_c) \approx 1.44(\lambda/L \cos \phi)$ for broadside ($\phi = 0$) through angles

near endfire ($\phi = \pi/2$), where $\lambda = c/f_c$ is the acoustic wavelength, c is the sound speed, f_c is the center frequency, and L is the array aperture length. At endfire, the angular resolution is $\beta(\phi = \pi/2, f_c) \approx 2.8\sqrt{\lambda/L}$. Permission for this National Oceanographic Partnership Program experiment was given in the Office of Naval Research document 5090 Ser 321RF/096/06.

3.2 Measurement and analysis of humpback whale vocalizations

Acoustic pressure time series measured by sensors across the receiver array were converted to two-dimensional (2D) beam-time series by time-domain beamforming [20,22,25,26], and further converted to spectrograms by temporal Fourier transform. Whale vocalizations were detected and characterized in time and frequency for each azimuth by visual inspection.

With our densely sampled, large-aperture array, multiple vocalizing humpback whale individuals could be tracked in beam-time and compared with the bearings of historic humpback whale habitats in the Gulf of Maine, including the Georges Bank, Stellwagen Bank, Great South Channel, and Northeast Channel as shown in Figure 16. Throughout our entire experiment, including the "before" and "during" periods discussed in Section 2.2, we measured roughly 2000 humpback whale vocalizations per day but none originated from Stellwagen Bank, as in the Figure 16 example.

As noted in Section 2.1, both humpback whale song [1,2,8,67,92–94] and non-song [6,7,9,54,66,95] vocalizations were measured, where non-song vocalizations contained repetitive and random calls. Songs [2] were composed of repeating themes, which could be sub-divided into phrases and units. A song session typically consisted of at least two themes and often lasted over tens of minutes, with gaps of silence not exceeding ten minutes between any two themes. An example of repeated song themes is shown in Figure 8. Repetitive non-song calls were defined as series of downsweep "meows" or "bow-shaped" calls, which contained at least two similarly structured "meows" or "bow-shaped" calls that were uttered within a short time interval of roughly 31 seconds or 58 seconds, respectively. Random non-song calls, were primarily composed of individual "meows", "bow-shaped" calls, and "feeding cries" that occurred at least one minute apart from any type of individually uttered non-song calls. We found that roughly 73% of the non-song vocalizations were "meows," roughly 22% were "bow-shaped" calls, and roughly 5% were "feeding cries." These non-song calls were observed in the frequency range of 250–700 Hz (Table 5). The standard and primary method of using spectral and temporal characteristics of the vocalizations to identify whale species [6,34,35,95–100] is used here. The specific

Table 1. Percentage of time the Risch et al. statistical test [34] incorrectly finds whales respond to sonar when no sonar is present using annual humpback whale song occurrence data reported from single sensor detections at Stellwagen Bank [35] in time-dependent ambient noise.

Analysis period	Excluding "during" period[a]	Before "during" period[a]
% of time with false-positive response	98.0%(49/50)	100%(35/35)

Risch et al. statistical test [34] is applied to all continuous 33-day periods, as described in Section 3.4.1, in the annual humpback whale song occurrence reported from single sensor detections at Stellwagen Bank in 2006 and 2008 [35], with 2006 as the test year and 2008 as the control year. The test false-positively finds humpback whales react to sonar 98–100% of the time over a yearly period when no sonars are present. The fraction of time when the Risch et al. statistical test [34] false-positively finds whales react to sonar is given in the parenthesis. The parenthetical numbers in the denominator represent the total number of 33-day periods with no sonar present within the analysis period and the parenthetical numbers in the numerator represent the number of 33-day periods when the Risch et al. statistical test [34] false-positively finds whales react to sonar when no sonar is present.
[a]The "during" period is defined in Section 2.2.

Table 2. The Risch et al. statistical test is applied to the same humpback whale song occurrence data reported in Risch et al. [34] over the 33-day period from September 15 to October 17 for 2008 and 2009, with either of these two years as the test year and the other as the control year.

Risch et al. statistical test	Result
With 2008 as the test year and 2009 as the control year	**False positive response**
With 2009 as the test year and 2008 as the control year	**False positive response**

It false-positively finds that whales react to sonar 100% of the time when no sonar is present, indicating self-contradictions in the Risch et al. [34] approach.

spectral and temporal characteristics of calls we observed are provided in Table 5, following a standard approach for classifying calls established by Dunlop et al. [6]. Since all non-song calls or non-song call sequences we detected *consistently* originated or ended at the the same spatial position as song calls, to within our reported position error in Section 3.3, and occurred immediately after or before these co-located song calls, alternating with song calls, it is most likely that the same species and group of whales produced the song and non-song calls we report. Given this and the fact that humpback whales are the only species known to produce song in this region, season and frequency range, it is most likely that the non-song calls we report are also from humpback whales and extremely unlikely that they originate from other species. Furthermore, humpback whales are the most abundant, by 1–2 orders of magnitude, vocalizing whales in the 250–700 Hz frequency range [2,6,7,9,101,102] in the Gulf of Maine during the fall season [45]. While North Atlantic right whales, minke whales

and sei whales have been observed to rarely vocalize solely in the 250–700 Hz frequency range, it is also unlikely that the non-song calls we observed were produced by these whales because (1) right and minke whale tonal calls are roughly 4–8 times shorter in time duration or roughly a factor of 2 lower in frequency than the non-song calls we observed [103–107]; (2) the typical right whale "gunshot" calls are of a much broader frequency content than 250–700 Hz and are more than an order of magnitude shorter in time duration than the non-song calls we observed [103,104,106,108]; (3) the more typical minke whale "pulse trains" lasting tens of seconds are comprised of pulses that are more than an order of magnitude shorter in time duration and have a minimum frequency roughly a factor of 2 lower than that of the non-song calls we observed [109,110]; (4) right whales are 20 times less abundant, minke whales are 10 times less abundant, and sei whales are 60 times less abundant than humpback whales in the Gulf of Maine during the fall season [45]; (5) sei whales have not

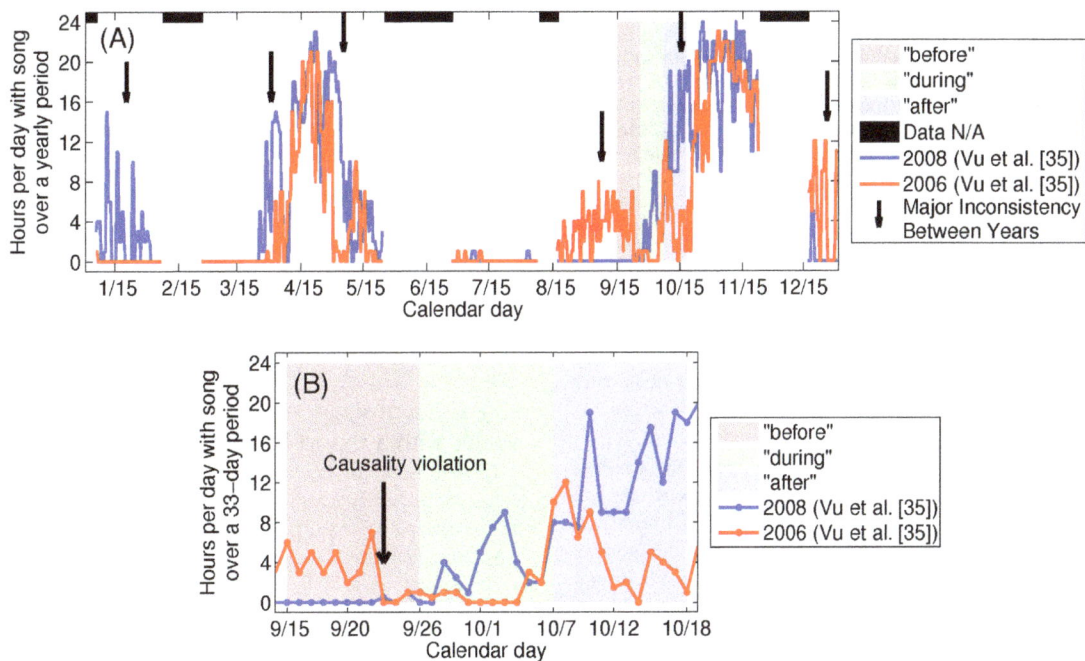

Figure 10. Reported humpback whale Stellwagen Bank song occurrence [35] **shows large natural variations within and across years.** Large natural variations in humpback whale song occurrence reported from single sensor detections at Stellwagen Bank [35] in time-dependent ambient noise within and across years are common in the absence of sonar. Line plots of reported single sensor daily humpback whale song occurrence at Stellwagen Bank in hours/day (A) for the entire year and (B) from September 15 to October 17, in 2006 and 2008 [35]. Many periods lasting roughly weeks where high song occurrence episodes are found in one year but not in another when no sonars are present are indicated by black arrows in (A). The reported reducing change in humpback whale song occurrence, to zero [34,35], occurred in the "before" period while the OAWRS vessels were inactive and docked on the other side of Cape Cod from Stellwagen Bank, at the Woods Hole Oceanographic Institution, due to severe winds for days before OAWRS transmissions for active surveying began on September 26, 2006, as marked by the black arrow in (B). This shows that Risch et al. [34] analysis violates temporal causality.

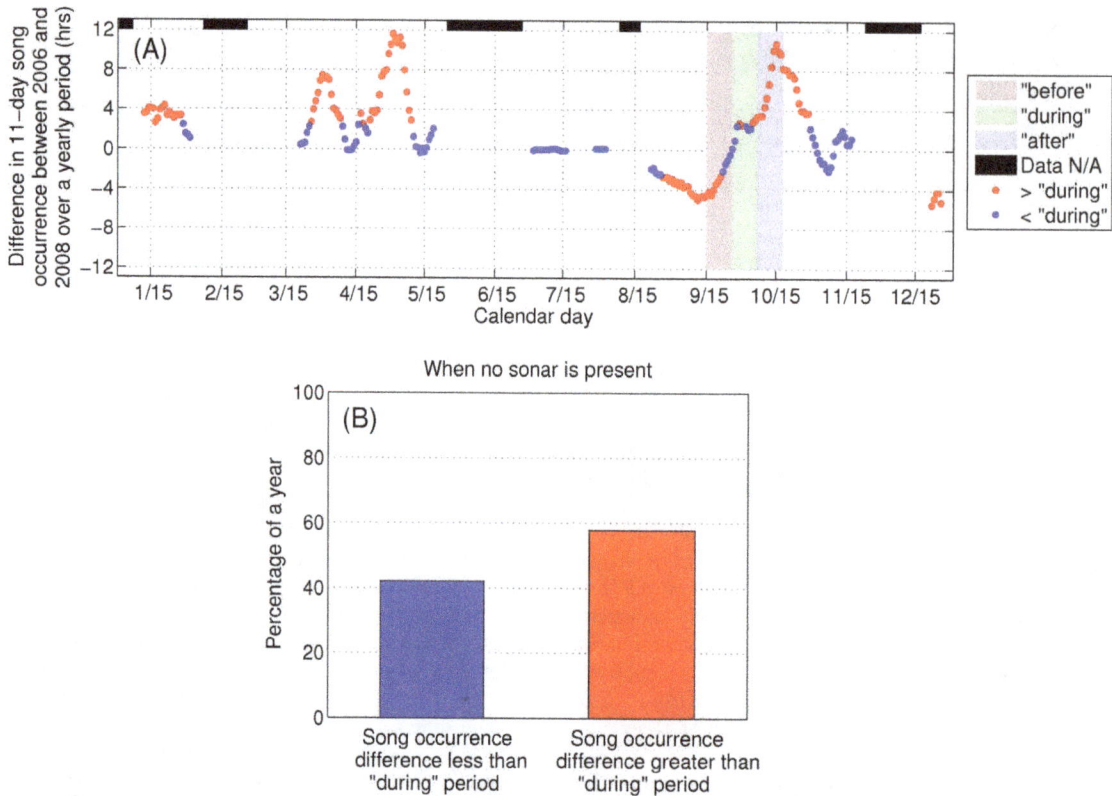

Figure 11. Quantifying large differences in the reported humpback whale song occurrence at Stellwagen Bank [35] **across years.** Difference in humpback whale song occurrence reported from single sensor detections at Stellwagen Bank [35] in time-dependent ambient noise across years exceeds that of the "during" period most of the time when no sonars are present. (A) Difference in mean humpback whale song occurrence at Stellwagen Bank over respective 11-day periods with 1-day increment in 2006 and 2008, (B) histogram of difference in mean humpback song occurrence over 11-day periods between 2006 and 2008 when no sonar is present, i.e. excluding the "during" period from September 26 to October 6. Periods when the difference in means of respective 11-day periods is greater than (red dots) and less than (blue dots) that of the "during" period are indicated in (A). The difference in means fluctuates randomly throughout the year, exceeding the "during" period 57.8% of the time (most of the time) when no sonars are present, indicating that there is nothing unusual about such differences, which are actually the norm.

been observed to vocalize in the 250–700 Hz frequency range in the North Atlantic and the North Pacific [111–114]; and (6) previous work shows humpback whales to be by far the dominant consumers of herring on Georges Bank of the whales that have been observed to vocalize in the 250–700 Hz range, where right and sei whales appear to consume negligible amounts of herring [115]. There were numerous sightings of humpback whales at Georges Bank during the 2006 Gulf of Maine experiment.

The diurnal humpback whale call rate (calls/min) time series of Figure 4(A) is obtained by averaging daily humpback whale call rate time series over the entire experiment. The daily humpback whale call rate time series is quantified in 15 minute bins over a diurnal cycle. We define a time period that (1) contains at least two song themes with (2) a gap of silence not exceeding 10 minutes between the adjacent song themes as the occurrence session of humpback whale songs. Similarly, a series of "meows" (Figure 7) or "bow-shaped" calls, and individually uttered non-song calls (Figure 6) constitute the occurrence sessions of repetitive non-song calls and random non-song calls, respectively. A time period longer than 10 minutes containing no calls is defined as the occurrence session of "No calls detected", and is mutually exclusive with the occurrence sessions of the other three categories. The percentage of time with songs, repetitive non-song calls and random non-song calls, as shown in Figure 5, are quantified using these defined occurrence sessions. The total percentage, the sum of

all four categories, may exceed 100% because different types of humpback whale calls may occur simultaneously in overlapping time windows. The number of whales singing at any given time within their detection ranges is found to be consistent with past observations [10,67,93,94,101,116–118].

3.3 Passive position estimation of vocalizing humpback whales with a towed horizontal receiver line-array

To determine the horizontal location of a vocalizing humpback whale, both bearing and range need to be estimated. With our densely sampled, large-aperture horizontal receiver array, bearings of vocalizing humpback whales are determined by time-domain beamforming. Synthetic aperture tracking [29] and the array invariant method [30] are applied to determine the range of vocalizing humpback whales from the horizontal receiver array center. The principle of the synthetic aperture tracking technique [29] is to form a synthetic array by combining a series of spatially separated finite apertures of a single towed horizontal line-array. The array invariant method [30] provides instantaneous source range estimation by exploiting the multi-modal arrival structure of guided wave propagation at the horizontal receiver array in a dispersive ocean waveguide. Position estimation error, or the root mean squared (RMS) distance between the actual and estimated location, is a combination of range and bearing errors. Range estimation error, expressed as the percentage of the range from the

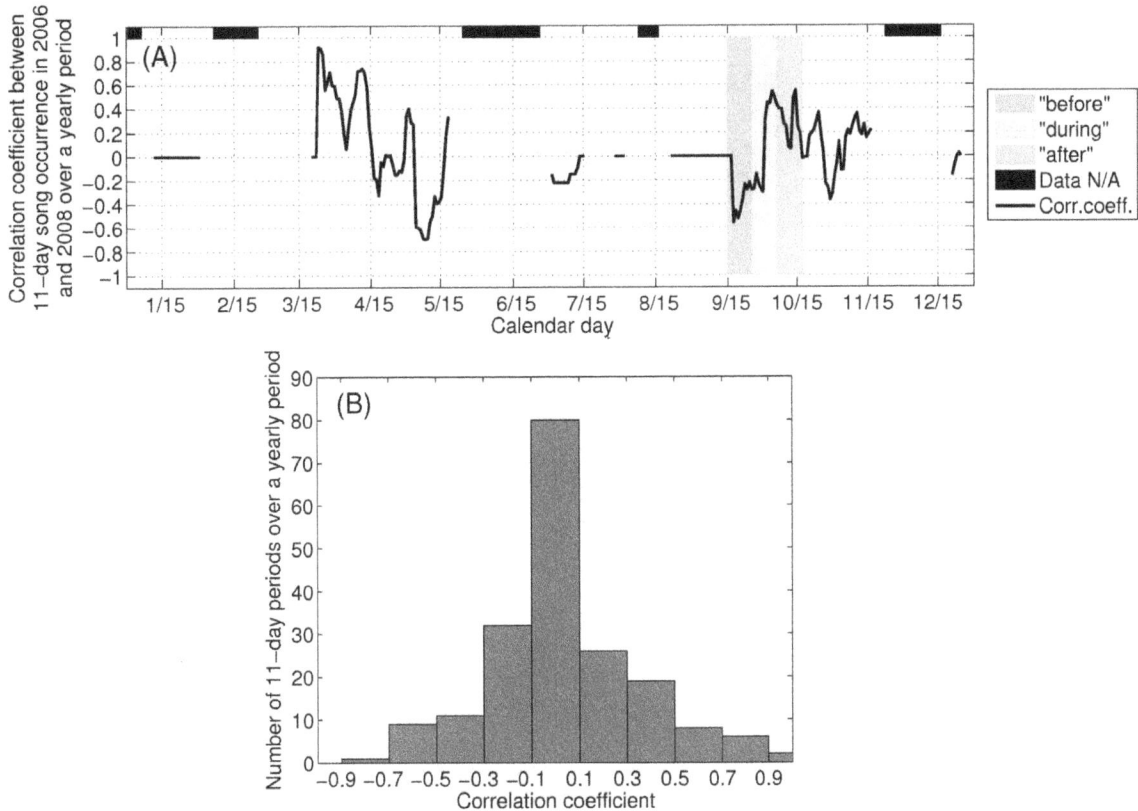

Figure 12. Reported annual humpback song occurrence at Stellwagen Bank [35] **are uncorrelated between years over 11-day periods.** Annual humpback whale song occurrence reported from single sensor detections at Stellwagen Bank [35] in time-dependent ambient noise are uncorrelated over 11-day periods across years. (A) Correlation coefficient between 2006 and 2008 humpback whale song occurrence time series over 11-day period with 1-day increment (B) histogram of the correlation coefficient in (A). The correlation coefficient of the annual humpback whale song occurrence time series over 11-day periods across years obeys a random distribution peaking at zero correlation about which it is symmetric, showing that correlation in trend between years is random and quantitatively expected to be zero with roughly as many negative correlations as positive ones. The correlation coefficient between the humpback whale song occurrence across years smoothly transitions from negative values in the "before" period, showing no similarity or relation in trend between years just before the 2006 OAWRS survey transmission period, to some of the highest positive correlations obtained between years in the "during" period. This demonstrates high similarity and relation in trend between years during the 2006 OAWRS active survey transmission period, which contradicts the results of the Risch et al. [34] study.

source location to the horizontal receiver array center, for the synthetic aperture tracking technique is roughly 2% at array broadside and gradually increases to 10% at 65° from broadside and 25% at 90° from broadside, i.e. near or at endfire [29]. Range estimation error for the array invariant method is roughly 4–8% [29] over all azimuthal directions. Bearing estimation error of the time domain beamformer is roughly 0.5° at broadside and gradually increases to 6.0° at endfire [29]. These errors are determined at the same experimental site and time period as the whale position estimates presented here, from thousands of controlled source signals transmitted by the same source array used to locate the herring shoals presented here [18] and are based on absolute Global Positioning System (GPS) ground truth measurements of the source array's position, which are accurate to within 3–10 meters [119]. More than 90% of vocalizing whales are found to be located 0–65° from the broadside direction of the horizontal receiver array. Position estimation error is then less than 2 km for most of the vocalizing whales localized in Figure 13 since they are found within roughly 40 km of the horizontal receiver array center. This error is over an order of magnitude smaller than the spatial scales of the whale concentrations shown in Figure 13, and consequently has negligible influence on the analyses and results. The measured source locations for all calls are

used to generate the whale call rate density maps shown in Figures 1–3 and 13. The source location of each call is characterized by a 2D Gaussian probability density function with mean equal to the measured mean position from synthetic aperture tracking or the array invariant method and standard deviations determined by the measured range and bearing standard deviations. The range standard deviation is 2% for sources located at and near array broadside and increases to 25% for sources located at and near array endfire, based on the range errors of both synthetic aperture tracking and the array invariant method [29]. The bearing standard deviation is 0.5° for sources located at or near array broadside and increases to 6.0° for sources located at or near array endfire [29]. The whale call rate density map is determined by superposition of the 2D spatial probability densities for the source location of each call, normalized by the total measurement time. Left-right ambiguity in determining the bearing of a sequence of source signals in this paper is resolved by changing the array's heading during the reception of the sequence of source transmissions, following the standard method for resolving left-right ambiguity in source bearing for line array measurements in the ocean [16,29,120–123]. For a far-field point source in free space, bearing ambiguity in line array measurements exists in a conical surface about the array's axis with cone angle

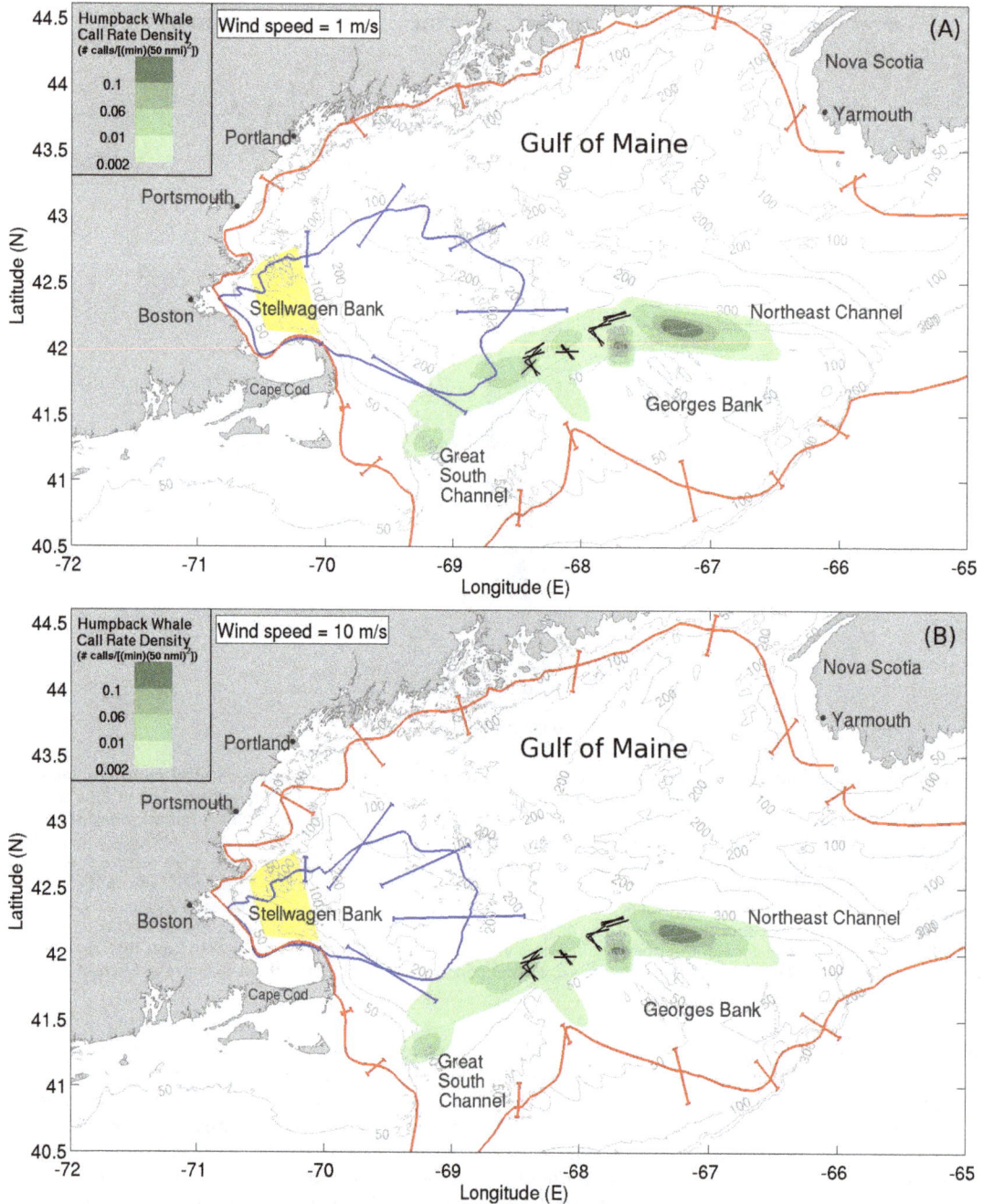

Figure 13. Wind-dependence of mean detection range for single sensor at Stellwagen Bank [34]**, and OAWRS receiver array.** The green shaded areas indicate the overall vocalizing humpback whale call rate densities (number of calls/[(min) (50 nmi)^2]) determined between September 22 and October 6, 2006 by our large aperture receiver array towed along several tracks (black lines). The mean detection ranges for the single sensor at Stellwagen Bank are in blue and for the OAWRS receiver array are in red, where Stellwagen Bank is marked by yellow shaded regions. These detection ranges are determined by the methods described in Section 3.5 given a humpback whale song unit source level of approximately 180 dB re 1 μPa and 1 m which is the median of all published humpback whale song source levels [93,101,102,152–154]. The error bars represent the spread in detection range due to typical humpback whale song source level variations (Section 3.5). Under (A) low wind speed conditions vocalizing whales are within the mean detection area for a single Stellwagen Bank sensor but for (B) higher wind speeds most vocalizing whales are outside the mean detection area of the same sensor, which results in reduction of detectable whale song occurrence by the single sensor [34] at Stellwagen Bank.

equal to the bearing of the source with respect to the array's axis, because the phase speed on the array is identical for far-field sources on this cone at any given frequency. When ambiguity is restricted to source locations in the ocean, only two ambiguous bearings remain, left and right in the horizontal plane about the

array's axis, for ranges large compared to the water depth of the source and receiver, as is the case in this paper. To resolve this ambiguity, array heading is varied by an amount $\Delta\theta$ with respect to an absolute coordinate system during the sequence of source transmissions. The true location of the source in absolute

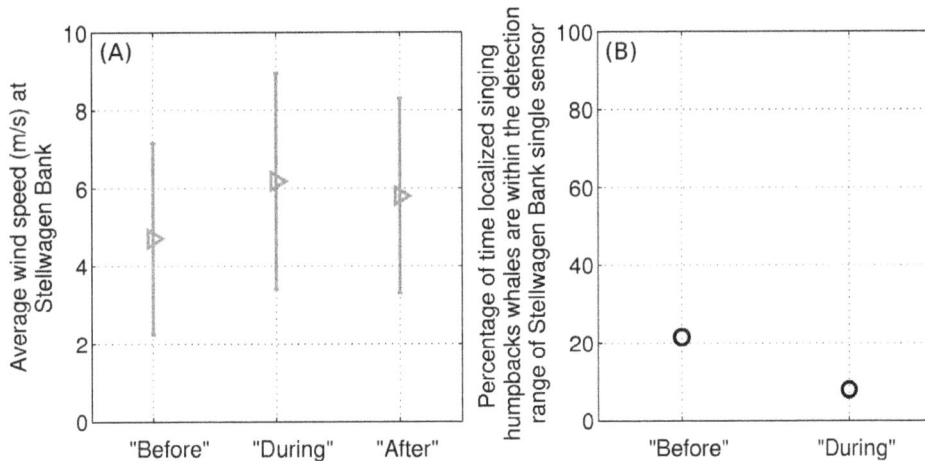

Figure 14. Wind-speed increase causes reduction in humpback song occurrence at Stellwagen Bank. Average wind speed increase from the "before" to the "during" period at Stellwagen Bank causes reduction in the percentage of time humpback whale songs are within mean detection range of a single Stellwagen Bank sensor. (A) Averaged wind speed measured at the NDBC buoy [72] closest to Stellwagen Bank over the "before," "during," and "after" 11-day periods; and (B) percentage of the time vocalizing humpback whales localized by our large aperture array are within the mean detection range of the single sensor [34] at Stellwagen Bank in the "before" and "during" periods, using waveguide propagation methods and whale song parameters described in Section 3.5. Since the OAWRS experiment was conducted only up to October 6, 2006, the humpback whale source distribution in the "after" period was not measured and we do not investigate the percentage of time that humpback whales are within the mean detection range of the single sensor at Stellwagen Bank [34] for the "after" period. The triangles represent the mean wind speed and the solid ticks represent the standard deviation of the wind speed over the respective 11-day periods.

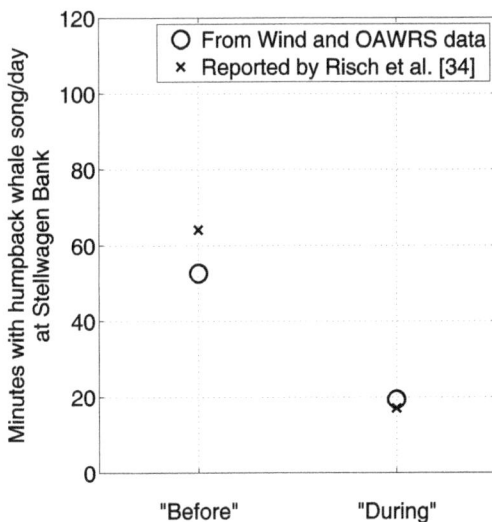

Figure 15. Humpback song occurrence detectable by single sensor matches reported humpback song occurrence at Stellwagen Bank [34]. Average humpback whale song occurrence detectable by a single hydrophone at Stellwagen Bank in time-dependent ambient noise in the "before" and the "during" periods matches the reported humpback whale song occurrence by Risch et al. [34]. Using the measured wind speeds at Stellwagen Bank [72] (Figure 14), the measured spatial distribution of vocalizing humpback whales (Figure 1), and constant song production rates (Figure 9) measured by our large-aperture array, the detectable song occurrence over the "before" and "during" period are found to be within ±18% of the reported means [34], much less than the standard deviations of reported song occurrence[34], using waveguide propagation methods and whale song parameters described in Section 3.5. Before and during OAWRS survey transmissions, this figure shows that reported variations in song occurrence at Stellwagen Bank by Risch et al. [34] are actually due to detection range changes caused by wind-dependent ambient noise, through well established physical processes [20,73].

coordinates is independent of the array heading, but the bearing of the virtual image source has a component that moves by $2\Delta\theta$ with the array heading. This is analogous to the case where a mirror is rotated by $\Delta\theta$, and the true source remains at an absolute position independent of the mirror's orientation but its virtual image in the mirror rotates by an apparent $2\Delta\theta$ with the mirror's rotation to maintain a specular angle with respect to the mirror's plane and satisfy Snell's Law [21,124]. The criterion used here to distinguish the virtual image bearing from the true source bearing is that established by Rayleigh [26,124,125], where ambiguity is robustly resolved by moving the array heading by an angular amount $\Delta\theta$ such that the change in virtual bearing $2\Delta\theta$ exceeds the array's angular resolution scale (the array beamwidth, Section 3.1) in the direction of the detected source. This Rayleigh resolved change in bearing of the virtual source of $2\Delta\theta$ with the array's heading change of $\Delta\theta$ is used to identify the virtual source and distinguish it from the true source, which has an absolute bearing independent of $\Delta\theta$. This procedure for ambiguity resolution with the Rayleigh criterion has been applied to all sequences of source transmissions used for source localization in this paper.

3.4 Risch et al. statistical test

To evaluate its bias and quantify the impact of this bias, the Risch et al. statistical test of Ref. [34] is applied to Stellwagen Bank humpback whale song occurrence data reported in Refs. [34,35], since the bias of this test has not been previously investigated, and the implications of a bias have not been previously analyzed or discussed for this test.

The Risch et al. statistical test [34] applies the Tukey method [126] for simultaneous pairwise multiple comparison with the quasi-Poisson generalized linear model (GLM) and log link in the statistical programming language 'R' [34,127,128] to humpback whale song occurrence over non-overlapping 11-day periods within a 33-day period across years, and tests the resulting pairwise comparisons following the statements of Table 6. The input to the statistical test of Ref. [34] is daily humpback whale song

Table 3. Typical anthropogenic noise sources at Stellwagen Bank.

Continuous anthropogenic noise source	Source level in dB re 1 μPa and 1 m	Frequency in Hz	Source range in km for received level above 120[a] dB re 1 μPa	Source range in km for received level between 88–110[b] dB re 1 μPa	Acoustic intensity in Watts/m² 1 m away from anthropogenic noise source
Cruise ship	219 [159]	10 to >1,000 [160]	<100	160 to >200	5,000
Cargo vessel	192 [20,161]	10 to >1,000 [20,161]	<10	30–200	10
Research vessel	166–195 [159]	40 to >1,000 [77,159]	<6	2–130	0.025–20
Outboard motor boat	176 [78,162]	100 to >1,000 [163,164]	<2	3–20	0.25
Whale watching boat	169 [165]	100 to >1,000 [165]	<1	3–25	0.05

[a]Recommended received pressure level in the NOAA guideline for continuous-type sources [84].

[b]Range of received pressure level at Stellwagen Bank single sensor reported by Risch et al. of OAWRS impulsive signal [34], of roughly 1–2 seconds duration and at least 75 seconds spacing between impulses. Source ranges are determined at the frequencies with maximum humpback whale vocalization energy, using the waveguide propagation methods described in Section 3.5. Humpback whale vocalizations are known to have source levels in the range of 175 to 188 dB re 1 μPa and 1 m [9,101,102,153], and have been reported to go up to 203 dB re 1 μPa and 1 m [166]. All data shown in the table is for sources and measurements in water where $L_{s,water} = L_w + 171$ based on the sound speed and density of water, L_w is the power level in dB re 1 Watt, and $L_{s,water}$ is the source level in dB re 1 μPa and 1 m. Underwater noise from a typical low flying jet airplane [26] can lead to underwater sound pressure levels exceeding 120 dB re 1 μPa in water at ranges less than 5 kilometers.

occurrence time series data over each 11-day period. Each pairwise comparison between the mean song occurrence in the jth 11-day period of the ith 33-day period in the kth year and that in the lth 11-day period of the ith 33-day period in the mth year is assigned a value of $p^i_{(j,k),(l,m)}$. The value of $p^i_{(j,k),(l,m)}$ is the probability that the absolute value of the Tukey test statistic [126] is greater than the observed value of the test statistic, conditioned on the null hypothesis, i.e. all mean humpback whale song occurrences over 11-day periods are the same, and is denoted by the variable P in Risch et al. [34]. If $p^i_{(j,k),(l,m)}$ is less than a threshold P_T set by the user, then the means are classified by the user to be significantly different, otherwise they are classified by the user to be not significantly different.

Suppose there are daily humpback whale song occurrence time series over M years, and for each year there are N 33-day periods. Let $\mu^i_{(j,k)}$ be the mean humpback whale song occurrence over the jth 11-day period of the ith 33-day period in the kth year, where $i=1,...,N, j=1,2,3$, and $k=1,...,M$. Let $k=1$ be the test year and let $k=2,...,M$ be the control years.

For a given 33-day period over M years, there are $_{3M}C_2 = \dfrac{(3M)!}{2!(3M-2)!}$ pairs of 11-day periods. Comparing the $p^i_{(j,k),(l,m)}$ with P_T for each of the $_{3M}C_2$ pairs, outcome $T^i_{(j,k),(l,m)}$ is assigned for the comparison between the mean song occurrence

pair $\mu^i_{(j,k)}$ and $\mu^i_{(l,m)}$. The possible outcomes $T^i_{(j,k),(l,m)}$ are (1) $X(\mu^i_{(j,k)} < \mu^i_{(l,m)})$, which is defined as: $\mu^i_{(j,k)}$ and $\mu^i_{(l,m)}$ are not significantly different and $\mu^i_{(j,k)} < \mu^i_{(l,m)}$; (2) $X(\mu^i_{(j,k)} \nless \mu^i_{(l,m)})$, which is defined as: $\mu^i_{(j,k)}$ and $\mu^i_{(l,m)}$ are not significantly different and $\mu^i_{(j,k)} \nless \mu^i_{(l,m)}$; (3) $Y(\mu^i_{(j,k)} < \mu^i_{(l,m)})$, which is defined as: $\mu^i_{(j,k)}$ and $\mu^i_{(l,m)}$ are significantly different and $\mu^i_{(j,k)} < \mu^i_{(l,m)}$; and (4) $Y(\mu^i_{(j,k)} \nless \mu^i_{(l,m)})$, which is defined as: $\mu^i_{(j,k)}$ and $\mu^i_{(l,m)}$ are significantly different and $\mu^i_{(j,k)} \nless \mu^i_{(l,m)}$, as given in Table 7.

The rate of false positive findings that whales respond to sonar when no sonar is present is

$$P_{FP} = \frac{\sum_{i=1}^{N_S} b_i}{N_S}, \tag{1}$$

where

$$b_i = \begin{cases} 1 & \text{when } \sum_{n=1}^{4} a_{i,n} \text{ is non-zero,} \\ 0 & \text{otherwise,} \end{cases} \tag{2}$$

N_S is the number of 33-day periods when no sonars are present,

Table 4. Received mean intensity of typical anthropogenic noise sources at Stellwagen Bank.

Continuous anthropogenic noise source	Received level in water in dB re 1 μPa (or corresponding mean intensity in Watts/m²) 500 m [a] away from an anthropogenic noise source over a minute or longer	How many decibels higher (or times greater) the mean intensity of the given anthropogenic noise source over a minute or longer at 500 m is than that reported for OAWRS at Stellwagen Bank [34]
Cruise ship	177 (0.33)	85 (300,000,000)
Cargo vessel	147 (0.00033)	55 (300,000)
Research vessel	121–144 (0.00000083–0.00017)	29–52 (750–150,000)
Outboard motor boat	131 (0.0000083)	39 (7,500)
Whale watching boat	124 (0.0000017)	32 (1,500)

[a]Whale watching vessels [167] are allowed to approach humpback whales at ranges much less than 500 m according to NOAA Whalewatching Guidelines [168].

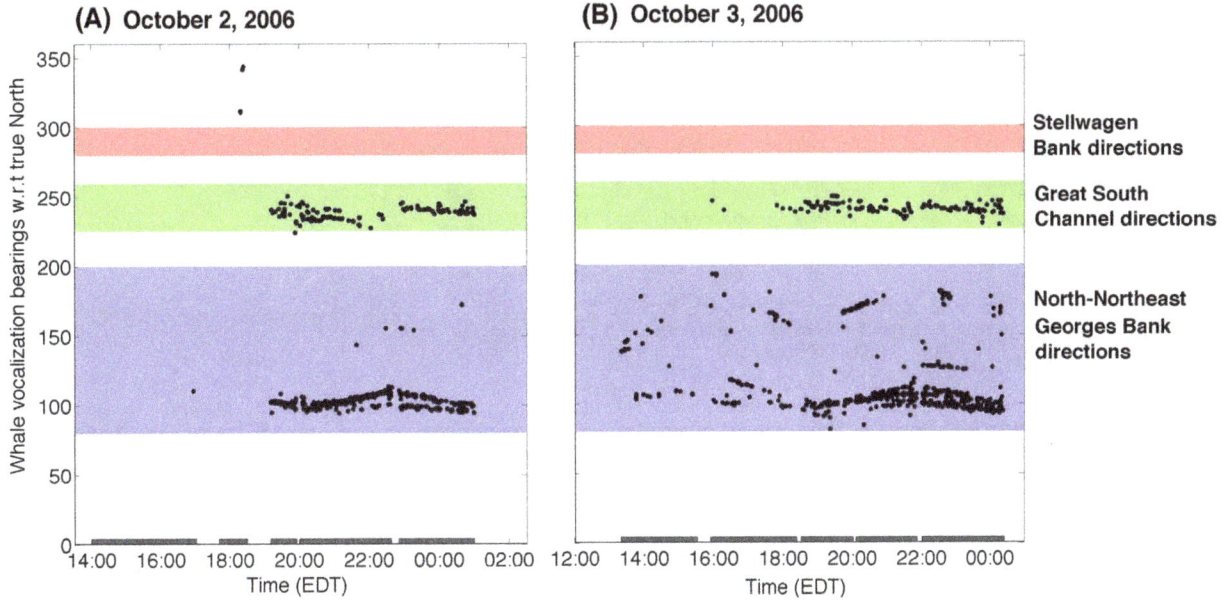

Figure 16. Vocalizing humpback whale bearings measured by our large-aperture receiver array. Examples of vocalizing humpback whale bearings measured on (A) October 2 and (B) October 3, 2006. Almost all humpback whale vocalizations are found to originate from North-Northeast Georges Bank directions (purple shaded areas) and the Great South Channel directions (green shaded areas), but none originates from Stellwagen Bank directions (red shaded areas). All vocalizing humpback whale bearings are measured from the true North in clockwise direction with respect to the instantaneous spatial locations of towed horizontal receiver array center. The techniques used here for resolving source bearing ambiguity about the horizontal line-array's axis are described in Section 3.3. The shaded bars on the x-axis indicate the operation time periods of the towed array.

the $a_{i,n}$ are defined in Table 6, and each ith 33-day period, for $i = 1,2,..,N_S$, has no sonar present.

3.4.1 False positive rate and statistical bias of the Risch et al. statistical test. When the Risch et al. statistical test [34], as described mathematically in Section 3.4 and Table 6, is applied

to the three 33-day humpback whale song occurrence time series data reported in Risch et al. [34], with 11-day time series indices $j = 1$ for the "before" period from September 15 to September 25, $j = 2$ for the "during" period from September 26 to October 6, and $j = 3$ for the "after" period from October 7 to October 17,

Table 5. Temporal and spectral characteristics of humpback whale non-song calls.

Non-song calls	Characteristics	Mean	Standard deviation	Minimum	Maximum
"Meows"	Overall call duration (s)	1.44	0.59	0.41	3.60
	Minimum frequency (Hz)	367	45	255	474
	Maximum frequency (Hz)	537	48	410	699
	Repetition interval (s)	31	8	3	50
Series of "Meows"	Overall series duration (s)	300	240	120	840
	Repetition interval (s)	510	288	270	1230
	Number of "Meows"	10	11	2	61
"Bow-shaped" calls	Overall call duration (s)	2.36	0.92	0.69	4.38
	Minimum frequency (Hz)	367	29	269	450
	Maximum frequency (Hz)	511	39	440	600
	Repetition interval (s)	58	2	55	62
"Feeding cries"	Overall call duration (s)	3.18	1.59	1.65	8.10
	Minimum frequency (Hz)	363	23	293	395
	Maximum frequency (Hz)	540	23	492	585
	Repetition interval (s)	692	464	78	1638

These calls include "meows" and "bow-shaped" calls, both of which are primarily uttered in series at night, and "feeding cries", which only occur at night but far less frequently than "meows" and "bow-shaped" calls. We find that roughly 73% of humpback whale non-song calls are "meows", roughly 22% are "bow-shaped" calls, and roughly 5% are "feeding cries".

Table 6. Risch et al. statistical test statements [34].

	Risch et al. Statement	Algorithmic representation
1	"While 'before' and 'after' periods differed significantly within the years 2008 and 2009 ($P < 0.001$), with more song recorded in the later period in both years, this increase was not significant in 2006 ($P = 0.2147$)."	If $T^i_{(1, 1),(3, 1)} = T^i_{(1, k),(3, k)}$ for all $k \neq 1$, then $a_{i, 1} = 0$, otherwise $a_{i, 1} = 1$.
2	"In 2006, the 'during' period was significantly different from the period 'after' ($P = 0.0093$), with more song recorded later. The 2006 'during' period was not detectably different from the period 'before' ($P = 0.5226$)."	If $T^i_{(2, 1),(3, 1)} = X(\mu^i_{(2, 1)} < \mu^i_{(3, 1)})$ or $T^i_{(2, 1),(3, 1)} = X(\mu^i_{(2, 1)} \nprec \mu^i_{(3, 1)})$ AND $T^i_{(1, 1),(2, 1)} = Y(\mu^i_{(1, 1)} < \mu^i_{(2, 1)})$ or $T^i_{(1, 1),(2, 1)} = Y(\mu^i_{(1, 1)} \nprec \mu^i_{(2, 1)})$, then $a_{i, 2} = 0$, otherwise $a_{i, 2} = 1$.
3	"When comparing the 'during' period across years, 2006 differed significantly from 2009 ($P = 0.0057$). The same time period did not differ significantly between 2006 and 2008 ($P = 0.1842$), or between 2008 and 2009 ($P = 0.4819$)."	If $T^i_{(2, 1),(2, k)} = Y(\mu^i_{(2, 1)} < \mu^i_{(2, k)})$ or $T^i_{(2, 1),(2, k)} = Y(\mu^i_{(2, 1)} \nprec \mu^i_{(2, k)})$ for all $k > 1$, then $a_{i, 3} = 1$, otherwise $a_{i, 3} = 0$.
4	"Yet, overall there was considerably less song recorded in the 11 'during' days in 2006 compared to both 2008 and 2009."	If $\mu^i_{(2, 1)} \nprec \mu^i_{(2, k)}$ for all $k > 1$, then $a_{i, 4} = 0$, otherwise $a_{i, 4} = 1$.

and indices $k = 1$ for year 2006, $k = 2$ for year 2008 and $k = 3$ for year 2009, we obtain the same P values and results reported in the 'Risch et al. Statement' column of Table 6. Specifically, daily humpback whale song occurrence time series denoted by Ψ_{2006} for year 2006, Ψ_{2008} for year 2008, and Ψ_{2009} for year 2009, from song occurrence data reported in Risch et al. [34] over the 33-day period from September 15 to October 17, are input to the Tukey tests of the statistical programming language 'R', as described in Section 3.4. Since there is only one 33-day period from September 15 to October 17, $i = 1$. This 33-day period consists of the three consecutive non-overlapping 11-day periods with indices j or $l = 1,2,3$ and year indices k or $m = 1,2,3$ for pairwise comparisons between periods within and across years. A value of $p^i_{(j,k),(l,m)}$, the P value, and a corresponding $T^i_{(j,k),(l,m)}$ outcome are determined for each pairwise comparison between the mean song occurrence in the jth 11-day period of the kth year and that in the lth 11-day period of the mth year from the Tukey tests, as described in Section 3.4.

We apply the Risch et al. statistical test [34] to the two-year humpback whale song occurrence daily time series data reported in Vu et al. [35] with the same statistical test settings used to obtain the P values and results reported in the 'Risch et al. Statement' column of Table 6. The Vu et al. [35] daily humpback whale song occurrence time series (Figure 3 of Ref. [35]) over the ith 33-day period, denoted by $\Xi_{i,2006}$ for year 2006 and $\Xi_{i,2008}$ for year 2008, are input to the Tukey tests of the statistical programming language 'R', as described in Section 3.4. For the ith 33-day period, consisting of three consecutive non-overlapping 11-day periods with indices j or $l = 1,2,3$, and year indices k or $m = 1$ for

the test year 2006 and k or $m = 2$ for the control year 2008, a value of $p^i_{(j,k),(l,m)}$, the P value, and a corresponding $T^i_{(j,k),(l,m)}$ outcome are determined for each pairwise comparison between the mean song occurrence in the jth 11-day period of the kth year and that in the lth 11-day period of the mth year from the Tukey tests, as described in Section 3.4. From the outcomes $T^i_{(j,k),(l,m)}$, the corresponding $a_{i,n}$ are determined based on Table 6. This is repeated for all continuous 33-day periods, where the $i + 1^{th}$ 33-day period begins 1-day after the ith 33-day period. Only 33-day periods that have 11-day periods with reported whale song occurrence are included. If data is missing in any day from a 33-day period, then that 33-day period is excluded from both years. False positive rates are then determined from $a_{i,n}$ via Equations (1) and (2). The Risch et al. statistical test [34] false-positively finds whales react to sonar in (a) 100% of the 35 continuous 33-day periods before the "during" period (Table 1) when no sonar is present; and (b) 98% of the 50 continuous 33-day periods excluding the "during" period (Table 1) when no sonar is present. No valid or meaningful conclusions can be drawn from such an overwhelmingly biased statistical test. This specific application of the Risch et al. statistical test [34] has not been previously reported.

When the Risch et al. statistical test [34] is applied to the same humpback whale song occurrence data, Ψ_{2008} and Ψ_{2009}, reported in Risch et al. [34] over the 33-day period between September 15 and October 17, with 11-day time series indices $j = 1$ for the "before" period, $j = 2$ for the "during" period, and $j = 3$ for the "after" period, and year indices $k = 1$ for the test year 2008 and $k = 2$ for the control year 2009, as well as with year indices $k = 2$

Table 7. Possible outcomes of each pairwise comparison between the mean humpback whale song occurrence in the jth 11-day period of the ith 33-day period in the kth year and that in the lth 11-day period of the ith 33-day period in the mth year in the Risch et al. statistical test [34].

Outcome $T^i_{(j, k),(l, m)}$	Description
$X(\mu^i_{(j, k)} < \mu^i_{(l, m)})$	Means are not significantly different and $\mu^i_{(j,k)} < \mu^i_{(l,m)}$
$X(\mu^i_{(j, k)} \nprec \mu^i_{(l, m)})$	Means are not significantly different and $\mu^i_{(j,k)} \nprec \mu^i_{(l,m)}$
$Y(\mu^i_{(j, k)} < \mu^i_{(l, m)})$	Means are significantly different and $\mu^i_{(j,k)} < \mu^i_{(l,m)}$
$Y(\mu^i_{(j, k)} \nprec \mu^i_{(l, m)})$	Means are significantly different and $\mu^i_{(j,k)} \nprec \mu^i_{(l,m)}$

for the control year 2008 and $k = 1$ for the test year 2009, the test false-positively finds that whales react to sonar 100% of the time when no sonar is present, indicating self-contradictions in the Risch et al. [34] approach, as shown in Table 2, which make their analysis and conclusions invalid. This specific application of the Risch et al. statistical test [34] has also not been previously reported.

3.5 Model for detectable humpback whale song occurrence

Detectable humpback whale song occurrence for a coherent sensor array can be quantified in terms of local wind-speed-dependent ambient noise for a given spatial distribution of vocalizing humpback whales. The humpback whale song occurrence depends on the presence of at least one singing humpback whale inside the mean wind-dependent detection range of the sensor array. The percentage of time in a day over which a humpback whale is within the mean detection area and is singing corresponds to the measured daily humpback whale song occurrence rate.

The detection range [20,23,25,39,71], r_d, is defined as the range from the center of the array at which signals, in this case humpback whale songs, can no longer be detected above the ambient noise, and is the solution of the sonar equation [20–24],

$$NL(v) + DT - AG = RL(r_d(v)) = SL - TL(r_d(v)), \quad (3)$$

where $NL(v)$ is the wind-speed-dependent ambient noise level, v is the wind speed, DT is the detection threshold, RL is the received sound pressure level due to a humpback whale song source level SL undergoing a transmission loss of $TL(r_d(v))$ at range $r_d(v)$ for some given source and receiver depths, and AG is the array gain equal to $10 \log_{10} N_0$ for a horizontal array, where N_0 is the number of coherent sensors spaced at half wavelength [20–24]. The capability of sensor arrays with high array gain such as ours to detect sources orders of magnitude more distant in range than a single sensor is standard, well established and well documented in many textbooks [20–24,27]. The array gain of our coherent horizontal OAWRS receiver array is 18 dB, which enables detection of whale vocalizations in an ocean acoustic waveguide [20,22,24,27] up to either two orders of magnitude lower in SNR or two orders of magnitude more distant in range than a single hydrophone [20–24,27], which has zero array gain [20–24,27], by direct inspection of Equation (3). We set the detection threshold, DT, such that the sum of signal and noise is detectable at least 5.6 dB [129–132] above the noise. The ambient noise and the received signal are filtered to the frequency band of the source. Further, the wind-speed-dependent ambient noise level is modeled as

$$NL(v) = 10 \log_{10} \left(\frac{\alpha v^n + \beta}{1 \mu Pa^2} \right) \quad (4)$$

where n is the power law coefficient of wind-speed-dependent ambient noise, α is the waveguide propagation factor [133] and β corresponds to the constant baseline sound pressure squared in the frequency band of the source. The coefficients n, α and β are empirically obtained by minimizing the root mean square error between the measured and the modeled ambient noise level as a function of measured wind speed during the OAWRS experiment in the Gulf of Maine [18]. We find $n \approx 1.2$ in the frequency range of the observed humpback song units, which is consistent with past ambient noise measurements in high shipping traffic regions [134–

137]. (A value of $n \approx 3$ would have been consistent with wind-dependent ambient noise with no significant shipping component [138–140] but a value of $n \approx 3$ was not obtained.) The noise levels obtained from Equation (4) in Stellwagen Bank are consistent with those reported in Risch et al. [34].

A standard parabolic equation model of the US Navy and the scientific community, Range-dependent Acoustic Model (RAM) [22,141–144], that takes into account range-dependent environmental parameters is used to calculate the transmission loss $TL(r_d(v))$ from the whale location to the sensor in a highly range-dependent continental-shelf environment in the Gulf of Maine including Stellwagen Bank. The model uses experimentally measured sound speed profiles acquired during the OAWRS 2006 experiment [19] and standard bathymetry data for the Gulf of Maine [145]. Expected transmission loss [146] is determined along any given propagation path from source to receiver by Monte-Carlo simulation over range-dependent bathymetry [145] and range-dependent sound speed structures measured from oceanographic data [19,55,147,148]. An estimate of detection range $\hat{r}_d(v)$ for a given humpback whale song unit source level can be obtained from Equation (3) by a minimum mean squared error method. Higher transmission loss occurs in shallower waters due to more intense and pervasive bottom interaction [20–24]. Transmission loss in deeper waters is typically significantly lower due to upward refraction [20,22] which leads to far less intense and pervasive bottom interaction, as is the case in the deeper waters surrounding Georges Bank [20–24]. Highly directional transmission loss may then occur when there are large depth variations about a receiver. Indeed, this effect makes the detection range of whales in directions to the North of our receiver and Georges bank much greater than in directions to its South where the relatively shallow waters of Georges Bank are found (Figure 13). The fact that we localized the sources of many whale calls at great distances along shallow water propagation paths on Georges Bank in directions where transmission loss was greater and found negligibly small vocalization rates much closer to the receiver in the deeper waters north of Georges Bank where transmission loss was much less greatly emphasizes the finding that the vocalization rates originating from north of Georges Bank were negligibly small. This indeed is expected based on general behavioral principles [33] since the whales' dominant prey was on Georges Bank, where the majority of whale vocalizations originated (Figures 1–3), and not in the deeper waters to the North, as we note in Section 2.1. This is also consistent with the historical distribution of humpback whales in the Gulf of Maine during the fall season [45]. The ranges and propagation paths from deep to shallow waters between our receiver array and Stellwagen Bank are very similar to those between our receiver array and the distant whale call sources localized along Georges Bank (Figure 13). The corresponding transmission losses have negligible differences. The fact that we localized the sources of many whale calls on Georges Bank but found negligibly small vocalization rates originating from Stellwagen Bank in the "before" or "during" periods, then emphasizes the fact that vocalization rates originating from Stellwagen Bank were negligibly small in these periods. As noted in Sections 2.1 and 2.2, this is consistent with the well documented findings that humpback whales migrate away from Stellwagen Bank where herring stocks have collapsed to feed at other locations that support large herring aggregations such as Georges Bank [33]. Our transmission loss calculations with the standard RAM parabolic equation model have been extensively and successfully calibrated and verified with (1) thousands of one-way transmission loss measurements made during the same 2006 Gulf of Maine experiment discussed here at the same time and at the same location [19,149]; (2) thousands of two-way transmission loss measurements made from herring shoal returns and verified by conventional fish finding sonar and ground truth trawl surveys during the same 2006 Gulf of Maine experiment discussed here at the same time and at the same location [18,19,150]; (3) roughly one hundred two-way transmission loss measurements made from calibrated targets with known scattering properties during the same 2006 Gulf of Maine experiment discussed here at the same time and at the same location [151]; and (4) thousands of one-way transmission loss measurements made during a past OAWRS experiment conducted in a similar continental shelf environment [147].

We find that the humpback whale song source levels measured from more than 4,000 song units recorded during the same 2006 Gulf of Maine experiment discussed here at the same time and at the same location approximately follow a Gaussian distribution and are in the range 155 to 205 dB re 1 μPa and 1 m (Figure 17) with a mean of 179.8 dB re 1 μPa and 1 m and a median of 179.4 dB re 1 μPa and 1 m. The high array gain [20–25] of our densely sampled, large aperture coherent OAWRS horizontal receiver array used here enables detection of whale songs two orders of magnitude lower in SNR than a single hydrophone, which has no array gain. Our measurements of humpback whale song source levels then have a high dynamic range and span the wide range of published source levels [9,93,101,102,152,153], except for those in Ref. [154], which appear to be anomalously low compared to the rest of the literature as has been previously noted in Ref. [9]. The mean and median of our measured source levels match very well (within 0.6 dB) with the median of all published humpback whale song unit source levels of 180 dB re 1 μPa and 1 m [93,101,102,152–154]. Our song unit source levels are determined given our estimated whale positions and waveguide propagation modeling. Results in Figures 14 and 15 are computed using our measured whale positions and the median of all published humpback song source levels of 180 dB re 1 μPa and 1 m [93,101,102,152–154], which has negligible difference from our measured median and mean song source levels, for the range of measured humpback singing depths of 2 m to 25 m [152,155]. Results in Figures 14 and 15 are insensitive to variations in whale position variations within the errors we report for our measured whale positions in Section 3.3, and so are insensitive to the whale position errors of our measurement system. Insensitivity here means the measured to modeled song occurrence match is within ±18% as in Figure 15.

The total humpback whale song occurrence in a day detectable by a sensor in varying wind speeds is

$$T_{song} = \int_0^{T_{day}} S(t)dt, \qquad (5)$$

where $S(t) = 1$ when $\hat{r}_d(v(t))$ is greater than or equal to the minimum of $r_i(t)$ over all i, and $S(t) = 0$ when $\hat{r}_d(v(t))$ is less than the minimum of $r_i(t)$ over all i, where $i = 1, 2, ..., N_w$, N_w is the total number of singing whales, $v(t)$ is the measured wind speed, $r_i(t)$ is the range of the ith singing humpback whale from the sensor at time t, and T_{day} is the full diurnal time period of 24 hours. The detectable humpback whale song occurrence rate is then $\dfrac{T_{song}}{T_{day}}$.

3.6 Autocorrelation of annual humpback whale song occurrence time series in 2006 and 2008

We calculated the normalized autocorrelation function [156] of the Vu et al. [35] 2006 and 2008 annual humpback whale song occurrence time series. The autocorrelation function at zero time

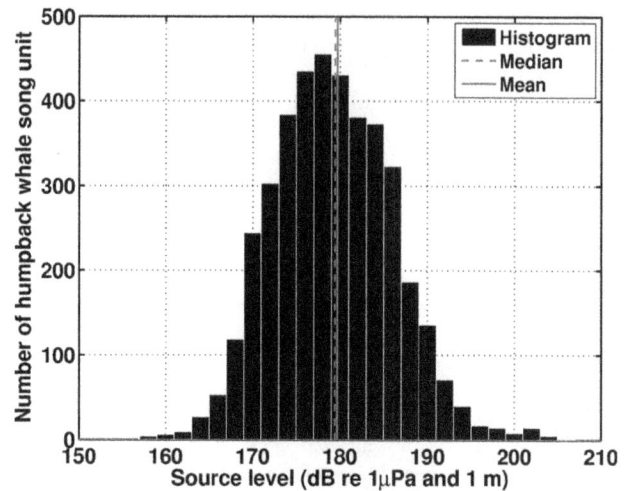

Figure 17. Histogram of the measured humpback whale song unit source levels. The humpback whale song unit source levels measured from more than 4,000 recorded song units during the same 2006 Gulf of Maine experiment discussed here at the same time and at the same location approximately follow a Gaussian distribution and are in the range 155 to 205 dB re 1 μPa and 1 m with a mean of 179.8 dB re 1 μPa and 1 m and a median of 179.4 dB re 1 μPa and 1 m, which are within 0.6 dB of the median of all published humpback whale song unit source levels of 180 dB re 1 μPa and 1 m [93,101,102,152–154]. The solid and dashed gray lines represent the mean and the median of the measured humpback song unit source levels, respectively.

lag, where perfect temporal correlation exists, is one. The time lag at which the autocorrelation function falls to $1/e$ is the e-folding time scale defining the width of the correlation peak, or coherence time scale, within which processes are conventionally taken to be correlated [156,157]. The e-folding time scale of the Vu et al. [35] annual humpback whale song occurrence time series is 18 days for 2006 and 21 days for 2008 (Figure 18). The roughly 20-day coherence time scale shows that the humpback song occurrence gradually changes over periods longer than the 11-day periods analyzed in Risch et al. [34]. This time is consistent with the smooth and gradual transition in Figure 12 of the correlation coefficient of 11-day periods across years from negative values in the "before" period to some of the highest positive correlations obtained between years in the "during" period, which contradicts the results of the Risch et al. [34] study and is consistent with a violation of temporal causality in the Risch et al. [34] study. It is noteworthy that (1) the humpback song occurrence dropped to zero in the "before" period, and (2) only after a time period consistent with the measured coherence time scale of song occurrence, within which temporal processes are correlated, did song occurrence begin to increase in the "during" period (Figure 10). The Risch et al. [34] analysis then also violates temporal causality because the correlated processes that caused the reduction in humpback song occurrence started days before the OAWRS survey transmissions began, yet the analysis and conclusions of Risch et al. [34] offer no other explanation than these OAWRS survey transmissions for the reduction, when only other causes are causally possible. Indeed as we have shown in Section 2.2 non-sonar causes regularly lead to such changes in song occurrence, and as we have shown in Section 3.5 standard detection range variations from measured wind speed dependent noise variations at Stellwagen and measured humpback whale song sources near Georges Bank completely account for the changes reported in Risch et al. [34].

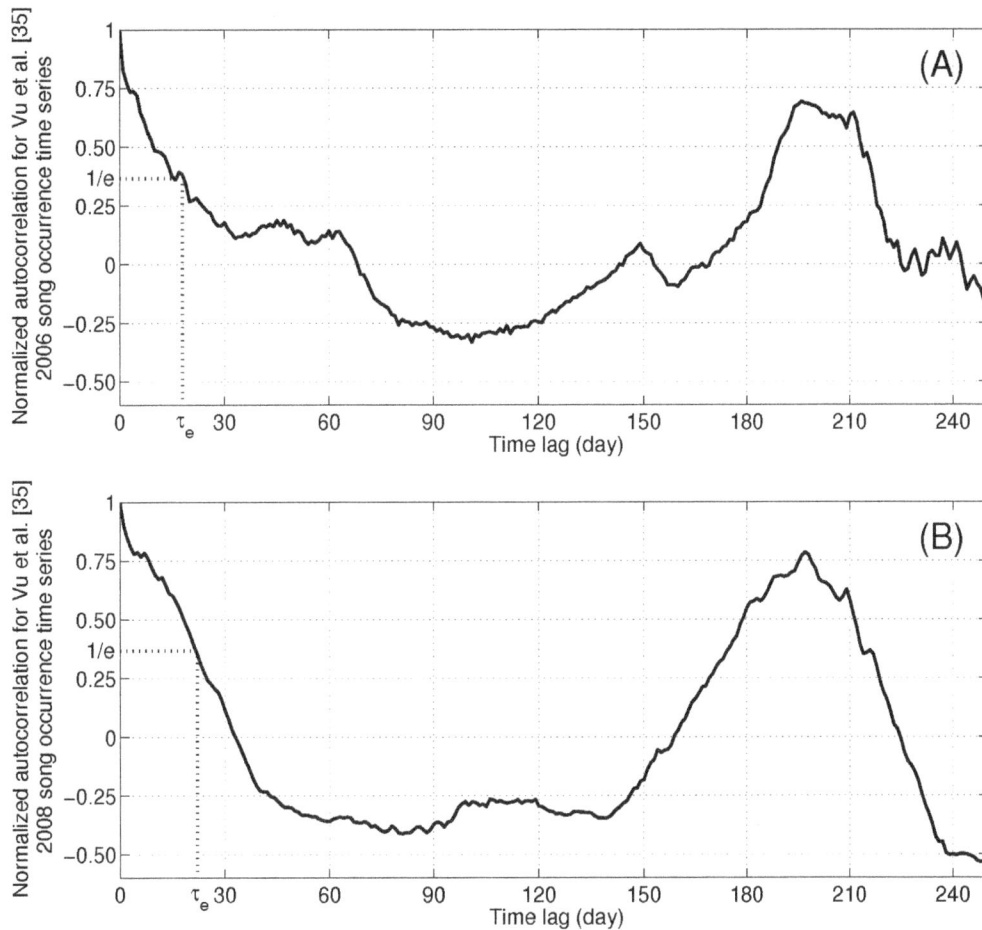

Figure 18. Autocorrelation of Vu et al. [35] humpback whale song occurrence time series in 2006 and 2008. The e-folding time scale τ_e of the Vu et al. [35] annual humpback whale song occurrence time series is (A) 18 days for 2006 and (B) 21 days for 2008. The roughly 20-day coherence time scale shows that the humpback song occurrence gradually changes over periods longer than the 11-day periods analyzed in Risch et al. [34]. It is noteworthy that (1) the humpback song occurrence dropped to zero in the "before" period, and (2) only after a time period consistent with the measured coherence time scale of song occurrence, within which temporal processes are correlated, did song occurrence begin to increase in the "during" period (Figure 10). The Risch et al. [34] analysis then violates temporal causality because the correlated processes that caused the reduction in humpback song occurrence started days before the OAWRS survey transmissions began, yet the analysis and conclusions of Risch et al. [34] offer no other explanation than these survey transmissions for the reduction. Both time series show high correlation at a time lag of roughly seven months due to increases in song occurrence during the spring and fall seasons (Figure 10), separated by roughly seven months.

Acknowledgments

This research was supported by the National Oceanographic Partnership Program, the Census of Marine Life, the Office of Naval Research, the Alfred P. Sloan Foundation, the National Science Foundation, the Presidential Early Career Award for Scientists and Engineers, Northeastern University, and Massachusetts Institute of Technology. We thank David Reed for providing technical assistance. The funders had no role in study design, data collection and analysis, decision to publish, or preparation of the manuscript.

Author Contributions

Contributed reagents/materials/analysis tools: ZG ADJ DT DHY FW AZ PR NCM. Wrote the paper: ZG ADJ DT DHY FW AZ PR NCM. Conceived and designed theory and experiments: ZG ADJ DT DHY PR NCM. Developed theory and performed experiments: ZG ADJ DT DHY PR NCM.

References

1. Noad M, Cato D, Bryden M, Jenner M, Jenner K (2000) Cultural revolution in whale songs. Nature 408: 537–537.
2. Cato D (1991) Songs of humpback whales: The Australian perspective. Mem Queensl Mus 30: 277–290.
3. Noad MJ, Cato DH, Stokes MD (2004) Acoustic tracking of humpback whales: measuring interactions with the acoustic environment. In: Proceedings of Acoustics. pp.353–358.
4. Cato D, McCauley R, Rogers T, Noad M (2006) Passive acoustics for monitoring marine animals - progress and challenges. In: Proceedings of Acoustics. pp.453–460.

5. Au WWL, Mobley J, Burgess WC, Lammers MO, Nachtigall PE (2000) Seasonal and diurnal trends of chorusing humpback whales wintering in waters off Western Maui. Mar Mamm Sci 16: 530–544.
6. Dunlop R, Noad M, Cato D, Stokes D (2007) The social vocalization repertoire of East Australian migrating humpback whales (*Megaptera novaeangliae*). J Acoust Soc Am 122: 2893–2905.
7. Dunlop RA, Cato DH, Noad MJ (2008) Non-song acoustic communication in migrating humpback whales (*Megaptera novaeangliae*). Mar Mamm Sci 24: 613–629.

8. Mattila DK, Guinee LN, Mayo CA (1987) Humpback whale songs on a North Atlantic feeding ground. J Mamm 68: 880–883.

9. Thompson PO, Cummings WC, Ha SJ (1986) Sounds, source levels and associated behavior of humpback whales, Southeast Alaska. J Acoust Soc Am 80: 735–740.

10. Noad M, Cato D (2001) A combined acoustic and visual survey of humpback whales off southeast Queensland. Mem Queensl Mus 47: 507–523.

11. Cato D (1998) Simple methods of estimating source levels and locations of marine animal sounds. J Acoust Soc Am 104: 1667–1678.

12. McDonald MA, Fox CG (1999) Passive acoustic methods applied to fin whale population density estimation. J Acoust Soc Am 105: 2643–2651.

13. Watkins W, Daher M, Reppucci G, George J, Martin D, et al. (2000) Seasonality and distribution of whale calls in the North Pacific. Oceanography 13: 62–67.

14. Stafford K, Nieukirk S, Fox C (2001) Geographic and seasonal variation of blue whale calls in the North Pacific. J Cetacean Res Manage 3: 65–76.

15. Watkins W, Daher M, George J, Rodriguez D (2004) Twelve years of tracking 52-hz whale calls from a unique source in the North Pacific. Deep-Sea Res Part I 51: 1889–1901.

16. Barlow J, Taylor B (2005) Estimates of sperm whale abundance in the northeastern temperate Pacific from a combined acoustic and visual survey. Mar Mamm Sci 21: 429–445.

17. Thode A (2004) Tracking sperm whale (*Physeter macrocephalus*) dive profiles using a towed passive acoustic array. J Acoust Soc Am 116: 245–253.

18. Makris NC, Ratilal P, Jagannathan S, Gong Z, Andrews M, et al. (2009) Critical Population Density Triggers Rapid Formation of Vast Oceanic Fish Shoals. Science 323: 1734–1737.

19. Gong Z, Andrews M, Jagannathan S, Patel R, Jech J, et al. (2010) Low-frequency target strength and abundance of shoaling Atlantic herring (*Clupea harengus*) in the Gulf of Maine during the Ocean Acoustic Waveguide Remote Sensing 2006 Experiment. J Acoust Soc Am 127: 104–123.

20. Urick RJ (1983) Principles of Underwater Sound. New York: McGraw Hill, 29–65 and 343–366 pp.

21. Clay CS, Medwin H (1977) Acoustical Oceanography, John Wiley Sons Inc, New York. pp.494–501.

22. Jensen FB, Kuperman WA, Porter MB, Schmidt H (2011) Computational Ocean Acoustics. New York: Springer-Verlag, 2nd edition, 708–713 pp.

23. Tolstoy I, Clay C (1966) Ocean Acoustics: Theory and Experiment in Underwater Sound. McGraw-Hill, 1–9 pp.

24. Burdic WS (1984) Underwater Acoustic System Analysis, Prentice-Hall Englewood Cliffs NJ, volume 2. pp.322–360.

25. Kay S (1998) Fundamentals of Statistical Signal Processing, Volume II: Detection Theory, volume 7. Upper Saddle River (New Jersey), 512 pp.

26. Crocker MJ (1998) Handbook of Acoustics. Wiley Interscience, 460 pp.

27. Rossing TD (2007) Springer Handbook of Acoustics. Springer Science + Business Media, New York, 179 pp.

28. Lurton X (2002) An Introduction to Underwater Acoustics. Springer-Verlag, 172–180 pp.

29. Gong Z, Tran D, Ratilal P (2013) Comparing passive source localization and tracking approaches with a towed horizontal receiver array in an ocean waveguide. J Acoust Soc Am 134: 3705–3720.

30. Lee S, Makris NC (2006) The array invariant. J Acoust Soc Am 119: 336–351.

31. Council NEFM (2011) Draft Amendment 5 to the Fishery Management Plan (FMP) for Atlantic Herring Including a Draft Environmental Impact Statement (DEIS), Volume 1. Technical report, New England Fishery Management Council in consultation with National Marine Fisheries Service, Atlantic States Marine Fisheries Commission and Mid-Atlantic Fishery Management Council.

32. King J, Camisa M, Manfredi V, Correia S (2011) Massachuseets Fishery Resource Assessment: 2010 Annual Performance Report. Technical report, Massachusetts Division of Marine Fisheries, United States Department of Interior Fish and Wildlife Service, Region 5 Wildlife and Sport Fish Restoration Program.

33. Weinrich M, Martin M, Bove J, Schilling M (1997) A shift in distribution of humpback whales, *Megaptra novaeangliae*, in response to prey in the southern Gulf of Maine. Fish Bull 95: 826–836.

34. Risch D, Corkeron PJ, Ellison WT, Van Parijs SM (2012) Changes in humpback whale song occurrence in response to an acoustic source 200 km away. PLoS ONE 7: e29741.

35. Vu ET, Risch D, Clark CW, Gaylord S, Hatch LT, et al. (2012) Humpback whale song occurs extensively on feeding grounds in the western North Atlantic Ocean. Aquat Biol 14: 175–193.

36. Moore D, McCabe G, Craig B (2010) Introduction to the Practice of Statistics. W. H. Freeman and Company, 268–294 pp.

37. Goldman R, Weinberg J (1985) Statistics, An Introduction, Prentice-Hall, chapter 3. pp.240–260.

38. Li Q (2012) Digital Sonar Design in Underwater Acoustics: Principles and Applications. New York: Springer-Verlag, 168–178 pp.

39. Au WWL (1993) The Sonar of Dolphins. Springer-Verlag, 8 pp.

40. Makris NC, Ratilal P, Symonds DT, Jagannathan S, Lee S, et al. (2006) Fish Population and Behavior Revealed by Instantaneous Continental Shelf-Scale Imaging. Science 311: 660–663.

41. Makris NC (2003) Geoclutter Acoustics Experiment 2003 Cruise Report. Technical report, MIT.

42. Hare J, Churchill J, Richardson D, Jech M, Deroba J, , et al. An evaluation of whether changes in the timing and distribution of Atlantic herring spawning on Georges Bank may have biased the NEFSC acoustic survey: Preliminary results from a NOAA FATE funded project. Technical report, Northeast Fisheries Science Center and Woods Hole Oceanographic Institution.

43. Jech J, Stroman F (2012) Aggregative patterns of pre-spawning Atlantic herring on Georges Bank from 1999–2010. Aquat Living Resour 25: 1–14.

44. Payne P, Wiley D, Young S, Pittman S, Clapham P, et al. (1990) Recent fluctuations in the abundance of baleen whales in the southern Gulf of Maine in relation to changes in selected prey. Fish Bull 88: 687–696.

45. Battista T, Clark R, Pittman S (2006) An ecological characterization of the Stellwagen Bank National Marine Sanctuary region. NOAA Technical Memorandum NCCOS 45, 1–20 pp.

46. Gong Z (2012) Remote Sensing of Marine Life and Submerged Target Motions with Ocean Waveguide Acoustics. PhD dissertation, Northeastern University, The Department of Electrical and Computer Engineering.

47. Introduction to NEFSC Fisheries Acoustic Survey Operations, Northeast Fisheries Science Center. National Oceanographic and Atmospheric Administration. Available: http://www.nefsc.noaa.gov/femad/ecosurvey/acoustics/pages/surveys3.htm (Accessed 2012 Oct 29).

48. Milinski M (1993) Predation risk and feeding behaviour. In: Pitcher TJ, editor, Behaviour of Teleost Fishes, Chapman & Hall. pp.285–305.

49. Mackinson S, Nøttestad L, Guénette S, Pitcher T, Misund O, et al. (1999) Cross-scale observations on distribution and behavioural dynamics of ocean feeding Norwegian spring-spawning herring (*Clupea harengus L.*). ICES J Mar Sci 56: 613–626.

50. Madsen P, Wahlberg M, Møhl B (2002) Male sperm whale (*Physeter macrocephalus*) acoustics in a high-latitude habitat: implications for echolocation and communication. Behav Ecol Sociobiol 53: 31–41.

51. Tyack P, Clark CW (2000) Communication and acoustic behavior of whales and dolphins. In: Au WWL, Popper AN, Fay RR, editors, Hearing by Whales and Dolphins, Handbook on Auditory Research, Springer, Berlin Heidelberge, New York. pp.156–224.

52. Edds-Walton PL (1997) Acoustic communication signals of mysticete whales. Bioacoustics 8: 47–60.

53. Johnson M, Hickmott L, Soto N, Madsen P (2008) Echolocation behaviour adapted to prey in foraging Blainville's beaked whale (*Mesoplodon densirostris*). Proc R Soc B 275: 133–139.

54. Stimpert A, Wiley D, Au W, Johnson M, Arsenault R (2007) Megapclicks: acoustic click trains and buzzes produced during night-time foraging of humpback whales (*Megaptera novaeangliae*). Biol Lett 3: 467–470.

55. Jagannathan S, Symonds D, Bertsatos I, Chen T, Nia H, et al. (2009) Ocean Acoustic Waveguide Remote Sensing (OAWRS) of marine ecosystems. Mar Ecol Prog Ser 395: 137–160.

56. Ratilal P, Lai Y, Symonds DT, Ruhlmann LA, Preston JR, et al. (2005) Long range acoustic imaging of the continental shelf environment: The Acoustic Clutter Reconnaissance Experiment 2001. J Acoust Soc Am 117: 1977–1998.

57. Chia CS, Makris NC, Fialkowski LT (2000) A comparison of bistatic scattering from two geologically distinct abyssal hills. J Acoust Soc Am 108: 2053–2070.

58. Makris NC, Chia CS, Fialkowski LT (1999) The bi-azimuthal scattering distribution of an abyssal hill. J Acoust Soc Am 106: 2491–2512.

59. Makris NC, Avelino LZ, Menis R (1995) Deterministic reverberation from ocean ridges. J Acoust Soc Am 97: 3547–3574.

60. Makris NC, Berkson JM (1994) Long-range backscatter from the mid-Atlantic ridge. J Acoust Soc Am 95: 1865–1881.

61. Makris NC (1993) Imaging ocean-basin reverberation via inversion. J Acoust Soc Am 94: 983–993.

62. Lai Y (2004) Acoustic scattering from stationary and moving targets in shallow water environments - with application of humpback whale detection and localization. PhD dissertation, Massachusetts Institute of Technology, Department of Ocean Engineering.

63. Jech J, Michaels W (2006) A multifrequency method to classify and evaluate fisheries acoustics data. Can J Fish Aquat Sci 63: 2225–2235.

64. Nero RW, Thompson CH, Jech JM (2004) *In situ* acoustic estimates of the swimbladder volume of Atlantic herring (*Clupea harengus*). ICES J Mar Sci: Journal du Conseil 61: 323–337.

65. Rose G (2007) Cod: The Ecological History of the North Atlantic Fisheries. Breakwater Books. Available: http://books.google.com/books?id=tDNe7GOOwfwC (Accessed 2014 Jul 21).

66. Cerchio S, Dahlheim M (2001) Variation in feeding vocalizations of humpback whales (*Megaptera novaeangliae*) from Southeast Alaska. Bioacoustics 11: 277–296.

67. Clark CW, Clapham PJ (2004) Acoustic monitoring on a humpback whale (*Megaptera novaeangliae*) feeding ground shows continual singing into late spring. Proc R Soc B 271: 1051–1057.

68. Altman D (1991) Practical Statistics for Medical Research. Chapman & Hall/CRC, 409–419 pp.

69. Chow S, Liu J (1998) Design and Analysis of Clinical Trials: Concept and Methodologies. Wiley-Interscience, 89–93 pp.

70. Van Trees HL (2001) Detection, Estimation, and Modulation Theory, Part I. Wiley-Interscience, 23–46 pp.

71. Council NR (1997) Oceanography and Naval Special Warfare: Opportunities and Challenges. Technical report, National Academy of Sciences.

72. National Data Buoy Center. National Oceanic and Atmospheric Administration. Available: http://www.ndbc.noaa.gov/(Accessed 2012 Oct 29).

73. Wenz GM (1962) Acoustic Ambient Noise in the Ocean: Spectra and Sources. J Acoust Soc Am 34: 1936–1956.

74. Abbott B, Ahmed R, Greene G, Kristanovich FC, Luchessa S, et al. (2011) Comments on Drakes Bay Oyster Company Special Use Permit Environmental Impact Statement: Point Reyes National Seashore. ENVIRON International Corporation, Seattle, Washington.

75. Ocean Studies Board, The Division on Earth and Life Studies, The National Academies (2012) Scientific Review of the Draft Environmental Impact Statement: Drakes Bay Oyster Company Special Use Permit. National Academies Press.

76. WhaleWatch. New England Aquarium. Available: http://www.neaq.org/visit_planning/whale_watch/index.php (Accessed 2012 Oct 29).

77. Hatch L, Clark C, Merrick R, Van Parijs S, Ponirakis D, et al. (2008) Characterizing the Relative Contributions of Large Vessels to Total Ocean Noise Fields: A Case Study Using the Gerry E. Studds Stellwagen Bank National Marine Sanctuary. Environ Manage 42: 735–752.

78. Greene CR, Moore SE (1995) Man-made noise. In: Richardson WJ, Greene CR, Malme CI, Thomson DH, editors, Marine Mammals and Noise, Academic Press: New York, chapter 12. pp.437–452.

79. Nowacek DP, Thorne LH, Johnston DW, Tyack PL (2007) Responses of cetaceans to anthropogenic noise. Mammal Rev 37: 81–115.

80. Pater LL, Grubb TG, Delaney DK (2009) Recommendations for improved assessment of noise impacts on wildlife. J Wildl Manage 73: 788–795.

81. Chapter 8.16, noise control. Code of Ordinances, Cambridge, Massachusetts. Available: http://library.municode.com/HTML/16889/level2/TIT8HESA_CH8. 16NOCO.html (Accessed 2014 Jul 21).

82. McCarthy E (2004) International Regulation of Underwater Sound: Establishing Rules and Standards to Address Ocean Noise Pollution, Springer, chapter 1 & 2.

83. Ocean Studies Board, National Research Council (2000) Marine Mammals and Low-Frequency Sound: Progress Since 1994, National Academies Press, chapter 1 & 2.

84. Endangered and Threatened Species: Designation of Critical Habitat for Cook Inset BelugaWhale. Proposed Rules, Federal Register. Technical report, National Oceanic and Atmospheric Administration. Department of Commerce National Oceanic & Atmospheric Administration.

85. Berger E (2003) The Noise Manual. AIHA, 26 pp.

86. Oster E (2004) Witchcraft, Weather and Economic Growth in Renaissance Europe. J Econ Perspect 18: 215–228.

87. Behringer W (1999) Climatic change and witch-hunting: the impact of the Little Ice Age on mentalities. Clim Chang 43: 335–351.

88. Pavlac BA (2009) Witch Hunts in the Western World: Persecution and Punishment from the Inquisition through the Salem Trials, Greenwood Pub Group. pp.1–188.

89. Ashforth A (2001) AIDS, witchcraft, and the problem of power in post-apartheid South Africa. School of Social Science. Princeton NJ: School of Social Science, Institute for Advanced Study.

90. Miguel E (2005) Poverty and witch killing. Rev Econ Stud 72: 1153–1172.

91. Becker K, Preston JR (2003) The ONR Five Octave Research Array (FORA) at Penn State. IEEE J Ocean Eng 5: 2607–2610.

92. Au W, Mobley J, Burgess W, Lammers M, Nachtigall P (2000) Seasonal and diurnal trends of chorusing humpback whales wintering in waters off western Maui. Mar Mamm Sci 16: 530–544.

93. Cato DH, Paterson RA, Paterson P (2001) Vocalisation rates of migrating humpback whales over 14 years. Mem Queensl Mus 47: 481–490.

94. Tyack PL (1981) Interactions between singing Hawaiian humpback whales and conspecifics nearby. Behav Ecol Sociobiol 8: 105–116.

95. Stimpert AK, Au WW, Parks SE, Hurst T, Wiley DN (2011) Common humpback whale (Megaptera novaeangliae) sound types for passive acoustic monitoring. J Acoust Soc Am 129: 476–482.

96. Mellinger D, Stafford K (2007) Fixed Passive Acoustic Observation Methods for Cetaceans. Oceanography 20: 36–45.

97. Swartz SL, Martinez A, Stamates J, Burks C, Mignucci-Giannoni A (2002) Acoustic and Visual Survey of Cetaceans in the Waters of Puerto Rico and the Virgin Islands, February-March 2001. US Department of Commerce, National Oceanic and Atmospheric Administration, NOAA Fisheries, Southeast Fisheries Science Center.

98. Baumgartner MF, Mussoline SE (2011) A generalized baleen whale call detection and classification system. J Acoust Soc Am 129: 2889.

99. Rebull OG, Cusí JD, Fernández MR, Muset JG (2006) Tracking fin whale calls offshore the Galicia Margin, North East Atlantic Ocean. J Acoust Soc Am 120: 2077.

100. Oswald JN, Barlow J, Norris TF (2003) Acoustic identification of nine delphinid species in the eastern tropical Pacific Ocean. Mar Mamm Sci 19: 20–037.

101. Winn HE, Perkins PJ, Poulter TC (1970) Sounds of the humpback whale. In: Proc 7th Annu Conf of Biol Sonar pp.39–52.

102. Au WWL, Andrews K (2001)Feasibility of using acoustic DIFAR technology to localize and estimate Hawaiian humpback whale population. Technical report, Hawaiian Islands Humpbak Whale National Marine Sanctuary, Office of National Marine Santuaries, NOAA, US Dept. of Commerce and Department of Land and Natural Resources, State of Hawaii, USA .

103. Parks S, Searby A, Celerier A, Johnson M, Nowacek D, et al. (2011) Sound production behavior of individual North Atlantic right whales: implications for passive acoustic monitoring. Endanger Species Res 15: 63–76.

104. Vanderlaan AS, Hay AE, Taggart CT (2003) Characterization of North Atlantic right-whale (Eubalaena glacialis) sounds in the Bay of Fundy. IEEE J Ocean Eng 28: 164–173.

105. Matthews J, Brown S, Gillespie D, Johnson M, McLanaghan R, et al. (2001) Vocalisation rates of the North Atlantic right whale (Eubalaena glacialis). J Cetacean Res Manage 3: 271–282.

106. Laurinolli MH, Hay AE, Desharnais F, Taggart CT (2003) Localization of North Atlantic right whale sounds in the Bay of Fundy using a sonobuoy array. Mar Mamm Sci 19: 708–723.

107. Edds-Walton PL (2000) Vocalizations of Minke Whales Balaenoptera acutorostrata in the St. Lawrence Estuary. Bioacoustics 11: 31–50.

108. Parks SE, Hamilton PK, Kraus SD, Tyack PL (2005) The gunshot sound produced by male North Atlantic right whales (Eubalaena glacialis) and its potential function in reproductive advertisement. Mar Mamm Sci 21: 458–475.

109. Mellinger DK, Carson CD, Clark CW (2000) Characteristics of minke whale (Balaenoptera acutorostrata) pulse trains recorded near Puerto Rico. Mar Mamm Sci 16: 739–756.

110. Nieukirk SL, Stafford KM, Mellinger DK, Dziak RP, Fox CG (2004) Low-frequency whale and seismic airgun sounds recorded in the mid-Atlantic Ocean. J Acoust Soc Am 115: 1832.

111. Rankin S, Barlow J (2007) Vocalizations of the sei whale Balaenoptera borealis off the Hawaiian Islands. Bioacoustics 16: 137–145.

112. Baumgartner MF, Fratantoni DM (2008) Diel periodicity in both sei whale vocalization rates and the vertical migration of their copepod prey observed from ocean gliders. Limono Oceanogr 53: 2197–2209.

113. Knowlton A, Clark CW, Kraus S (1991) Sounds recorded in the presence of sei whale, Balaenoptera borealis. J Acoust Soc Am 89: 1968.

114. Thompson TJ, Winn HE, Perkins PJ (1979) Mysticete Sounds. Springer.

115. Overholtz W, Link J (2007) Consumption impacts by marine mammals, fish, and seabirds on the Gulf of Maine–Georges Bank Atlantic herring (Clupea harengus) complex during the years 1977–2002. ICES J Mar Sci: Journal du Conseil 64: 83–96.

116. Winn H, Winn L (1978) The song of the humpback whale Megaptera novaeangliae in the West Indies. Mar Biol 47: 97–114.

117. Darling JD, Jones ME, Nicklin CP (2006) Humpback whale songs: Do they organize males during the breeding season? Behaviour 143: 1051–1102.

118. Glockner DA, Venus S (1983) Determining the sex of humpback whales (Megaptera novaeangliae) in their natural environment. In: Communication and Behavior of Whales. AAAS Selected Symposia Series, Westview Press, Boulder, CO, USA. pp.447–464.

119. FURUNO Marine GPS Navigator Model GP-90 brochure, FURUNO Electric CO., LTD. Available: http://techserv.gso.uri.edu/DownLoad/FurunoGP90.pdf (Accessed 2013 Apr 18).

120. Thode A, Skinner J, Scott P, Roswell J, Straley J, et al. (2010) Tracking sperm whales with a towed acoustic vector sensor. J Acoust Soc Am 128: 2681–2694.

121. Hinich MJ, Rule W (1975) Bearing estimation using a large towed array. J Acoust Soc Am 58: 1023–1029.

122. Mukhopadhyay N, Datta S, Chattopadhyay S (2013) Applied sequential methodologies: real-world examples with data analysis, CRC Press. pp.11–16.

123. Greening MV, Perkins JE (2002) Adaptive beamforming for nonstationary arrays. J Acoust Soc Am 112: 2872–2881.

124. Born M, Wolf E (1999) Principles of Optics: Electromagnetic Theory of Propagation, Interference and Diffraction of Light. CUP Archive, 462 pp.

125. Rayleigh L (1879) Xxxi. Investigations in optics, with special reference to the spectroscope. Lond Edinb Dubl Philos Mag 8: 261–274.

126. Tukey J (1991) The philosophy of multiple comparisons. Stat Sci 6: 100–116.

127. Bretz F, Hothorn T, Westfall P (2010) Multiple comparisons using R. Chapman & Hall/CRC, 82–93 pp.

128. R: a language and environment for statistical computing. R Development Core Team. Available: http://www.R-project.org (Accessed 2012 Oct 29).

129. Makris NC (1996) The effect of saturated transmission scintillation on ocean acoustic intensity measurements. J Acoust Soc Am 100: 769–783.

130. Dyer I (1970) Statistics of Sound Propagation in the Ocean. J Acoust Soc Am 48: 337–345.

131. Medwin H, Blue JE (2005) Sounds in the Sea: From Ocean Acoustics to Acoustical Oceanography. Cambridge University Press.

132. Pierce AD (1989) Acoustics: An Introduction to its Physical Principles and Applications. Acoust Soc Am.

133. Wilson JD, Makris NC (2006) Ocean acoustic hurricane classification. J Acoust Soc Am 119: 168–181.

134. Cato D (1976) Ambient sea noise in waters near Australia. J Acoust Soc Am 60: 320–328.

135. Burgess A, Kewley D (1983) Wind-generated surface noise source levels in deep water east of Australia. J Acoust Soc Am 73: 201–210.

136. Crouch WW, Burt PJ (1972) The Logarithmic Dependence of Surface-Generated Ambient-Sea-Noise Spectrum Level on Wind Speed. J Acoust Soc Am 51: 1066–1072.

137. Piggott C (1964) Ambient sea noise at low frequencies in shallow water of the Scotian Shelf. J Acoust Soc Am 36: 2152–2163.

138. Cato DH, Tavener S (1997) Ambient sea noise dependence on local, regional and geostrophic wind speeds: implications for forecasting noise. Appl Acoust 51: 317–338.
139. Kewley D, Browning D, Carey W (1990) Low-frequency wind-generated ambient noise source levels. J Acoust Soc Am 88: 1894–1902.
140. Wilson JD, Makris NC (2008) Quantifying hurricane destructive power, wind speed, and air-sea material exchange with natural undersea sound. Geophys Res Lett 35: L10603.
141. Collins MD (1993) Generalization of the Split-Step Padé solution. J Acoust Soc Am 93: 1736–1742.
142. RAM to Navy Standard Parabolic Equation: Transition from Research to Fleet Acoustic Model. R.A. Zingarelli and D.B. King, Acoustics Division, Naval Research Laboratory. Available: http://www.nrl.navy.mil/research/nrl-review/2003/simulation-computing-modeling/zingarelli/(Accessed 2013 Nov 07).
143. Etter PC (2013) Underwater Acoustic Modeling and Simulation. CRC Press.
144. Brekhovskikh LM, Lysanov Y (2003) Fundamentals of Ocean Acoustics. Springer.
145. Argo database. International Argo Project. Available: www.argo.ucsd.edu, http://argo.jcommops.org (Accessed 2012 Oct 29).
146. Simmen J, Flatté S, Wang G (1997) Wavefront folding, chaos, and diffraction for sound propagation through ocean internal waves. J Acoust Soc Am 102: 239–255.
147. Andrews M, Chen T, Ratilal P (2009) Empirical dependence of acoustic transmission scintillation statistics on bandwidth, frequency, and range in New Jersey continental shelf. J Acoust Soc Am 125: 111–124.
148. Andrews M, Gong Z, Ratilal P (2011) Effects of multiple scattering, attenuation and dispersion in waveguide sensing of fish. J Acoust Soc Am 130: 1253–1271.
149. Tran D, Andrews M, Ratilal P (2012) Probability distribution for energy of saturated broadband ocean acoustic transmission: Results from Gulf of Maine 2006 experiment. J Acoust Soc Am 132: 3659–3672.
150. Northeast Fisheries Science Center (2012) 54th Northeast Regional Stock Assessment Workshop (54th SAW): Assessment Report. US Dept Commer, Northeast Fish Sci Cent Ref Doc. 12–18; 600 p.
151. Jagannathan S, Küsel ET, Ratilal P, Makris NC (2012) Scattering from extended targets in range-dependent fluctuating ocean-waveguides with clutter from theory and experiments. J Acoust Soc Am 132: 680.
152. Au WWL, Pack AA, Lammers MO, Herman LM, Deakos MH, et al. (2006) Acoustic properties of humpback whale songs. J Acoust Soc Am 120: 1103–1110.
153. Cato D (1992) The biological contribution to the ambient noise in waters near Australia. Acoust Aust 20: 76–80.
154. Levenson C (1972) Characteristics of sound produced by humpback whales (*Megaptera novaeangliae*). NAV-OCEANO Technical Note: 7700–7706.
155. Thode AM, Gerstoft P, Burgess WC, Sabra KG, Guerra M, et al. (2006) A Portable Matched-Field Processing System Using Passive Acoustic Time Synchronization. IEEE J Ocean Eng 31: 696–710.
156. Zar JH (1984) Biostatistical Analysis. 2nd edition, pp.1–688.
157. Archer D (2007) Global Warming: Understanding the Forecast, Cambridge Univ Press. p. 48.
158. Gerry E. Studds Stellwagen Bank National Marine Sanctuary: Research Programs. National Oceanic and Atmospheric Administration. Available: http://stellwagen.noaa.gov/about/location.html (Accessed 2012 Oct 31).
159. Allen J, Peterson M, Sharrard G, Wright D, Todd S (2012) Radiated noise from commercial ships in the Gulf of Maine: Implications for whale/vessel collisions. J Acoust Soc Am 132: EL229–EL235.
160. Kipple B, Gabriele C (2004) Underwater noise from skiffs to ships. In: Proc. of Glacier Bay Science Symposium. pp.172–175.
161. Arveson P, Vendittis D (2000) Radiated noise characteristics of a modern cargo ship. J Acoust Soc Am 107: 118–129.
162. Vasconcelos RO, Amorim MCP, Ladich F (2007) Effects of ship noise on the detectability of communication signals in the Lusitanian toadfish. J Exp Biol 210: 2104–2112.
163. Pol M, Carr AH (2002) Developing a low impact sea scallop dredge, Final Report NOAA/NMFS Saltonstall-Kennedy Program NA96FD0072. Technical report, Massachusetts Division of Marine Fisheries, Pocasset, MA, USA.
164. Lesage V, Barrette C, Kingsley M, Sjare B (1999) The effect of vessel noise on the vocal behavior of belugas in the St. Lawrence River Estuary, Canada. Mar Mamm Sci 15: 65–84.
165. Erbe C (2002) Underwater noise of whale-watching boats and potential effects on killer whales (*Ornicus orca*), based on an acoustic impact model. Mar Mamm Sci 18: 394–418.
166. Pack AA, Potter J, Herman LM, Hoffmann-Kuhnt M, Deakos MH (2001) Determining source levels, sound fields, and body sizes of singing humpback whales (*Megaptra novaeangliae*) in the Hawaiian Winter Grounds. Technical report, The Dolphin Institute, Honolulu, Hawaii.
167. Gerry E. Studds Stellwagen Bank National Marine Sanctuary Condition Report 2007. U.S. Department of Commerce, National Oceanic and Atmospheric Administration, National Marine Sanctuary Program, Silver Spring, MD.
168. Gerry E. Studds Stellwagen Bank National Marine Sanctuary: Whalewatching Guidelines, Northeast Region including Stellwagen Bank. National Oceanic and Atmospheric Administration. Available: http://stellwagen.noaa.gov/visit/whalewatching/guidelines.html (Accessed 2012 Oct 29).

2

Carbon and Nitrogen Isotopes from Top Predator Amino Acids Reveal Rapidly Shifting Ocean Biochemistry in the Outer California Current

Rocio I. Ruiz-Cooley[1]*, Paul L. Koch[2], Paul C. Fiedler[3], Matthew D. McCarthy[1]

1 Ocean Sciences Department, University of California Santa Cruz, Santa Cruz, California, United States of America, 2 Earth and Planetary Sciences Department, University of California Santa Cruz, Santa Cruz, California, United States of America, 3 Southwest Fisheries Science Center, National Marine Fisheries Service, National Oceanic and Atmospheric Administration, La Jolla, California, United States of America

Abstract

Climatic variation alters biochemical and ecological processes, but it is difficult both to quantify the magnitude of such changes, and to differentiate long-term shifts from inter-annual variability. Here, we simultaneously quantify decade-scale isotopic variability at the lowest and highest trophic positions in the offshore California Current System (CCS) by measuring $\delta^{15}N$ and $\delta^{13}C$ values of amino acids in a top predator, the sperm whale (*Physeter macrocephalus*). Using a time series of skin tissue samples as a biological archive, isotopic records from individual amino acids (AAs) can reveal the proximate factors driving a temporal decline we observed in bulk isotope values (a decline of ≥ 1 ‰) by decoupling changes in primary producer isotope values from those linked to the trophic position of this toothed whale. A continuous decline in baseline (i.e., primary producer) $\delta^{15}N$ and $\delta^{13}C$ values was observed from 1993 to 2005 (a decrease of ~ 4‰ for $\delta^{15}N$ source-AAs and 3‰ for $\delta^{13}C$ essential-AAs), while the trophic position of whales was variable over time and it did not exhibit directional trends. The baseline $\delta^{15}N$ and $\delta^{13}C$ shifts suggest rapid ongoing changes in the carbon and nitrogen biogeochemical cycling in the offshore CCS, potentially occurring at faster rates than long-term shifts observed elsewhere in the Pacific. While the mechanisms forcing these biogeochemical shifts remain to be determined, our data suggest possible links to natural climate variability, and also corresponding shifts in surface nutrient availability. Our study demonstrates that isotopic analysis of individual amino acids from a top marine mammal predator can be a powerful new approach to reconstructing temporal variation in both biochemical cycling and trophic structure.

Editor: Wei-Chun Chin, University of California, Merced, United States of America

Funding: Funding was provided by Marine Mammal and Turtle Division, Southwest Fisheries Science Center-National Oceanographic Atmospheric Administration for data collection and isotope analysis and National Science Foundation(Division of Ocean Sciences(OCE)-1155728, and OCE-0623622) for analysis of amino acids and data. Funding for Open Access provided by the University of California, Santa Cruz, Open Access Fund. The funders had no role in study design, data collection and analysis, decision to publish, or preparation of the manuscript.

Competing Interests: The authors have declared that no competing interests exist.

* Email: rcooley@ucsc.edu

Introduction

The California Current System (CCS) contains one of the five major coastal upwelling zones in the world's oceans, and hosts a great diversity and abundance of marine life [1]. The oceanographic state of this large ecosystem is dynamic. Natural climate variation and anthropogenic stressors alter biochemical cycling, food web dynamics, and the fitness of species [1–3]. Known interannual and decadal changes are related both to the El Niño-Southern Oscillation (ENSO) and to basin-scale processes associated with the Pacific Decadal Oscillation (PDO) [1].The latter, is an index of interannual sea surface temperature (SST) variability in the North Pacific, that is related to physical and biochemical variations and influences community changes in plankton, fish and other taxa [4,5]. In addition to this natural variability, humans have perturbed climate by increasing atmospheric CO_2 concentrations, which have increased ocean temperatures, water column stratification, hypoxia, and water column

anoxia and have decreased surface ocean pH [6,7]. These environmental factors may negatively impact populations of species, increasing mortality and decreasing reproductive success due to habitat compression and metabolic constraints [8]. Other anthropogenic pressures, such as intensive fisheries and the past whaling industry (which principally targeted sperm whales, *Physeter macrocephalus*) might have triggered top-down effects. Given the lack of detailed proxy records to trace simultaneously biochemical baselines and length of food webs, assessing the extent to which biogeochemical cycling and community structure in pelagic ecosystems have changed over the past century is difficult, as is attributing change to natural cycles versus anthropogenic disturbances.

The isotopic values of marine primary producers are sensitive to environmental variation, such as change in temperature, and CO_2 or nitrate concentrations, as well as biological differences such as physiology and growth rate [9–11]. Hence, the carbon and nitrogen isotope values ($\delta^{13}C$ and $\delta^{15}N$ values, respectively) of primary producers, also known as "baseline isotope values", vary in space and time as a function of these fundamental ecosystem properties [12]. Baseline isotope values are then integrated into consumers' tissues through diet, typically with metabolic fractionation leading to enrichment in the heavier isotope (especially ^{15}N) in consumers [13,14]. Therefore, isotopic values of marine consumers could be used to reconstruct changes in diet and/or ecosystem biogeochemistry. The $\delta^{13}C$ and $\delta^{15}N$ values from a resident animal can potentially provide an integrated record of the biogeochemical characteristics of its habitat, as well as its trophic position [15]. However, because multiple factors influence the bulk $\delta^{13}C$ and $\delta^{15}N$ values ultimately recorded in consumer tissues, it is often difficult to disentangle the effects of changing trophic position from shifts in baseline values.

Studies in different ocean basins have shown that bulk tissue $\delta^{13}C$ or $\delta^{15}N$ values have declined over the last century, but interpretations of these trends have varied widely [16]. For example, declining bulk tissue $\delta^{15}N$ values are sometimes attributed to a drop in consumer trophic level [17,18] or to baseline shifts due to either changes in foraging zone or biogeochemical cycles [16]. In particular, two recent studies in the Pacific have revealed pervasive declines in $\delta^{15}N$ values in the offshore Central Pacific [18] and North Pacific Subtropical Gyre (NPSG) [19], but offered diametrically opposing interpretations as to underlying mechanism. In the highly productive CCS, despite accumulating evidence for oceanographic changes since the 1950s [2], isotopic data from plankton species have been contradictory. Bulk $\delta^{15}N$ values from three zooplakton species have exhibited no long-term trends, whereas data for a specialized zooplankton feeder decreased by approximately 3‰ [20,21]. Declines in $\delta^{13}C$ values over the 20th century are expected due to the combustion of fossil fuels (i.e., the Suess effect), and have been observed in many records and ecosystems [22]. However, variability in the magnitude and timing of $\delta^{13}C$ declines has suggested that other factors, such as declining primary productivity, could also contribute in some regions [16]. In the offshore CCS, there are currently no $\delta^{13}C$ time series for organic or inorganic material.

Isotopic analysis of individual amino acids (AAs) can effectively separate trophic effects from shifts in baseline isotope values [23,24]. Regardless of an animal's trophic position, the original $\delta^{15}N$ and $\delta^{13}C$ values from primary producers are relatively well preserved within the group of 'source-AAs' for nitrogen [25] and the 'essential-AAs' for carbon [26]. In contrast, isotopic values from the 'trophic-AAs' for nitrogen, and 'non-essential-AAs' for carbon, undergo significant metabolic fractionation, and vary in association with a consumer's diet [23,24], tissue turnover rates, and possibly metabolism [27]. Hence, isotopic analysis of amino acids from apex marine mammal predators offers a unique opportunity to simultaneously investigate temporal variation at the lowest and highest trophic levels of their food web. Sperm whales are top predators of the mesopelagic ocean. Mark-recapture studies, morphology, and acoustic analysis indicate that female sperm whales forage within the same oceanic region year round [28]. Consequently, they can function as natural biological samplers, broadly integrating biogeochemical information from their home ecosystem. In this study, we use sperm whale skin as a novel biological archive of time series data. Our data combine bulk tissue and AA isotope analysis to examine temporal variation in baseline values (reflecting ecosystem biogeochemistry) and whale trophic position (indicating trophic structure) from offshore waters of the California Current ecosystem.

Results and Discussion

Foraging zone of sperm whales sampled in CCS

In the CCS off the US west coast, sperm whales are found in oceanic waters from California to Washington [29]. Their habitat therefore excludes the coastal upwelling system that exhibits strong latitudinal isotopic gradients [30]. Mitochondrial and nuclear markers reveal that the CCS whales are an independent population and a single genetic stock [31]. Our isotopic data from skin biopsies (Figure 1) indicate that whales fed homogenously within the offshore northern and central CCS. First, the variation in bulk isotope values (n = 18; SD = 1.2‰ for $\delta^{13}C$ and 1.2‰ for $\delta^{15}N$) is similar to the variation observed in other sperm whale populations that are considered to be resident (i.e., Gulf of Mexico and Gulf of California, SD≤0.8‰ for both $\delta^{15}N$ and $\delta^{13}C$ [15]; SE Pacific, SD = 3.5‰ for $\delta^{15}N$ and 0.7‰ for $\delta^{13}C$ [32]). In addition, the $\delta^{15}N$ values for phenylalanine (Phe; n = 12; mean (SD) = 10.9‰ (0.9)) are relatively consistent with expected nitrate and particulate organic matter $\delta^{15}N$ values from the oceanic northern CCS (~6 to 10 ‰) [12], and also with published Phe $\delta^{15}N$ values from muscle of the jumbo squid (*Dosidicus gigas*; potential prey of sperm whales) [33]. Phe $\delta^{15}N$ values are a proxy for primary producer values [25] as they exhibit only minor ^{15}N-enrichment with trophic transfer [23]. In top predators (such as sperm whales), this likely results in slightly higher Phe $\delta^{15}N$ values versus baseline inorganic N sources. Lastly, because latitudinal trends in the $\delta^{15}N$ values from predator source-AAs can indicate their geographic residency [24,33], the lack of any latitudinal variation in Phe $\delta^{15}N$ values ($r^2 = 0$; n = 12) strongly suggests that the individual sperm whales sampled here were not foraging in different localized regions, but rather foraged over a broad latitudinal range within the northern and central CCS. While the isotopic incorporation rate for extremely large animals like whales is not well known, the thick skin of sperm whales likely integrates information for at least three and possibly more than six months prior to sampling [34]. Our data set encompasses information mainly from the fall and winter, except for the samples collected in 2001 and 2003, which also integrate information from the summer.

Coupled decadal declines in $\delta^{15}N$ and $\delta^{13}C$ values

Bulk $\delta^{13}C$ and $\delta^{15}N$ values in whale skin decreased from 1993 to 2005 by 1.1‰ and 1.7‰, respectively. These decreases were statistically significant at the alpha = 0.05 level. Inclusion of a single sample available from 1972 further suggests possible longer-term temporal declines for both $\delta^{13}C$ and $\delta^{15}N$ values by ≥4‰ and >3‰, that are also statistically significant (Table 1). Together, these coupled time-series declines in bulk $\delta^{13}C$ and $\delta^{15}N$ values suggest coincident biogeochemical or trophic system perturbation (Table 1, Figure 2). In particular, the rate of decrease for bulk $\delta^{15}N$ values since the 1970's is at least five times greater than the rate for the long-term $\delta^{15}N$ decrease recently documented in the central Pacific from proteinaceous corals (2.3‰ in 150 years; annual decrease calculated at 0.015‰) [19], and it is more similar to the rate of change observed for a single zooplankton $\delta^{15}N$ record from southern California (~3‰ in 50 years) [21].

To disentangle the factors driving the declines in bulk isotope values, we analyzed individual AA isotope values, focusing on AAs that have been demonstrated to track baseline changes (as noted above, essential AA for $\delta^{13}C$ values, source AA for $\delta^{15}N$ values). Linear regression models for average $\delta^{13}C$ and $\delta^{15}N$ values from the most accurately measured essential- and source-AAs both exhibited strong negative temporal trends across all the data (i.e. for both 1972 and 1993 to 2005; Table 1, Figure 2), with drops of

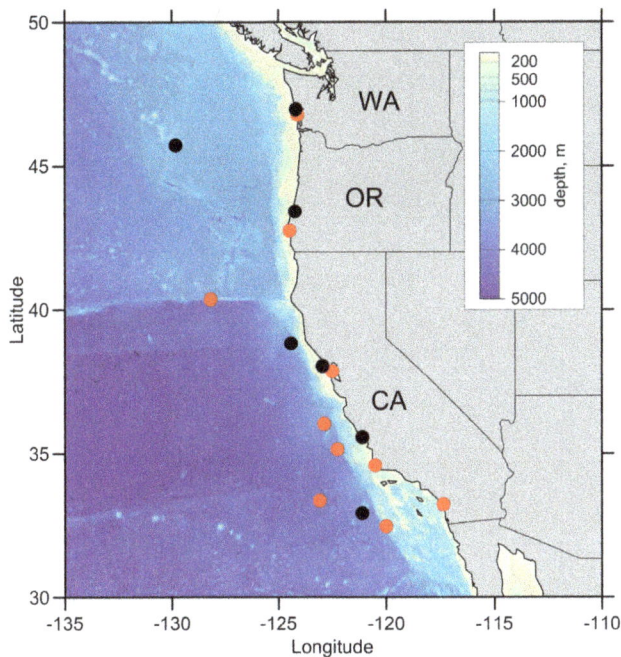

Figure 1. Sperm whales are distributed year-round in offshore deep waters (~>150 km off the US west coast [29]). Skin samples (○) from free-ranging sperm whales were collected together with skin from stranded individuals. Tissue samples were used for bulk (in black) and amino acid (in red) stable isotope analysis.

$\geq 3‰$ and $>4‰$ respectively indicated by compound-specific isotope data. Residuals for all regressions exhibited a random pattern. In contrast, average $\delta^{15}N$ values for the trophic-AAs were much variable, resulting in a lower r^2, but overall they paralleled the source-AA trend (Table 1, Figure 2A). These results are not consistent with any significant drop in sperm whale trophic level as the primary driver of decreases in bulk isotope ratios, and instead strongly implicate coupled changes in baseline $\delta^{15}N$ and $\delta^{13}C$ values.

These negative trends in baseline $\delta^{15}N$ and $\delta^{13}C$ values might relate to changes in biochemical cycling, rates of primary production, or primary producer species composition. In particular, the decline in average essential-AA $\delta^{13}C$ values (Figure 2B), which are a direct proxy for primary producers, is far too high to be explained solely by the Seuss effect (~0.2 ‰ per decade since 1960 [35]), and it also coincides with the decline in average source-AA $\delta^{15}N$ values. This suggests that the mechanism explaining a drop in primary producer $\delta^{15}N$ values should be consistent with a concurrent large decline in $\delta^{13}C$ values. One possiblity, which would represent a direct analogy to changes in other ocean regions, would be a shift towards more oligotrophic conditions for the outer CCS. This explanation would be consistent with coupled declines in both isotopes, linked to decreased primary production and a shift in species composition that is typically associated with warmer and more stratified ocean conditions [36]. Oligotrophy in the world ocean is increasing due to climate shifts [37] and is projected to continue increasing in the North Pacific [38]. Recent isotopic records from deep sea proteinaceous corals, for example, provide strong support for such linked trends associated with warming of the NPSG [39]. The nitrogen isotope record from deep sea coral indicate that the long-term declines in baseline $\delta^{15}N$ values are likely linked to progressive increases in seasonal gyre extent, leading to steady increases in N contribution from diazotrophy [19]. Therefore, an analogous explanation would imply that oceanographic conditions in the offshore CCS region (which have conditions more similar to the open ocean and represent the base of sperm whales' food web) might have shifted toward more "gyre-like" conditions, driving baseline isotope values toward those more typical of the oligotrophic open ocean.

However, to our knowledge, there is currently no evidence for substantially increasing SST and diazotrophy in the CCS itself. Instead, recent analyses suggest largely the opposite: overall, the thermocline weakened and shoaled in the offshore CCS between 1950 and 1993 [7], possibly increasing nutrient availability in the euphotic zone despite increased stratification [40]. Additionally, the offshore CCS has cooled (not heated) since the early 1990s (Figure 3), and this trend is also reflected in the present "cool" PDO regime. Furthermore, the generalization that global warming will universally increase stratification and thus decrease surface nutrient supply has been recently challenged for some regions including the CCS [41]. For example, one recent model projects increases in nitrate supply and productivity in the CCS during the 21st century despite increases in stratification and limited change in wind-driven upwelling [42]. In the southern CCS, coastal surface nutrients have increased possibly linked to a general shoaling of the nutricline [43]. In the Southern California Bight, the most intensively monitored region of the CCS, nutrients in source waters have also increased over the last three decades, but the N:P and Si:N ratios were greatly reduced, possibly shifting phytoplankton species composition and abundance [44]. Whether or not these trends in nutrient dynamics extend to other regions of the CCS is unclear, because the oceanographic state of this ecosystem varies regionally [1,45].

In particular, shifts in offshore and onshore oceanographic conditions appear to be decoupled. Coastal upwelling has recently increased, as expected for enhanced alongshore winds [46], but has decreased offshore where upwelling is driven by wind-stress curl [47]. Since 1997, trends in satellite chlorophyll estimates, an index of phytoplankton biomass, have been positive in coastal upwelling waters but tend to be zero or negative in offshore waters [48]. Together, this current evidence indicates cooling, but not increases in productivity, in the offshore CCS concurrent with the observed 1993–2006 trends in sperm whale $\delta^{15}N$ and $\delta^{13}C$ values. Lower temperatures increase the solubility of CO_2 and change the fractionation associated with carbon fixation, often resulting in lower phytoplankton $\delta^{13}C$ values [49]; lower temperatures might have also changed phytoplankton growth and species composition. If surface nitrate also increased along the outer CCS region sampled by these whales, then the degree of nitrate utilization by primary producers (and so their $\delta^{15}N$ values) could have also changed, since phytoplankton preferentially assimilate $^{14}NO_3^-$ [50]. In general, proportional nitrate utilization is lower where surface NO_3^- concentrations are higher [50]. Therefore, lower NO_3^- utilization during seasonal upwelling might also be expected to depress the $\delta^{15}N$ values of primary producers, propagating the ^{15}N-depleted signal into food webs during their most productive periods. At present, there simply are not enough detailed data on nutrient concentrations and other oceanographic factors in the outer CCS to deduce a mechanism. However, the observed declining baseline values revealed by sperm whales do indicate a recently progressive shift in primary producer dynamics, likely associated with changes in SST, average state of surface nutrients and/or primary production.

Table 1. Temporal variation in $\delta^{15}N$ and $\delta^{13}C$ values from the offshore California Current System in sperm whale skin samples.

Time period	Tracer	Linear Regression	n	r^2	p-value	Isotopic shift (‰)	Annual decrease
1993–2005	$\delta^{15}N$						
	Bulk	y = 302−0.143 * year	17	0.25	<0.05	1.7	0.14
	Mean Source-AA	y = 717−0.354 * year	11	0.52	= 0.01	4.2	0.35
	Mean Trophic-AA	y = 311−0.143 * year	11	0.12	>0.05	1.7	
1972–2005	$\delta^{15}N$						
	Bulk	y = 218−0.101 * year	18	0.37	<0.05	3.3	0.10
	Mean Source-AA	y = 298−0.145 * year	12	0.39	<0.05	4.7	0.14
	Mean Trophic-AA	y = 88−0.031 * year	12	0.03	>0.05	1.0	
1993–2005	$\delta^{13}C$						
	Bulk	y = 174−0.095 * year	17	0.24	<0.05	1.1	0.09
	Mean Essential-AA	y = 474−0.250 * year	8	0.62	<0.05	3.0	0.25
1972–2005	$\delta^{13}C$						
	Bulk	y = 242−0.129 * year	18	0.67	<0.01	4.2	0.12
	Mean Essential-AA	y = 184−0.105 * year	9	0.58	<0.05	3.4	0.10

For mean calculations: Source-AAs are phenylalanine, glycine, lysine, tyrosine; Trophic-AA: glutamic acid, alanine, isoleucine, leucine, proline; Essential-AA: phenylalanine, valine, leucine. Isotopic shifts were calculated using the corresponding linear regression equations listed in this table. The annual decrease was calculated for shifts that exhibited a p-value≤0.05.

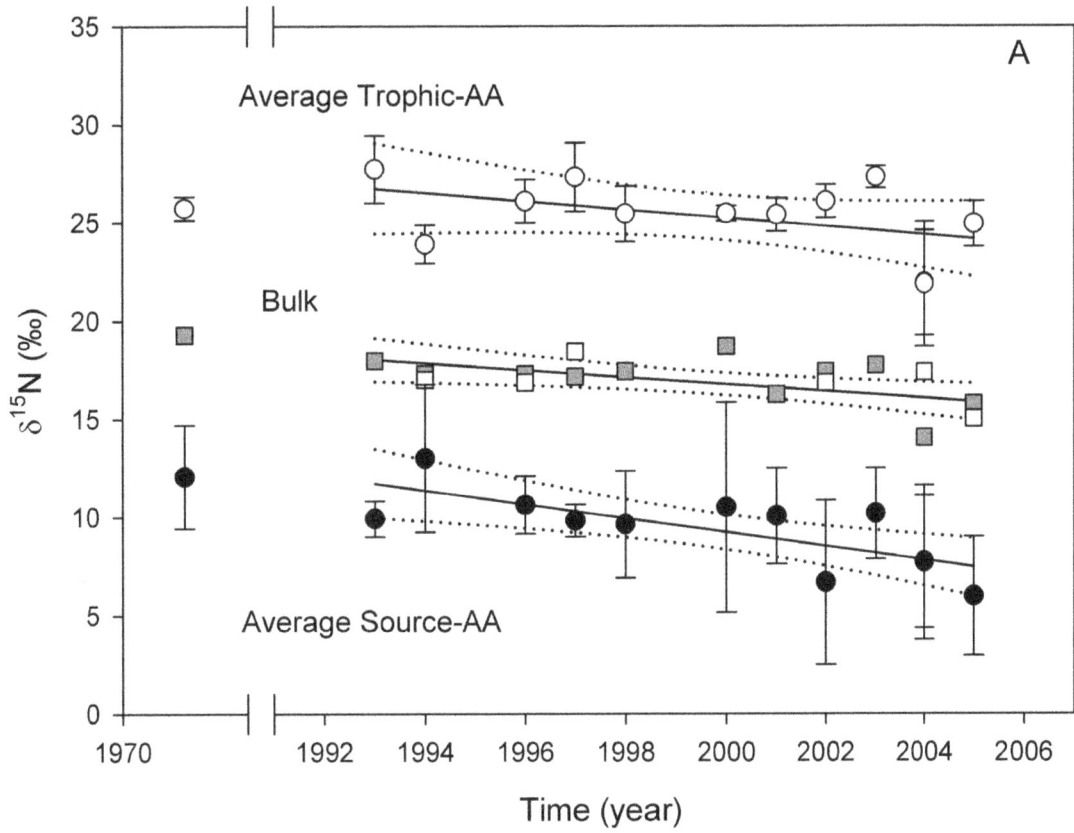

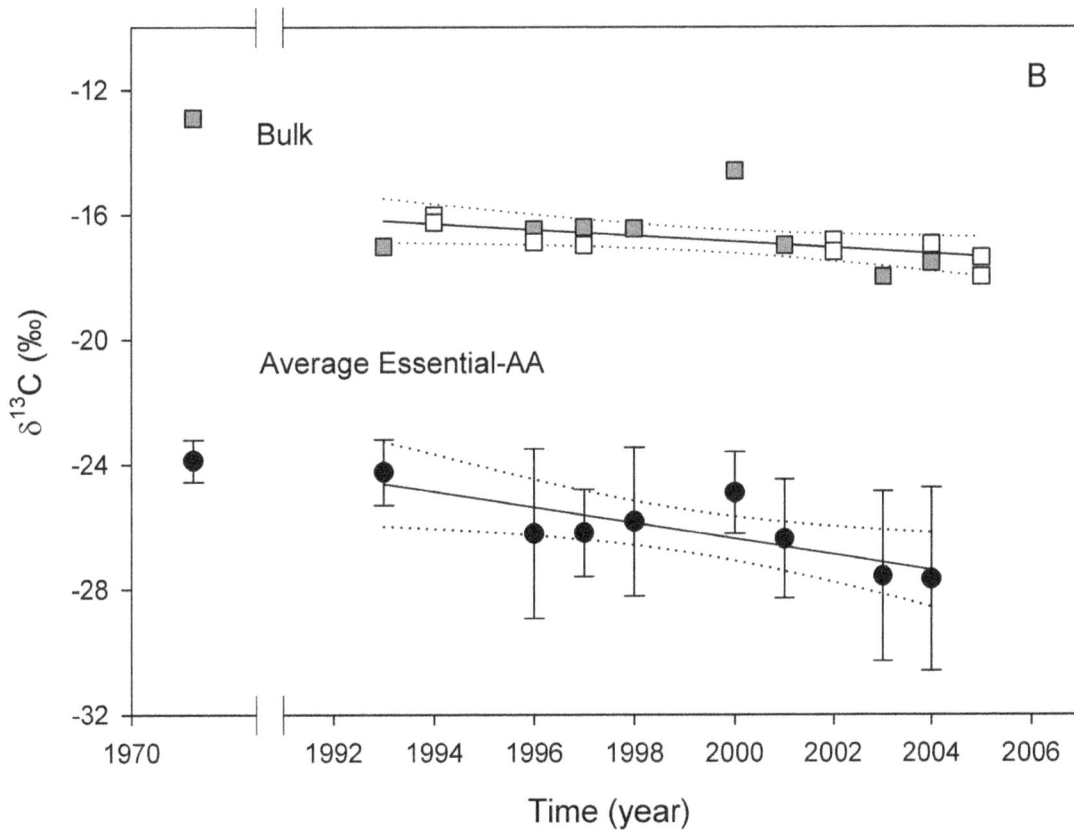

Figure 2. Time series of isotopic data from sperm whale skin. (A) $\delta^{15}N$ values from bulk skin, average source-AAs and average trophic- AAs ($\pm$ SD); and (B) $\delta^{13}C$ values from bulk skin and average essential-AAs ($\pm$SD). Bulk isotope data are plotted with a square symbol ($\square$), filled grey squares indicate the samples that were also analyzed for amino acid stable isotope analysis. The corresponding linear regression equations are provided in Table 1, as are the amino acids included within each AA-group.

Implications of Rapid Change for offshore CCS Biogeochemistry

Although our time series data are limited for both elements, the compound-specific AA data identify a parallel decline in both baseline $\delta^{15}N$ and $\delta^{13}C$ values in the outer CCS from 1992 to 2005, likely indicative of major recent shifts in biochemical cycling. At the same time, however, the overall similarity in whale trophic position signifies that the broad trophic structure is realtively unaffected. We note that in comparison with the recent deep sea coral data from the gyre offshore of this region [19], our data suggest that both the rate and scale of biochemical change on the CCS margin may be far greater than in the open Pacific Ocean. The coral record from the NPSG indicates a fairly steady $\delta^{15}N$ annual decrease of $\sim$0.015‰ over the last 150 years with a total drop of 2.3 ‰ in exported primary production $\delta^{15}N$ values over that period. In contrast, our molecular-level proxies for $\delta^{15}N$ values at the base of the food chain (the source AAs) indicate more rapid annual declines of 0.35 ‰ since the 1990's. The independent molecular proxies for primary production $\delta^{13}C$ values (the essential AAs) indicate relatively similar declines.

Together with the CCS observations discussed above, the contrast with the NPSG coral data (while not directly comparable in terms of time scale), suggests that despite the fact that baseline $\delta^{15}N$ declines are observed in both data sets, different biogeo-chemical mechanisms may underlie the changes in these very different oceanographic regions. Climate variability likely affects the biochemistry of ecosystems differently depending on the oceanographic properties, microbial and phytoplankton commu-nities, and species assemblages. In the eastern Pacific Ocean, the structure of the pycnocline varies strongly among the known biogeochemical provinces [51]. This likely influences geographic variation in surface nutrient availability, and therefore stable isotope ratios in POM, primary producers [12] and consumers [14]. Temporal trends in pycnocline depth, SST, stratification, and mixed layer depths also differ between these biogeochemical provinces [40]. For example, while SST decreased overall since 1958 in many parts of the California Current, SST increased in

the easternmost southern subtropical gyre and equatorial Pacific [40]. Ultimately, more detailed data that couple integrated measures of ecosystem baseline with oceanographic state will be required to understand the substantial biogeochemical changes our data indicate.

Our work highlights that detailed time-series of biochemical baseline and trophic structure records among different ecosystems will be crucial to identify rapid ecosystem shifts in response to climate change. In particular, in the face of uncertain coupling of natural and anthropogenic climate forcing, understanding the timing, extent and especially the mechanistic basis for baseline shifts now represents an urgent challenge. However, despite many efforts to unravel the linkage and feedback controls between the carbon and nitrogen cycles, and the effect of their variability on primary production and food-web dynamics, they are still not well understood. This study has demonstrated the great potential in coupling molecular isotopic tools with the unique bioarchive of sperm whales (or other top predators), as sentinels of offshore ecosystems. This may allow, for the first time, decoding of the factors that underlie temporal trends in bulk isotopic records, while simultaneosly monitoring changes at both the highest and lowest trophic levels. We suggest that integrating this approach with detailed oceanographic data will be a major new tool to identify the effects of natural climate variability versus anthropo-genic global warming on ecosystem biochemistry and primary production. Elucidating such patterns from this and other ocean margin regions, in particular their relationships with oceano-graphic and climatic variations and shifts in primary production, will be an essential part of the critical task of predicting future trends in both ecosystem biochemistry and trophic dynamics.

Material and Methods

A total of 18 skin samples (Figure 1) were analyzed for bulk stable isotope analysis. Skin tissue samples with enough material (3.5 mg) were selected for CSIA-AA. Data from 12 samples were obtained for individual AA $\delta^{15}N$ values, and 9 samples for $\delta^{13}C$ values. The Southwest Fisheries Science Center/Pacific Islands

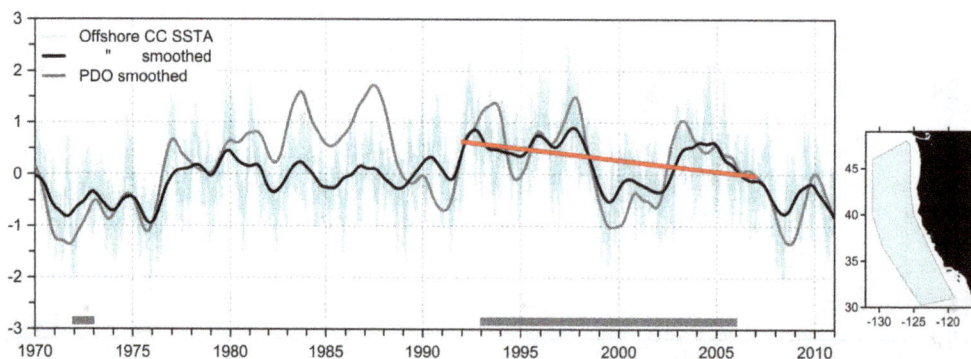

Figure 3. Time series data of sea surface temperature anomaly (SSTA) from the offshore California Current (inset map) and the Pacific Decadal Oscillation (PDO). Monthly SSTA was computed in 0.5-deg fields from the Simple Ocean Data Assimilation version 2.2.4 reanalysis (http://coastwatch.pfeg.noaa.gov/erddap/griddap/hawaii_d90f_20ee_c4cb.html), and then averaged in the offshore area (the plot shows $\pm$1sd). Monthly SSTA (°C) and PDO values were smoothed with a 25-month lowess smooth. The linear fit is for 1992–2006 (red line, slope −0.044°C y-1). Sample periods are indicated along the time axis.

Fisheries Science Center Institutional Animal Care and Use Committee (IACUC) approved the original animal work that produced the samples. Sex was determined genetically using qPCR sexing assay by the PRD-Genetic Lab at NOAA [52]. These samples consisted of 5 females, 2 males and 2 unidentified individuals possibly corresponding to females or juvenile males. Large adult males were not included. Bulk isotope values were analyzed by continuous flow isotope ratio mass spectrometry (IRMS; Thermo Finnigan) and standardized relative to Vienna-Pee Belemnite (V-PDB) for carbon and atmospheric N_2 for nitrogen. Results are expressed in part per thousand (‰) and standard notation: $\delta^H X = [(R_{sample}/R_{standard})-1] \times 1000$, where H is the mass number of the heavy isotope, X is either C or N, and R_{sample} and $R_{standard}$ are the ratio of $^{13}C/^{12}C$ or $^{15}N/^{14}N$ in the sample and standard, respectively.

We hydrolyzed and prepared approximately 3.5 mg of skin as well as a control (Cyanno; bacteria tissue) [53] to quantify $\delta^{15}N$ values from source- and trophic-AAs and $\delta^{13}C$ values from essential- and non-essential-AAs. All derivatives were injected with an AA control, N-leucine, to verify accuracy during each run, and analyzed via gas chromatography-IRMS to obtain $\delta^{15}N$ and $\delta^{13}C$ values from individual AAs. Each sample was run 3–4 times to maximize accuracy among chromatograms. The associated analytical error among replicates was <1.0 ‰. For all samples, $\delta^{15}N$ values were obtained from a total of four source-AAs (phenylalanine, glycine, lysine, tyrosine), and five trophic-AAs (glutamic acid, alanine, isoleucine, leucine, proline) (Figure S1A). For $\delta^{13}C$ values, the essential-AAs that we consistently determined were phenylalanine, valine and leucine, and the non-essential-AA were alanine, proline, aspatic acid, glutamic acid and tyrosine (Figure S1B).

The relative pattern of AA $\delta^{15}N$ and $\delta^{13}C$ values was highly consistent with past work from other organisms and tissues [23,25,54]. We grouped data as source- or trophic-AAs for $\delta^{15}N$ values, and essential- or non-essential-AAs for $\delta^{13}C$ values to increase power in the analysis and evaluate temporal variation. We calculated average values for each AA group and they are reported in Table S1. Regression analyses were conducted to evaluate linear relationship between time and each isotopic tracer for both bulk and individual-AA $\delta^{15}N$ and $\delta^{13}C$ values (Table 1).

There was a weak correlation between average source-AA and trophic-AA ($r^2 = 0.13$; $p = 0.67$), indicating that trophic-AA $\delta^{15}N$ values could not be predicted by the variability in source-AAs, and vice versa. However, the correlation between average essential-AA and non-essential-AA $\delta^{13}C$ values was moderate ($r^2 = 0.63$, $p = 0.06$). Since the controls on isotopic patterns for non-essential-AA $\delta^{13}C$ values are complex and dependent on diet quality and quantity, including *de novo* synthesis and routing of AAs from diet-to-tissue, this group was not considered in the linear regression analysis.

Supporting Information

Figure S1 Stable isotope values of individual amino acids (AAs) in skin samples of sperm whales (*Physeter macrocephalus*). (A) Four $\delta^{15}N$ Source-AAs: phenylalanine (phe), glycine (gly), lysine (lys), tyrosine (tyr), and five Trophic-AAs: glutamic acid (glx), alanine (ala), isoleucine (ile), leucine (leu), proline (Pro); and (B) Three $\delta^{13}C$ essential-AAs: phe, leu, and valine (val).

Table S1 Average values and one standard deviations (SD) were calculated for Source-AAs (phenylalanine, glycine, lysine, tyrosine), Trophic-AAs (glutamic acid, alanine, isoleucine, leucine, proline) and Essential-AAs (phenylalanine, valine, leucine).

Acknowledgments

We thank L. T. Ballance, J. Barlow, K. Robertson (SWFSC/NMFS/NOAA) and J. Calambokidis (Cascadia Research) for facilitating the use of tissues samples, and the genetic SWFSC lab for molecular whale sex identification.

Author Contributions

Conceived and designed the experiments: RIRC MDM. Performed the experiments: RIRC MDM. Analyzed the data: RIRC PCF. Contributed reagents/materials/analysis tools: MDM. Contributed to the writing of the manuscript: RIRC PLK PCF MDM.

References

1. Checkley JDM, Barth JA (2009) Patterns and processes in the California Current System. Prog Oceanogr 83: 49–64.
2. Bograd SJ, William JS, Barlow J, Booth A, Brodeur RD, et al. (2010) Status and trends of the California Current region, 2003–2008. PICES Special Publication. 106–141 p.
3. McGowan JA, Bograd SJ, Lynn RJ, Miller AJ (2003) The biological response to the 1977 regime shift in the California Current. Deep Sea Res II 50: 2567–2582.
4. Brinton E, Townsend A (2003) Decadal variability in abundances of the dominant euphausiid species in southern sectors of the California Current. Deep Sea Res II 50: 2449–2472.
5. Chavez FP, Ryan J, Lluch-Cota SE, Ñiquen CM (2003) From anchovies to sardines and back: Multidecadal change in the Pacific Ocean. Science 299: 217–221.
6. Chan F, Barth JA, Lubchenco J, Kirincich A, Weeks H, et al. (2008) Emergence of anoxia in the California Current Large Marine Ecosystem. Science 319: 920.
7. Palacios DM, Bograd SJ, Mendelssohn R, Schwing FB (2004) Long-term and seasonal trends in stratification in the California Current, 1950–1993. J Geophy Res 109: C10016.
8. Bograd SJ, Castro CG, Di Lorenzo E, Palacios DM, Bailey H, et al. (2008) Oxygen declines and the shoaling of the hypoxic boundary in the California Current. Geophys Res Lett 35: L12607.
9. Farrell JW, Pedersen TF, Calvert SE, Nielsen B (1995) Glacial-interglacial changes in nutrient utilization in the equatorial Pacific Ocean. Nature 377: 514–517.
10. Rau GH, Sweeney RE, Kaplan IR (1982) Plankton $^{13}C:^{12}C$ ratio changes with latitude: differences between northern and southern oceans. Deep Sea Res 29: 1035–1039.
11. Goericke R, Fry B (1994) Variations of marine plankton $\delta^{13}C$ with latitude, temperature, and dissolved CO_2 in the World Ocean. Global Biogeochem Cy 8: 85–90.
12. Somes CJ, Schmittner A, Galbraith ED, Lehmann MF, Altabet MA, et al. (2010) Simulating the global distribution of nitrogen isotopes in the ocean. Global Biogeochem Cy 24.
13. Peterson BJ, Fry B (1987) Stable Isotopes in Ecosystem Studies. Annu Rev Ecol Evol Syst 18: 293–320.
14. Ruiz-Cooley RI, Gerrodette T (2012) Tracking large-scale latitudinal patterns of $\delta^{13}C$ and $\delta^{15}N$ along the eastern Pacific using epi-mesopelagic squid as indicators. Ecosphere 3: 63.
15. Ruiz-Cooley R, Engelhaupt D, Ortega-Ortiz J (2012) Contrasting C and N isotope ratios from sperm whale skin and squid between the Gulf of Mexico and Gulf of California: effect of habitat. Mar Biol: 1–14.
16. Schell DM (2001) Carbon isotope ratio variations in Bering Sea biota: The role of anthropogenic carbon dioxide. Limnol Oceanogr Methods 46: 999–1000.
17. Emslie SD, Patterson WP (2007) Abrupt recent shift in $\delta^{13}C$ and $\delta^{15}N$ values in Adélie penguin eggshell in Antarctica. Proc Natl Acad Sci USA 104: 11666–11669.
18. Wiley AE, Ostrom PH, Welch AJ, Fleischer RC, Gandhi H, et al. (2013) Millennial-scale isotope records from a wide-ranging predator show evidence of recent human impact to oceanic food webs. Proc Natl Acad Sci USA 110: 8972–8977.
19. Sherwood OA, Guilderson TP, Batista FC, Schiff JT, McCarthy MD (2013) Increasing subtropical North Pacific Ocean nitrogen fixation since the Little Ice Age. Nature 505: 78–81.

20. Rau GH, Ohman MD, Pierrot-Bults A (2003) Linking nitrogen dynamics to climate variability off central California: a 51 year record based on $^{15}N/^{14}N$ in CalCOFI zooplankton. Deep Sea Res Part II 50: 2431–2447.

21. Ohman MD, Rau GH, Hull PM (2012) Multi-decadal variations in stable N isotopes of California Current zooplankton. Deep Sea Res Part I 60: 46–55.

22. Sonnerup RE, Quay PD, McNichol AP, Bullister JL, Westby TA, et al. (1999) Reconstructing the oceanic ^{13}C Suess Effect. Global Biogeochem Cy 13: 857–872.

23. Chikaraishi Y, Ogawa NO, Kashiyama Y, Takano Y, Suga H, et al. (2009) Determination of aquatic food-web structure based on compound-specific nitrogen isotopic composition of amino acids. Limnol Oceanogr Methods 7 740–750.

24. Popp BN, Graham BS, Olson RJ, Hannides CCS, Lott MJ, et al. (2007) Insight into the trophic ecology of yellowfin tuna, Thunnus albacares, from compound-specific nitrogen isotope analysis of proteinaceous amino acids. In: Dawson TD, Siegwolf, R. T W., editor. Stable isotopes as indicators of ecological change. New York: Elsevier Academic Press. pp. 173–190.

25. McClelland JW, Montoya JP (2002) Trophic relationships and the nitrogen isotopic composition of amino acids in plankton. Ecology 83: 2173–2180.

26. O'Brien DM, Fogel ML, Boggs CL (2002) Renewable and nonrenewable resources: Amino acid turnover and allocation to reproduction in Lepidoptera. Proc Natl Acad Sci USA 99: 4413–4418.

27. Germain LR, Koch PL, Harvey JT, McCarthy MD (2013) Nitrogen isotopic fractionation of amino acids in harbor seals (Phoca vitulina): Differential trophic enrichment factors based on ammonia vs. urea excretion. Mar Ecol Prog Ser 482: 265–277.

28. Default S, Whitehead H, Dillon M (1999) An examination of the current knowledge on the stock structure of sperm whales (Physeter macrocephalus) worldwide. J Cetac Res Manage 1: 1–10.

29. Carretta JV, Forney KA, Lowry MS, Barlow J, Baker J, et al. (2010) U.S. Pacific marine mammal stock assessments: 2009. California, USA. 336 p.

30. Sigman DM, Casciotti KL (2001) Nitrogen Isotopes in the Ocean. In: Editor-in-Chief: John HS, editor. Encyclopedia of Ocean Sciences. Oxford: Academic Press. pp. 1884–1894.

31. Mesnick SL, Taylor BL, Archer FI, Martien KK, TreviÑO SE, et al. (2011) Sperm whale population structure in the eastern and central North Pacific inferred by the use of single-nucleotide polymorphisms, microsatellites and mitochondrial DNA. Mol Ecol Resour 11: 278–298.

32. Marcoux M, Whitehead H, Rendell L (2007) Sperm whale feeding variation by location, year, social group and clan: Evidence from stable isotopes. Mar Ecol Prog Ser 333: 309–314.

33. Ruiz-Cooley RI, Ballance LT, McCarthy MD (2013) Range expansion of the jumbo squid in the NE Pacific: $\delta^{15}N$ decrypts multiple origins, migration and habitat Use. PLoS ONE 8: e59651.

34. Ruiz-Cooley RI, Gendron D, Aguiniga S, Mesnick S, Carriquiry JD (2004) Trophic relationships between sperm whales and jumbo squid using stable isotopes of C and N. Mar Ecol Prog Ser 277: 275–283.

35. Francey RJ, Allison CE, Etheridge DM, Trudinger CM, Enting IG, et al. (1999) A 1000-year high precision record of $\delta^{13}C$ in atmospheric CO_2. Tellus B 51: 170–193.

36. Karl DM, Bidigare RR, Letelier RM (2001) Long-term changes in plankton community structure and productivity in the North Pacific Subtropical Gyre: The domain shift hypothesis. Deep Sea Res Part II 48: 1449–1470.

37. Polovina JJ, Howell EA, Abecassis M (2008) Ocean's least productive waters are expanding. Geophys Res Lett 35: L03618.

38. Polovina JJ, Dunne JP, Woodworth PA, Howell EA (2011) Projected expansion of the subtropical biome and contraction of the temperate and equatorial upwelling biomes in the North Pacific under global warming. ICES J Mar Sci 68: 986–995.

39. Guilderson TP, McCarthy MD, Dunbar RB, Englebrecht A, Roark EB (2013) Late Holocene variations in Pacific surface circulation and biogeochemistry inferred from proteinaceous deep-sea corals. Biogeosciences 10: 3925–3949.

40. Fiedler PC, Mendelssohn R, Palacios DM, Bograd SJ (2012) Pycnocline Variations in the Eastern Tropical and North Pacific, 1958–2008. J Climate 26: 583–599.

41. Dave AC, Lozier MS (2013) Examining the global record of interannual variability in stratification and marine productivity in the low-latitude and mid-latitude ocean. J Geophysi Res-Oceans 118: 3114–3127.

42. Rykaczewski RR, Dunne JP (2010) Enhanced nutrient supply to the California Current Ecosystem with global warming and increased stratification in an earth system model. Geophys Res Lett 37: L21606.

43. Aksnes DL, Ohman MD (2009) Multi-decadal shoaling of the euphotic zone in the southern sector of the California Current System. Limnol Oceanogr 54: 1272–1281.

44. Bograd SJ, Buil MP, Lorenzo ED, Castro CG, Schroeder ID, et al. (2014) Changes in source waters to the Southern California Bight. Deep Sea R Part II. Available: http://dx.doi.org/10.1016/j.dsr2.2014.04.009.

45. McClatchie S (2013) Regional fisheries oceanography of the California Current System: the CalCOFI Program. Dordrecht: Springer. 253 p.

46. García-Reyes M, Largier J (2010) Observations of increased wind-driven coastal upwelling off central California. J Geophy Res 115.

47. Jacox MG, Moore AM, Edwards CA, Fiechter J (2014) Spatially resolved upwelling in the California Current System and its connections to climate variability. Geophysl Res Lett 41: 3189–3196.

48. Kahru M, Kudela RM, Manzano-Sarabia M, Greg Mitchell B (2012) Trends in the surface chlorophyll of the California Current: Merging data from multiple ocean color satellites. Deep Sea Res Part II 77–80: 89–98.

49. Rau GH, Takahashi T, Marais DJD (1989) Latitudinal variations in plankton $\delta^{13}C$: implications for CO_2 and productivity in past oceans. Nature 341: 516–518.

50. Wada E, Hattori A (1991) Nitrogen in the sea: forms, abundances, and rate processes. Boca Raton: CRC Press. 208 p.

51. Longhurst AR (2007) Ecological Geography of the Sea; Press. EA, editor. 542 p.

52. Morin PA, Nestler A, Rubio-Cisneros NT, Robertson KM, Mesnick S (2005) Interfamilial characterization of a region of the ZFX and ZFY genes facilitates sex determination in cetaceans and other mammals. Mol Ecol 14: 3275–3286.

53. McCarthy MD, Benner R, Lee C, Fogel M (2007) Amino acid nitrogen isotopic fractionation patterns as indicators of heterotrophy in plankton, particulate, and dissolved organic matter. Geochim Cosmochim Acta 71: 4727–4744.

54. Sherwood OA, Lehmann MF, Schubert CJ, Scott DB, McCarthy MD (2011) Nutrient regime shift in the western North Atlantic indicated by compound-specific $\delta^{15}N$ of deep-sea gorgonian corals. Proc Natl Acad Sci USA.

3

The Whale Pump: Marine Mammals Enhance Primary Productivity in a Coastal Basin

Joe Roman[1]*, James J. McCarthy[2]

1 Gund Institute for Ecological Economics, University of Vermont, Burlington, Vermont, United States of America, **2** Museum of Comparative Zoology, Harvard University, Cambridge, Massachusetts, United States of America

Abstract

It is well known that microbes, zooplankton, and fish are important sources of recycled nitrogen in coastal waters, yet marine mammals have largely been ignored or dismissed in this cycle. Using field measurements and population data, we find that marine mammals can enhance primary productivity in their feeding areas by concentrating nitrogen near the surface through the release of flocculent fecal plumes. Whales and seals may be responsible for replenishing 2.3×10^4 metric tons of N per year in the Gulf of Maine's euphotic zone, more than the input of all rivers combined. This upward "whale pump" played a much larger role before commercial harvest, when marine mammal recycling of nitrogen was likely more than three times atmospheric N input. Even with reduced populations, marine mammals provide an important ecosystem service by sustaining productivity in regions where they occur in high densities.

Editor: Peter Roopnarine, California Academy of Sciences, United States of America

Funding: Funding was supported by Stellwagen Bank National Marine Sanctuary, Office of Naval Research (ONR) grant N00014-08-1-0630, National Oceanographic Partnership Program (NOPP) grant N00014-07-1-1029, and the Museum of Comparative Zoology, Harvard University. The funders had no role in study design, data collection and analysis, decision to publish, or preparation of the manuscript.

Competing Interests: The authors have declared that no competing interests exist.

* E-mail: jroman@uvm.edu

Introduction

The biological pump mediates the removal of carbon and nitrogen from the euphotic zone through the downward flux of aggregates, feces, and vertical migration of invertebrates and fish [1]. Copepods and other zooplankton produce sinking fecal pellets and contribute to downward transport of dissolved and particulate organic matter by respiring and excreting at depth during migration cycles, thus playing an important role in the export of nutrients (N, P, and Fe) from surface waters [2,3]. Perhaps because of the prevalence of this flux of zooplankton biomass and detritus, it has often been presumed that the fecal matter of top predators such as marine mammals is also lost rapidly to deep waters and the benthos [4]. Yet predators such as whales, pinnipeds, and seabirds must rise to the surface to breathe, and so may play a different role in nutrient cycling.

There is a growing body of evidence supporting the important role of large vertebrates in many ecosystem processes. Grazing animals in the Serengeti, for example, stimulate net primary productivity and carbon sequestration [5,6]. Changes in vertebrate density and composition can have local and even global impacts: the decline of Pleistocene megafauna may have impacted methane production and thus atmospheric temperature [7]. Similarly, the removal of sperm whales from the Southern Ocean may have diminished this region's role as a reservoir for carbon [8].

Several lines of evidence indicate that most of the nitrogen released by marine mammals is expected to be in the shallower portion of their depth range: attachment to the surface for respiration, reduced metabolism at depth, physiological response to hydrostatic pressure, a decrease in glomular filtration rate and urine flow during forced diving studies, and observations of buoyant fecal plumes at the surface [9,10,11]. As early as 1983, Kanwisher and Ridgway noted that cetaceans could play an analogous role to upwelling, "lifting nutrients from deep waters" and releasing fecal material "that tends to disperse rather than sink when it is released." [12] Whale foraging dives are characterized by rapid descents and ascents to reduce transit time to prey aggregations [13,14], and high metabolic rates in gray seals while motionless at the surface support the idea that marine mammals process food during extended surface intervals following deep-water foraging [15]. Even if defecation occurred randomly, it would on average occur higher in the water column than where these animals feed, since they are unlikely to dive deeper than foraging efforts require.

Thus opposing the contribution of zooplankton, such as copepods, to the downward biological pump, cetaceans feeding deep in the water column effectively create an upward pump, enhancing nutrient availability for primary production in locations where whales gather to feed (Figure 1). Released nitrogenous compounds that can be used by primary producers are likely to remain in the euphotic zone, either as urea (the primary mammalian N-excretory product in urine), or as amino-N and NH_4^+ as the fecal plume material is consumed and metabolized. Pinnipeds that breed on shore and seaside ledges are also a source of nitrogenous nutrients in coastal waters [16].

We examined the relative importance of the whale pump in the Gulf of Maine, a partially isolated, highly productive basin in the western North Atlantic Ocean where nitrogen is generally considered to be the limiting nutrient for phytoplankton growth [17]. Townsend observed that the advective flux of nitrogen from

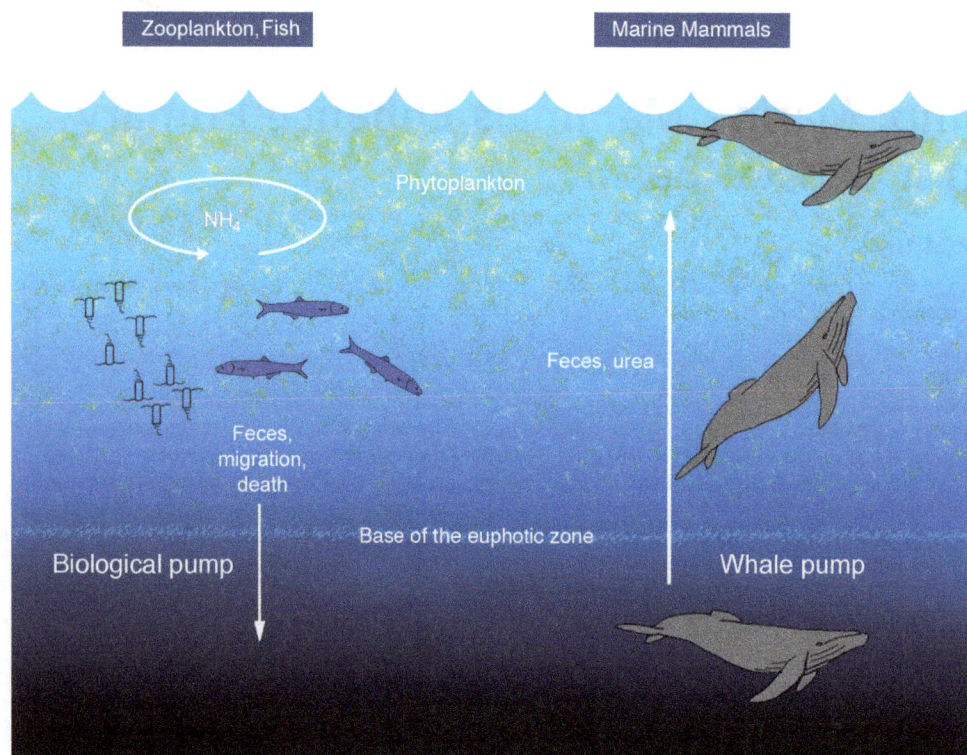

Figure 1. A conceptual model of the whale pump. In the common concept of the biological pump, zooplankton feed in the euphotic zone and export nutrients via sinking fecal pellets, and vertical migration. Fish typically release nutrients at the same depth at which they feed. Excretion for marine mammals, tethered to the surface for respiration, is expected to be shallower in the water column than where they feed.

deep and adjacent waters could not sustain primary production in this basin, noting that the "construction of carbon and nitrogen budgets that consider only fluxes into and out of the Gulf, and not internal recycling, will be in error" [18].

Results and Discussion

Field Measurements

We collected and analyzed 16 fecal plume samples during two whale-tagging cruises on Stellwagen Bank. PON concentrations of the humpback fecal plume samples were elevated by as much as two orders of magnitude above typical mixed-layer concentrations for summer in this area [19]. Concentrations of NH_4^+ in fecal plumes ranged from 0.4 to 55.5 μmol kg^{-1}. All reference samples collected away from visible fecal plumes had concentrations <0.1 μmol kg^{-1} (the nominal limit of detection), which is typical for summer surface waters [19]. Hence, nearly all of the samples taken near whale fecal plumes had dramatically elevated NH_4^+. The results of shipboard incubation time-course experiments are plotted in Figures 2a and 2b. These fecal plume samples contain phytoplankton and microbes capable of utilizing NH_4^+. Thus any change over time would be the net difference between what was produced by microbial activity associated with the feces (presumably gut flora) and the constituent microbial plankton minus the consumption of NH_4^+ by plankton and microbes. No samples showed a net loss of NH_4^+ during these experiments.

The measured NH_4^+ production rates in incubated samples were strongly correlated with sample PON concentration (Figure 2a), which implicates fecal particulate material as the

source of this nitrogen. The highest observed production rate was equivalent to about 50 times a typical plankton assimilation rate during summer in Massachusetts Bay [19]. Rates of increase in NH_4^+ show no relationship to initial NH_4^+ concentrations (Figure 2b), suggesting that the source is the fecal particulate material rather than another dissolved compound (amino-N or urea) that was co-released with NH_4^+.

Ecosystem Effects

We propose that marine mammals play an important role in the delivery of recycled nitrogen to surface waters (Table 1). Over the course of a year, marine mammals release approximately 2.3×10^4 metric tons (1.7×10^9 mol N) per year to the surface of the Gulf of Maine, more than all rivers combined and approximately the same as current coastal point sources (Figure 3a, Table 2, [20]). Although atmospheric deposition delivers more nitrogen to the Gulf than rivers or marine mammals, it is important to note that the atmospheric source is currently much higher than the estimated preindustrial levels (Figure 3b) [21].

The release of nutrients at the ocean surface is a pattern common to many air-breathing vertebrates, however, in the Gulf of Maine, and presumably in many other systems, it is dominated by whales, especially baleen whales. Currently cetaceans deliver approximately 77% of the nutrients released to the gulf by mammals and birds (Table 2); their biomass in the North Pacific and Southern Oceans indicate that they also play a dominant role in these systems [22,23]. For some marine ecosystems it may be appropriate to expand this term beyond one that emphasizes whales to acknowledge greater importance of pinnipeds or seabirds. In the gulf, the whale pump

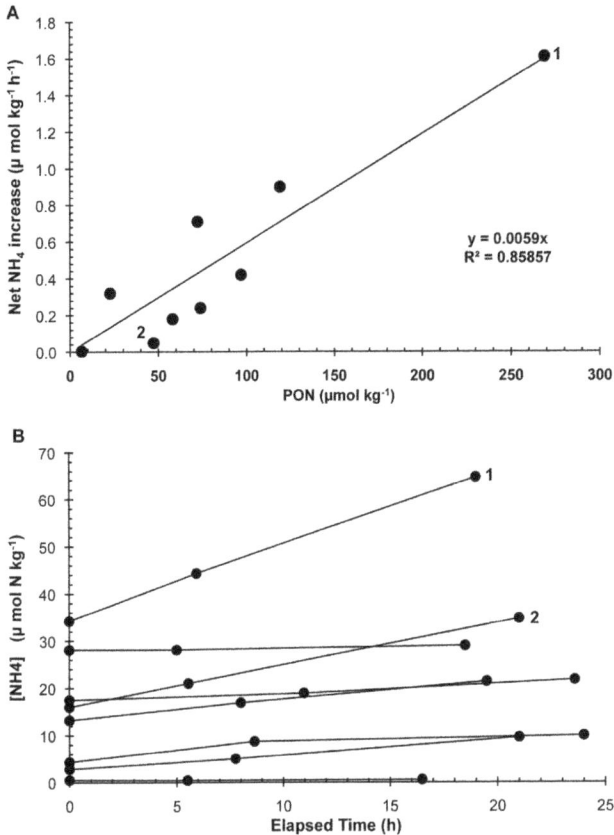

Figure 2. Shipboard incubation time-course experiments on Humpback whale samples collected on Stellwagen Bank, Gulf of Maine. (a) Net NH_4^+ production vs. fecal PON concentration in time course incubations of material collected in whale fecal plumes. Samples 1 and 2 had the highest initial NH_4^+ concentrations, yet their rates of NH_4^+ production ranged from the second lowest to the highest in the entire data set. **(b)** NH_4^+ concentration vs. incubation time.

will be most active in spring and summer, when feeding whales are present and when nitrate levels are low (Figure 4). Concentrations are ~8 µmol kg^{-1} in winter but approach undetectable levels in summer [18]. Kenney et al. have estimated that 30% of the annual prey consumed by cetaceans in the Gulf of Maine occurs in spring and 48% in summer [24]. Surface excretion may extend seasonal plankton productivity during these seasons, after a thermocline has formed. The effects of the pump are also expected to be much greater in highly productive areas such as Stellwagen and Georges Banks and the Bay of Fundy, where diving and surfacing transcends warm-season stratification and can markedly increase surface nitrogen levels.

The whale pump provides a positive plankton nutrition feedback. On Stellwagen Bank, humpback whales bottom feed on sand lance (*Ammodytes* spp), especially at night when these forage fish burrow into the sandy substrate [25]. In the Grand Manan Basin, right whales feed beneath the thermocline, on concentrated bands of diapausing copepods, in direct proportion to the abundance and quality of food available [14,26]. The density of copepods in this layer is orders of magnitude greater than average estimates of water-column prey density [27]. The average dive depth (113–130 m) for right whales is strongly correlated with peak prey abundance (fifth copepodites of *Calanus finmarchicus*) and the thermocline [14]. Fin whale foraging dives often exceed 100 m to locate dense concentrations of euphausiids [13].

Not all feeding occurs along or below the pycnocline. Right whales surface feed on copepods in Cape Cod Bay and the Great South Channel in the spring [28]. On Stellwagen, humpbacks tend to surface feed during daylight hours, when their prey is most abundant in the upper portion of the water column [25]. Several species have diel patterns in foraging behavior: sei whales feed on aggregations of *C. finmarchicus* when they migrate to the surface at night, reducing transit time for the whales and maximizing foraging efficiency [29]. Although the upward movement of nutrients is essential to our conception of the whale pump, the feeding of marine mammals at the surface, especially on prey that migrate across the pycnocline themselves, and the subsequent excretion of nutrients at the surface are important parts of the overall pattern of the pump.

Because of their large size and the high energetic cost of foraging, baleen whales require dense patches of food [13]. Production of phytoplankton stocks that support copepods, euphasiids, and fish consumed by whales will benefit most immediately from the release of nitrogenous excreta in nutrient-limited waters during stratified summer conditions. The whale pump could also reinforce the aggregative behavior and cooperative foraging of some cetaceans. The predictability of finding food in regions of high productivity is critical to individual survival and reproductive success: many species return to the same locations year after year, using the same feeding grounds across generations [30,31]. Another possible concentration-enhancing mechanism of the whale pump is the attraction of zooplankton to fecal material. The initial observation that led Hamner and Hamner to study the use of scent trails by zooplankton was an aggregation of copepods on the regurgitated meal of a seasick dive-boat tender [32]. At least one of the fecal plumes we collected—suspended just below the surface, about the size of our inflatable sampling boat, and the color of oversteeped green tea—had high numbers of copepods. Consumption of the fine particulate fraction in the fecal plume by zooplankton would provide further nutrition for the lower trophic levels that nourish these mammals.

Any attempt to study the role of marine mammals in coastal ecosystems must consider that many species now occur only in remnant populations, drastically reduced by commercial exploitation, incidental mortality, and habitat destruction (Figure 3b). Three species of mammals (sea mink, Atlantic walrus, and possibly Atlantic gray whale) are now extinct or absent in the Gulf of Maine, along with several marine birds, including the great auk. In the Bay of Fundy, humans have reduced the biomass of the upper trophic level of vertebrates by at least an order of magnitude [33]. One unanticipated consequence of this depletion of deep-diving mammals is a likely decline in the carrying capacity for higher trophic levels in coastal ecosystems.

Looking beyond the Gulf of Maine, it is important to consider the roles of present and past stocks of large air-breathing predators in the nutrient cycle of marine ecosystems. In the North Pacific, whale populations consume approximately 26% of the average daily net primary productivity; pre-exploitation populations may have required more than twice this sum [34]. Might primary productivity have been higher in the past as a result of a stronger whale pump? One recent study provides evidence that phytoplankton abundance has declined in 8 of 10 oceanic regions over the past century, and the authors suggest that this can be explained by ocean warming over this period [35]. Yet declines in both the Arctic and Southern Ocean regions, areas with especially high harvests of whale and seal populations over the past century, are in excess of the mean global rate. Full recovery from one serious anthropogenic impact on marine ecosystems, namely the dramatic depletion of whale populations, can help to counter the impacts of

Table 1. Effect of common and historically important marine mammals on the nitrogen cycle in the Gulf of Maine ecosystem.

Species	N excreted (kg day^{-1})	Population (M)	N flux (10^8 mol N yr^{-1})
Cetaceans			
Baleen			
Right whale	15.9	345	1.2
Humpback whale	9.42	902	1.8
Fin whale	15.0	2,065	6.7
Sei whale	8.32	91	0.16
Minke whale	2.94	3,497	2.3
Toothed			
Pilot whale	0.63	219	0.036
White-sided dolphin	0.15	20,400	0.78
Common dolphin	0.09	139	0.0034
Harbor Porpoise	0.05	89,700	1.2
Pinnipeds			
Harbor seal	0.09	99,340	2.4
Gray seal	0.22	1,731	0.10
Total			16.7

Total annual nitrogen released is 365 x N excreted day^{-1} for resident toothed whales and pinnipeds; for baleen whales, which migrate seasonally out of the study area, the total nitrogen released is expected to be 83% of annual excretion [48].

another now underway—the decline in nutrients for phytoplankton growth caused by ocean warming. The whale pump may have even played a role in helping to support a greater number of apex consumers. In the Southern Hemisphere, Willis has noted that a decrease in krill abundance followed the near elimination of large whales [36]. He hypothesized that one factor in this counterintuitive decline is a shift in krill behavior. Another factor could be the diminished whale pump, which would have affected productivity by reducing the recycling of nutrients to near-surface waters: Smetacek and Nicol et al. have shown that whales recycle iron in surface waters of the Southern Ocean [23,37]. The fertilization events of the whale pump can apply to nitrogen, iron, or other limiting nutrients.

These findings have important implications for the management of ocean resources. As marine mammal populations recover, it has been suggested that whales and other predators should be culled to limit competition with human fishing efforts, an idea that has been championed to challenge international restrictions on whaling [38]. Yet no data have been forthcoming to support the logic of this assertion. Furthermore, recent studies suggest that marine mammals have a negligible effect on fisheries in the North Atlantic [39,40]; simulated reductions in large whale abundance in the Caribbean did not produce any appreciable increase in biomass of commercially important fish species [41]. On the contrary, marine mammals provide important ecosystem services. On a global scale, they can influence climate, through fertilization events and the export of carbon from surface waters to the deep sea through sinking whale carcasses [42]. In coastal areas, whales retain nutrients locally, increasing ecosystem productivity and perhaps raising the carrying capacity for other marine consumers, including commercial fish species. An unintended effect of bounty programs and culls could be reduced availability of nitrogen in the euphotic zone and decreased overall productivity.

Methods

Ammonium analysis

An important question in this research was whether elevated NH_4 could be detected in whale fecal plumes, and whether rates of NH_4^+ production could be measured when freshly sampled feces are held in experimental chambers in the shipboard laboratory. Humpback whale fecal plumes were sampled with a 30-cm diameter, 150-μm mesh plankton net from small boats engaged in whale-tagging operations on Stellwagen Bank during July 2008 and 2009. The large greenish plumes, typically suspended just below the surface and at times as big as the collecting boat, were visibly heterogeneous and did not allow for quantitative sampling. Surface-water controls away from visible fecal plumes were collected both in close proximity (~20 m) to groups of surfacing whales and distant (>1 km) from any visible activity.

One-liter samples were placed in a cooler and returned to the support ship (NOAA Ship *Nancy Foster*) within 1–6 hours of collection, at which time a 200-ml aliquot of the fecal suspension was filtered (combusted Whatman GF/F). The filtrate was analyzed for initial NH_4^+-N concentration [43]. The filter was dried at 50°C, then sealed in a glass vial and retained for later particulate organic nitrogen (PON) analysis onshore [44]. The remaining unfiltered sample was placed in a dark refrigerator (12°C) to monitor changes in NH_4^+ over time. (Mean surface water temperature during the study period was ~18°C.)

At approximately 10 and 20 hours from the time the samples were onboard, subsamples were drawn from the refrigerated sample, filtered, and the filtrate was analyzed for NH_4^+-N concentration. In addition, single point NH_4^+-N and PON determinations were made on the control water samples described above, as well as samples from eight additional distinct fecal plumes sampled during this period and a similar operation in July 2008. Extremely dense aggregations of copepods were observed in a few fecal samples. We were unable to satisfactorily remove animals in these samples for analysis of fecal PON, and thus data from these samples are not reported here. We did not determine if the copepods were coprophagous.

Marine Mammal Consumption

To calculate the effect of marine mammals on the nitrogen cycle, we used estimates of daily consumption employing standard metabolic models scaled for assimilation, activity, and migratory fasting. This consumption rate has traditionally been estimated as 2–3% of body mass for rorqual whales, representing a daily average for summer consumption in Antarctica [45]. We employed more conservative estimates, as considered by Barlow and colleagues [46], using mass (M) to calculate the basal metabolic rate (BMR), where BMR $= 293.1\ M^{0.75}$. Rather than relying on a factor of 2.5 x BMR to calculate the field metabolic rate (FMR) we used 3 x BMR, in light of recent studies by Kjeld and colleagues, who derived consumption rates of 3.5% per day for fin whales and 4.6% for sei whales—about 30% higher than previously estimated [47]. Lockyer also found higher levels of consumption, calculating that baleen whales increase consumption rates ten fold in the summer [48]. The average daily ration was calculated as FMR divided by $(0.8[3900\mathcal{Z} + 5450(1-\mathcal{Z})])$, where $\mathcal{Z}$ is the fraction of crustaceans in the diet [46]. Values for $\mathcal{Z}$ are from the dietary composition table in Kenney et al. [24]. See Table 3 for daily consumption rates.

We employed an average daily consumption rate of 6.9% for seals in the Gulf of Maine, based on data from gray seals collected by Sparling et al. [15]. This aligns well with data from other pinnipeds, such as sea lions, which require daily consumption of

a)

b)

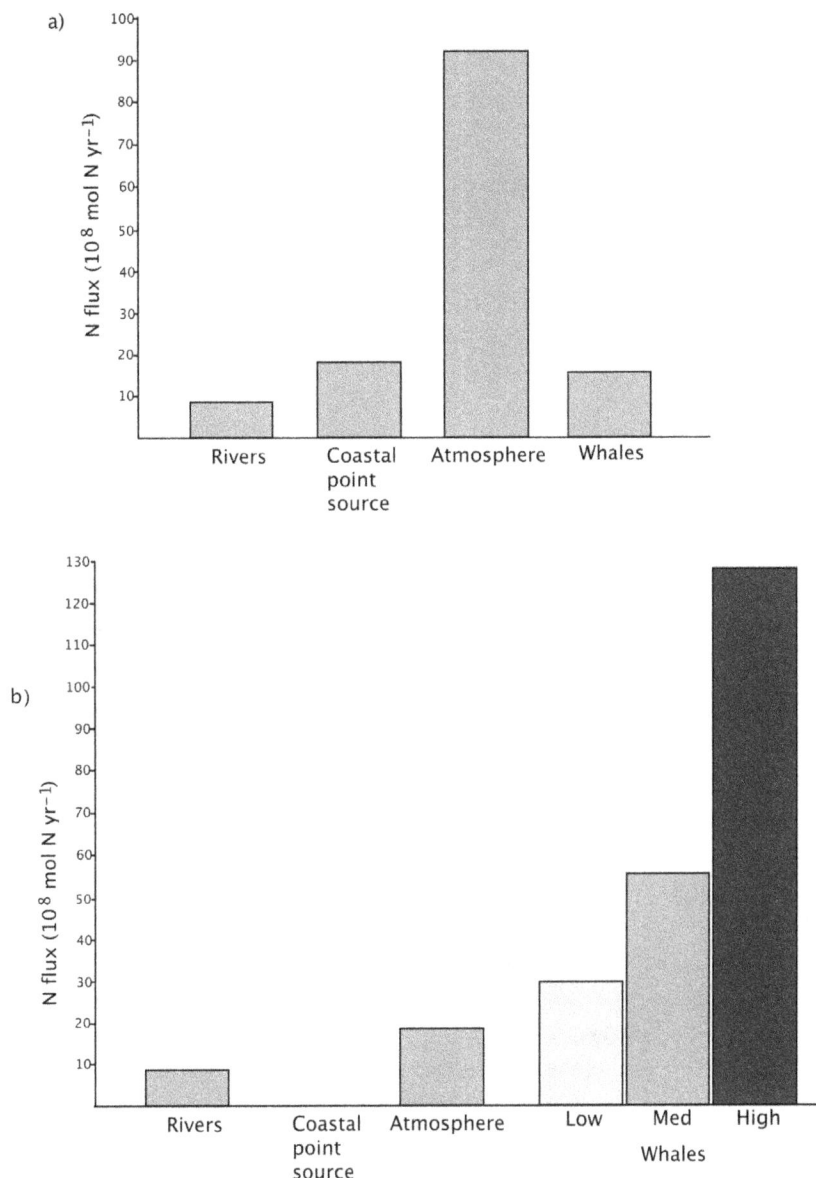

Figure 3. The flux of nitrogen in the Gulf of Maine (a) at present and (b) before commercial hunting. Point-source pollution, industrial emissions of nitrogen, and allochthonous sources from Townsend [18]. The range of historical estimates are adapted from Lotze [66]. Sources that are not expected to be influenced by anthropogenic change, such as offshore transport from Scotian Shelf water, are not included in this graph.

between 5% (adult males) and 13% (young females) of their body mass, with lactating females increasing their consumption by 70% [49]. Carlini et al. estimated a consumption rate of 6.8% during the post-breeding aquatic phase for southern elephant seals [50].

Marine Mammal Nitrogen Excretion

Fish and crustaceans such as euphasiids are approximately 15% protein [45] (about 17% nitrogen by weight) or 2.5% nitrogen. Nitrogen consumption = feces + urine + storage. Feces and urine are egested; stored nitrogen is retained for growth, energy reserve, eggs, sperm, and embryos. We assume that approximately 80% of ingested nitrogen is metabolized and 20% is retained [51]. Although the great majority of fecal matter is expected to stay in the euphotic zone, we employed this conservative estimate to account for the fact that no quantitative analysis has been performed to account for potential sinking. Although prey consumption and body weight vary

according to age and reproductive status, we employed average adult weights for all marine mammals.

Pinnipeds excrete approximately 87% of ingested nitrogen [16,52]. We employed an estimate of 80% to account for potential exported nitrogen. We recognize that seal feces can be important to the coastal ecosystem, but assume that the amount retained by terrestrial systems would be negligible in relation to the total nitrogen flux. Even during the breeding period, pinnipeds such as sea lions spend more than 80% of their time at sea [53]. Rookeries are rarely far from the sea, and it is assumed that most nutrients are returned to the ocean during storms [16]. Approximately 3% of the excretion from pinniped colonies is expected to be volatilized as NH_3 into the atmosphere [16], with some of this nitrogen returned to the sea via wet atmospheric deposition.

Urinary nitrogen from marine mammals would disperse diffusively and advectively, and the amount released would be

Table 2. Contemporary nitrogen flux in the Gulf of Maine.

Source	N flux per year (10^8 mol N)
Biological	
Cetaceans	14
Pinnipeds	2.5
Seabirds	1.2–2.3
Influx	
Offshore	1,479
Rivers	8
Coastal point sources	18
Atmosphere	93
Loss	
Denitrification	331
Burial	44

Influx and loss from Townsend [18].
Coastal point sources from Sowles [20].

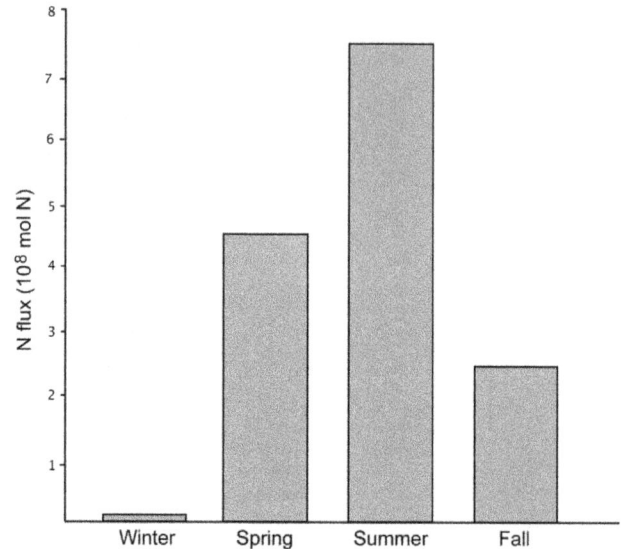

Figure 4. The role of cetaceans in the nitrogen cycle by season. Seasonal estimates based on the percentage of total consumption in the Gulf of Maine [24].

difficult to sample quantitatively. Particulate and dissolved nitrogen associated with flocculent fecal plumes can, however, be sampled because the plumes are visible from ships. Microbial proteolitic and deaminating processes will liberate NH_4^+ from the released particulate material, and these processes may have begun in the animal's gut.

Seabirds

Seabird estimates were unavailable for the entire Gulf of Maine. Huettmann estimated that the total marine food consumption of the 10 most common seabirds along the western Scotian Shelf was approximately 84,000 tons per year [54]. As the Scotian Shelf forms the eastern boundary of the Gulf of Maine, we used this annual consumption estimate of 0.87 tons km^{-2} yr^{-1} to determine the total effect of seabirds on the nitrogen cycle in the Gulf of Maine. Powers & Backus estimated an annual consumption rate of 1.6 tons km^{-2} yr^{-1} for the seabirds of Georges Bank [55]. We employed these two rates to estimate a reasonable range of the role that seabirds play in this basin.

For seabirds, foraging effort may be targeted at the zone below the thermocline [56], and nutrient cycling is expected to be quick. In birds, nitrogen is excreted primarily as uric acid, which is unstable in seawater, undergoing rapid conversion to urea [57]. We estimated that approximately 80% of nitrogen consumed was excreted at the surface, with 20% stored for fat and reproduction or exported to terrestrial systems and the seafloor. The entire area of the Gulf of Maine is 1.03×10^5 km^2 [18], yielding a total nitrogen flux of $1.2–2.3 \times 10^8$ mol N yr^{-1}, or about 10% of the current nutrient contribution from marine mammals.

Body Mass, Residence Time, and Population Size for Marine Mammals

Body mass is from Trites and Pauly [58], using mean mass of males and females assuming a 1:1 sex ratio. Right whale body mass is from Kenney et al [24]. Population size for cetaceans is also from Kenney et al., employing an average of the summer and spring estimates of abundance, except for humpback whales [59], harbor porpoises [60], white-sided dolphins [61], and gray and harbor seals [61,62]. Right and fin whale populations are from NOAA stock assessments [61] Estimates for fin whales come

from a survey conducted in 2006 from the southern Gulf of Maine to the Gulf of Saint Lawrence. Although part of this survey took place outside of our study area, the numbers are lower than previous studies for just the Gulf of Maine. We applied this abundance estimate as a reasonable, and conservative, estimate. Seal estimates are also probably conservative: many harbor seals are year-round residents, and we only account for the spring and summer seasons when they are pupping along the Maine coast (assuming that 50% of their yearly ration comes from the gulf). Both harbor and gray seal populations have likely grown since the last estimates were made (harbor seals in 2001, gray seals in 1999).

Total annual nitrogen flux was estimated as the product of the mean annual flux (365 x N excreted day^{-1}) and the estimated abundance of each species. For baleen whales, which migrate outside of the study area, we used Lockyer's estimate that 83% of the annual intake occurs in summer feeding areas [46,48].

Seasonal variation

We expect seasonal variation in feeding, as has been observed in captive adult gray seals [63] and many other marine mammals [64]. Periods of fasting in pinnipeds, for example, are assumed to be balanced by periods of more intensive feeding over the course of the year [65]. Because feeding is likely to decline in the winter, we suspect that our estimates are conservative for the many of the organisms included in this study.

Historic Estimates

We used data from Lotze et al. [66] to estimate historical numbers of cetaceans in the Gulf of Maine. Large whales in Massachusetts Bay are 10% of their historical numbers and small cetaceans 50%. In the Bay of Fundy, large whales were estimated to have a relative abundance of 45% compared to pre-exploitation numbers and small cetaceans 50%. We took estimates for Massachusetts Bay as the upper end for past population sizes and estimates from the Bay of Fundy in the lower end. It is worth noting that several ocean-wide studies support the higher end of this range [67,68]. As a medium estimate, we took an approximate

Table 3. Body mass and consumption rates for cetaceans and seals in the Gulf of Maine.

Species	Body mass (kg)	Percent of zooplankton in diet	Wet weight consumed (kg day^{-1})
Cetaceans			
Baleen			
Right whale	40,000	100	797
Humpback whale	30,408	5	471
Fin whale	55,590	10	751
Sei whale	16,811	100	416
Minke whale	6,566	5	149
Toothed			
Pilot whale	850	0	32
White-sided dolphin	120	0	7.3
Common dolphin	65	0	4.6
Harbor porpoise	31	0	2.6
Pinnipeds			
Harbor seal	67	0	4.6
Gray seal	160	0	11

average of these percentages, assuming that large whales constitute 25% of historical numbers and small cetaceans 50%.

Acknowledgments

We thank J. Nevins for experimental design and nitrogen analyses; E. Norse for suggesting the term "whale pump"; C. Campbell for help with analysis; M. Raila for helping design Figure 1; L. Farrell, S. Kraus, and R. Rolland for reviews and comments; and D. Wiley, A. Friedlaender, P. Halpin, and colleagues aboard NOAA RVs *Nancy Foster* and *Auk* for help in the field. Research conducted under NMFS permits 775-1875 & 605-1904.

Author Contributions

Conceived and designed the experiments: JR JJM. Performed the experiments: JR. Analyzed the data: JR. Contributed reagents/materials/analysis tools: JJM. Wrote the paper: JR JJM.

References

1. Longhurst AR, Harrison WG (1989) The biological pump: Profiles of plankton production and consumption in the upper ocean. Prog Oceanog 22: 47–123.
2. Hutchins D, Wang, W-X, Fisher NS (1995) Copepods grazing and the biogeochemical fate of diatom iron. Limnol Oceanogr 40: 989–994.
3. Steinberg DK, Goldthwait SA, Hansell DA (2002) Zooplankton vertical migration and the active transport of dissolved organic and inorganic nitrogen in the Sargasso Sea. Deep-Sea Res I 49: 1445–1461.
4. Huntley ME, Lopez MDG, Karl DM (1991) Top predators in the Southern Ocean: a major leak in the biological carbon pump. Science 253: 64–66.
5. McNaughton SJ (1985) Ecology of a Grazing Ecosystem: The Serengeti. Ecological Monographs 55: 259–294.
6. Holdo R, Sinclair A, Dobson A, Metzger K, Bolker B, et al. (2009) A disease-mediated trophic cascade in the Serengeti and its implications for ecosystem C. PLoS Biol 7: e1000210. doi:1000210.1001371/journal.pbio.1000210.
7. Smith FA, Elliott SM, Lyons SK. Methane emissions from extinct megafauna. Nature Geoscience 3: 374–375.
8. Lavery TJ, Roudnew B, Gill P, Seymour J, Seuront L, et al. (2010) Iron defecation by sperm whales stimulates carbon export in the Southern Ocean. Proceedings of the Royal Society B: doi:10.1098/rspb.2010.0863.
9. Ortiz RM (2001) Osmoregulation in marine mammals. J Exp Biol 204: 1831–1844.
10. Katona S, Whitehead H (1988) Are cetacea ecologically important? Oceanogr Mar Biol Annu Rev 26: 553–568.
11. Kooyman G, Castellini MA, Davis RW (1981) Physiology of diving in marine mammals. Annu Rev Physiol 43: 343–356.
12. Kanwisher JW, Ridgway SH (1983) The physiological ecology of whales and porpoises. Scientific American 248: 110–120.
13. Croll DA, Acevedo-Gutierrez A, Tershy BR, Urban-Ramirez J (2001) The diving behavior of blue and fin whales: is dive duration shorter than expected based on oxygen stores? Comp Biochem Physiol A 129: 797–809.
14. Baumgartner MF, Mate BR (2003) Summertime foraging ecology of North Atlantic right whales. Mar Ecol Prog Ser 264: 123–135.
15. Sparling CE, Fedak MA, Thompson D (2007) Eat now, pay later? Evidence of deferred food-processing costs in diving seals. Biol Lett 3: 94–98.
16. Theobald MR, Crittenden PD, Hunt AP, Tang YS, Dragosits U, et al. (2006) Ammonia emissions from a Cape fur seal colony, Cape Cross, Namibia. Geophys Res Lett 33: 1–4.

17. Rabalais NN (2002) Nitrogen in aquatic ecosystems. Ambio 31: 102–112.
18. Townsend DW (1998) Sources and cycling of nitrogen in the Gulf of Maine. J Mar Syst 16: 283–295.
19. Moreno P thesis, Harvard University.
20. Sowles J (2001) Nitrogen in the Gulf of Maine: Sources, Susceptibility and Trends. NOAA/UNH Cooperative Institute for Coastal and Estuarine Technology, Gulf of Maine Council on the Marine Environment, & NOAA Ocean Service.
21. Prospero JM, Barrett K, Church T, Dentener F, Duce RA, et al. (1996) Atmospheric deposition of nutrients to the North Atlantic Basin. Biogeochemistry 35: 27–73.
22. Pfister B, DeMaster DP (2006) Changes in marine mammal biomass in the Bering Sea/Aleutian Islands region before and after the period of commercial whaling. In: Estes JA, DeMaster DP, Doak DF, Williams TM, BrownellJr RL, eds. Whales, whaling, and ocean ecosystems. Berkeley: University of California Press. pp 116–133.
23. Nicol S, Bowie A, Jarmon S, Lannuzel D, Meiners KM, et al. (2010) Southern Ocean iron fertilization by baleen whales and Antarctic krill. Fish and Fisheries 11: 203–209.
24. Kenney RD, Scott GP, Thompson TJ, Winn HE (1997) Estimates of prey consumption and trophic impacts of cetaceans in the USA northeast continental shelf ecosystem. J Northw Atl Fish Sci 22: 155–171.
25. Friedlaender AS, Hazen EL, Nowacek DP, Halpin PN, Ware C, et al. (2009) Diel changes in humpback whale *Megaptera novaeangliae* feeding behavior in response to sand lance *Ammodytes* spp. behavior and distribution. Mar Ecol Prog Ser 395: 91–100.
26. Michaud J, Taggart CT (2007) Lipid and gross energy content of North Atlantic right whale food, *Calanus finmarchicus*, in the Bay of Fundy. End Species Res 3: 77–94.
27. Michaud J (2005) The prey field of the North Atlantic right whale in the Bay of Fundy: spatial and temporal variation. Halifax, NS: thesis, Dalhousie University.
28. Mayo CA, Marx MK (1990) Surface foraging behaviour of the North Atlantic right whale, *Eubalaena glacialis*, and associated zooplankton characteristics. Can J Zool 68: 2214–2220.
29. Baumgartner MF, Fratantoni DM (2008) Diel periodicity in both sei whale vocalization rates and the vertical migration of their copepod prey observed from ocean gliders. Limnol Oceanogr 53: 2197–2209.

30. McConnell BJ, Fedak MA, Lovell P, Hammond PS (1999) Movements and foraging areas of grey seals in the North Sea. J Appl Ecol 36: 573–590.

31. Weinrich MT (1998) Early experience in habitat choice by humpback whales (*Megaptera novaeangliae*). J Mammal 79: 163–170.

32. Hamner P, Hamner W (1977) Chemosensory tracking of scent trails by the planktonic shrimp *Acetes sibogae australis*. Science 4281: 886–888.

33. Lotze HK, Milewski I (2004) Two centuries of multiple human impacts and successive changes in a North Atlantic food web. Ecol Appl 14: 1428–1447.

34. Croll DA, Kudela R, Tershy BR (2006) Ecosystem impact of the decline of large whales in the North Pacific. In: Estes JA, DeMaster DP, Doak DF, Williams TM, BrownellJr RL, eds. Whales, Whaling, and Ocean Ecosystems. Berkeley: University of California Press. pp 202–214.

35. Boyce DG, Lewis MR, Worm B (2010) Global phytoplankton decline over the past century. Nature 466: 591–596.

36. Willis J (2007) Could whales have maintained a high abundance of krill? Evol Ecol Res 9: 651–662.

37. Smetacek V, Nicol S (2005) Polar ocean ecosystems in a changing world. Nature 437: 362–368.

38. Holt SJ (2003) The tortuous history of "scientific" Japanese whaling. BioScience 53: 205–206.

39. Hansen B, Harding K (2006) On the potential impact of harbour seal predation on the cod population in the eastern North Sea. J Sea Res 56: 329–337.

40. Corkeron PJ (2009) Marine mammals' influence on ecosystem processes affecting fisheries in the Barents Sea is trivial. Biology Letters 5: 204–206.

41. Morissette L, Kaschner K, Gerber LR. 'Whales eat fish'? Demystifying the myth in the Caribbean marine ecosystem. Fish and Fisheries: DOI: 10.1111/j.1467-2979.2010.00366.x.

42. Pershing AJ, Christensen LB, Record NR, Sherwood GD, Stetson PB (2010) The impact of whaling on the ocean carbon cycle: Why bigger was better. PLoS ONE 5(8): e12444. doi:12410.11371/journal.pone.0012444.

43. Parsons TR, Maita Y, Lalli CM (1984) A Manual of Chemical and Biological Methods for Seawater Analysis. New York: Pergamon.

44. McCarthy JJ, Garside C, Nevins JL (1999) Nitrogen dynamics during the Arabian Sea northeast Monsoon. Deep-Sea Res Part II 46: 1623–1664.

45. Gaskin DE (1982) The Ecology of Whales and Dolphins. London: Heinemann. 459 p.

46. Barlow J, Kahru M, Mitchell BG (2008) Cetacean biomass, prey consumption, and primary production requirements in the California Current ecosystem. Mar Ecol Prog Ser 371: 285–295.

47. Kjeld M (2003) Salt and water balance of modern baleen whales: rate of urine production and food intake. Can J Zool 81: 606–616.

48. Lockyer C (1981) Growth and energy budgets of large baleen whales from the Southern Hemisphere. FAO Fish Ser 3: 379–487.

49. Winship AJ, Trites AW, Rosen DAS (2002) A bioenergetic model for estimating the food requirements of Steller sea lions *Eumetopias jubatus* in Alaska, USA. Mar Ecol Prog Ser 229: 291–312.

50. Carlini G, Marquez MEI, Bornemann H, Panarello H, Casaux R, et al. (2005) Food consumption estimates of southern elephant seal females during their post-breeding aquatic phase at King George Island. Polar Biol 28: 769–775.

51. Boyd IL, Lockyer C, Marsh HD (1999) Reproduction in marine mammals. In: Reynolds III JE, Rommel SA, eds. Biology of marine mammals. Washington, DC: Smithsonian Institution Press. pp 218–286.

52. Ronald K, Keiver KM, Beamish FWH, Frank R (1984) Energy requirements for maintenance and faecal and urinary losses of the grey seal (*Halichoerus grypus*). Can J Zool 62: 1101–1105.

53. Trites AW, Porter B (2001) Attendance patterns of Stellar sea lions (*Eumetopias jubatus*) and their young during winter. J Zool 256: 547–556.

54. Huettmann F (2001) Estimates of abundance, biomass, and prey consumption for selected seabird species for the eastern and western Scotian Shelf, 1966-1992: Canadian Department of Fisheries and Oceans, contract F5245-000520.

55. Powers KD, Brown RGB (1987) Seabirds. In: Backus RH, ed. Georges Bank. CambridgeMassachusetts: MIT Press. pp 359–371.

56. Daunt F, Wanless S, Peters G, Benvenuti S, Sharples J, et al. (2006) Impacts of oceanography on the foraging dynamics of seabirds in the North Sea. In: Boyd IL, Wanless S, Camphuysen CJ, eds. Top Predators in Marine Ecosystems: Their Role in Monitoring and Management. Cambridge: Cambridge University Press. pp 177–190.

57. Antia NJ, Landymore AF (1974) Physiological and ecological significance of the chemical instability of the uric acid and related purines in sea water and marine algal culture medium. J Fish Res Board Can 31: 1327–1335.

58. Trites AW, Pauly D (1998) Estimating mean body masses of marine mammals from maximum body lengths. Can J Zool 76: 886–896.

59. Clapham P, Barlow J, Bessinger M, Cole T, Mattila D, et al. (2003) Abundance and demographic parameters of humpback whales from the Gulf of Maine, and stock definition relative to the Scotian Shelf. J Cetacean Res Manag 5: 13–22.

60. Palka D (2000) Abundance of the Gulf of Maine/Bay of Fundy harbor porpoise based on shipboard and aerial surveys during 1999. NOAA/NMFS/NEFSC-00-07.

61. Waring GT, Josephson E, Fairfield CP, Maze-Foley K, eds (2006) US. Atlantic and Gulf of Mexico marine mammal stock assessments. NMFS-NE-194.

62. Gilbert JR, Waring GT, Wynne KM, Guldager N (2005) Changes in abundance of harbor seals in Maine, 1981-2001. Mar Mamm Sci 23: 519–535.

63. Sparling CE, Speakman JR, Fedak MA (2006) Seasonal variation in the metabolic rate and body composition of female grey seals: fat conservation prior to high-cost reproduction in a capital breeder? J Comp Physiol B 176: 505–512.

64. Millar JS, Hickling GJ (1990) Fasting endurance and the evolution of mammalian body size. Funct Ecol 4: 5–12.

65. Cornick LA, Neill W, Grant WE (2006) Assessing competition between Steller sea lions and the commercial groundfishery in Alaska: A bioenergetics modelling approach. Ecol Model 199: 107–114.

66. Lotze HK, Lenihan HS, Bourque BJ, Bradbury RH, Cooke RG, et al. (2006) Depletion, degradation, and recovery potential of estuaries and coastal seas. Science 312: 1806–1809.

67. Roman J, Palumbi SR (2003) Whales before whaling in the North Atlantic. Science 301: 508–510.

68. Aguilar A (1986) A review of old Basque whaling and its effect on the right whales (*Eubalaena glacialis*) of the North Atlantic. Report of the International Whaling Commission 10: 191–199.

4

Investigating Population Genetic Structure in a Highly Mobile Marine Organism: The Minke Whale *Balaenoptera acutorostrata acutorostrata* in the North East Atlantic

María Quintela[1,2], **Hans J. Skaug**[1,3], **Nils Øien**[4], **Tore Haug**[5], **Bjørghild B. Seliussen**[1], **Hiroko K. Solvang**[4], **Christophe Pampoulie**[6], **Naohisa Kanda**[7], **Luis A. Pastene**[7], **Kevin A. Glover**[1,8]*

1 Dept. of Population Genetics, Institute of Marine Research, Bergen, Norway, **2** BIOCOST Research Group, Dept. of Animal Biology, Plant Biology and Ecology, University of A Coruña, A Coruña, Spain, **3** Department of Mathematics, University of Bergen, Bergen, Norway, **4** Dept. of Marine Mammals, Institute of Marine Research, Bergen, Norway, **5** Dept. of Marine Mammals, Institute of Marine Research, Tromsø, Norway, **6** Marine Research Institute of Iceland, Reykjavik, Iceland, **7** Institute of Cetacean Research, Tokyo, Japan, **8** Department of Informatics, Faculty of Mathematics and Natural Sciences, University of Bergen, Bergen, Norway

Abstract

Inferring the number of genetically distinct populations and their levels of connectivity is of key importance for the sustainable management and conservation of wildlife. This represents an extra challenge in the marine environment where there are few physical barriers to gene-flow, and populations may overlap in time and space. Several studies have investigated the population genetic structure within the North Atlantic minke whale with contrasting results. In order to address this issue, we analyzed ten microsatellite loci and 331 bp of the mitochondrial D-loop on 2990 whales sampled in the North East Atlantic in the period 2004 and 2007–2011. The primary findings were: (1) No spatial or temporal genetic differentiations were observed for either class of genetic marker. (2) mtDNA identified three distinct mitochondrial lineages without any underlying geographical pattern. (3) Nuclear markers showed evidence of a single panmictic population in the NE Atlantic according STRUCTURE's highest average likelihood found at K = 1. (4) When K = 2 was accepted, based on the Evanno's test, whales were divided into two more or less equally sized groups that showed significant genetic differentiation between them but without any sign of underlying geographic pattern. However, mtDNA for these individuals did not corroborate the differentiation. (5) In order to further evaluate the potential for cryptic structuring, a set of 100 *in silico* generated panmictic populations was examined using the same procedures as above showing genetic differentiation between two artificially divided groups, similar to the aforementioned observations. This demonstrates that clustering methods may spuriously reveal cryptic genetic structure. Based upon these data, we find no evidence to support the existence of spatial or cryptic population genetic structure of minke whales within the NE Atlantic. However, in order to conclusively evaluate population structure within this highly mobile species, more markers will be required.

Editor: Valerio Ketmaier, Institute of Biochemistry and Biology, Germany

Funding: Funding was provided by Norwegian Ministry of Fisheries and Coastal Affairs. The funders had no role in study design, data collection and analysis, decision to publish, or preparation of the manuscript.

Competing Interests: The authors have declared that no competing interests exist.

* Email: kevin.glover@imr.no

Introduction

Anthropogenic activities are key factors affecting wildlife populations, including altering population structure and distribution patterns [1–4]. Overexploitation by the whaling industry led to serious declines in many of the world's populations of whales. Currently, the IUCN conservation status "least concern" is applicable to only ~20% of whale species and only 8% show increasing population trends [5]. Marine mammals are highly mobile and may travel large distances (*e.g.* Stevick *et al.* [6]). A number of factors are thought to play a role in shaping the genetic structuring of cetacean populations such as the complex social structure (*e.g.* matrilineal based groups), the resource specialization and the great capacity for learning [7–9].

Minke whales, the second smallest baleen whales (about 10 m in length), are currently considered as two species [10]: the cosmopolitan common minke whale (*Balaenoptera acutorostrata*, Lacepede, 1804) and the Antarctic minke whale (*B. bonaerensis*, Burmeister, 1867), which is confined to the Southern hemisphere with the exceptions of rare inter-oceans migration events [11,12]. The former is further divided into three sub-species: the North Atlantic (*B.a. acutorostrata*), the North Pacific (*B.a. scammoni*), and the dwarf common minke whale (*B.a.* unnamed sub-species), which is thought to be restricted to the Southern hemisphere.

B.a. acutorostrata occurs in the entire North Atlantic during the Northern hemisphere summer months, limited in the northern range by the ice [13]. Although their winter distribution and thus the location of breeding areas is unknown, they probably fit the general ecological pattern of large cetaceans in the Northern hemisphere and migrate to lower latitudes, inhabiting temperate and tropical waters where pairing and birth of calves takes place

[14]. Calves are born between November and March after a gestation period of ten months [15,16]. In the western North Pacific, two *B.a. scammoni* breeding populations on either side of Japan are known to mix on feeding grounds in the Okhotsk Sea [17].

The minke whale is still harvested in significant numbers, and the management of *B.a. acutorostrata* in the North Atlantic is regulated under the Revised Management Procedure (RMP) developed by the Scientific Committee of the International Whaling Commission which also regularly reviews the species status through Implementation Reviews, the last one completed in 2009 [18]. The RMP implements the concept of Management Areas, which are currently outlined by taking into account different factors including distribution, life history parameters, local conservation threats such as bycatch, pollution, direct human exploitation and competition with fisheries, as well as differences in national legislation [19]. Five Management Areas have been established in the Eastern North Atlantic (*i.e.* "IWC Small Areas" [20]). The main concern of this outline is that, for minke whales, which is a migratory species, the small Management Areas would not reflect the real population boundaries but instead temporary mixed assemblages [21]. Therefore, careful assessment of the genetic diversity and genetic structure of the populations is essential to enable any successful conservation strategy.

Distinct breeding populations have not been identified for North Atlantic minke whales. Hence, the assessment of genetic structuring of minke whales within the North Atlantic has been based upon samples collected in the feeding grounds and stranded individuals. The question of population genetic structure within this species remains unresolved with partially conflicting results. There seems to be a general agreement regarding the absence of any clear spatial genetic structuring at mitochondrial level [22–25] although the possibility of co-existence of two breeding popula- tions of common minke whales in the North Atlantic was proposed by Palsbøll [26] after finding two main groups of genotypes when analyzing restriction fragment length polymorphism on mtDNA. Likewise, whereas some studies based on nuclear markers [24,27,28] failed to reveal any genetic differentiation between individuals from the central and north-eastern parts of the North Atlantic; some other insights based on stable isotopes and heavy metals [29], levels of radioactive caesium ^{137}Cs [30], persistent organochlorines [30], microsatellites [23] or isozymes [31–33] suggest a geographic substructuring across different areas of the North Atlantic. Recently, Anderwald *et al.* [24], using a set of ten microsatellites, reported the possible existence of two cryptic stocks across the North Atlantic. All these uncertainties regarding population identification and assessment have further increased the scientific and political controversy that whaling already poses [34] and therefore, the need to elucidate population genetic structure within this species.

Norway conducts a commercial harvest of minke whale, *B.a. acutorostrata* in the Northeast Atlantic, and each year, approxi- mately 500 whales are captured across five IWC Management Areas (Fig. 1). In order to enforce domestic regulation and compliance within this harvest, an individual-based DNA register (NMDR) has been maintained since 1996 [12]. This register contains genetic data of ten microsatellites and mtDNA for approximately 8000 whales harvested during the period 1996– 2011. In addition, the register includes biometric information together with the geographic position of captures, what provides a powerful database to investigate the potential genetic structure of this species in the NE Atlantic.

The main objective of this study was to investigate spatial and temporal genetic structure of *B. a. acutorostrata* harvested in the NE Atlantic IWC Management Areas during the period 2007– 2011. Secondly, we examined the possible existence of cryptic populations distributed across the North Atlantic as proposed by Anderwald *et al.* [24]. Therefore, the present study also included a set of samples from 2004 to match their sampling time frame and hence to enable comparisons. To achieve this second objective, conventional genetic analyses as well as simulation studies were conducted.

Material and Methods

Sampling, genotyping, and mtDNA sequencing

The time frame of the present study circumscribes to the period 2007–2011 when the genotyping of the individuals was performed by the Institute of Marine Research in Bergen, following very strict procedures to ensure the data quality [35,36]. In addition, samples collected in 2004 have been included for comparative purposes with former studies. Thus, the present data consists of genetic data from 2990 whales (2156 females and 834 males) that were harvested in the period from April to September. The distribution of individuals per sex, year and Management Areas is shown in Table 1. No animals were killed to provide samples for the present study as all the samples analyzed existed prior to it and were included in the NMDR [12] from which all the information used in the present work has been obtained; *i.e.* the biometrics, the position of the catches, the microsatellite genotypes and the mtDNA sequences of each of the 2990 individuals that were analyzed. The analytical approaches used for nuclear and mitochondrial markers at the NMDR were the following: DNA was extracted twice from muscle stored in ethanol using Qiagen DNeasy Blood & Tissue Kit following manufacturer's instructions and DNA concentration was measured on a Nanodrop. Ten microsatellite loci: EV1*Pm*, EV037*Mn* [37]; GATA028, GATA098, GATA417 [38]; GT023, GT211, GT310, GT509, GT575 [39] were amplified in three multiplex reactions based on a 2 minute hot start at 94°C, denaturing for 20 seconds at 94°C, annealing for 45 seconds, elongation at 72°C for 1 minute and a final hold at 4°C. Multiplex specific conditions are detailed in Glover *et al.* [12]. Individuals were sexed using specific primers for the ZFY/ZFX gene [40].

The D-loop region of mtDNA was amplified by performing forward sequencing of one DNA isolate and reverse sequencing of the second one. The first PCR reaction yielded a 1066 bp amplification product, which was forward strand sequenced. The second PCR reaction entailed the amplification of a 331 bp product that was sequenced in the reverse direction. PCR conditions for the two directions were identical, thus containing 0.5 units Go Taq polymerase (Promega), 1.5 mM MgCl$_2$, 0.2 mM dNTP and 0.2 µM of each primer. Forward product used primers MT4(M13F) and MT3(M13Rev) modified from Árnason *et al.* [41], whereas primers for the reverse product were: BP15851(M13F), modified from Larsen *et al.* [42] and MN312(M13R), modified from Palsbøll *et al.* [43]. PCR conditions were: hot start at 94°C for 2 min, followed by 30 cycles of denaturizing at 94°C for 50 seconds annealing at 53°C for 50 seconds and elongation at 72°C for 3 min 30 seconds, and finally a 10 min elongation at 72°C and a 4°C hold.

Genetic structure according to microsatellites

Total number of alleles, allelic richness and the inbreeding coefficient F_{IS} per population and per year were calculated with MSA [44], whereas observed (H_O) and unbiased expected heterozygosity (UH_E) were computed with GenAlEx [45]. The genotype distribution of each locus per year class and its direction

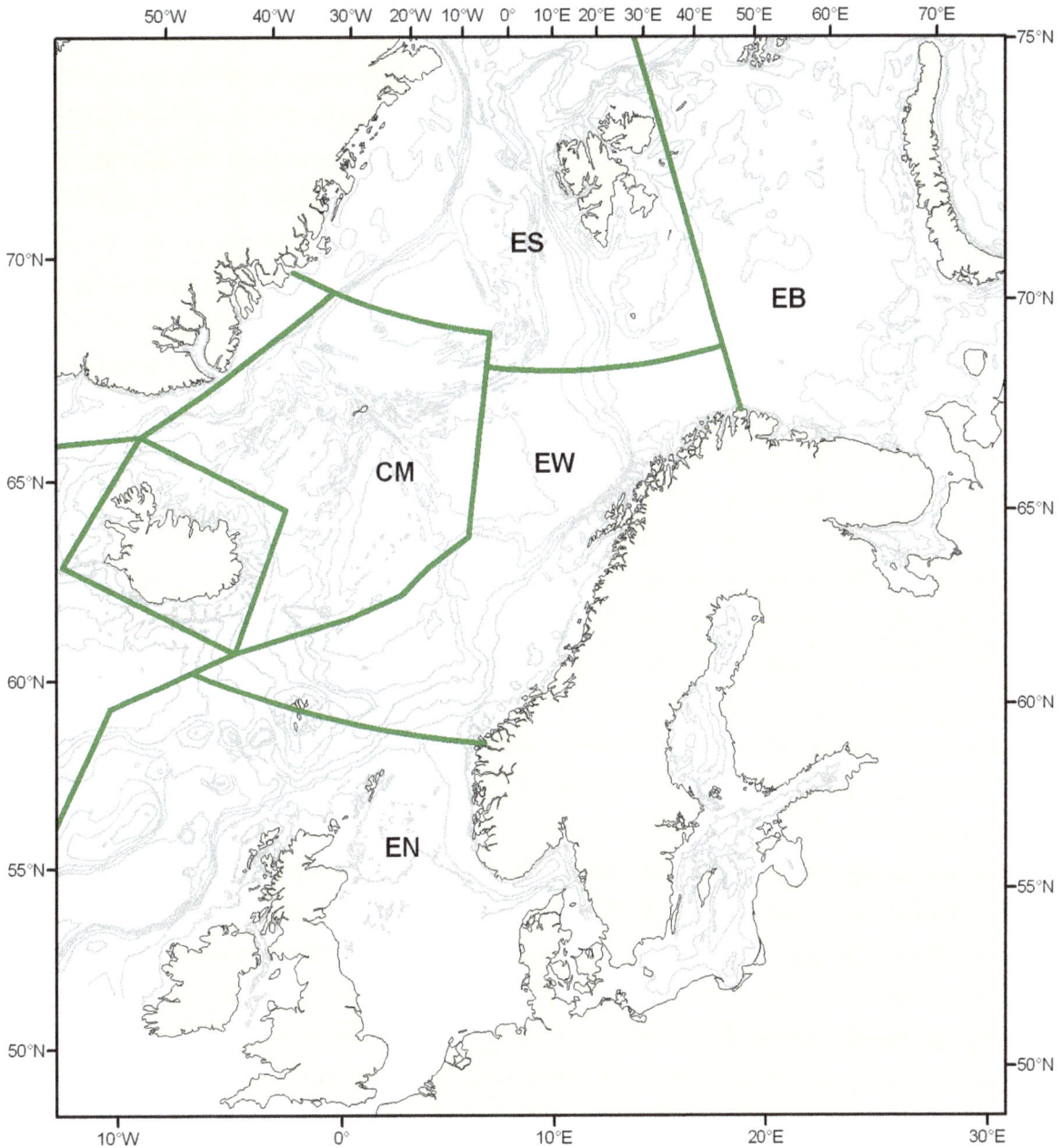

Figure 1. Geographic distribution of the five International Whaling Commission (IWC) Management Areas: ES (Svalbard-Bear Island area), EB (Eastern Barents Sea), EW (Norwegian Sea and coastal zones off North Norway, including the Lofoten area), EN (North Sea), and CM (Western Norwegian Sea-Jan Mayen area).

(heterozygote deficit or excess) was compared with the expected Hardy-Weinberg distribution using the program GENEPOP 7 [46] as was the linkage disequilibrium. Both were examined using the following Markov chain parameters: 10000 steps of dememorisation, 1000 batches and 10000 iterations per batch.

We used several methods to estimate population structure, including STRUCTURE [47], BAPS [48], and traditional F_{ST} [49] and R_{ST} analyses [50]. Slatkin's R_{ST} is an analogue of Wright's F_{ST} [51], adapted to microsatellite loci by assuming a high-rate stepwise mutation model instead of a low-rate K- or infinite-allele mutation model.

Both genetic differentiation among Management Areas per year class, and the level of temporal population genetic differentiation were tested using the Analysis of Molecular Variance (AMOVA) implemented in ARLEQUIN v.3.5.1.2 [52]. We also calculated the pairwise F_{ST} between populations from year class 2004 to 2011.

Both STRUCTURE [47,53,54] and BAPS [48,55] conduct a Bayesian analysis to identify hidden population structure, the former using allele frequency and linkage disequilibrium information from the data set directly, the latter identifying populations with different allele frequencies. Thus, BAPS first infers the most likely individual clusters in the sample population and then

Table 1. Distribution of females (F) and males (M) per Management Area (Fig. 1) on a per year class basis.

Year	Period	MANAGEMENT AREAS										
		EW		ES		EB		EN		CM		Total
		F	M	F	M	F	M	F	M	F	M	
2004	25th April – 23rd September	102	83	107	2	100	23	52	29	17	0	515
2007	22nd April – 22nd August	89	83	265	11	8	20	44	47	0	0	567
2008	30th April – 5th September	52	90	212	8	9	11	47	39	25	5	498
2009	11th April - 15th September	84	87	229	14	3	0	24	25	0	0	466
2010	29th April - 11th September	60	80	252	12	11	6	23	4	1	0	449
2011	1st May – 27th August	93	110	160	24	78	18	9	3	0	0	495
	Total	480	533	1225	71	209	78	199	147	43	5	2990

performs the most likely admixture of genotypes [55]; an approach that is more powerful in identifying hidden structure within populations [56].

We used the Bayesian model-based clustering algorithms implemented in STRUCTURE v. 2.3.4 to identify genetic clusters under a model assuming admixture and correlated allele frequencies without using population information. Ten runs with a burn-in period consisting of 100000 replications and a run length of 1000000 Markov chain Monte Carlo (MCMC) iterations were performed for a number of clusters ranging from $K = 1$ to $K = 5$. If applicable, we then used STRUCTURE Harvester [57] to calculate the Evanno *et al.* [58] *ad hoc* summary statistic ΔK, which is based on the rate of change of the 'estimated likelihood' between successive K values. The usual scenario where this approach is appropriate are those cases where once the real K is reached, L(K) at larger Ks plateaus or continues increasing slightly and the variance between runs increases. Hence, the estimated 'log probability of data' does not provide a correct estimation of the number of clusters and instead, ΔK accurately detects the uppermost hierarchical level of structure [58]. Runs were automatized with the program ParallelStructure [59] that controls the program STRUCTURE and distributes jobs between parallel processors in order to significantly speed up the analysis time. Afterwards, runs were averaged with CLUMPP version 1.1.1 [60] using the LargeKGreedy algorithm and the G′ pairwise matrix similarity statistics. Averaged runs were graphically displayed using barplots on a per year class basis.

Secondly, we used BAPS 6.0 [48] for a number of clusters ranging between $K = 1$ and $K = 5$ (10 runs per K), and then we performed the most likely admixture of genotypes [55], again on a per year class basis.

Mitochondrial DNA

Estimates of genetic diversity were calculated with DnaSP [61] and consisted of number of segregating sites, average number of pairwise nucleotide differences, nucleotide diversity and haplotype diversity.

Demographic changes were examined using three different approaches: Tajima's D [62], Fu's F_S [63] and by comparing mismatch distributions of pairwise nucleotide differences between haplotypes to those expected under a sudden population expansion model [64–66]. The analyses were implemented in the program ARLEQUIN v.3.5.1.2, and P-values were generated using 10000 simulations.

We used Tajima's D and Fu's Fs to test for shift in the allele frequency spectrum compared to a neutral Wright-Fisher model consistent with population expansion under neutral evolution. The neutrality test Fs [63] has been shown to be a powerful test to detect population growth when large sample sizes are available [67]. Large and negative significant values of Fs indicate an excess of recent mutations (haplotypes at low frequency) compared to those expected for a stable population, which can be interpreted as a signature of recent population growth, genetic hitchhiking or population expansion following a bottleneck event [63]. Demographic changes were also investigated by calculating the raggedness index of the observed mismatch distribution for each of the populations according to the population expansion model. This measure quantifies the smoothness of the observed mismatch distribution. Small raggedness values represent a population which has experienced sudden expansion (possibly following a bottleneck) whereas higher values suggest stationary populations [68,69]. Unimodal distributions are expected for populations that recently expanded or experienced a bottleneck, as individuals within a population will present similar haplotype divergence (in terms of nucleotide differences) [64,66]. In contrast, a multimodal or 'ragged' distribution is expected for a stable or slowly declining population [64]. Statistical significance for the mismatch distributions was obtained using a goodness-of-fit test based on the sum of squared deviations between the observed and expected distributions [70] and the Harpending's raggedness index, rg [68] after 10000 simulations using the estimated parameters of the expected distribution for a population expansion.

The evolutionary relationships between haplotypes were examined with the software Network [71] using the median-joining algorithm to build an unrooted cladogram. Networks were built separately for every year class and also for the full data set ranging from 2004 to 2011. Singletons were removed, transitions weights were changed into 10 whereas tranversions and gaps were changed into 30; epsilon was set at 10, and the MP option [72] was enabled to delete redundant links and median vectors.

BAPS clustering was used to validate Network results and thus the program was run 100 times for the number of clusters reported by the median-joining tree.

Testing the hypothesis of cryptic stock clustering of North Atlantic minke whale

Anderwald et al. [24] identified genetic sub-structuring of North Atlantic minke whales and proposed the existence of two putative cryptic stocks. In their paper, the detection of genetic differentiation among minke whale individuals was enhanced by the use of an outgroup in STRUCTURE, and thus they included 30 individuals of B.a. scammoni from the Sea of Japan as an outgroup. We added this approach to our former STRUCTURE analyses using two different outgroups: firstly, 95 individuals of the subspecies Pacific minke whale (B. a. scammoni); secondly, 93 individuals of the Antarctic minke whale (B. bonaerensis) and, thirdly, both outgroups simultaneously. STRUCTURE and BAPS analyses together with the assessment of genetic differentiation between groups of individuals after clustering procedures are exhaustively detailed in the File S1 in Supporting Information.

In addition to the above analyses, and to test the alternative hypothesis of minke whales constituting a panmictic population, we created a set of 100 in silico generated panmictic populations based on the allele frequencies observed in our samples. Hence, at each of the ten loci, the allelic values (two per individual) were put in a pool, and then randomly re-assigned to individuals, thereby preserving the original allele frequencies. The resulting in silico simulated panmictic populations were analysed automatizing STRUCTURE with the program ParallelStructure under a model assuming admixture and correlated allele frequencies without using population information. Ten runs per K ranging from 1 to 5, a burn-in period of 100000 replications, and a run length of 1000000 MCMC iterations were followed by Evanno's test. For the sake of the comparison, ten of the populations yielding K = 2 after Evanno's test were averaged with CLUMPP and pairwise F_{ST} between resulting clusters was performed as above. In addition, BAPS analyses were also conducted for K ranging from 1 to 5.

Detection of sex-biased dispersal

The potential for sex-biased dispersal was investigated using the microsatellite data with the methods described by Goudet et al. [73] and implemented in GenAlEx [45]. The statistics used were: F_{IS}, F_{ST}, Ho, Hs (the within group gene diversity), the mean corrected assignment index (mAIc) and the variance around the assignment index (vAIc) [74,75]. When comparing allele frequencies between individuals of the dispersing sex and those of the more philopatric one, a greater similarity is expected among the

more dispersing sex. Likewise, expectations would be mAIc to be higher in the more philopatric sex, while vAIc should be lower [73]. Female philopatry and male dispersal are the expected patterns for mammalian species based on the expectation that partuating females will be more dependent on local resources [76]. Thus, a one-tailed Mann–Whitney U-test was used to test if dispersal was biased toward males, as in most marine mammals.

Results

Spatial and temporal genetic structure according to microsatellites

The sex distribution per Management Area across years was biased towards females in 73% of the cases (Table 1). Namely, in ES, females were 7–54 fold more abundant than males, whereas in CM no males were reported with the exception of 5 individuals in 2008. The spatial distribution showed that females and males overlapped in latitudes below 71°N but hardly any males occurred in the northernmost regions.

The microsatellite data set contained no missing data and both the number of alleles (116–124), and allelic richness (11.5–12.3) were stable across the six year classes analysed (Table 2). Observed heterozygosity ranged between 0.757 and 0.795, and unbiased expected heterozygosity, between 0.768 and 0.801. Analysis of HWE revealed that at the significance level of α 0.05, 4.8% of loci by sample combinations displayed significant deviations; whereas this number decreased to 0.4% at the significance level of α 0.001. LD was detected 68 times (5.6%) at α 0.05 and 9 (0.7%) at α 0.001.

The analysis of geographic genetic structuring among Management Areas revealed no differentiation over time. Thus, AMOVA performed separately for each of the year classes reported no significant F_{ST} in any case (Table 3). Likewise, all the pairwise comparisons between Management Areas were non-significant for all year classes analysed with the only exception of EB-EW in 2010 being marginally significant at $R_{ST} = 0.024$ ($P = 0.046$) but not after Bonferroni correction ($P = 0.0017$).

Similar to results of spatial genetic structure above, AMOVA reported high temporal genetic stability, and the pairwise comparison between the different year classes (Table 4) yielded only one significant albeit very weak pairwise F_{ST} between years 2007–2008 ($F_{ST} = 0.0004$, $P = 0.0270$); which was no longer significant after Bonferroni correction.

The individual analysis of every Management Area also showed temporal genetic stability as a general picture. Hence, the AMOVA performed separately in each of the five IWC zones yielded a non-significant F_{ST} that exhibited 0.003 as the highest value. Likewise, the pairwise comparisons between years within each area were also non-significant with the exception of area EN. The pairwise matrix for EN reached significance in three out of the total fifteen cases corresponding to the comparisons between year 2008 and years 2004, 2007 and 2009 respectively.

STRUCTURE showed that the highest average likelihood was found to be K = 1 in all year classes together with a decreasing trend of LnP(D) across consecutive values of K (Table A in File S1). In these situations, although there is no need to perform Evanno's test; we found that ΔK took its highest value for K = 2 in all the year classes with the exception of 2008 where K = 3. The common feature to all the sampling years were the low values reported for ΔK, which reached its maximum in 2011 (ΔK = 16.7) but otherwise ranged between 3 and 8. Barplots for each year class are shown in Fig. A in File S1. For K≥2, every individual showed that the membership to each cluster was evenly distributed among groups producing flag-like barplots.

Table 2. Minke whale microsatellites.

Year	No alleles	Ar	No private alleles	Ho	uHe	F_{IS}
2004	119	11.7	2	0.757±0.015	0.768±0.010	0.0057±0.0243
2007	120	11.7	0	0.776±0.012	0.770±0.010	−0.0050±0.0185
2008	123	12.1	3	0.791±0.012	0.777±0.010	−0.0150±0.0205
2009	124	12.3	4	0.790±0.018	0.775±0.013	−0.0057±0.0183
2010	116	11.6	0	0.787±0.024	0.801±0.022	0.0093±0.0229
2011	116	11.5	1	0.795±0.017	0.778±0.010	−0.0057±0.0151

Summary statistics per year showing total number of alleles, allelic richness (based on minimum sample size of 449 diploid individuals), number of private alleles, observed heterozygosity (average ± SE), unbiased expected heterozygosity (average ± SE), and inbreeding coefficient (F_{IS}) (average ± SD).

BAPS showed that most likely K was 3 for year classes 2008, 2010 and 2011 and 4 for the remaining ones. No admixture was detected in any of the sampling years but in 2004 with one admixed individual.

Mitochondrial DNA

A total of 92 haplotypes were found in the complete dataset (*i.e.* year classes 2004 and 2007–2011), 25 of them unique (0.8% of the individuals). Six of the haplotypes were shared by 4–9% of the individuals whereas the most abundant one was present in 806 whales (27%). The number of haplotypes found per year class ranged mostly between 36 and 38 (Table 5), and took its maximum value in 2004 (N_H = 62). Both haplotype (H_D) and nucleotide diversity (π) showed high and stable values across the years (Table 5).

The distribution of the most common haplotypes was even across Management Areas and AMOVA revealed that no differentiation was observed among them in any of the sampling years (Table 3), with the exception of the significant pairwise comparison EB-ES (F_{ST} = 0.008, P = 0.035) in 2004. Similarly, high temporal stability (Table 4) was reported with one weak but marginally significant pairwise comparison: 2010–2011 (F_{ST} = 0.0019, P = 0.043), although not significant after Bonferroni correction.

A median-joining tree (Fig. 2) was built for the total data set (*i.e.* 2004 and 2007–2011 excluding singletons) given the temporal stability detected across year classes. A central ancestral haplotype was reported in 27% of the individuals, whereas none of the remaining ones exceeded a frequency of 9%. This ancestral haplotype did not show any phylogeographic structure, *i.e.* it was

not linked to any of the Management Areas as it was present in relatively even proportions in each of them. The MJ-tree suggests the existence of three lineages: a central one that evolved through two episodes of expansion, with haplotypes connected between them via single mutational steps in the vast majority of cases. Two of the lineages gathered 85% of the haplotypes in a quite even distribution whereas the third one accounted for the 15% remaining. The haplotype composition per lineages revealed by Network perfectly matched BAPS clustering for K = 3, with 100% coincidence. The same individuals that showed significant differentiation in three lineages at mtDNA yielded F_{ST} = 0.00018 (P = 0.8966) when analysed for microsatellites.

Different insights (Table 6) invoke population size expansion such as: a) large negative and significant Fu's Fs, b) negative albeit non-significant Tajima's D values (except in 2010), c) small raggedness values, d) unimodal mismatch distributions (not shown), and e) the star-shape of haplotype network.

Examining potential cryptic population structure using clustering methods

Evanno's test revealed that K = 2 was the most likely scenario in all the clustering approaches performed per year class, either with or without outgroups, and even regardless of the number of outgroups included in the analysis (with the exception of year class 2010 without outgroup that showed K = 3). Thus, when using outgroups, NE Atlantic minke whales (*B.a. acutorostrata*) constituted a compact cluster whereas the outgroups (i.e *B.a. scammoni* or/and *B. bonaerensis*) constituted a second compact one (see Fig. B in File S1). Therefore, in such a case, we needed to explore K = 3 to have NE Atlantic minke whales divided into two groups

Table 3. Genetic differentiation into Management Areas per year class.

Year	Microsatellites		mtDNA	
	F_{ST}	R_{ST}	F_{ST} (Haplotype frequency)	F_{ST} (Tamura-Nei)
2004	0.0000 (0.6836)	0.0000 (0.6827)	0.0012 (0.2524)	0.0010 (0.4364)
2007	0.0000 (0.6964)	0.0000 (0.8940)	0.0000 (0.6157)	0.0000 (0.7972)
2008	0.0000 (0.9613)	0.0000 (0.9442)	0.0000 (0.8310)	0.0000 (0.5750)
2009	0.0004 (0.2406)	0.0000 (0.7125)	0.0000 (0.5240)	0.0006 (0.3684)
2010	0.0000 (0.9374)	0.0000 (0.4030)	0.0000 (0.4723)	N.C.[NOTE]
2011	0.0000 (0.9867)	0.0000 (0.6003)	0.0012 (0.2652)	0.0000 (0.4919)

Summary of AMOVA (F_{ST} and *P*-value) conducted with ARLEQUIN with 10000 permutations at microsatellites and mtDNA.
[NOTE]N.C. not calculated. Nucleotide composition too unbalanced for Tamura-Nei correction.

Table 4. Temporal genetic differentiation: Pairwise F_{ST} between year classes calculated with ARLEQUIN for microsatellites (lower diagonal) and mtDNA (upper diagonal).

	2004	2007	2008	2009	2010	2011
2004		0.0005	0.0012	0.0005	0.00000	0.0037
2007	0.00000		0.0002	0.00000	0.00000	0.0024
2008	0.00023	**0.00043***		0.00000	0.00000	0.00000
2009	0.00000	0.00000	0.00016		0.00000	0.0005
2010	0.00000	0.00000	0.00015	0.00000		**0.0019***
2011	0.00006	0.00000	0.00000	0.00000	0.00018	

Significance calculated after 10000 permutations. Values highlighted in boldface type as significant at P<0.05 (*) lost significance after Bonferroni correction.

(*e.g.* Fig. 3c–f, and Fig. A3 in File S1). The distribution of those individuals into clusters was conditioned to overcoming a threshold of membership of 0.50. Although an exhaustive report of all the results regarding cryptic clustering can be found in the File S1 in Supporting Information, the main findings of each assignment procedure were as follows:

a) **STRUCTURE without outgroup (K = 2).-** Although individuals showed very narrow ranges of membership (0.51–0.63, average 0.59) to clusters (Table B a,b in File S1); there was a significant albeit weak genetic differentiation between groups per year class demonstrated by pairwise F_{ST} (Table B a,b in File S1), Fisher's exact test ($\chi^2 =$ infinity, df = 20, P<0.0001) and Factorial Correspondence Analyses (despite a low percentage of total variation explained by the two first axes ranging between 3.97 and 4.22%). The same individuals genotyped at mtDNA did not produce any significant F_{ST} in any sampling year (Table B a,b in File S1). GeneClass corroborated clustering with an average percentage of correct assignment of 86% (ranging from 83.5 to 90.2%, Table J in File S1).

b) **STRUCTURE with outgroups (K = 3).**

   a. Pacific minke whale (*B.a. scammoni*) as an outgroup (Fig. C in File S1, left column).- The average membership to cluster was higher than when using STRUCTURE without outgroups (0.88). Individuals were divided into two clusters that showed weak albeit significant genetic differentiation (average F_{ST} between clusters was 0.0130). The same individuals genotyped at mtDNA did not show any genetic differentiation between clusters (Table D in File S1).

   b. Antarctic minke whale (*B. bonaerensis*) as an outgroup (Fig. C in File S1, right column).- The average membership to cluster was higher, 0.95, whereas the average F_{ST} between clusters was slightly lower, 0.0122. Again, the individuals from both clusters genotyped at mtDNA did not reveal any genetic differentiation (Table E in File S1).

   c. We used a conservative approach and divided the NE Atlantic minke whales into two groups after taking the consensus of the results of the analyses with Antarctic and Pacific outgroups together. This means that individuals were assigned to cluster 1 or 2 after comparing the assignment obtained after Antarctic and Pacific analyses. Likewise, a number of individuals was left unassigned and this comprised those that did

not reach the inferred ancestry 0.5 threshold plus the mismatches between both procedures (*e.g.* individuals that belonged to cluster 1 with Antarctic outgroup and to cluster 2 in the Pacific clustering). Again a weak but significant differentiation between groups per year class was shown by pairwise F_{ST} (Table F in File S1), Fisher's exact test ($\chi^2 =$ infinity, df = 20, P<0.0001) and Factorial Correspondence Analyses (albeit the low percentage of the total variation explained by the two first axes ranging from 3.9 to 4.2%). Once more, the same individuals genotyped at mtDNA did not show any evidence of genetic differentiation (Table F in File S1). GeneClass corroborated STRUCTURE consensus clustering with a high percentage of correct assignment (97–98.6%) across all year classes with the exception of 2008 that was slightly lower (85.6%), Table J in File S1.

c) **BAPS for K = 2.-** BAPS divided individuals of each sampling year class into two groups of even size for year classes 2008, 2009 and 2011 whereas for the remaining ones, the size ratio was around 1.4–1.5 (Table H in File S1). No admixed individuals were detected in any of the sampling years. A weak albeit significant F_{ST} (Table H in File S1) was found between groups per year class, a differentiation that was further confirmed by Fisher's exact test ($\chi^2 =$ infinity, df = 20, P<0.0001) and Factorial Correspondence Analyses (percentage of the total variation explained by the two first axes ranging between 3.73 and 4.10%). The same individuals genotyped at mtDNA only produced significant F_{ST} for year classes 2007 and 2008 based on Tamura-Nei distance and haplotype frequencies, respectively (Table H in File S1). GeneClass corroborated BAPS clustering with a percentage of correct assignment of 100% in all the cases (Table J in File S1).

The geographic distribution of individuals after both procedures of clustering was slightly different. Hence, STRUCTURE-clustered individuals were evenly distributed among Management Areas per year class whereas for the BAPS-clustered ones, this distribution was less homogeneous in some of the cases (Fig. D and Table K in File S1).

The analyses of the 100 *in silico* generated panmictic populations with STRUCTURE revealed that, again and like the real data: a) the highest average likelihood was detected at K = 1, and b) a decreasing trend of LnP(D) across consecutive values of K was found in all the cases. As formerly reported, even if Evanno's test is not applicable in this situation, we wanted to test

Table 5. Minke whale mtDNA.

Year	N	N_H	N_{UH}	S	k	$\Pi \times 10^2$	H_D
2004	515	62	29	21	3.071	0.969	0.908±0.008
2007	567	49	22	26	2.863	0.906	0.895±0.008
2008	498	38	14	21	2.854	0.900	0.886±0.010
2009	466	36	11	24	2.824	0.891	0.884±0.010
2010	449	38	14	18	2.940	0.931	0.897±0.009
2011	495	37	10	20	2.781	0.877	0.864±0.011

Summary of diversity statistics: Sample size (N), number of haplotypes (N_H), number of unique haplotypes (N_{UH}), number of segregating sites (S), average number of pairwise nucleotide differences (k), nucleotide diversity (π) and haplotype diversity (H_D mean ± SD).

its outputs and thus we found that the most likely number of clusters showed different values: $K = 2$ in 58% of the cases, $K = 3$ in 33% and $K = 4$ in the remaining 9%. Similarly, low values of ΔK (ranging from 1 to 15) were also reported for the 100 panmictic populations. CLUMPP was performed on a set of ten randomly chosen simulated populations that showed $K = 2$ after Evanno's method, and individuals were distributed into clusters after overcoming a threshold of 0.50. In all cases, both clusters showed similar size (ratio 1–1.3) and 7–19% of individuals were left unassigned (Table M in File S1). Although the range of membership to cluster was very low (0.51–0.64), pairwise F_{ST} between groups exhibited low (0.012–0.020) but significant values ($P<0.0001$). Importantly, these values were equal in magnitude to the observations based upon the real data reported above. BAPS analyses showed that, in spite of dealing with panmictic populations, in no case the most likely K was found to be 1. Instead, the number of putative populations took the following values: 3 (4% of the cases), 4 (38%) and 5 (58%) respectively.

Detection of sex-biased dispersal

According to the expectations that dispersal should be biased towards males, as in most of mammals, mAIc was lower in males than in females (-0.051 *vs.* 0.020) and vAIc was higher (2.90 *vs.* 2.35) Fig. G in File S1, whereas the rest of the statistics (F_{IS}, F_{ST}, Ho and Hs) took almost identical values in both sexes. However, the Mann–Whitney U-test proved to be non-significant ($P>0.5$) therefore we were unable to detect sex-biased dispersal in North Atlantic minke whales. Furthermore, when performing a two-tailed U-test we found a non-significant result ($P = 0.953$) that would not support a higher female dispersal either.

Discussion

Overall, the total data set ($N = 2990$) consisted of 28% males and 72% females; proportions that exactly coincide with Anderwald *et al.* [24] and are very similar to the 21% males 79% females reported by Andersen *et al.* [23]. This uneven presence of sexes was also reflected in the sex composition across Management Areas (Table 1), which also agrees with Andersen *et al.* [23] and Anderwald *et al.* [24] and corresponds to the known segregational behaviour with respect to sex and age during summer as mature females tend to occur further north than males [77–79].

Microsatellite loci used here exhibited a range of variation of genetic diversity comparable to what has been formerly reported for the same species [23–25] as well as for other balaenopterids such as Bryde's whales, *B. brydei* [80]; fin whales, *B. physalus* [81,82]; sei whales, *B. borealis* [83]; bowhead whales, *B. mysticetus* [84,85] and gray whales, *Eschrichtius robustus* [86]. Likewise, a similar magnitude of mtDNA genetic diversity, measured either as nucleotide (average of 0.009) or haplotype diversity (average of 0.9), was formerly reported for *B.a. acutorostrata* within the same geographic area [22–25,43] and resembles what has been described for other whale species [83,86–91]. However, the nucleotide diversity reported for the Antarctic minke whale (*B. bonaerensis*) is higher (0.0159) and this was interpreted as the Antarctic species having larger long-term effective population size than *B.a. acutorostrata* [22]. It has been proposed that the current size of the Antarctic minke whale population is unusually high as an indirect result of the whaling that killed more than 2 million of large whales leading to competitive release for smaller krill-eating species [92].

The mismatch distribution analyses were consistent with exponential population expansion suggesting that populations of

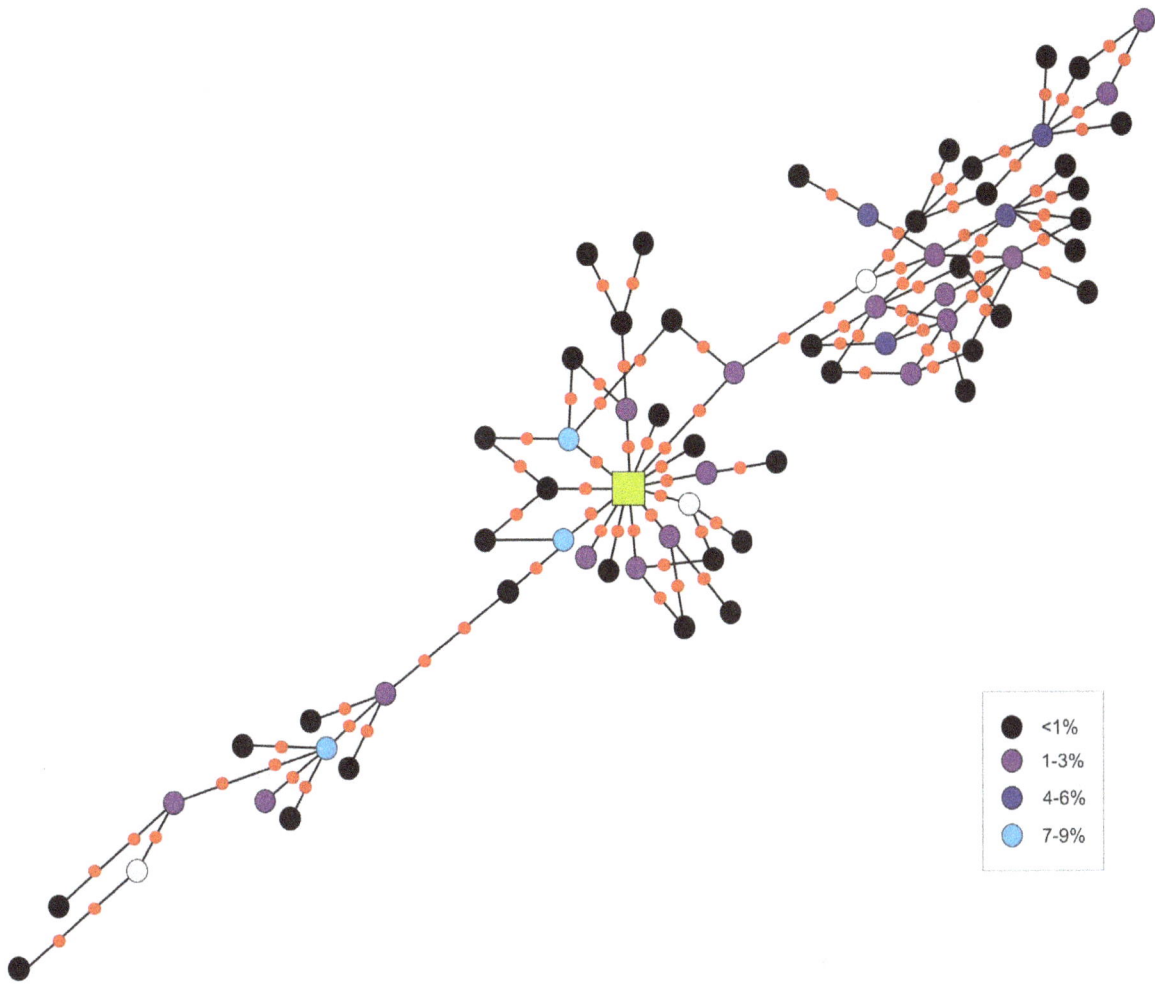

Figure 2. Median-joining network of mtDNA haplotypes corresponding to the period 2004 and 2007–2011. Haplotypes are represented as circles which area is not proportional to its relative frequency for simplicity. Instead, the frequency of haplotypes is depicted through the color code detailed in the legend. The green square represents the ancestral and more abundant haplotype (present in 27% of the individuals). The minimum number of steps connecting parsimoniously two haplotypes is indicated as a red dot, and the open circles represent extinct or missing haplotype that might have not been sampled (mv).

North Atlantic minke whale are not at equilibrium, something that had already been reported in the literature for this species in the same geographic area [24,25]. Earlier studies proposed that this expansion followed the last glacial maximum, as seen for various other cetacean species in the North Atlantic [24].

Spatial genetic structure

The present molecular markers (ten microsatellite loci and 331 bp of mtDNA D-loop) studied on 2990 individuals congruently failed to reveal any genetic differentiation among Management Areas during the period 2004 and 2007–2011. Only two

Table 6. Minke whale mtDNA.

Year	Fu's F_S	Tajima's D	SSD (P-value)	rg (P-value)
2004	**−24.916 (0.0002)**	−0.0062 (0.5859)	0.0098 (0.4413)	0.0153 (0.6398)
2007	**−23.861 (0.0000)**	−0.6057 (0.3004)	0.0105 (0.4586)	0.0159 (0.6831)
2008	**−13.279 (0.0119)**	−0.1924 (0.4987)	0.0115 (0.4099)	0.0158 (0.6806)
2009	**−10.710 (0.0279)**	−0.5333 (0.3516)	0.0128 (0.4439)	0.0193 (0.6339)
2010	**−13.711 (0.0104)**	0.2234 (0.6487)	0.0136 (0.3308)	0.0192 (0.5360)
2011	**−12.002 (0.0188)**	−0.1403 (0.5213)	0.0117 (0.4639)	0.0178 (0.6527)

Analyses of population stability (Tajima's D and Fu's F_S tests) and population expansion (sum of squared deviations, SSD and raggedness, rg mismatch distribution tests). Significant values are indicated with boldface type.

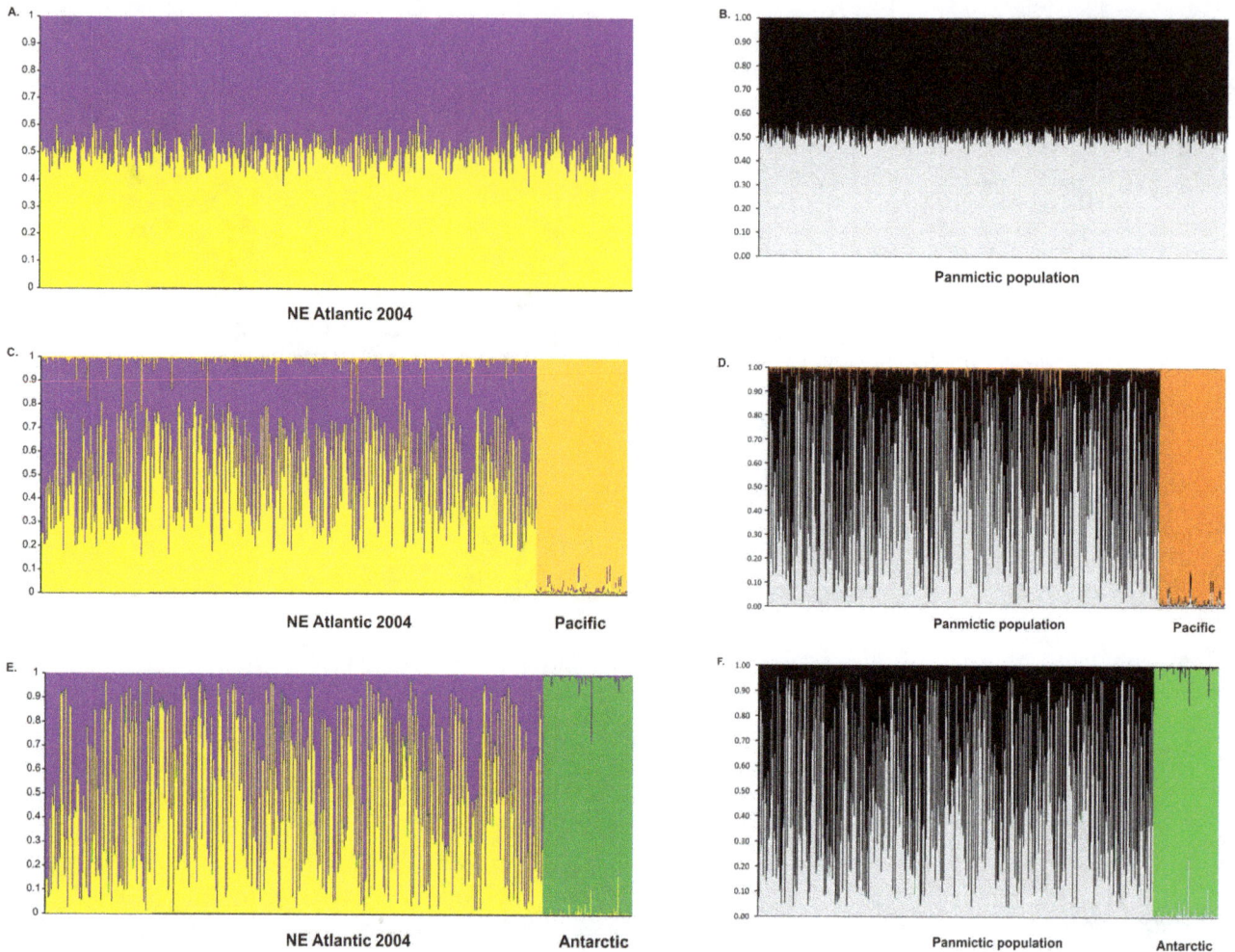

Figure 3. Example of comparison between real populations and the simulated panmictic ones. Bayesian clustering of North East Atlantic minke whale corresponding to year class 2004 (left column) and to a randomly chosen simulated panmictic population (right column). Inferred ancestry of individuals was calculated after averaging ten STRUCTURE runs with CLUMPP for K=2 (barplots a,b) and K=3 (barplots c–f). The outgroups were 95 individuals of the Pacific subspecies (*B. a. scammoni*) and 93 individuals of the Antarctic species (*B. bonaerensis*).

weak and marginally significant pairwise comparisons were recorded: EB-EW for microsatellites in 2010 ($R_{ST} = 0.024$, $P = 0.046$) and EB-ES for mtDNA in 2004 ($F_{ST} = 0.008$, $P = 0.035$). This translates to 1% of pairwise tests showing some spatial and temporal divergence, and neither were significant following Bonferroni correction for multiple testing. This lack of spatial genetic differentiation was the case when analyzing each of the six year classes separately, and also for all the specimens combined in the same AMOVA analysis, both for mitochondrial ($F_{ST} = 0.0005$, $P = 0.1134$) and nuclear ($F_{ST} = 0.0000$, $P = 0.9473$) DNA. Likewise, none of the resulting pairwise comparisons between areas were significant. The analyses using the joint data set seems legitimate given the temporal genetic stability found at both markers, stability that had already been reported within similar time frames [23,25,27].

The lack of geographic genetic differentiation as revealed in the present study is in agreement with some former studies of minke whales in the N Atlantic that were based on nuclear [24,25,27,28] and mitochondrial DNA markers [22–25]. However, the consensus about this issue is far from being commonplace, as the opposite scenario has also been reported for nuclear markers [23,31–33]. In particular, Andersen *et al.* [23] suggested the existence of four

genetically discrete subpopulations in the Atlantic (*i.e.* West Greenland, NE Atlantic, North Sea and Central North Atlantic) and indicated that this could be the result of the profound ecological differences between feeding areas (environmental conditions, prey availability) posing different selective pressures, coupled with a strong affiliation between mother and calf to the feeding site. Seasonal site fidelity that had been already reported for minke whales [93–95] as well as for other species such as humpback whales [96].

We also tested Tiedemann's [97] thesis that states that, for marine large mammals, the F_{ST} obtained for females would reflect the maximum spatial genetic differentiation of the species. Through the population structure observed in the maternally inherited mtDNA, Baker *et al.* [96] demonstrated that humpback whales show strong fidelity to migration destinations such as feeding grounds. Following a similar approach, we performed AMOVA across Management Areas by pooling all females sampled between 2007 and 2011 and, once more, we found no genetic differentiation for microsatellites ($F_{ST} = 0.00009$, $P = 0.326$) or mtDNA ($F_{ST} = -0.005$, $P>0.05$). This result disagrees, again, with Andersen *et al.* [23] who reported a significant F_{ST} at both markers for females.

In conclusion, our data set of ten microsatellites and 331 bp of mtDNA control region failed to reveal any spatial genetic variation across 2990 individual whales harvested in the five management areas in the NE Atlantic for the years 2004 and 2007–2011.

Cryptic population genetic structure

The division of North Atlantic minke whale into two mtDNA lineages had already been reported [25,26], and Palsbøll [26] suggested the presence of two potential breeding populations coexisting at feeding grounds in the North Atlantic. The division of mtDNA showing a lack of concordance between haplotypes and geographic regions was first mentioned by Bakke *et al.* [22] who proposed the existence of two or more differentiated populations sharing the same feeding grounds. However, the pattern observed in mtDNA might also reflect a residual ancestral polymorphism or a "recent" isolation of two populations at breeding sites, which roam through large parts of the North Atlantic Ocean during the feeding migration, as proposed by Palsbøll [26], Bakke *et al.* [22] and Pampoulie *et al.* [25]. Our results also agree with the discordance between haplotypes and geographic areas; however, we support the division of mtDNA into three distinct lineages (Fig. 2), with a central group that evolved through two different expansion episodes. This possible expansion was further corroborated by large negative and significant Fu's Fs, negative Tajima's D (except in 2010), small raggedness values and unimodal mismatch distributions (Table 6). The lack of connections among lineages further suggested genetic differentiation. Importantly, microsatellites did not corroborate this result.

Nuclear markers provided no evidence to reject the hypothesis that North Atlantic minke whales constitute a single panmictic population. This is in spite of certain insights from STRUCTURE analyses conducted both in this study and in Anderwald *et al.* [24] that appeared to spuriously suggest the existence of cryptic subpopulations. First, LnP(D) obtained after the STRUCTURE analyses conducted here revealed that $K = 1$ was the most likely number of clusters, both for the real data distributed in six year classes and for the 100 simulated panmictic populations. In all cases, a clear decreasing trend of LnP(D) along consecutive values of K was recorded. The *ad hoc* statistic ΔK based on the rate of change in the log probability of data between successive K values obtained through Evanno's test can, unfortunately, never validate $K = 1$ [58]. Furthermore, this test is not even applicable in situations of decreasing pattern of LnP(D) [58]. However, when ignoring this limitation, Evanno's test showed that the highest ΔK was found at values ranging between $K = 2$ and $K = 4$. Thus, in the real data, 5 out of the 6 cases yielded $K = 2$ at the Evanno criterion, whereas year class 2008 reported $K = 3$. Likewise, the 100 *in silico* generated panmictic populations revealed $K = 2$ in a majority of cases (58%) whereas the remaining ones were distributed between $K = 3$ (33%) and $K = 4$ (9%). Hence, both LnP(D) pointed at 1, together with the fact that the highest ΔK was found mainly at $K = 2$ in the real and simulated panmictic populations, supports that North Atlantic minke whale constitutes one single panmictic population.

In addition, when Evanno's test is computed in non-pertinent situations and seems to reveal substructuring in the population ($K = 2$), there are multiple features that strongly suggest a false result. The first hint to be considered is the low values of ΔK, which is an indication that the strength of the signal detected by STRUCTURE is weak [58]. In our case, both the 100 simulated panmictic populations and the six real ones showed extremely low values of ΔK (ranging mainly from 1 to 10). In contrast, when the differentiation signal between two populations is strong, *i.e.* when the number of clusters is unequivocally two, ΔK exhibits

significantly higher values. Thus, for instance, when we conducted STRUCTURE analyses for Atlantic minke whales including Antarctic or Pacific whales as an outgroup (Fig. B in File S1), the value of ΔK at $K = 2$ was higher by three orders of magnitude than the one found when running STRUCTURE without outgroups. Secondly, when $K = 1$ but the model is forced for $K = 2$, most individuals will have a probability around 0.5 and 0.5 of belonging to cluster 1 and cluster 2 respectively. Our results also corroborated this extent as the inferred membership to clusters ranged from 0.51 to 0.64 in the real and the ten *in silico* generated panmictic populations. However, even in this situation, a weak albeit significant F_{ST} between clusters (average value of 0.010 in the real data and of 0.016 in the panmictic populations) can still be found (see Tables A2, A13 in File S1). The fact that values of F_{ST} are of similar magnitude in the real data and in the 100 panmictic populations sheds important doubts about the reliability of such genetic structure.

When running STRUCTURE with outgroups to enhance the genetic differentiation, the resulting barplots for $K = 3$ (Fig. C in File S1) showed that the subdivision of NE Atlantic minke whales revealed a higher inferred membership to cluster compared to when no outgroups were used. Furthermore, when the outgroup was the Pacific subspecies (*B.a. scammoni*), the averaged inferred membership was 0.89 but when the outgroup was the Antarctic species (*B. bonaerensis*), this value was even higher (average 0.95) and the percentage of non-assigned individuals was slightly lower. This higher inferred membership to cluster could be expected to result in a higher genetic differentiation. However, the resulting F_{ST} values between these clusters were virtually identical in the following cases: real data without outgroups, real data with outgroups, simulated panmictic populations without outgroups, one randomly chosen simulated panmictic population without outgroups (Tables A2, A4, A5, A6, A13 in File S1). Additionally, all of these values overlap with the level of genetic differentiation observed using a similar approach in Anderwald et al. 2011 [24]. The fact that both real and simulated panmictic populations showed the same patterns further increased the doubts upon the reliability of the clustering analyses upon which subdivision of North East Atlantic minke whales into cryptic populations has been suggested [24]. Furthermore, in most cases, the distribution of the individuals belonging to clusters 1 and 2 across Management Areas was surprisingly similar for the six year classes sampled (Table K in File S1), and in the *in silico* generated panmictic populations.

Anecdotally, North Atlantic minke whales have been suggested to follow an annual migration cycle between Arctic feeding grounds and Southern breeding grounds. The information on sightings of minke whales in the Southern North Atlantic is however very scarce [78] and one or more breeding grounds have so far not been demonstrated. Also, foetuses in different stages of development have been found in catches from the northern feeding grounds [78], indicating that mating may take place even there. The hypothesis of panmixia could therefore be well supported by these observations, also implying that separate breeding grounds may not exist.

As a general picture, the data analysed here show that while nuclear markers suggest panmixia, mitochondrial markers reveal the existence of three distinct lineages in North East Atlantic minke whales; which can be a reflection of the different time scales both type of markers represent. Besides, due to maternal inheritance, mitochondrial genes have lower effective migration rates than nuclear genes [98], and random drift is faster for the haploid, maternally inherited mt genome compared to a diploid, biparentally inherited nuclear locus [99]. Furthermore, an accelerated substitution rate of the mitochondrial genome contributes to faster differentiation [100]. Thus, the aforementioned discordance of higher population subdivision in mtDNA than in nuclear DNA is indicative of migration and breeding sex ratios not being biased [101]. Accordingly, and in agreement with Pampoulie *et al.* [25], we did not reach statistical support for the hypothesis of male-biased dispersal in this species. In contrast, male-biased dispersal has been reported for other whale species such as sperm whales [102,103] or gray whales [86].

Conclusions

The population structure of North Atlantic minke whale, *B.a. acutorostrata*, has been the subject of a long debate with contrasting results, partially driven by the fact that most previous studies have been limited by low numbers of samples, or genetic markers, or a combination of both. In order to shed further light on this topic, we conducted a spatial, temporal and cryptic population analysis of 2990 whales harvested in the North East Atlantic during the period 2004 and 2007–2011. This large data set, which has been genotyped according to strict protocols upon which the NMDR is based [12], and is thus of very high data quality [36], failed to reveal any indication of geographical or temporal population genetic structure within the NE Atlantic based upon the analysis of ten microsatellites and 331 bp of the mitochondrial D-loop. Furthermore, while three mtDNA lineages were revealed in the data, these did not show any underlying geographic pattern, and possibly represent an ancestral signal. In order to address the possibility of cryptic population structure as suggested by Anderwald *et al.* [24], we run STRUCTURE using a similar approach. However, while Evanno's test might seem to suggest the existence of two genetically differentiated clusters per year class, there were a number of facts strongly suggesting that these results were potentially an artefact. Firstly, as this approach can never validate K = 1, it shows K = 2 instead but with a very low value of ΔK, which is an indication of a very weak genetic signal. Furthermore, there was a lack of corroboration with mtDNA, in each case there was close to a 50/50 division between individuals into groups 1 and 2, and there was an absence of any clear geographic pattern underlying the clusters. The suspicion that these analyses would spuriously reveal population substructure was subsequently confirmed when it was possible to falsely create two cryptic populations in our *in silico* generated panmictic populations. These displayed more or less identical genetic characteristics both as in the real data in this study, and in the study by Anderwald *et al.* [24]. Therefore, we conclude that there is at present no or very little evidence to suggest that the minke whale displays spatial or cryptic population genetic structure throughout the North East Atlantic. However, it is also duly acknowledged that all studies conducted thus far have been limited by low numbers of genetic markers. Therefore, in order to conclusively evaluate the potential for spatial or cryptic population genetic structure within this highly mobile species, significantly larger numbers of markers will be required. Recent publication of the minke whale genome [104] will represent a major resource to identify the numbers of markers needed to address this issue in the future.

Supporting Information

File S1 Supporting Information. File S1 contains detailed information on the following issue: "Testing the hypothesis of cryptic stock clustering in North East Atlantic minke whales": including Material and Methods, and Results. This appendix also comprises eight figures (Fig. A–G) and thirteen tables (Table A–

M). **Figures. Fig. A1.** Bayesian clustering of North East Atlantic minke whales genotyped at 8 microsatellites for the six sampled year classes. Inferred ancestry of individuals was calculated after averaging ten STRUCTURE runs with CLUMPP after Evanno's test. **Fig. A2.** Bayesian clustering of North East Atlantic minke whales genotyped at 10 microsatellites for the six sampled year classes. Inferred ancestry of individuals was calculated after averaging ten STRUCTURE runs with CLUMPP after Evanno's test. **Fig. B.** Bayesian clustering of North East Atlantic minke whale year class 2004 with outgroups: a) 95 individuals of the subspecies Pacific minke whale (*B. a. scammoni*); b) 93 individuals of the Antarctic minke whale (*B. bonaerensis*), and c) both former outgroups together. The number of clusters that best fitted the data was K = 2 after Evanno's [58] test in each case. This scenario was consistent across year classes. **Fig. C.** Bayesian clustering of North East Atlantic minke whale with outgroups in each year class. In the column to the left, the outgroup are 95 individuals of the subspecies Pacific minke whale (*B. a. scammoni*) whereas in the column to the right, the outgroup are 93 individuals of the Antarctic minke whale (*B. bonaerensis*). The number of clusters that best fitted the data was distinctively K = 2 after Evanno's [58] test in each case. **Fig. D.** Geographic distribution of individuals after different clustering methods: BAPS and STRUCTURE for microsatellites. Pie charts represent the percentage of individuals belonging to clusters 1 (dark grey) and 2 (light grey) per Management Area taking year class 2008 as an example (the full data for all the year classes is available in Table K in File S1). **Fig. E.** Bayesian clustering of individuals of ten of the simulated panmictic populations that showed K = 2 after Evanno's test. Inferred ancestry of individuals was calculated after averaging ten STRUCTURE runs with CLUMPP. **Fig. F.** Distribution of pairwise F_{ST} after 10000 random clustering of North Atlantic minke whale individuals per year class into two groups. **Fig. G.** Frequency distributions of the corrected assignment index (AIc) for 2156 females (light grey bars above axis) and 834 males (dark grey bars below axis). AIc values differed among sexes, males having on average negative values (−0.051) and higher variance (2.90) and females positive values (0.020) with lower variance (2.35). However, Mann–Whitney U-test proved sex-biased dispersal to be non-significant ($P > 0.5$). **Tables. Table A.** Summary result of STRUCTURE without outgroups: a) Data set of 8 microsatellites. b) Data set of 10 microsatellites. **Table B.** STRUCTURE without outgroups: Clustering of individuals per year class after Evanno's test (the two cases that showed the highest Evanno's ΔK at K = 3 are depicted in italics and analysed for K = 2 for comparison): Number of individuals per cluster and range of inferred membership to each of them (in brackets). Summary of the results of the AMOVA (F_{ST} and *P*-value) conducted with Arlequin with 10000 permutations. Analyses were performed for the same sets of individuals genotyped at mtDNA. Statistically significant values were highlighted in boldface type. Negative F_{ST} values found at mtDNA were transformed into 0. a) Data set of 8 microsatellites. b) Data set of 10 microsatellites. **Table C.** Summary statistics after STRUCTURE clustering showing total number of alleles, number of private alleles, observed heterozygosity (average ± SE), unbiased expected heterozygosity (average ± SE), and inbreeding coefficient (F_{IS}) (average ± SD). We show in italics the distribution of the individuals for K = 2 for the two year classes that showed the highest Evanno's ΔK at K = 3. a) Data set of 8 microsatellites. b) Data set of 10 microsatellites. **Table D.** STRUCTURE with the Pacific minke whale subspecies (*B. a. scammoni*) as an outgroup. Clustering of individuals per year class and one randomly chosen simulated panmictic population after Evanno's test and CLUMPP averaging: Number of individuals per cluster and range of inferred membership to each of them (in brackets). Summary of the results of the AMOVA (F_{ST} and P-value) conducted with Arlequin with 10000 permutations. Analyses were performed for the same sets of individuals genotyped at mtDNA. Statistically significant values were highlighted in boldface type. Negative F_{ST} values found at mtDNA were transformed into 0. **Table E.** STRUCTURE with Antarctic minke whale species (*B. bonaerensis*) as an outgroup. Clustering of individuals per year class and one randomly chosen simulated panmictic population after Evanno's test and CLUMPP averaging: Number of individuals per cluster and range of inferred membership to each of them (in brackets). Summary of the results of the AMOVA (F_{ST} and P-value) conducted with Arlequin with 10000 permutations. Analyses were performed for the same sets of individuals genotyped at mtDNA. Statistically significant values were highlighted in boldface type. Negative F_{ST} values found at mtDNA were transformed into 0. **Table F.** STRUCTURE consensus clustering of individuals (*i.e.* agreement between Antarctic and Pacific outgroup clustering) into two groups per year class. Summary of the results of the AMOVA (F_{ST} and P-value) conducted with Arlequin with 10000 permutations. Analyses were performed for the same sets of individuals at mtDNA. Statistically significant values are highlighted in boldface type. **Table G.** Summary statistics after STRUCTURE consensus clustering (*i.e.* consensus between Antarctic and Pacific outgroup clustering) showing total number of alleles, allelic richness (minimum sample size), number of private alleles, observed heterozygosity (average $\pm$ SE), unbiased expected heterozygosity (average $\pm$ SE), and inbreeding coefficient (F_{IS}) (average $\pm$ SD). **Table H.** BAPS clustering of individuals genotyped with microsatellites into two groups per year class. Summary of the results of the AMOVA (F_{ST} and P-value) conducted with ARLEQUIN with 10000 permutations. Analyses were performed for the same sets of individuals at mtDNA. Statistically significant values were highlighted in boldface type. **Table I.** Summary statistics after BAPS clustering showing total number of alleles, allelic richness (minimum sample size), number

of private alleles, observed heterozygosity (average $\pm$ SE), unbiased expected heterozygosity (average $\pm$ SE), and inbreeding coefficient (F_{IS}) (average $\pm$ SD). **Table J.** GeneClass self-assignment: Percentage of individuals genotyped at microsatellites that were correctly assignment after clustering procedures. **Table K.** Number of individuals genotyped at microsatellites per Management Areas after clustering with BAPS and STRUCTURE (with and without outgroup). ND = No data. **Table L.** Matrix of numbers and percentage of coincident individuals when comparing the three clustering methods: BAPS, STRUCTURE without outgroup (STR), and STRUCTURE with outgroup (STR consensus). The percentage of coincident individuals was calculated by dividing the number of by the lowest number of individuals in the corresponding cluster. STRUCTURE analyses were performed with 8 microsatellites. **Table M.** STRUCTURE clustering of individuals in the 10 randomly selected simulated panmictic populations showing K = 2 after Evanno's test. Number of individuals per cluster and range of inferred membership to each of them (in brackets); number of non-assigned individuals (and % of the total). Summary of the results ofthe AMOVA (F_{ST} and P-value) conducted with Arlequin with 10000 permutations. Statistically significant values were highlighted in boldface type.

Acknowledgments

We thank François Besnier for help with ParallelStructure and simulations, and Lúa López and Rodolfo Barreiro for insightful discussions. We are grateful to Rus Hoelzel and Pia Anderwald for constructive criticism to earlier drafts of this manuscript.

Author Contributions

Conceived and designed the experiments: MQ HJS NØ TH KAG. Performed the experiments: BBS KAG. Analyzed the data: MQ HJS HKS. Contributed reagents/materials/analysis tools: HJS NØ TH NK LAP KAG. Wrote the paper: MQ HJS NØ TH BBS HKS CP NK LAP KAG.

References

1. Laikre L, Schwartz MK, Waples RS, Ryman N (2010) Compromising genetic diversity in the wild: unmonitored large-scale release of plants and animals. Trends in Ecology & Evolution 25: 520–529.
2. Allendorf FW, Luikart G (2006) Conservation and the Genetics of Populations: Blackwell Publishing.
3. Frankham R (2005) Stress and adaptation in conservation genetics. Journal of Evolutionary Biology 18: 750–755.
4. Glover KA, Quintela M, Wennevik V, Besnier F, Sørvik AGE, et al. (2012) Three decades of farmed escapees in the wild: A spatio-temporal analysis of Atlantic salmon population genetic structure throughout Norway. PLoS ONE 7: e43129.
5. IUCN (2013) International union for the conservation of nature and natural resources. In: species Irlot, editor: www.iucnredlist.org.
6. Stevick PT, McConnell BJ, Hammond PS (2002) Patterns of Movement. In: Hoelzel AR, editor. Marine Mammal Biology: An Evolutionary Approach. Oxford, U.K.: Blackwell Publishing. pp. 185–216.
7. Hoelzel A (1998) Genetic structure of cetacean populations in sympatry, parapatry, and mixed assemblages: implications for conservation policy. Journal of Heredity 89: 451–458.
8. Whitehead H, Dillon M, Dufault S, Weilgart L, Wright J (1998) Non-geographically based population structure of South Pacific sperm whales: dialects, fluke-markings and genetics. Journal of Animal Ecology 67: 253–262.
9. Hoelzel AR (2009) Evolution of population genetic structure in marine mammal species In: Bertorelle G, Bruford MW, Hauffe HC, Rizzoli A, Vernesi C, editors. Population Genetics for Animal Conservation. Cambridge: Cambridge University Press. pp. 410.
10. Rice DW (1998) Marine Mammals of the World. Systematics and Distribution. Society for Marine Mammalogy 4: 1–231.
11. Glover K, Kanda N, Haug T, Pastene L, Oien N, et al. (2013) Hybrids between common and Antarctic minke whales are fertile and can back-cross. BMC Genetics 14: 25.
12. Glover KA, Haug T, Øien N, Walløe L, Lindblom L, et al. (2012) The Norwegian minke whale DNA register: a data base monitoring commercial harvest and trade of whale products. Fish and Fisheries 13: 313–332.
13. Johnsgård Å (1966) The distribution of Balaenopteridae in the North Atlantic Ocean. In: Norris KS, editor. Whales, Dolphins and Porpoises. California: University of California Berkely Press. pp. 114.
14. Stewart BS, Leatherwood S (1985) Minke Whale, *Balaenoptera acutorostrata* Lacépède, 1804. In: Ridgway SH, Harrison SR, editors. Handbook of Marine Mammals, Volume 3: The Sirenians and Baleen Whales: Academic Press. pp. 91–136.
15. Jonsgård Å (1951) Studies on the little piked whale or minke whale (*Balaenoptera acutorostrata* Lacépède). Norsk Hvalfangsttid 40.
16. Sergeant DE (1963) Minke whales, *Balaenoptera acutorostrata* Lacépède, of the Western North Atlantic. Journal of the Fisheries Research Board of Canada 20: 1489–1504.
17. Wada S (1991) Genetic distinction between two minke whale stocks in the Okhotsk Sea coast of Japan.
18. IWC (2010) Report of the Scientific Committee. Journal of Cetacean Research and Management 11: 1–98.
19. Donovan GP (1991) A review of IWC stock boundaries. Cambridge, U.K. 39–68 p.
20. IWC (1992) Annex K. Report of the working group on North Atlantic minke trials. Rep Int Whal Comm 42: 246–251.
21. Hoelzel AR (1991) Whaling in the dark. Nature 352: 481–481.
22. Bakke I, Johansen S, Bakke Ø, El-Gewely MR (1996) Lack of population subdivision among the minke whales (*Balaenoptera acutorostrata*) from Icelandic and Norwegian waters based on mitochondrial DNA sequences. Marine Biology 125: 1–9.
23. Andersen LW, Born EW, Dietz R, Haug T, Øien N, et al. (2003) Genetic population structure of minke whales *Balaenoptera acutorostrata* from Greenland, the North East Atlantic and the North Sea probably reflects different ecological regions. Marine Ecology Progress Series 247: 263–280.

24. Anderwald P, Daníelsdóttir AK, Haug T, Larsen F, Lesage V, et al. (2011) Possible cryptic stock structure for minke whales in the North Atlantic: Implications for conservation and management. Biological Conservation 144: 2479–2489.

25. Pampoulie C, Daníelsdóttir AK, Víkingsson GA (2008) Genetic structure of the North Atlantic common minke whale (*Balaenoptera acutorostrata*) at feeding grounds: a microsatellite loci and mtDNA analysis. SC/F13/SP17 SC/60/PFI10 XXX: 1–17.

26. Palsbøll PJ (1989) Restriction fragment pattern analysis of mitochondrial DNA in minke whales, *Balaenoptera acutorostrata*, from the Davis Strait and the Northeast Atlantic: Copenhagen.

27. Martínez I, Pastene LA (1999) RAPD-typing of Central and Eastern North Atlantic and Western North Pacific minke whales, *Balaenoptera acutorostrata*. ICES Journal of Marine Science 56: 640–651.

28. Martínez I, Elvevoll EO, Haug T (1997) RAPD typing of north-east Atlantic minke whale (*Balaenoptera acutorostrata*). ICES Journal of Marine Science 54: 478–484.

29. Born EW, Outridge P, Riget FF, Hobson KA, Dietz R, et al. (2003) Population substructure of North Atlantic minke whales (*Balaenoptera acutorostrata*) inferred from regional variation of elemental and stable isotopic signatures in tissues. Journal of Marine Systems 43: 1–17.

30. Hobbs KE, Muir DCG, Born EW, Dietz R, Haug T, et al. (2003) Levels and patterns of persistent organochlorines in minke whale (*Balaenoptera acutorostrata*) stocks from the North Atlantic and European Arctic. Environmental Pollution 121: 239–252.

31. Daníelsdóttir AK, Sverrir DH, Sigfríður G, Alfreð Á (1995) Genetic variation in northeastern Atlantic minke whales (*Balaenoptera acutorostrata*). Developments in Marine Biology 4: 105–118.

32. Daníelsdóttir AK, Duke EJ, Árnason A (1992) Genetic variation at enzyme loci in North Atlantic minke whales, *Balaenoptera acutorostrata*. Biochemical Genetics 30: 189–202.

33. Árnason A (1995) Genetic markers and whale stocks in the North Atlantic ocean: a review. Developments in Marine Biology Volume 4: 91–103.

34. Skåre M (1994) Whaling: A sustainable use of natural resources or a violation of animal rights? Environment: Science and Policy for Sustainable Development 36: 12–31.

35. Skaug HJ, Bérubé M, Palsbøll PJ (2010) Detecting dyads of related individuals in large collections of DNA-profiles by controlling the false discovery rate. Molecular Ecology Resources 10: 693–700.

36. Haaland O, Glover K, Seliussen B, Skaug H (2011) Genotyping errors in a calibrated DNA register: implications for identification of individuals. BMC Genetics 12: 36.

37. Valsecchi E, Amos W (1996) Microsatellite markers for the study of cetacean populations. Molecular Ecology 5: 151–156.

38. Palsbøll PJ, Bérubé M, Larsen AH, Jørgensen H (1997) Primers for the amplification of tri- and tetramer microsatellite loci in baleen whales. Molecular Ecology 6: 893–895.

39. Bérubé M, Jørgensen H, McEwing R, Palsbøll PJ (2000) Polymorphic dinucleotide microsatellite loci isolated from the humpback whale, *Megaptera novaeangliae*. Molecular Ecology 9: 2181–2183.

40. Bérubé M, Palsbøll P (1996) Identification of sex in cetaceans by multiplexing with three ZFX and ZFY specific primers. Molecular Ecology 5: 283–287.

41. Árnason U, Gullberg A, Widegren B (1993) Cetacean mitochondrial DNA control region: sequences of all extant baleen whales and two sperm whale species. Molecular Biology and Evolution 10: 960–970.

42. Larsen AH, Sigurjonsson J, Øien N, Vikingsson G, Palsboll P (1996) Populations genetic analysis of nuclear and mitochondrial loci in skin biopsies collected from Central and Northeastern North Atlantic humpback whales (*Megaptera novaeangliae*): Population identity and migratory destinations. Proceedings of the Royal Society of London Series B: Biological Sciences 263: 1611–1618.

43. Palsbøll PJ, Clapham PJ, Mattila DK, Larsen F, Sears R, et al. (1995) Distribution of mtDNA haplotypes in North-Atlantic humpback whales: The influence of behaviour on population structure. Marine Ecology Progress Series 116: 1–10.

44. Dieringer D, Schlötterer C (2003) MICROSATELLITE ANALYSER (MSA): A platform independent analysis tool for large microsatellite data sets. Molecular Ecology Notes 3: 167–169.

45. Peakall R, Smouse PE (2006) GenAlEx 6: genetic analysis in Excel. Population genetic software for teaching and research. Molecular Ecology Notes 6: 288–295.

46. Rousset F (2008) GENEPOP'007: a complete re-implementation of the genepop software for Windows and Linux. Molecular Ecology Resources 8: 103–106.

47. Pritchard JK, Stephens M, Donnelly P (2000) Inference of population structure using multilocus genotype data. Genetics 155: 945–959.

48. Corander J, Waldmann P, Marttinen P, Sillanpaa MJ (2004) BAPS 2: enhanced possibilities for the analysis of genetic population structure. Bioinformatics 20: 2363–2369.

49. Weir BS, Cockerham CC (1984) Estimating F-statistics for the analysis of population structure. Evolution 38: 1358–1370.

50. Slatkin M (1995) A measure of population subdivision based on microsatellite allele frequencies. Genetics 139: 457–462.

51. Wright S (1969) Evolution and the Genetics of Populations. Chicago: University of Chicago Press. 295 p.

52. Excoffier L, Laval G, Schneider S (2005) Arlequin ver. 3.0: An integrated software package for population genetics data analysis. Evolutionary Bioinformatics Online 1: 47–50.

53. Falush D, Stephens M, Pritchard JK (2003) Inference of population structure using multilocus genotype data: Linked loci and correlated allele frequencies. Genetics 164: 1567–1587.

54. Hubisz M, Falush D, Stephens M, Pritchard J (2009) Inferring weak population structure with the assistance of sample group information. Molecular Ecology Resources 9: 1322–1332.

55. Corander J, Waldmann P, Sillanpaa MJ (2003) Bayesian analysis of genetic differentiation between populations. Genetics 163: 367–374.

56. Corander J, Marttinen P (2006) Bayesian identification of admixture events using multilocus molecular markers. Molecular Ecology 15: 2833–2843.

57. Earl DA, von Holdt BM (2012) STRUCTURE HARVESTER: a website and program for visualizing STRUCTURE output and implementing the Evanno method. Conservation Genetics Resources 4: 359–361.

58. Evanno G, Regnaut S, Goudet J (2005) Detecting the number of clusters of individuals using the software STRUCTURE: a simulation study. Molecular Ecology 14: 2611–2620.

59. Besnier F, Glover KA (2013) ParallelStructure: a R package to distribute parallel runs of the population genetics program STRUCTURE on multi-core computers. PLoS ONE 8: e70651.

60. Jakobsson M, Rosenberg NA (2007) CLUMPP: a cluster matching and permutation program for dealing with label switching and multimodality in analysis of population structure. Bioinformatics 23: 1801–1806.

61. Rozas J, Sánchez-del Barrio JC, Messeguer X, Rozas R (2003) DnaSP, DNA polymorphism analyses by the coalescent and other methods. Bioinformatics 19: 2496–2497.

62. Tajima F (1989) Statistical Method for Testing the Neutral Mutation Hypothesis by DNA Polymorphism. Genetics 123: 585–595.

63. Fu YX (1997) Statistical tests of neutrality of mutations against population growth, hitchhiking and background selection. Genetics 147: 915–925.

64. Slatkin M, Hudson RR (1991) Pairwise comparisons of mitochondrial DNA sequences in stable and exponentially growing populations. Genetics 129: 555–562.

65. Rogers AR (1995) Genetic evidence for a Pleistocene population explosion. Evolution 49: 608–615.

66. Rogers AR, Harpending H (1992) Population growth makes waves in the distribution of pairwise genetic differences. Molecular Biology and Evolution 9: 552–569.

67. Ramos-Onsins SE, Rozas J (2002) Statistical properties of new neutrality tests against population growth. Molecular Biology and Evolution 19: 2092–2100.

68. Harpending HC (1994) Signature of ancient population growth in a low-resolution mitochondrial DNA mismatch distribution. Human Biology 66: 591–600.

69. Rogers AR, Harpending HC (1983) Population structure and quantitative characters. Genetics 105: 985–1002.

70. Schneider S, Excoffier L (1999) Estimation of past demographic parameters from the distribution of pairwise differences when the mutation rates vary among sites: Application to human mitochondrial DNA. Genetics 152: 1079–1089.

71. Bandelt HJ, Forster P, Rohl A (1999) Median-joining networks for inferring intraspecific phylogenies. Molecular Biology and Evolution 16: 37–48.

72. Polzin T, Vahdati Daneshmand S (2003) On Steiner trees and minimum spanning trees in hypergraphs. Operations Research Letters 31: 12–20.

73. Goudet J, Perrin N, Waser P (2002) Tests for sex-biased dispersal using bi-parentally inherited genetic markers. Molecular Ecology 11: 1103–1114.

74. Favre L, Balloux F, Goudet J, Perrin N (1997) Female-biased dispersal in the monogamous mammal *Crocidura russula*: Evidence from field data and microsatellite patterns. Proceedings of the Royal Society of London Series B: Biological Sciences 264: 127–132.

75. Mossman CA, Waser PM (1999) Genetic detection of sex-biased dispersal. Molecular Ecology 8: 1063–1067.

76. Greenwood PJ (1980) Mating systems, philopatry and dispersal in birds and mammals. Animal Behaviour 28: 1140–1162.

77. Øien N (1988) Length distributions in catches from the north-eastern Atlantic stock of minke whales. Rep Int Whal Comm 38: 289–295.

78. Horwood J (1990) Biology and exploitation of the minke whale. Boca Raton, Florida: CRC Press, Inc.

79. Laidre KL, Heagerty PJ, Heide-Jørgensen MP, Witting L, Simon M (2009) Sexual segregation of common minke whales (*Balaenoptera acutorostrata*) in Greenland, and the influence of sea temperature on the sex ratio of catches. ICES Journal of Marine Science: Journal du Conseil 66: 2253–2266.

80. Kanda N, Goto M, Kato H, McPhee MV, Pastene LA (2007) Population genetic structure of Bryde's whales (*Balaenoptera brydei*) at the inter-oceanic and trans-equatorial levels. Conservation Genetics 8: 853–864.

81. Bérubé M, Urbán J, Dizon AE, Brownell RL, Palsbøll PJ (2002) Genetic identification of a small and highly isolated population of fin whales (*Balaenoptera physalus*) in the Sea of Cortez, México. Conservation Genetics 3: 183–190.

82. Bérubé M, Aguilar A, Dendanto D, Larsen F, Notarbartolo Di Sciara G, et al. (1998) Population genetic structure of North Atlantic, Mediterranean Sea and

Sea of Cortez fin whales, *Balaenoptera physalus* (Linnaeus 1758): analysis of mitochondrial and nuclear loci. Molecular Ecology 7: 585–599.

83. Kanda N, Goto M, Yoshida H, Pastene LA (2009) Stock structure of sei whales in the North Pacific as revealed by microsatellite and mitochondrial DNA analyses.

84. Jorde PE, Schweder T, Bickham JW, Givens GH, Suydam R, et al. (2007) Detecting genetic structure in migrating bowhead whales off the coast of Barrow, Alaska. Molecular Ecology 16: 1993–2004.

85. Morin PA, Archer FI, Pease VL, Hancock-Hanser BL, Robertson KM, et al. (2012) Empirical comparison of single nucleotide polymorphisms and microsatellites for population and demographic analyses of bowhead whales. Endangered Species Research 19: 129–147.

86. Lang AR, Weller DW, Leduc RG, Burdin AM, Brownell RLj (2010) Genetic differentiation between Western and Eastern (*Eschrichtius robustus*) gray whale populations using microsatellite markers. University of Nebraska - Lincoln. 19 p.

87. Sremba AL, Hancock-Hanser B, Branch TA, LeDuc RL, Baker CS (2012) Circumpolar diversity and geographic differentiation of mtDNA in the critically endangered Antarctic blue whale (*Balaenoptera musculus intermedia*). PLoS ONE 7: e32579.

88. LeDuc RG, Weller DW, Hyde J, Burdin AM, Rosel PE, et al. (2002) Genetic differences between western and eastern gray whales (*Eschrichtius robustus*). Journal of Cetacean Research and Management 4: 1–5.

89. LeDuc RG, Dizon AE, Pastene LA, Kato H, Nishiwaki S, et al. (2007) Patterns of genetic variation in Southern Hemisphere blue whales and the use of assignment test to detect mixing on the feeding grounds. Journal of Cetacean Research and Management 9: 73–80.

90. Olavarría C, Baker CS, Garrigue C, Poole M, Hauser N, et al. (2007) Population structure of South Pacific humpback whales and the origin of the eastern Polynesian breeding grounds. Marine Ecology Progress Series 330: 257–268.

91. Patenaude NJ, Portway VA, Schaeff CM, Bannister JL, Best PB, et al. (2007) Mitochondrial DNA diversity and population structure among Southern right whales (*Eubalaena australis*). Journal of Heredity 98: 147–157.

92. Ruegg KC, Anderson EC, Scott Baker C, Vant M, Jackson JA, et al. (2010) Are Antarctic minke whales unusually abundant because of 20th century whaling? Molecular Ecology 19: 281–291.

93. Gill A, Fairbairrns RS (1995) Photo-identification of the minke whale *Balaenoptera acutorostrata* off the Isle of Mull, Scotland. In: Arnoldus Schytte

Blix LW, Øyvind U, editors. Developments in Marine Biology: Elsevier Science. pp. 129–132.

94. Dorsey EM, Stern SJ, Hoelzel AR, Jacobsen J (1990) Minke whales (*Balaenoptera acutorostrata*) from the west coast of North America: individual recognition and small-scale site fidelity. 357–368 p.

95. Dorsey EM (1983) Exclusive adjoining ranges in individually identified minke whales (*Balaenoptera acutorostrata*) in Washington state. Canadian Journal of Zoology 61: 174–181.

96. Baker CS, Slade RW, Bannister JL, Abernethy RB, Weinrich MT, et al. (1994) Hierarchical structure of mitochondrial DNA gene flow among humpback whales *Megaptera novaeangliae*, world-wide. Molecular Ecology 3: 313–327.

97. Tiedemann R, Hardy O, Vekemans X, Milinkovitch MC (2000) Higher impact of female than male migration on population structure in large mammals. Molecular Ecology 9: 1159–1163.

98. Birky CW, Maruyama T, Fuerst P (1983) An approach to population and evolutionary genetic theory for genes in mitochondria and chloroplasts, and some results. Genetics 103: 513–527.

99. Palumbi SR, Baker CS (1994) Contrasting population structure from nuclear intron sequences and mtDNA of humpback whales. Molecular Biology and Evolution 11: 426–435.

100. Avise JC, Ellis D (1986) Mitochondrial DNA and the evolutionary genetics of higher animals [and Discussion]. Philosophical Transactions of the Royal Society of London B, Biological Sciences 312: 325–342.

101. Birky CW, Fuerst P, Maruyama T (1989) Organelle gene diversity under migration, mutation, and drift: equilibrium expectations, approach to equilibrium, effects of heteroplasmic cells, and comparison to nuclear genes. Genetics 121: 613–627.

102. Engelhaupt D, Rus Hoelzel A, Nicholson C, Frantzis A, Mesnick S, et al. (2009) Female philopatry in coastal basins and male dispersion across the North Atlantic in a highly mobile marine species, the sperm whale (*Physeter macrocephalus*). Molecular Ecology 18: 4193–4205.

103. Lyrholm T, Leimar O, Johanneson B, Gyllensten U (1999) Sex–biased dispersal in sperm whales: contrasting mitochondrial and nuclear genetic structure of global populations. Proceedings of the Royal Society of London Series B: Biological Sciences 266: 347–354.

104. Yim H-S, Cho YS, Guang X, Kang SG, Jeong J-Y, et al. (2014) Minke whale genome and aquatic adaptation in cetaceans. Nat Genet 46: 88–92.

5

Current and Future Patterns of Global Marine Mammal Biodiversity

Kristin Kaschner[1]*, **Derek P. Tittensor**[2¤a,¤b], **Jonathan Ready**[3], **Tim Gerrodette**[4], **Boris Worm**[2]

1 Evolutionary Biology and Ecology Lab, Institute of Zoology, Albert-Ludwigs-University, Freiburg, Germany, **2** Department of Biology, Dalhousie University, Halifax, Nova Scotia, Canada, **3** Institute of Estuarine and Coastal Studies, Universidade Federal do Pará – Campus de Bragança, Bragança, Pará, Brazil, **4** Protected Resources Division, Southwest Fisheries Science Center, National Marine Fisheries Service, La Jolla, California, United States of America

Abstract

Quantifying the spatial distribution of taxa is an important prerequisite for the preservation of biodiversity, and can provide a baseline against which to measure the impacts of climate change. Here we analyse patterns of marine mammal species richness based on predictions of global distributional ranges for 115 species, including all extant pinnipeds and cetaceans. We used an environmental suitability model specifically designed to address the paucity of distributional data for many marine mammal species. We generated richness patterns by overlaying predicted distributions for all species; these were then validated against sightings data from dedicated long-term surveys in the Eastern Tropical Pacific, the Northeast Atlantic and the Southern Ocean. Model outputs correlated well with empirically observed patterns of biodiversity in all three survey regions. Marine mammal richness was predicted to be highest in temperate waters of both hemispheres with distinct hotspots around New Zealand, Japan, Baja California, the Galapagos Islands, the Southeast Pacific, and the Southern Ocean. We then applied our model to explore potential changes in biodiversity under future perturbations of environmental conditions. Forward projections of biodiversity using an intermediate Intergovernmental Panel for Climate Change (IPCC) temperature scenario predicted that projected ocean warming and changes in sea ice cover until 2050 may have moderate effects on the spatial patterns of marine mammal richness. Increases in cetacean richness were predicted above 40° latitude in both hemispheres, while decreases in both pinniped and cetacean richness were expected at lower latitudes. Our results show how species distribution models can be applied to explore broad patterns of marine biodiversity worldwide for taxa for which limited distributional data are available.

Editor: Steven J. Bograd, National Oceanic and Atmospheric Administration/National Marine Fisheries Service/Southwest Fisheries Science Center, United States of America

Funding: This research was supported by the Sloan Foundation (Census of Marine Life, Future of Marine Animal Populations Project). Partial funding for the main author was provided through the Future Oceans Cluster Project D 1067/88 "Winners and Losers in the Future Oceans" project, IFM-Geomar Kiel. The funders had no role in study design, data collection and analysis, decision to publish, or preparation of the manuscript.

Competing Interests: The authors have declared that no competing interests exist.

* E-mail: Kristin.Kaschner@biologie.uni-freiburg.de

¤a Current address: United Nations Environment Programme World Conservation Monitoring Centre, Cambridge, United Kingdom
¤b Current address: Microsoft Research Computational Science Laboratory, Cambridge, United Kingdom

Introduction

The global distribution of species diversity and richness has been of interest to naturalists for centuries and remains an important research topic in ecology today [1]. More recently, this quest has been further motivated by systematic conservation planning efforts, which require detailed data on the distribution of biodiversity in space and time [2]. Quantifying patterns of biodiversity can be costly and challenging, particularly in the oceans where most taxa cannot easily be seen and many species are highly mobile with large ranges that extend far into the open oceans [3].

In terms of species number, marine mammals are a relatively small taxonomic group, yet given their biomass and position in the food web they represent an ecologically important part of marine biodiversity [4,5,6] Furthermore they are of significant conservation concern, with 23% of species currently threatened by extinction [6]. Therefore, marine mammals often feature prominently in marine conservation planning and protected area design [7,8,9]. Their large-scale patterns of biodiversity have only recently been analyzed using

expert knowledge [6] or regional observations [10]. Using expert knowledge, Schipper and colleagues [6] delineated the known, or suspected, range of individual species and then overlaid maps to produce global patterns of marine mammal species richness. This approach can accommodate all species on a global scale, but represents a relatively coarse approach that does not distinguish between core and marginal habitats, attributing the same probability of occurrence for a species throughout its range [6]. In addition, resulting patterns remain to be quantitatively validated and cannot be used directly to investigate shifts in distributions under different environmental conditions, since distributions are based on expert knowledge, rather than predictive models that take into account environmental forcings. In contrast, due to the lack of occurrence records for most marine mammal species, existing empirical attempts using sighting surveys to estimate realized cetacean richness have been restricted in taxonomic and spatial coverage, and resulting global predictions may suffer from undersampling [10]. Similar to the trade-offs of different habitat prediction modeling approaches [11], these two methods lie on opposite ends of a spectrum from potentially overpredicting expert-derived (range maps) to potentially underpredicting (empirical sighting surveys) range sizes. Here we present a complementary modelling approach that combines both types of data to make predictions of large-scale marine mammal species distributions using a relative environmental suitability (RES) model [12]; an environmental niche model developed specifically to deal with the prevailing paucity of data for many marine mammal species [12]. The RES model delineates the environmental tolerances of all species with respect to basic parameters known to determine marine mammal distributions directly or indirectly. It does so by combining available data on species occurrence and habitat usage, supplemented by expert knowledge [12]. The relative environmental suitability of different habitats for a given species can then be computed and used to predict long-term mean annual species distributions. Here we superimpose individual species predictions to generate global patterns of species richness, defined as the number of species present in a given area [13], which we subsequently validated using independent survey data.

Bioclimatic envelope models such as the RES models are based on the relationship between species occurrence and environmental proxies, and have been used to explore possible range shifts of marine and terrestrial species under changing environmental conditions [14,15], although results tend to be sensitive to model assumptions and uncertainties [16,17,18]. Global warming is imposing environmental changes on a large scale, and empirical observations indicate shifts in the distributional ranges of many species; these shifts are often consistent with global warming as a driving mechanism [19]. In the oceans, many taxa, ranging from benthic invertebrates to plankton and fish, have shown such range shifts (reviewed in [20]). There is much concern about climate change impacts on marine mammals [21,22], but the assessment of impacts has mostly been restricted to theoretical considerations [23,24,25]. The quantification of possible effects on health [26], food availability [27] and migration [28] remains difficult and impacts are expected to vary for different species [27]. However, species distributions are expected to be affected by temperature and ice cover changes [23], with changes in community structure [29], range expansions into higher-latitude waters [10,30], and decreases in suitable habitat [31] among the probable outcomes. Here we apply species-specific RES models to explore the possible consequences of temperature change for the global distribution of marine mammal richness in the near future.

Methods

Mapping marine mammal richness

We explored marine mammal species richness by overlaying predictions of the relative probability of occurrence for 115 marine mammal species. These included 68 toothed whales (Order: Odontocetii), 15 baleen whales (Mysticetii), and 32 seals and sea lions (Pinnipedia), but excluded all freshwater species, dugongs and manatees (Sirenia), sea otter (*Enhydra lutis*), and polar bear (*Ursus maritimus*). Individual species' ranges were derived from an environmental niche model that predicted distributions and the relative environmental suitability (RES) for different species on a 0.5°x0.5° global grid. Predicted results represent mean annual geographic ranges defined as the maximum area between the known outer-most limits of a species' regular or periodic occurrence [12]. While this definition is inclusive of all areas covered during annual migrations, dispersal of juveniles etc., it specifically excludes extralimital sightings, which are sometimes difficult to distinguish from the core range [32]. The RES modeling approach was developed because of the paucity of marine mammal data available for standard species distribution modelling approaches, and well-known spatial biases in the available

data: point occurrence records are currently only available for <60% of known marine mammals [33], and 70% of all available sighting records come from continental shelf waters of the Northern Hemisphere, according to the Ocean Biogeographic Information System (OBIS, www.iobis.org, 05/2010). Unlike other species distribution models, the RES model therefore is based primarily on expert knowledge, compiled through extensive literature review, supplemented by occurrence data (where possible). This synthesized information is used to assign species to pre-defined habitat use categories, represented by simple trapezoid response curves, with respect to three basic environmental predictors [12]. For migratory species with known shifts in habitat usages during different seasons, habitat categories were selected to reflect both winter and summer usage [12]. Generic environmental predictors, including bathymetry, sea surface temperature and sea ice were selected *a priori* as predictive variables for all species, based on their documented importance in determining marine mammal occurrences directly or indirectly, e.g. through influencing prey availability. For example, strong correlations between bathymetry and patterns of species' occurrences have been noted for cetaceans and pinnipeds in different regions and ocean basins [34,35,36,37]. Sea surface temperature (SST) changes may be indicative of oceanographic processes that ultimately determine predator occurrence across multiple temporal scales [38] and significant correlations of SST with marine mammal presence and species richness of different predator groups have been demonstrated across regions and taxa [e.g. 3,34]. Another key environmental parameter that has been demonstrated to determine marine mammal species presence is sea ice concentration [39,40], since the edge of the pack ice represents an important feeding ground for many species [41]. The environmental data sets used for range predictions include gridded bathymetry data (from the ETOPO2 dataset, National Geophysical Data Center, www.ngdc.noaa.gov/products/ngdc_products.html) as well as mean SST extracted from the World Ocean Atlas [42] for the 1990s [12]; mean annual sea ice concentration data (United States National Snow & Ice Data Center (NSIDC) [43] was used instead of the formerly used data on distance to ice edge.

The RES model generates an index of species-specific relative environmental suitability of each individual half degree grid cell by scoring how well its physical attributes matched the known aspects of species' habitat use. RES values range between 0 (not suitable) to 1 (highly suitable) and represent the product of the suitability scores assigned for the individual environmental attributes (bottom depth, SST, sea ice concentration, and distance from land in some cases), which were calculated using pre-defined trapezoidal functional response curves. Model-predicted ranges and parameter settings for all species were summarized by Kaschner [44]. RES predictions for data-rich species have been successfully validated across different areas and time periods using independent data sets from dedicated marine mammal surveys [12] Validation analyses showed a strong positive relationship between the effort-corrected sighting rates of individual species and the corresponding predicted relative environmental suitability. Similarly, long-term habitat usage of species derived from effort-corrected whaling data provided support for the shape of pre-defined habitat categories used for RES input. Nevertheless, RES predictions often included parts of a species fundamental niche as well as its realized niche and suitability thresholds beyond which predicted presences were matched with observed occurrences varied by species. Since there is insufficient data to determine such presence thresholds empirically for all species, we used an alternative approach to generate species richness maps. Using a uniform presence threshold for all species, we generated richness maps across a range of different RES thresholds (RES>0 to RES = 1). To investigate how different assumptions about environmental suitability and species presence might affect predicted patterns of marine mammal richness, we then validated predictions against survey data.

Validation with survey data

We validated our predictions of marine mammal richness using available cetacean sighting data sets collected during dedicated surveys. To avoid circularity, we only used data which had not contributed extensively to assign species to specific habitat categories in the RES model. Validation data sets included a) the IWC-IDCR circumpolar cruises conducted regularly in the Southern Ocean between 1978–2001 [45], b) four NASS surveys, conducted in 1987, 1989, 1995 and 2001 in the Northeastern Atlantic [46], and c) seven SWFSC-ETP surveys conducted across the Eastern Tropical Pacific from 1986–1990 and 1992–93 [47,48] (Table 1). These three data sets likely represent the largest existing efforts to date to survey cetacean populations. Since pinniped observations were not reported consistently, validation analyses were limited to cetaceans only.

Comparison of predicted cetacean species richness from the RES model with observed richness from cetacean surveys was performed on a $5°\times5°$ grid to ensure sufficient sightings to estimate species richness for each cell empirically. Sightings data per $5°\times5°$ cell were combined across all years for each survey. Only records with high certainty in species identification were included. We used rarefaction to standardise for varying survey effort in different cells [49,50]. The rarefaction model is based on the hypergeometric distribution, sampling without replacement from a parent distribution. It is widely used to compare the number of species in a collection of samples with uneven sample sizes [51]. Species richness is expressed as the expected number of species from a standardized subsample of size n, which is computed as

$$E(S_n) = \sum_{i=1}^{S} \left[1 - \binom{N-m_i}{n} \Big/ \binom{N}{n} \right],$$

where N is the total number of individual sightings in the sample (here a $5°\times5°$ cell), S is the total number of species in the sample, and m_i is the number of individuals of species i in the sample. We calculated rarefied richness estimates for different n [52], namely the expected number of species per $n = 20$ sightings (ES_{20}), 50 sightings (ES_{50}) and 100 sightings (ES_{100}). Selection of an appropriate n represents a trade-off since the range of potential diversity per sampling unit will increase with increasing n, but sample size (i.e. the number of cells with enough effort to produce

rarefied estimates) and consequently geographic coverage will decrease.

We modeled the relationship between predicted and observed species richness using spatial eigenvector mapping (SEVM) to account for the effect of spatial autocorrelation on model results [53]. We fit Gaussian generalized linear models across all combinations of n and RES presence thresholds using the SEVMs package spdep v. 0.4–52 [54] in R [55] (version 2.8.1). Goodness of model fit (corrected for spatial autocorrelation) was assessed for each survey area separately as well as for all surveys combined using the coefficient of determination (adjusted r^2), and the most parsimonious model was identified using the Akaike Information Criterion (AIC) (Table S1). To ensure the broadest geographic representation, we based the selection of a RES threshold for forward projections of species richness patterns on the model combining data from all three survey areas and including the largest possible number of cells covered by enough survey effort to be included in the rarefaction analysis. Threshold selection was thus based on lowest AIC of the combined data set model for the lowest possible rarefaction basis, but excluded all n values and RES threshold combinations for which models did not produce significant relationships with validation data at the level of individual surveys (Table S1).

Since the validation analysis did not allow the unequivocal identification of a single best RES threshold model across all rarefaction bases n and survey areas, we also calculated the variation in predicted species richness for different ranges of RES thresholds for each survey cell. Mean standard deviation and coefficients of variations computed across all cells covered by a given survey and for all surveys combined can then provide an indication of the uncertainty in predicted species richness associated with the threshold selection process (Table 2).

Forward projections

To assess potential effects of climate warming and sea ice change on marine mammal biodiversity, we projected future distributions using mean temperatures and ice concentrations derived from the IPCC climate change scenario A1B for the years 2040–2049. This 'intermediate' scenario assumes very rapid economic but low population growth, rapid introduction of new and more efficient technologies, and moderate use of resources with a balanced use of technologies [56]. Assuming that species would maintain the same environmental preferences with respect to SST and sea-ice concentrations, we generated predictions of future species distribution for all species and superimposed them, applying the best presence threshold as determined by our validation with survey data. To assess changes in species richness

Table 1. Summary of validation data sets.

Survey Acronym	IWC-IDCR	NASS	SWFSC-ETP
Survey Name	International Whaling Commission - International Decade of Cetacean Research	North Atlantic Sightings Survey	Southwest Fisheries Science Centre - Eastern Tropical Pacific Surveys
Agency/Source	IWC Member State collaboration	North Atlantic Marine Mammal Commission (NAMMCO)	US National Marine Fisheries Service (NMFS) - SWFSC
Time period	1978–2001	1987, 1989, 1995 & 2001	1986–1990 and 1992–93
Ocean basin	Antarctica (S of 60° S)	NE Atlantic	Eastern Tropical Pacific
No. of sighting events	~35000	~7500	~8800
No. of identified species reported	31	17	34

Table 2. Effects of RES threshold selection on predicted species richness.

| Survey Area | Variation in predicted species richness (number of species) | | | | | |
| | 0.00< RES ≤1.00 | | 0.25≤ RES ≤0.75 | | 0.55≤ RES ≤0.65 | |
	Mean SD	Mean CV	Mean SD	Mean CV	Mean SD	Mean CV
SWFSC-ETP	5.68	0.26	2.84	0.13	1.02	0.05
NASS	3.04	0.24	1.60	0.13	0.35	0.03
IWC-IDCR	3.42	0.36	1.63	0.17	0.72	0.11
All Surveys	3.98	0.32	1.95	0.16	0.76	0.08

Variation (expressed as standard deviation, SD and coefficient of variation, CV) in number of species predicted to occur in each surveyed 5° cells between different assumed RES thresholds, averaged across all cells covered by a given survey. Estimates correspond to the level of uncertainty associated with predicted species richness in different survey areas that is introduced by the threshold selection process to generate species richness maps.

and distribution over time, we compared future patterns with current ones produced from a control data set (mean modeled 1990–99 environmental data) from the same climate scenario. Changes in species richness were shown in terms of absolute loss of native species from a given area or the absolute number of species that were newly predicted to occur in a given cell relative to the 1990–99 scenario. Similarly, we computed proportional increases and decreases in net biodiversity for each cell and the expected total and relative change in the number of species for different taxonomic groups. To assess potential effects of climate change on individual species, we also calculated the change in the size of distributions for each species between 1990–99 and 2040–49. Following the approach of a similar study [23], we then divided species into those that were predicted to expand, contract, or show no change in their range size. To provide an indication of the extent of the expected effect across different taxa, we computed the mean proportional change in size of distribution across all species falling into a specific category in each taxon.

Results

1990s species richness

Predicted patterns of marine mammal biodiversity were relatively consistent across all assumed RES thresholds, showing broad bands of high species richness in temperate waters of both hemispheres (Figure S1). Patterns based on a presence threshold RES>0.6 (Fig. 1) were most strongly supported by empirical species richness data (see 'Validation of species richness', below). The largest concentrations of marine mammal biodiversity were found in temperate waters between 20–50°S where up to 30% of all species may co-occur (Fig. 1A). Southern-hemisphere hotspots of high species richness were predicted in waters surrounding New Zealand, some Sub-Antarctic and Southeastern Pacific islands, and offshore waters along the coasts of southern South America. Biodiversity was also predicted to be high in subtropical and temperate waters of the Northern Hemisphere, although hotspots tended to be fewer and smaller in size. These included the waters surrounding Japan and Korea, Northwest Africa, the Southeastern U.S., parts of the mid-Atlantic ridge, Baja California, the Galapagos Islands and Hawaii. Overall, hotspots were relatively small: the total area of hotspots containing more than the 75th percentile of the maximum predicted species richness amounted to less than 5% of the oceans. Areas of high diversity were more abundant in the southern hemisphere where many species are more wide ranging and distributions tend to be less restricted by land barriers.

The comparison of species richness maps for different subgroups (Fig. 1B–D) with the overall species richness pattern shows that hotspots are probably mostly influenced by predicted odontocete species occurrence. Both odontocetes (Fig. 1B) and mysticetes (Fig. 1C) showed a band of high species richness in temperate waters of the Southern Hemisphere. However, while odontocete species richness was also high along ocean ridges in warmer waters (Fig. 1B), mysticetes concentrated in mid-latitudes (Fig. 1C). Distributional ranges for both groups were relatively large on average, resulting in large areas of overlap where many species co-occur. In contrast, pinniped species richness was mostly concentrated in subpolar and polar waters, and the lower degree of overlap in distribution between species resulted in 'weaker' hotspots with only up to six co-occurring species (Fig. 1D). Pinniped hotspots were located around the Sub-Antarctic islands and the Antarctic Peninsula, in the Bering Sea and the Sea of Okhotsk (Fig. 1D).

Latitudinal gradients of predicted marine mammal richness showed a bimodal distribution, with total species richness lowest in polar regions, highest between 30–60° N or S, and intermediate in tropical waters (Fig. 2). This basic pattern was shared across groups, although peaks in species richness occurred more polewards and tropical richness was much lower in pinnipeds compared with cetaceans. Small odontocetes (dolphins and porpoises) had the highest number of species of all groups, particularly at subtropical and tropical latitudes (Fig. 2).

Validation of 1990s species richness

For all three cetacean surveys we observed a strong linear relationship between the number of species seen in a given area and effort, expressed as total number of sightings in that area (Fig. 3 A–C). This suggests that the use of rarefaction is necessary to account for uneven effort across cells. None of the rarefaction curves calculated for each survey reached a full asymptote, which suggested that surveys are still incomplete in terms of marine mammal species detection (Fig. 3D). This may in part be explained by the difficulty to distinguish some closely related species, such as the numerous *Mesoplodon* spp. (beaked whales) at sea, sightings of which are often reported at a higher taxonomic level and thus would not be considered in this analysis. Survey effort was greatest in Antarctic waters in terms of sightings, but the total number of species observed in this region was still 30% lower than in the Eastern Tropical Pacific (Fig. 3D).

We found significant linear relationships between predicted (Fig. 4A) and observed rarefied richness (Fig. 4B) for all three survey areas individually as well as combined (Fig. 4C).

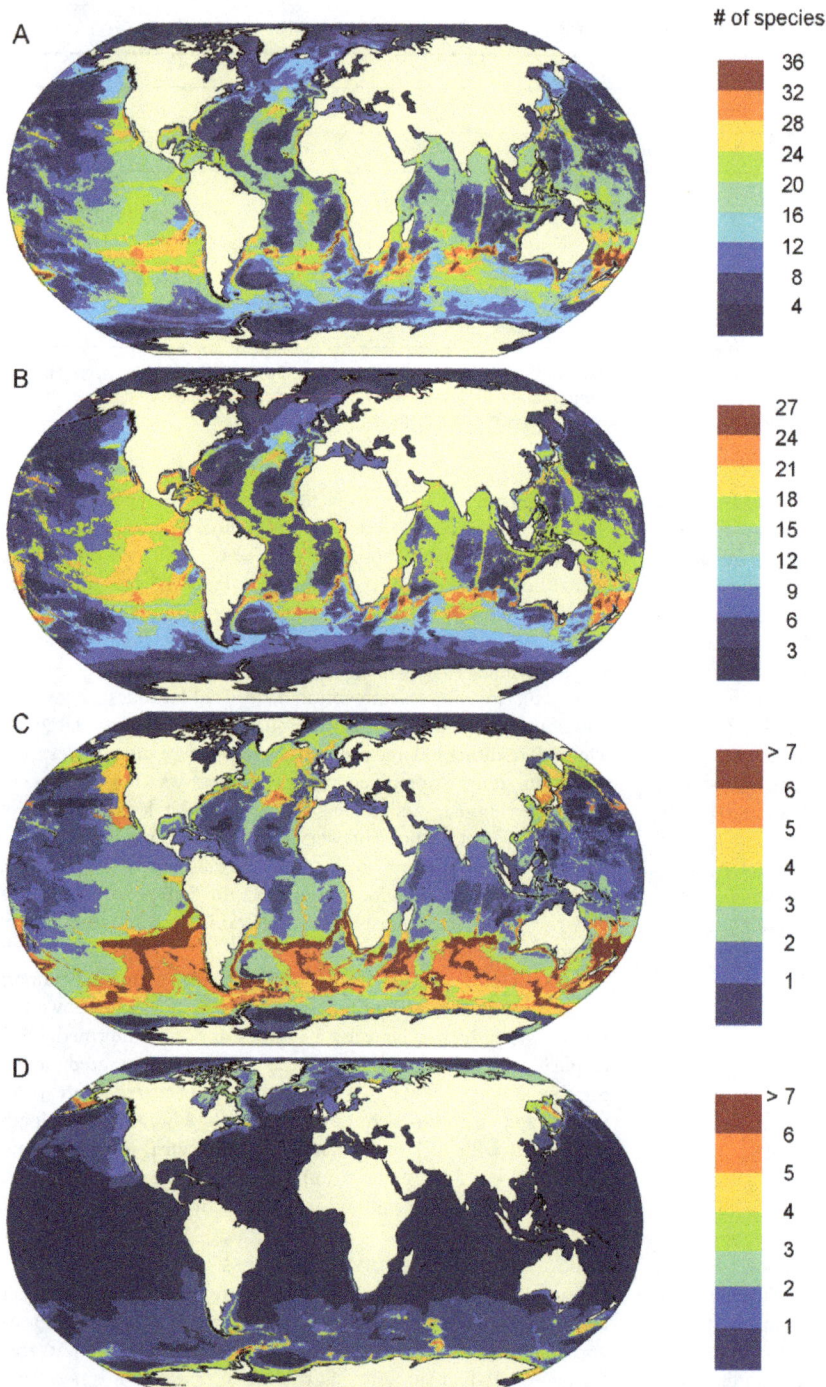

Figure 1. Predicted patterns of marine mammal species richness. A. All species (n = 115), **B.** Odontocetes (n = 69), **C.** Mysticetes (n = 14), **D.** Pinnipeds (n = 32). Colors indicate the number of species predicted to occur in each 0.5°x0.5° grid cell from a relative environmental suitability (RES) model, using environmental data from 1990–1999, and assuming a presence threshold of RES>0.6.

Significant linear relationships were seen across a wide range of different rarefaction bases and RES presence thresholds (Table S1). Presence thresholds associated with the most parsimonious models (lowest AIC) varied among survey areas and rarefaction bases, ranging between 0.25<RES<0.75, making the selection of a single best threshold somewhat subjective. However, the variation in species richness predicted for each 5° cell across this range of RES thresholds was relatively small on average,

amounting to only 16% of the total number of predicted species or ±2 species on average (Table 2). This indicates that predicted estimates of absolute species richness appear to be relatively robust across a range of thresholds. For display purposes we used the RES threshold >0.6, associated with the second lowest AIC for all surveys combined at rarefaction basis n = 50 (i.e. expected species per 50 sightings). This threshold was associated with the lowest possible ES basis to ensure the widest possible geographic

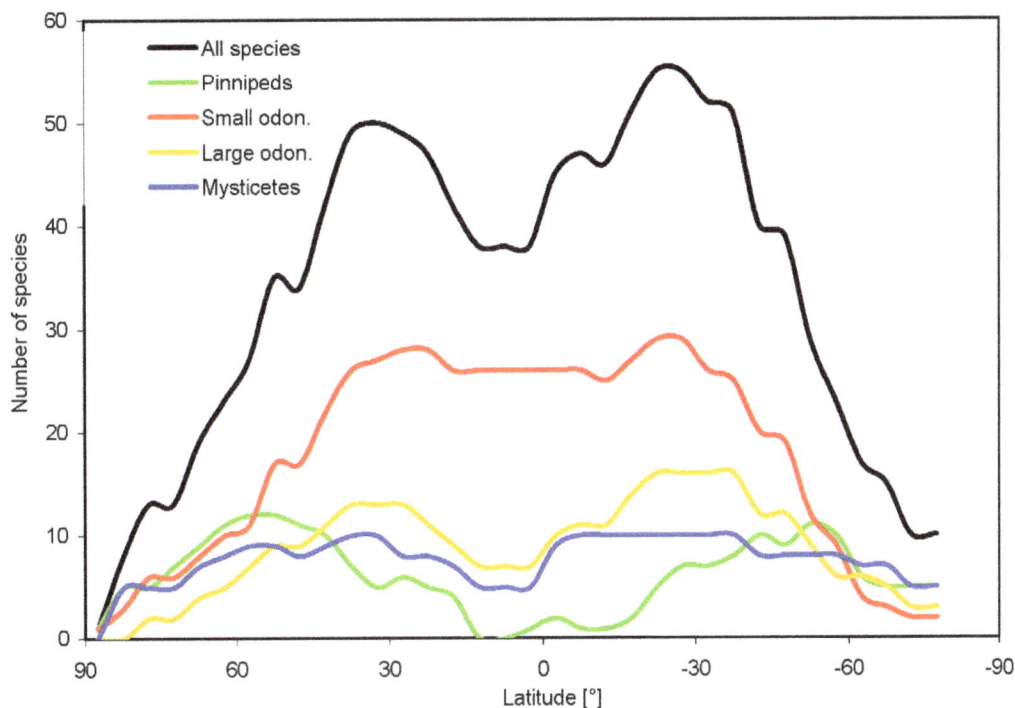

Figure 2. Marine mammal species richness by latitude. Number of predicted species was summed over 5° latitudinal bands for all species, mysticetes, small odontocetes, large odontocetes (beaked whales and sperm whale), and pinnipeds.

coverage, while at the same time consistently producing significant relationships and good model fits at the level of individual surveys (Table S1). We note, that the species richness maps based on the RES>0.6 threshold correspond to areas of overlap in highly suitable habitat across many species, but species may also occur in habitat predicted to be less suitable than the selected threshold].

Based on the estimated regression slope of our best model, only between 10–50% of all species predicted to occur in a given 5°x5° cell had actually been observed in any of the survey areas given the effort of survey data sets included in the analysis (Fig. 4C).

Forward projections

Projecting environmental change according to the intermediate IPCC-A1B climate change scenario for the years 2040–49, the predicted effects on global marine mammal biodiversity based on RES >0.6 were moderate (Fig. 5). Although the absolute loss in optimal habitat for native species might regionally affect as many as 11 species, this is predicted only in relatively small areas (Fig. 5A). In the Northern Hemisphere, the areas most likely to experience a decrease in the number of native species were the Barents Sea, parts of the North Atlantic ridge, and the Northern Indian Ocean as well as waters surrounding Japan (Fig. 5A). In addition, species loss was predicted to occur along coastlines or across continental shelves (Fig. 5A). In the Southern Hemisphere, decreases in native species richness were predicted mostly along 30° south, but also around the Galapagos Islands and in the Coral Triangle (Fig. 5A). At the same time, increases in biodiversity, mostly through the invasion of new species in polar waters, might also be substantial, particularly in the Northern Hemisphere (Fig. 5B). Areas most likely to experience an increase in the number of species due to invasion, were the Northern Greenland Sea, the Barents Sea, and the central Bering Sea as well the high

Arctic waters (Fig. 5B), where temperature increases might enable colonization of up to 10 new species. In the Southern Hemisphere, as sea ice melts and retracts, species richness might also increase substantially in parts of the Weddell Sea (Fig. 5B). Roughly 84% of all areas in which marine mammals were predicted to occur may experience only small changes in species composition and richness due to projected changes in temperature or sea ice concentration (dark blue areas in Fig. 5A, B, where predicted changes are within the bounds of uncertainty associated with the RES threshold selection process, see above).

With respect to individual taxa, pinniped biodiversity in tropical and temperate waters was predicted to decrease substantially (Fig. 6A), with the Galapagos fur seal (*Arctocephalus galapagonensis*) and the Hawaiian monk seal (*Monachus schauinslandii*) being most affected, but not the Galapagos sea lion (*Zalophus wollebaeki*). In contrast, the number of mysticete species at high latitudes, of the northern hemisphere in particular, was predicted to increase substantially (Fig. 6A). Overall, changes in species composition in terms of absolute number of species in different taxa were predicted to be highest in tropical waters, but taxonomically, the proportional composition of marine mammal communities was predicted to change most drastically in Arctic waters (Fig. 6B).

Increases in range size were predicted for 54% of all species, while 45% might experience a net loss in range size, and 1% of all species may not change (Table 3). However, these changes will typically be small, i.e. less than 10% for most taxa (Table 3). Notable exceptions include a substantial increase of predicted suitable habitat, as defined by our model, for endangered North Pacific (*Eubalena japonica*) and Atlantic (*E. glacialis*) right whales (15% and 27% increase respectively), the gray whale (*Eschrichtius robustus*) (40%) and Steller sea lions (*Eumetopias jubatus*) (85%).

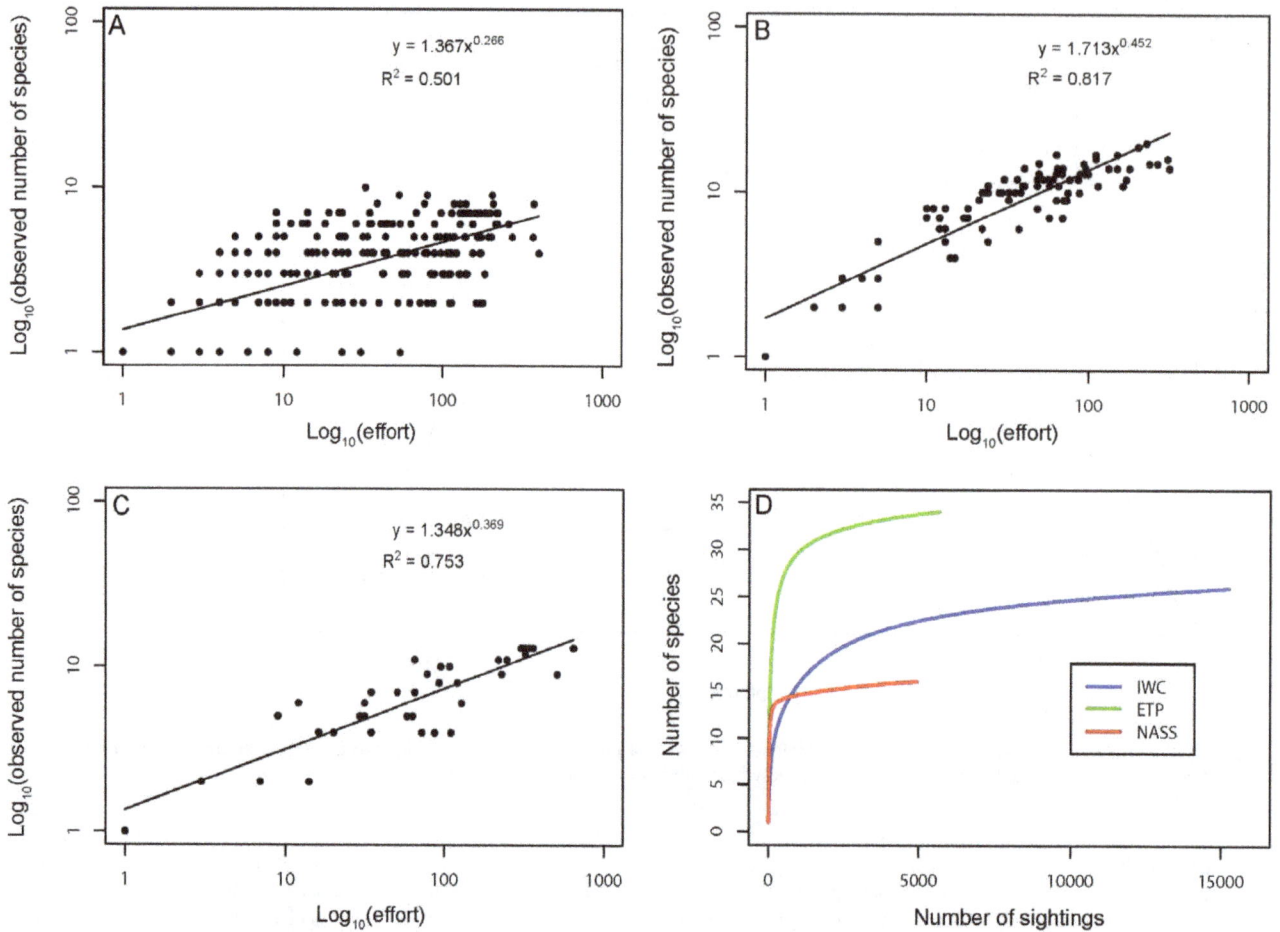

Figure 3. Rate of species discovery with survey effort. Number of species detected with increasing sampling effort in each 5°x5° cell in **A.** Antarctic waters, **B.** The Northeastern Atlantic, **C.** The Eastern Tropical Pacific, **D.** Species accumulation curves in different survey areas.

Discussion

We used a model combining empirical observations, expert-derived range maps, and environmental niche associations to predict present-day and future distributions of marine mammals. Validation with available survey data sets indicated that broad patterns of species richness are reproduced reasonably well, and lend confidence to the global approach taken here. Forward projections based on expected changes in temperature and sea ice concentration alone suggested modest changes over the course of the next 40 years, with possible declines in marine mammal species richness at lower latitudes and increases at higher latitudes, assuming an intermediate IPCC climate change scenario.

Low spatial coverage, relative to the global distribution of marine mammals, is a problem of marine mammal surveys in general (our Fig. 4B), and shows how much is still unknown with respect to the distribution of these animals. Yet even in our best-surveyed regions, there was still evidence of an incomplete inventory of species richness (Fig. 3D). However, continuing survey efforts, such as those that have been conducted in the ETP since 1993, are expected to improve species inventories over time. Nevertheless, fully complete survey-based inventories will likely remain a challenge, given, for example, the rarity and low detectability of numerous beaked whale species, combined with difficulties to distinguish species at sea. Similarly, seasonal

coverage of existing surveys rarely exceeds the summer months, and seasonal occurrences of migratory species may thus be missed, but will be included by a model that predicts long-term annual average occurrences. Consequently, the use of environmental suitability models might be viewed as a complementary tool to explore patterns of biodiversity, particularly in less well-surveyed regions around the world. Modeled species ranges can then be refined and validated as new survey data continue to be collected.

Our predictive maps show distinct peaks of marine mammal species richness in temperate waters of both hemispheres, similar to those that have been found for other marine predators and zooplankton [3,57]. These areas represent highly productive oceanographic transition zones (e.g. [58]), where range extents of tropical and temperate species overlap. The much stronger peak in the Southern hemisphere might be explained by macro-evolutionary patterns of speciation in the absence of geographic barriers – this may have resulted in a much greater number of panglobal species in the Southern compared with the Northern hemisphere.

Our results compare well with patterns of global marine mammal diversity reported by the International Union for the Conservation of Nature (IUCN) mammal specialist group [6]. Hotspots of species richness and latitudinal patterns reported by the IUCN are similar to those reported here, although species richness in the IUCN maps appears to be higher in tropical waters

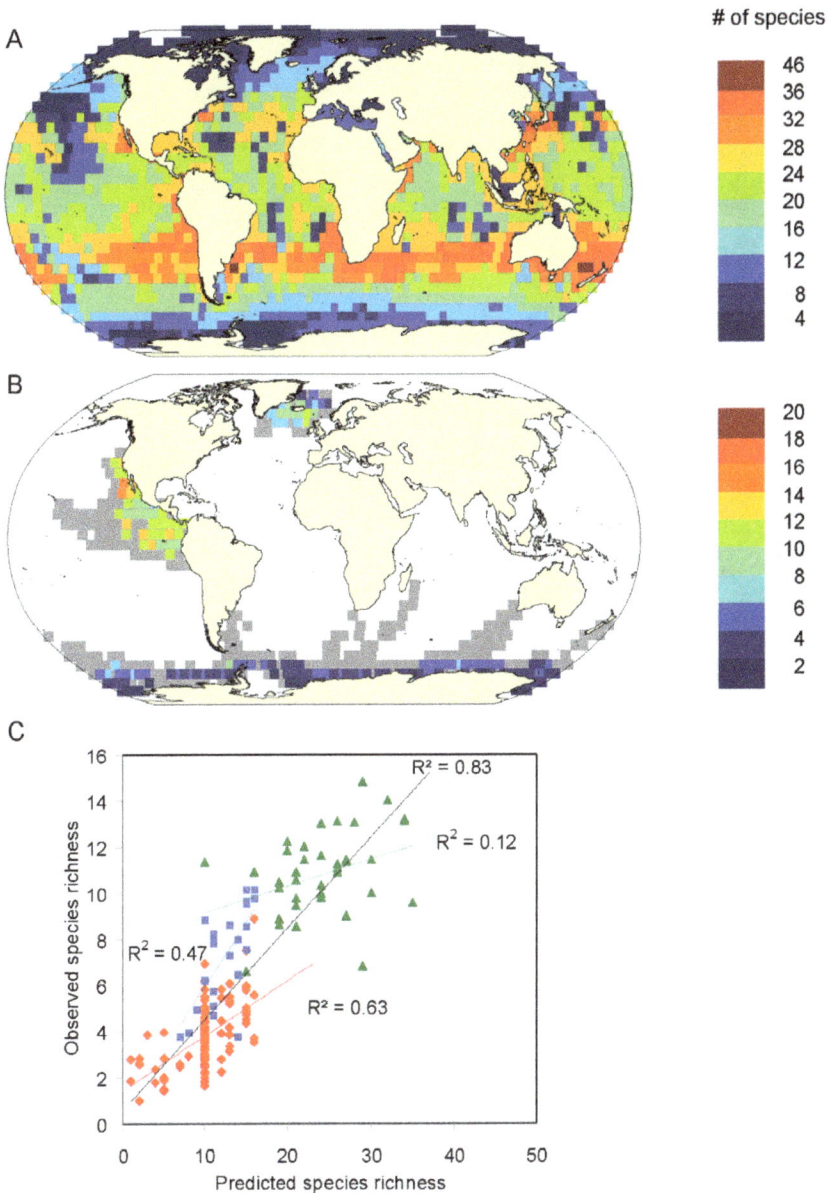

Figure 4. Validation with empirically observed marine mammal occurrences (5°x5° cells, 1990–1999). A. Predicted species richness of all cetaceans (RES presence threshold >0.6), **B.** Observed cetacean species richness per standardized sample of 50 sightings (grey cells have been covered by surveys but had insufficient effort for analysis), **C.** Relationship between observed and predicted species richness in the Antarctic (red), North Atlantic (blue) and Eastern Tropical Pacific (green) and across all three surveys (black). Data points correspond to individual 5°grid cells, regression lines to best linear fits, r^2 values were corrected for spatial autocorrelation.

than in our analysis, and the latitudinal bands of high biodiversity in temperate waters are less pronounced. Most of these differences can be explained by the higher level of spatial detail provided by the RES models in terms of the relative probabilities of occurrence of species. Our approach relaxes the assumption of equal probability of occurrence throughout the range, which is implicit in the IUCN range maps of species. This assumption effectively translates into large proportions of the oceans to be represented as almost homogenous in terms of species richness, given the high number of cosmopolitan or pantropical species. For instance, in the IUCN study [6] many of the baleen whales with maximum ranges extending from pole to pole contribute to the equatorial band in high species richness, even though the occurrence of

baleen whales in tropical waters is limited to a few species, such as the Bryde's whales (*Balaenoptera brydei* and *B. edeni*), humpback whales (*Megaptera novaeangliae*) or some resident blue whale populations (*B. musculus*) [59,60]. As a further consequence of this approach, predicted biodiversity hotspots are likely determined largely by overlapping species with restricted ranges, while possible concentrations of cosmopolitan species may be masked by the assumed uniform global occurrences. In contrast, our non-binary predictions of species-specific relative environmental suitability combined with the selected threshold of >0.6 effectively describe geographic areas of predicted co-occurrence of highly suitable habitat for many species. As a consequence RES-model derived hotspots, for instance, were more concentrated in temperate

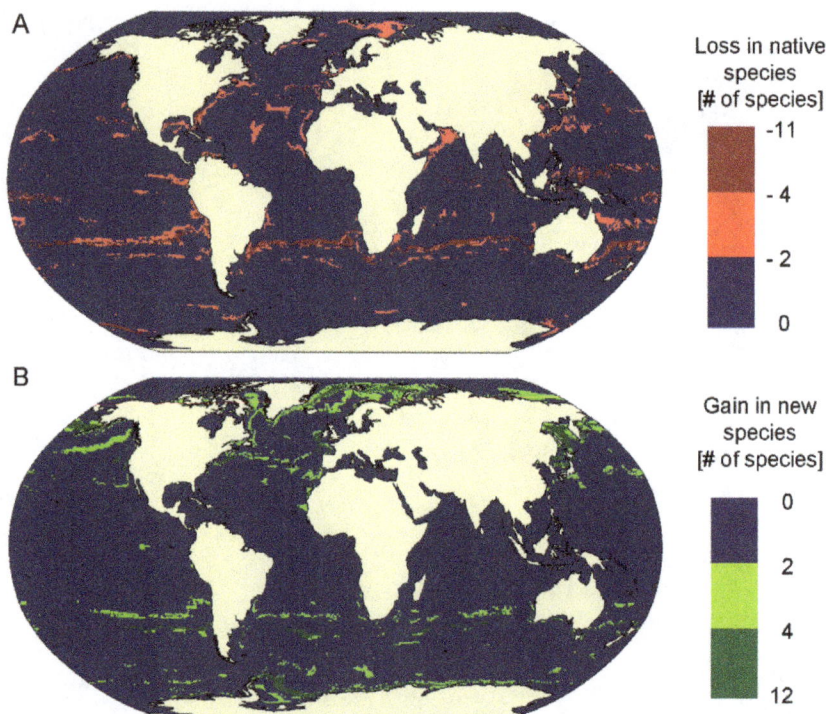

Figure 5. Projected effects of climate change on marine mammal species richness. Projected changes in overlap of optimal habitat across all species from 1990–1999 to 2040–2049 using the IPCC-A1B climate change scenario (0.5°×0.5° grid cells) **A.** Loss in number of native species, **B.** Gain in number of new species. Biodiversity changes are expressed relative to species richness predicted for the 1990s, and assuming a presence threshold of RES>0.6.

waters and around topographical features such as the mid-Atlantic ridge and seamounts [see also 61] than suggested by the IUCN maps. In the context of marine spatial planning, information on the relative importance of areas throughout a species range can help identify areas where the implementation of conservation measures will be most beneficial, to ensure the protection of both individual species and marine mammal biodiversity.

It should be noted, however, that the importance of bathymetry in determining species occurrence might be overestimated by our approach, and that observed patterns of species richness might bear less resemblance to bathymetric maps if additional environmental parameters were taken into consideration. For instance, a modified version of the RES model, the AquaMaps model (available: www.aquamaps.org, [62]), has incorporated primary production, shown to be an important driver of global species richness patterns for marine mammals [3], and salinity as additional optional environmental proxies. Resulting distributions are very similar to RES based maps, but appear less dominated by bathymetry. AquaMaps outputs have also been successfully validated for some species [63], but not all marine mammal species have been fully reviewed and the relative importance of the additional parameters still remains to be thoroughly investigated. We have therefore opted for conducting the present analysis based on RES predictions for the time being.

Another study of marine mammal biodiversity analyzed empirical sightings data for deep-water cetaceans [10]. While this approach provided important insights, the disadvantage is that only a subset of species can be included, and spatial coverage is necessarily low. Yet when we compare broad latitudinal patterns of species richness, these empirical results match well with our predictions. This previous study also supported a

correlation between ocean temperature and patterns of marine mammal richness [10], an assumed driving factor in our RES models.

A frequent application of bioclimatic envelope models is to project changes in distributional ranges using modeled climate change scenarios [14,64,65]. Recent studies provide some support that such modeled shifts in species distribution match observed range expansions towards higher latitudes [66,67]. The results presented here match relatively well with the broad-scale predictions derived independently by Whitehead et al. [10], who also explored various climate change scenarios, and their possible effects on deepwater cetacean richness. Our approach provides more detail, as projections are based on the ranges for individual species, rather than total species richness. This allows for differing responses among species. Furthermore we do not need to assume a consistent relationship between richness and temperature. Finally, we also included ice cover, which determines food availability and breeding habitat for many of the polar species [68,69], and is expected to influence marine mammal species at higher latitudes [30,70,71,72]. Nevertheless, predictions of latitudinal patterns in species richness agree among the two studies.

Overall, our findings provide support for hypothesized impacts of climate change on cetacean ranges based on recently developed theoretical framework [23]. Therein it was proposed that the majority of cetacean species will experience some climate-change-driven range expansion or contraction [23], which matches our results qualitatively (Table 3). Our modeling approach, however, indicates that, over the course of the next 40 years, negative effects such as net range contractions may be modest for most species, while a number of species might benefit from substantial increases in optimal habitat.

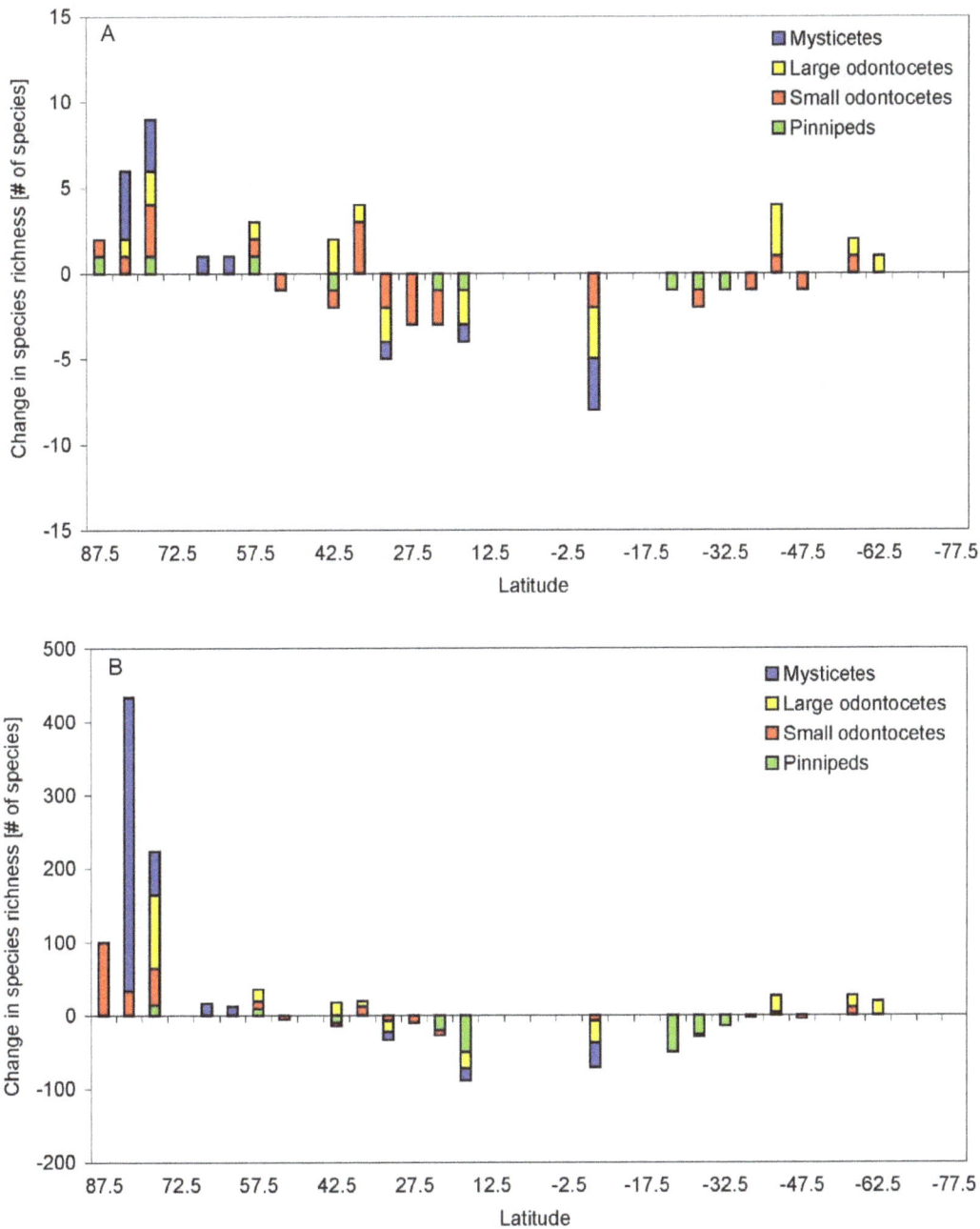

Figure 6. Projected absolute and proportional changes in marine mammal species richness and community composition at different latitudes. Changes were calculated relative to predicted species richness for the 1990s summed over 5° latitudinal bands for mysticetes, small odontocetes, large odontocetes, and pinnipeds.

Despite the encouraging empirical validations of RES predictions [this paper and 12,63] as well as the observed agreement with findings from two independent studies [6,10], we emphasize that there are obvious limitations to our approach. For example, assumed static habitat usage of species over time is a strong assumption in a highly dynamic marine environment. Backwards validation of predicted temporal changes using historic data sets would be one potential avenue for assessing the robustness of our predictions beyond the time period of data collection. Another limitation of this study lies in our focus on a single snapshot projection of species richness into the future. Further research

using time series projections based on intermediate intervals would allow the assessment of possible effects of intermittent temperature fluctuations on species distribution that could result in local extinction of populations or species not detectable in this analysis.

Our approach also cannot reproduce the full range of factors that affect marine mammal distributions today or in the future. Most important among the variables not considered by our model are the distribution of food supply, and the availability of breeding habitat, both of which could change under various climate change scenarios [30,70,71,72]. Although it has been proposed that prey distributions may also shift to higher latitudes [14,73], there is

Table 3. Predicted changes in mean size of optimal ranges due to climate change by the years 2040–49 across different taxa (IPCC-A1B scenario & RES>0.6).

Suborder	Family	Range expansion		Range contraction		Stable range size		Total
		Mean increase [%]	# of species	Mean decrease [%]	# of species	No change	# of species	
Pinnipedia	Otariidae	35.47	3	−8.82	12			15
Pinnipedia	Phocidae	28.02	11	−4.77	5			16
Pinnipedia	Odobenidae			−3.15	1			1
Mysticeti	Balaenidae	20.68	2	−3.98	2			4
Mysticeti	Balaenopteridae	1.70	3	−3.21	5			8
Mysticeti	Eschrichtiidae	40.12	1					1
Mysticeti	Neobalaenidae			−4.49	1			1
Odontoceti	Delphinidae	6.44	24	−4.55	11			35
Odontoceti	Kogiidae	0.68	1	−14.19	1			2
Odontoceti	Monodontidae	5.70	1	−8.95	1			2
Odontoceti	Phocoenidae	15.14	3	−3.20	2	0	2	7
Odontoceti	Physeteridae	0.48	1					1
Odontoceti	Pontoporiidae			−5.85	1			1
Odontoceti	Ziphiidae	4.27	12	−3.87	9			21
	Total		62		51		2	115

some evidence that overall prey abundance and biomass may decline in some areas [74]. It is difficult to assess how marine mammal species, which are often opportunistic foragers, will respond to shifts or reductions in prey distributions caused by increasing temperatures, but this could have an equal or greater effect on marine mammal distributions than the direct effects of temperature modeled here. Similarly, indirect effects such as changes in species interactions [27] or population dynamics [75] cannot be captured by our approach. Finally, ocean chemistry, also changing due to the uptake of anthropogenically produced carbon dioxide [76], will potentially impact calcareous organisms, the effects of which may propagate up the food-web. The general paucity of relevant data for the majority of species will likely preclude the consideration of these more complex factors in our models for the foreseeable future.

In conclusion, our models should be interpreted as minimally realistic models that generate testable predictions about the distribution of individual species and biodiversity and how these might be impacted by climate change. These predictions must be further scrutinized with independent empirical data as they become available, in order to be useful for conservation planning or management purposes. Using forward projection, RES models can be usefully applied to investigate potential future effects of climate change, which we have illustrated here using the 2040–2049 snapshot as an arbitrary reference point. However, given that climate models predict effects of global warming to become more pronounced during the later half of the 21st century, the investigation of long-term changes in marine mammal biodiversity patterns in smaller time intervals and beyond the year 2050 are needed to more comprehensively assess the effect of climate change on marine mammals. With these caveats in mind, however, we conclude that the RES model represents a powerful exploratory tool to investigate the large scale occurrences patterns of taxa for which global distributional data are still remarkably incomplete.

Supporting Information

Figure S1 Predicted current patterns of global marine mammal species richness based on different presence thresholds. Relative environmental suitability (RES) threshold for assumed species presence in 0.5° grid cells (1990s). A. RES>0, B. RES>0.2, C. RES>0.4, D. RES>0.6, E. RES>0.8. Biodiversity hotspots in maps based on higher assumed RES thresholds represent areas of overlap in predicted optimal habitat of many species.

Table S1 Summary of validation results comparing observed versus predicted species occurrence per 5° grid cell in different survey areas. Red values represent models with lowest AIC, yellow values correspond to models falling into the range of ΔAIC <2 and grey cells represent models with non-significant relationships.

Acknowledgments

We thank J.M. Frometin for discussions about spatial autocorrelation, C. F. Dormann for statistical advice, C. Muir for comments on the manuscript, and gratefully acknowledge the use of environmental data sets as provided through the Sea Around Us Project (www.seaaroundus.org) and AquaMaps project (www.aquamaps.org). We are indebted to C. Allison and the Secretariat of the International Whaling Commission as well as L. Burt and the NAMMCO secretariat for access to the IWC-IDCR and NASS cetacean survey data sets. In addition, we would like to thank the two reviewers whose comments helped to greatly improve this manuscript.

Author Contributions

Conceived and designed the experiments: KK BW. Analyzed the data: KK DPT. Contributed reagents/materials/analysis tools: KK DPT JR TG. Wrote the paper: KK BW DPT JR TG.

References

1. Gaston KJ (2000) Global patterns in biodiversity. Nature 405: 220–227.
2. Margules CR, Pressey RL (2000) Systematic conservation planning. Nature 405: 243–253.
3. Tittensor DP, Mora C, Jetz W, Ricard D, Vanden Berghe E, et al. (2010) Global patterns and predictors of marine biodiversity. Nature 466: 1098–1011.
4. Katona S, Whitehead H (1988) Are cetacea ecologically important? Oceanography and Marine Biology 26: 553–568.
5. Pauly D, Trites AW, Capuli E, Christensen V (1998) Diet composition and trophic levels of marine mammals. ICES Journal of Marine Science 55: 467–481.
6. Schipper J, Chanson JS, Chiozza F, Cox NA, Hoffmann M, et al. (2008) The status of the world's land and marine mammals: Diversity, threat, and knowledge. Science. pp 225–230.
7. Grech A, Marsh H (2008) Rapid assessment of risks to a mobile marine mammal in an ecosystem-scale marine protected area Conservation Biology 22: 711–720.
8. Zacharias MA, Gerber LR, Hyrenbach KD (2006) Review of the Southern Ocean Sanctuary: Marine protected areas in the context of the International Whaling Commission Sanctuary Programme. Journal of Cetacean Research and Management 8: 1–12.
9. Hoyt E (2005) Marine Protected Areas for Whales, Dolphins and Porpoises. A World Handbook for Cetacean Habitat Conservation. London, UK: Earthscan. pp 492.
10. Whitehead H, McGill B, Worm B (2008) Diversity of deep-water cetaceans in relation to temperature: Implications for ocean warming. Ecology Letters 11: 1198–1207.
11. Guisan A, Zimmermann N (2000) Predictive habitat distribution models in ecology. Ecological Modelling 135: 147–186.
12. Kaschner K, Watson R, Trites AW, Pauly D (2006) Mapping worldwide distributions of marine mammals using a Relative Environmental Suitability (RES) model. Marine Ecology Progress Series 316: 285–310.
13. Sanjit L, Bhatt D (2005) How relevant are the concepts of species diversity and species richness? Journal of Biosciences 30: 557–560.
14. Cheung WWL, Lam VWY, Sarmiento JL, Kearney K, Watson R, et al. (2009) Projecting global marine biodiversity impacts under climate change scenarios. Fish and Fisheries 10: 235–251.
15. Araújo MB, New M (2007) Ensemble forecasting of species distributions. Trends in Ecology and Evolution 22: 42–47.
16. Pearson RG, Thuiller W, Araujo MB, Martinez-Meyer E, Brotons L, et al. (2006) Model-based uncertainty in species range prediction. Journal of Biogeography 33: 1704–1711.
17. Thuiller W, Araujo MB, Pearson RG, Whittaker RJ, Brotons L, et al. (2004) Biodiversity conservation - Uncertainty in predictions of extinction risk. Nature 430: 145–148.
18. Araujo MB, Pearson RG, Thuiller W, Erhard M (2005) Validation of species-climate impact models under climate change. Global Change Biology 11: 1504–1513.
19. Parmesan C, Yohe G (2003) A globally coherent fingerprint of climate change impacts across natural systems. Nature 421: 37–42.
20. Worm B, Lotze HK (2009) Changes in marine biodiversity as an indicator of climate and global change. In: Letcher T, ed. Climate and global change: observed impacts on planet earth: Elsevier. pp 263–279.
21. Simmonds MP, Isaac SJ (2007) The impacts of climate change on marine mammals: Early signs of significant problems. Oryx 41: 19–26.
22. Learmonth JA, MacLeod CD, Santos MB, Pierce GJ, Crick HQP, et al. (2006) Potential effects of climate change on marine mammals. Oceanography and Marine Biology - an Annual Review 44: 431–464.
23. MacLeod CD (2009) Global climate change, range changes and potential implications for the conservation of marine cetaceans: a review and synthesis. Endangered Species Research 7: 125–136.
24. Harwood J (2001) Marine mammals and their environment in the twenty-first century. Journal of Mammalogy 82: 630–640.
25. Laidre KL, Stirling I, Lowry LF, Wiig Ø, Heide-Jørgensen MP, et al. (2008) Quantifying the sensitivity of Arctic marine mammals to climate-induced habitat change. Ecological Applications 18: S97–S125.
26. Burek KA, Gulland FMD, O'Hara TM (2008) Effects of climate change on Arctic marine mammal health. Ecological Applications 18: S126–S134.
27. Moore S, Huntington HP (2008) Arctic marine mammals and climate change: Impacts and resilience. Ecological Applications 18: 157–165.
28. Robinson RA, Crick HQP, Learmonth JA, Maclean IMD, Thomas CD, et al. (2009) Travelling through a warming world: Climate change and migratory species. Endangered Species Research 7: 87–99.
29. MacLeod CD, Bannon SM, Pierce GJ, Schweder C, Learmonth JA, et al. (2005) Climate change and the cetacean community of north-west Scotland Biological Conservation 124: 477–483.
30. Moore SE (2008) Marine mammals as ecosystem sentinels. Journal of Mammalogy 89: 534–540.
31. Kovacs KM, Lydersen C (2008) Climate change impacts on seals and whales in the North Atlantic Arctic and adjacent shelf seas. Science Progress 92: 117–150.
32. Gaston KJ (1994) Measuring geographic range sizes. Ecography 17: 198–205.
33. Read AJ, Halpin PN, Crowder LB, Best BD, Fujioka E OBIS-SEAMAP: mapping marine mammals, birds and turtles. World Wide Web electronic publication. http://seamap.env.duke.edu.
34. Baumgartner MF, Mullin KD, May LN, Leming TD (2001) Cetacean habitats in the Northern Gulf of Mexico. Fishery Bulletin 99: 219–239.
35. Hamazaki T (2002) Spatiotemporal prediction models of cetacean habitats in the mid-western North Atlantic ocean (from Cape Hatteras, North Carolina, U.S.A. to Nova Scotia, Canada). Marine Mammal Science 18: 920–939.
36. Moore SE, DeMaster DP, Dayton PK (2000) Cetacean habitat selection in the Alaskan Arctic during summer and autumn. Arctic 53: 432–447.
37. Payne PM, Heinemann DW (1993) The distribution of pilot whales (Globicephala spp.) in shelf/shelf-edge and slope waters of the Northeastern United States, 1978–1988. In: Donovan GP, Lockyer CH, Martin AR, eds. Biology of Northern Hemisphere Pilot Whales - Reports of the International Whaling Commission (Special Issue 14). Cambridge, UK: IWC. pp 51–68.
38. Au DWK, Perryman WL (1985) Dolphin habitats in the Eastern Tropical Pacific. Fishery Bulletin 83: 623–644.
39. Moore SE, DeMaster DP (1997) Cetacean habitats in the Alaskan Arctic. Journal of Northwest Atlantic Fishery Science 22: 55–69.
40. Ribic CA, Ainley DG, Fraser WR (1991) Habitat selection by marine mammals in the marginal ice zone. Antarctic Science 3: 181–186.
41. Murase H, Matsuoka K, Ichii T, Nishiwaki S (2002) Relationship between the distribution of euphausiids and baleen whales in the Antarctic (35°E - 145°W). Polar Biology 25: 135–145.
42. NOAA/NODC (1998) World Ocean Atlas 1998. Ocean Climate Laboratory, National Oceanographic Data Center.
43. Cavalieri D, Parkinson C, Gloersen P, Zwally HJ (1996) Sea ice concentrations from Nimbus-7 SMMR and DMSP SSM/I passive microwave data (1979-2002). National Snow and Ice Data Center, Boulder, Colorado USA updated 2008.
44. Kaschner K (2004) Modelling and mapping of resource overlap between marine mammals and fisheries on a global scale [Ph.D.]. Vancouver, Canada: University of British Columbia. 184 p.
45. IWC (2001) IDCR-DESS SOWER Survey data set (1978–2001). IWC.
46. NAMMCO (2001) North Atlantic Sightings Surveys (NASS) data: 1987, 1989, 1995, 2001. North Atlantic Marine Mammal Commission.
47. Gerrodette T, Forcada J (2005) Non-recovery of two spotted and spinner dolphin populations in the eastern tropical Pacific Ocean. Marine Ecology Progress Series 291: 1–21.
48. Kinzey D, Olson P, Gerrodette T (2000) Marine mammal data collection procedures on research ship line-transect surveys by the Southwest Fisheries Science Center. La Jolla, California: Southwest Fisheries Science Center (SWFSC), National Marine Fisheries Service, (NMFS), National Oceanic and Atmospheric Administration (NOAA)., Admin. Rept. LJ-00-08. 22 p.
49. Sanders HL (1968) Marine benthic diversity - a comparative study. American Naturalist 102: 243–282.
50. Hurlbert SH (1971) Nonconcept of species diversity - critique and alternative parameters. Ecology 52: 577–585.
51. Gotelli NJ, Graves GR (1996) Null models in ecology. Washington, D.C.: Smithsonian Institution Press. 388 p.
52. Colwell RK, Mao CX, Chang J (2004) Interpolating, extrapolating, and comparing incidence-based species accumulation curves. Ecology 85: 2717–2727.
53. Dormann CF, McPherson JM, Araujo MB, Bivand R, Bolliger J, et al. (2007) Methods to account for spatial autocorrelation in the analysis of species distributional data: a review. Ecography 30: 609–628.
54. Bivand R, Anselin L, Assunção R, Berke O, Bernat A, et al. (2009) spdep: Spatial dependence: weighting schemes, statistics and models. R package version 0.4-56 ed.
55. Team RDC (2009) R: A language and environment for statistical computing. ViennaAustria: R Foundation for Statistical Computing.
56. Junghaus J (2006) IPCC-AR4 MPI-ECHAM5_T63L31 MPI-OM_GR1.5L40 SRESA1B run no.1: ocean monthly mean values MPImet/MaD Germany. World Data Center for Climate, CERA-DB "OM-GR1.5L40_EH5-T63L31_A1B_1_MM".
57. Worm B, Sandow M, Oschlies A, Lotze HK, Myers RA (2005) Global patterns of predator diversity in the open oceans. Science 309: 1365–1369.
58. Longhurst A (1995) Seasonal cycles of pelagic production and consumption. Progress in Oceanography 36: 77–167.
59. Rice DW (1998) Marine Mammals of the World - Systematics and Distribution (Special publication 4); Wartzok D, ed. Lawrence, KS: Allen Press, Inc.. 231 p.
60. Branch TA, Stafford KM, Palacios DM, Allison C, Bannister JL, et al. (2007) Past and present distribution, densities and movements of blue whales Balaenoptera musculus in the Southern Hemisphere and northern Indian Ocean. Mammal Review 37: 116–175.
61. Kaschner K (2007) Air-breathing visitors to seamounts. Section A: Marine mammals. In: Pitcher TJ, Morato T, Hart PJB, Clark MR, Haggan N, et al. (2007) Seamounts: Ecology, Fisheries & Conservation, Oxford, UK Blackwell. pp 230–238.
62. Kaschner K, Ready JS, Agbayani E, Rius J, Kesner-Reyes K, et al. (2008) AquaMaps: Predicted range maps for aquatic species. World wide web electronic publication, www.aquamaps.org, Version 08/2010.

63. Ready J, Kaschner K, South AB, Eastwood PD, Rees T, et al. (2010) Predicting the distributions of marine organisms at the global scale. Ecological Modelling 221: 467–478.

64. Araujo MB, Thuiller W, Pearson RG (2006) Climate warming and the decline of amphibians and reptiles in Europe. Journal of Biogeography 33: 1712–1728.

65. Skov F, Svenning J-C (2004) Potential impact of climatic change on the distribution of forest herbs in Europe. Ecography 27: 366–380.

66. Walther G-R, Berger S, Sykes MT (2005) An ecological "footprint" of climate change. Proceedings of the Royal Society of London (Series B): Biological Sciences 272: 1427–1432.

67. Root TL, Price JT, Hall KR, Schneider SH, Rosenzweig C, et al. (2003) Fingerprints of global warming on wild animals and plants. Nature 421: 57–60.

68. Lunn NJ, Stirling I, Nowicki SN (1997) Distribution and abundance of ringed (*Phoca hispida*) and bearded seals (*Erignathus barbatus*) in western Hudson Bay. Canadian Journal of Fisheries & Aquatic Sciences 54: 914–921.

69. Loeb V, Siegel V, Holm-Hansen O, Hewitt R, Fraser W, et al. (1997) Effects of sea-ice extent and krill or salp dominance on the Antarctic food web. Nature 387: 897–900.

70. Tynan CT, Demaster DP (1997) Observations and predictions of Arctic climatic change: Potential effects on marine mammals. Arctic 50: 308–322.

71. Cotté C, Guinet C (2007) Historical whaling records reveal major regional retreat of Antarctic sea ice. Deep Sea Research Part I: Oceanographic Research Papers 54: 243–252.

72. Krafft BA, Kovacs KM, Frie AK, Haug T, Lydersen C (2006) Growth and population parameters of ringed seals (*Pusa hispida*) from Svalbard, Norway, 2002-2004. ICES Journal of Marine Science 63: 1136–1144.

73. Nicol S, Worby A, Leaper R (2008) Changes in the Antarctic sea ice ecosystem: Potential effects on krill and baleen whales. Marine and Freshwater Research 59: 361–382.

74. Greene CR, Pershing AJ (2004) Climate and the conservation biology of North Atlantic right whales: The right whale at the wrong time? Frontiers in Ecology and the Environment 2: 29–34.

75. Leaper R, Cooke J, Trathan P, Reid K, Rowntree V, et al. (2006) Global climate drives southern right whale (*Eubalaena australis*) population dynamics. Biology Letters 2: 289–292.

76. Feely RA, Sabine CL, Lee K, Berelson W, Kleypas J, et al. (2004) Impact of anthropogenic CO2 on the CaCO3 system in the oceans. Science 305: 362–366.

6

A Systematic Health Assessment of Indian Ocean Bottlenose (*Tursiops aduncus*) and Indo-Pacific Humpback (*Sousa plumbea*) Dolphins Incidentally Caught in Shark Nets off the KwaZulu-Natal Coast, South Africa

Emily P. Lane[1]*, Morné de Wet[2], Peter Thompson[2], Ursula Siebert[3], Peter Wohlsein[4], Stephanie Plön[5¤]

1 Department of Research and Scientific Services, National Zoological Gardens of South Africa, Pretoria, South Africa, 2 Epidemiology Section, Department of Production Animal Studies, Faculty of Veterinary Science, University of Pretoria, Pretoria, South Africa, 3 Institute for Terrestrial and Aquatic Wildlife Research, University of Veterinary Medicine, Hannover, Foundation, Germany, 4 Department of Pathology, University of Veterinary Medicine, Hannover, Foundation, Germany, 5 South African Institute for Aquatic Biodiversity, c/o Port Elizabeth Museum/Bayworld, Port Elizabeth, South Africa

Abstract

Coastal dolphins are regarded as indicators of changes in coastal marine ecosystem health that could impact humans utilizing the marine environment for food or recreation. Necropsy and histology examinations were performed on 35 Indian Ocean bottlenose dolphins (*Tursiops aduncus*) and five Indo-Pacific humpback dolphins (*Sousa plumbea*) incidentally caught in shark nets off the KwaZulu-Natal coast, South Africa, between 2010 and 2012. Parasitic lesions included pneumonia (85%), abdominal and thoracic serositis (75%), gastroenteritis (70%), hepatitis (62%), and endometritis (42%). Parasitic species identified were *Halocercus* sp. (lung), *Crassicauda* sp. (skeletal muscle) and *Xenobalanus globicipitis* (skin). Additional findings included bronchiolar epithelial mineralisation (83%), splenic filamentous tags (45%), non-suppurative meningoencephalitis (39%), and myocardial fibrosis (26%). No immunohistochemically positive reaction was present in lesions suggestive of dolphin morbillivirus, *Toxoplasma gondii* and *Brucella* spp. The first confirmed cases of lobomycosis and sarcocystosis in South African dolphins were documented. Most lesions were mild, and all animals were considered to be in good nutritional condition, based on blubber thickness and muscle mass. Apparent temporal changes in parasitic disease prevalence may indicate a change in the host/parasite interface. This study provided valuable baseline information on conditions affecting coastal dolphin populations in South Africa and, to our knowledge, constitutes the first reported systematic health assessment in incidentally caught dolphins in the Southern Hemisphere. Further research on temporal disease trends as well as disease pathophysiology and anthropogenic factors affecting these populations is needed.

Editor: Lloyd Vaughan, Veterinary Pathology, Switzerland

Funding: Pathological investigations on cetaceans caught in shark nets in South Africa was funded by the German Science Foundation (SI 1542/4-1) as part of a Research Cooperation Programme with the South African National Research Foundation (Grant number 707140), as well as by a National Research Foundation SEAChange grant (Grant number 74241). The funders had no role in study design, data collection and analysis, decision to publish, or preparation of the manuscript.

Competing Interests: The authors have declared that no competing interests exist.

* Email: emily@nzg.ac.za

¤ Current address: Coastal and Marine Research Unit, Nelson Mandela Metropolitan University, Port Elizabeth, South Africa

Introduction

Surveillance and research on diseases in wildlife populations present many challenges but are important tools to identify changes in ecosystem health and emerging threats to human and animal health [1]. Health assessments in coastal cetaceans can be used to indirectly monitor marine ecosystem health, investigate the effects of human activities on animal health, and identify risks to humans utilizing the same habitat for food or recreation [2,3]. Marine mammal researchers over the past 40 years have raised concerns about deteriorating ocean health. Although increased surveillance and improved diagnostic techniques may account for a portion of the recent proliferation of disease reports [4], mortality events due to harmful algal blooms and morbillivirus outbreaks are thought to be increasingly common in the North Atlantic [4–6]. However, lack of baseline data precludes accurate recording of temporal changes in the prevalence of many diseases [4,7,8]. Expected increasing effects of climate change, inter- and intra-specific competition and habitat degradation as well as exposure to pollutants, lend new urgency to understanding the causes of marine mammal disease outbreaks [3,7–9].

Coastal cetaceans are particularly vulnerable to anthropogenic impacts including net entanglement [10], boat strike [11], disturbances due to boat traffic [10], pollution [7], nutrient enrichment [10], novel pathogens [12], habitat degradation [10], and prey depletion through fishing [10,12]. Dolphins have long life spans [12,13], feed at a high trophic level [13], and their fat stores accumulate chemical pollutants [13–15]. Increased mortalities in polluted waters during morbillivirus epidemics suggest that pollutants may impair disease defense mechanisms [12]. Habitat destruction and prey depletion increase inter- and intra-species competition and stress that further undermine host defense mechanisms [7,12]. Nutrient enrichment with sewage and fertilizers has been implicated in an increase in the occurrence of devastating toxic algal blooms [16,17]. River runoff from urban areas may be responsible for the introduction of new marine pathogens such as *T. gondii* [18,19].

Both *Tursiops aduncus* (Indian Ocean bottlenose dolphin) and *Sousa plumbea* (Indo-Pacific humpback dolphin) occur along the Southern African coast within 10 km of the shore, [20–23]. Gill nets are deployed off the South African east coast by the KwaZulu-Natal Sharks Board (KZNSB) to reduce the risk of shark-human interactions [22,24]. Approximately 20 dolphins, mainly *T. aduncus* and *S. plumbea*, are incidentally caught (by-caught) annually in the shark nets [25]. This paper reports the results of the first systematic health assessment of incidentally caught coastal dolphins, based on 40 animals examined between 2010 and 2012. Pathological findings are analyzed in relation to species, catch location, age, sex, and body condition. This survey provides valuable baseline data for assessing the health status of these dolphin populations and for future monitoring of temporal and spatial health trends.

Materials And Methods

Ethics Statement

Evaluation of dolphins incidentally caught in the shark nets was performed under research permits issued to the Port Elizabeth Museum/Bayworld (PEM) by the South African Departments of Environmental Affairs and Agriculture, Forestry and Fisheries (RES2012/40 and RES2013/19). The protocol for this study was approved by the Research Committee of the Faculty of Veterinary Science; the Animal Use and Care Committee of the University of Pretoria (Protocol V011/12) and the Ethics and Scientific Committee of the National Zoological Gardens of South Africa (P10/23). Formalin-fixed tissues are stored at the PEM; paraffin embedded tissues and glass slides are stored at the National Zoological Gardens of South Africa.

From April 2010 to April 2012, dead dolphins were retrieved from the shark nets, weighed and frozen at -20°C by the KZNSB. Every 6–8 months, carcasses were defrosted and morphological measurements taken [26]. Of the 46 dolphins retrieved, 35 *T. aduncus* and five *S. plumbea* were deemed sufficiently fresh for necropsy and histopathological examination [27]. Age was estimated by total body length in *T. aduncus* [21] and by counting the annual growth layers in a mandibular tooth in *S. plumbea*. Animals were classified as unweaned calves (<2 years), juveniles (2–12 years), or sexually mature adults (>12 years) [21,28]. Blubber thickness measurements were used (ventral, lateral and dorsal midline cranial to the dorsal fin) to assess nutritional condition [29].

Using a standard necropsy and sampling protocol [30], all organs were examined macroscopically and representative samples fixed in 10% buffered formalin. Paraffin wax embedded tissues were sectioned (5 µm) and stained with haematoxylin and eosin (HE). Selected tissues were also stained with Gram, Von Kossa (VK), Stamps, Masson's Trichrome (MT), Ziehl-Neelsen (ZN), Gomori's methenamine silver (GMS), Perl's prussian blue, Hall's bile, periodic acid-Schiff (PAS), Fontana Masson's and Bielschowsky's modified silver stains [31]. Immunohistochemical reactions for

Toxoplasma gondii (Department of Pathology, University of Pretoria) and dolphin morbillivirus (Department of Pathology, University of Veterinary Medicine, Hannover) [33] were performed on sections where lymphoplasmacytic inflammation was present in the brain, lung, muscle or heart.

Parasites found during necropsy were preserved in 70% ethanol and identified according to published methods [34]. Lung tissue samples from all 40 dolphins were frozen, until the end of the collection period, thawed in the laboratory and cultured using standard bacteriological methods.

Statistical analyses

Animals were divided into two groups based on capture location region: North and South of Ifafa beach (Figure 1), since population and genetic studies of *T. aduncus* indicate that these are different subpopulations [35,36]. Too few *S. plumbea* were sampled for statistical analysis; all statistical comparisons are for *T. aduncus* only, unless otherwise stated. Blubber thickness was compared between age classes and sample sites using a linear mixed model adjusted for sex and region with Bonferroni correction for multiple comparisons. Occurrence of selected lesions with possible biological significance was compared between species, and for *T. aduncus*, between age classes, sexes and capture location region using Fisher's exact test. For univariable associations with p<0.25, adjustment for possible confounding between age class, sex and region was done using multivariable exact logistic regression models. Associations between the occurrence of selected lesions within the same animals was tested using McNemar's test. Due to the exploratory nature of the analysis and the relatively small sample size, significance was assessed at p<0.1. Statistical analysis was done using Stata 12.1 (StataCorp, College Station, TX, U.S.A.).

Results

More *T. aduncus* (35; 88%) were caught in the nets than *S. plumbea* (5; 12%) (Figure 1). Most *T. aduncus* (25; 71%) and all five *S. plumbea* were sampled from the northern region nets; and seven of the 35 *T. aduncus* (20%) were from the nets off Durban. Most *T. aduncus* in all age classes were females (24; 69%); and more juveniles (16; 46%) and calves (11; 31%) were caught than adults (8; 22%) of both sexes.

Blubber was thicker at the dorsal and thinner at the lateral sampling site for each age class (p<0.05; Figure 2). Blubber thickness did not differ between the sexes or between dolphins from different regions. Blubber was thicker in juveniles and adults compared to calves, at the dorsal (p<0.001) and ventral (p<0.05) sites.

Moderate to severe autolysis, putrefaction and freezing artefact were present histologically in most organs, particularly in the respiratory and intestinal mucosae, pancreas, brain and eye. Eosinophils were relatively well preserved compared to other inflammatory cells. Freezing distorted tissue architecture and caused lysis of erythrocytes. In addition, variable numbers of variably sized, round to oval, vacuoles (<0.1 cm diameter) with no associated nuclei or saprophytic bacteria were found in blood vessel lumina and the parenchyma of various organs. Mild to severe, acute congestion was present in most organs in all the dolphins.

Dolphin number, species, sex, age, sampling region, lesion severity and health status for *T. aduncus* and *S. plumbea* are listed in Table S1. Common and newly reported lesions and lesions that may have affected organ function are described below, along with their prevalence in *T. aduncus* and *S. plumbea* (Table 1). Exact

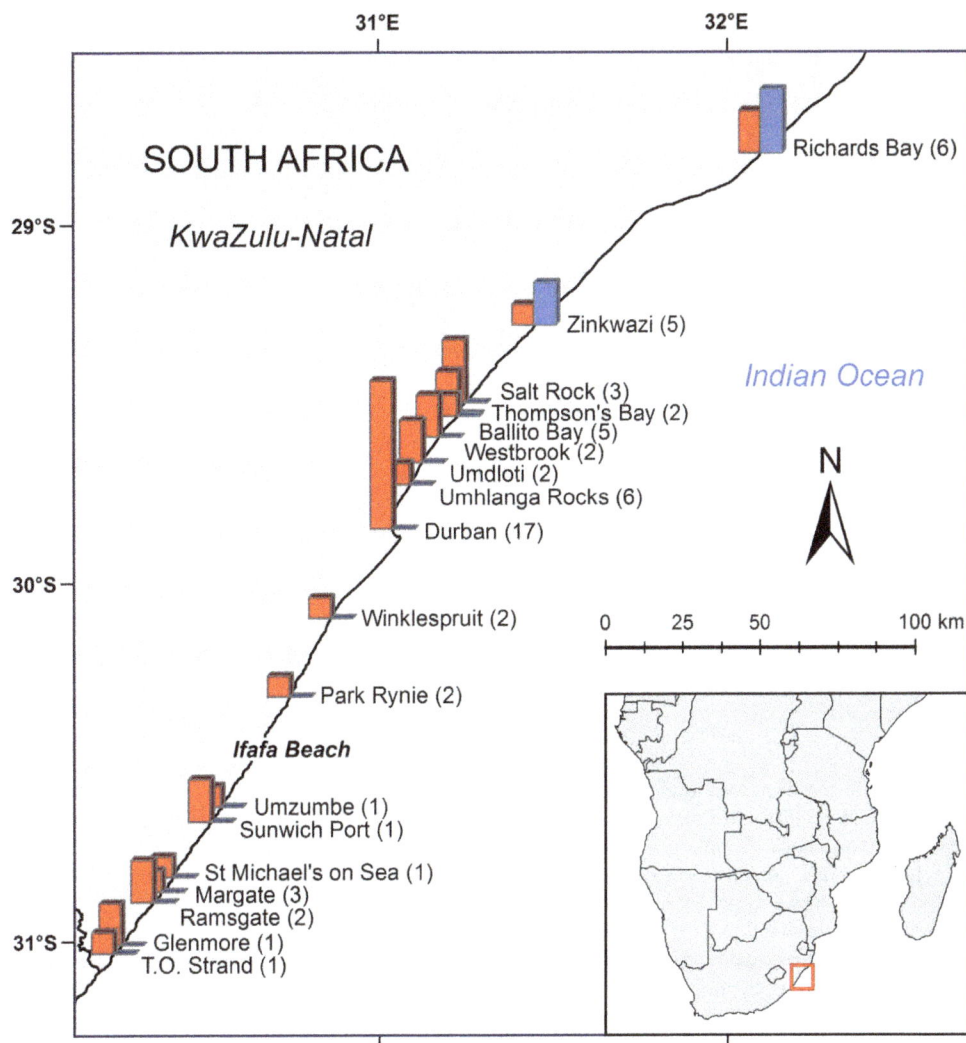

Figure 1. Location (beach name), number of shark nets per beach (in parenthesis) and number of *T. aduncus* **(red) and** *S. plumbea* **(blue) sampled along the KwaZulu-Natal coast, South Africa.** Gill nets are 110 m long and 10 m deep. Adapted from [87].

logistic regression models for lesions significantly associated (p< 0.1) with age class, sex and region are given in Table 2. Supplementary materials include a complete list, with prevalence by species, age class and region, of all pathological findings (Table S2) and common pathology observed in *T. aduncus* by age class, sex and region (Table S3).

Mild to severe, multifocal to diffuse, acute pulmonary congestion, oedema and emphysema were common, characterized by lungs that were heavy, poorly collapsed, mottled pink to deep red and contained air-filled bullae (1–4 mm diameter) beneath the pleura and throughout the lung parenchyma. White foam filled airways of affected lungs. Variable numbers of fine white round helminths (<50×1×1 mm, *Halocercus* sp.) were present in multiple firm, white to tan, unencapsulated pulmonary nodules (<2 cm diameter) and ectatic bronchi (<8 cm diameter) in 37 animals (93%), in all ages and both sexes and species (Figure 3). Affected bronchi were lined by discontinuous attenuated epithelium, with large amounts of necrotic cellular and inflammatory debris and medium number of filarial larvae. Similar inflammation often extended into and disrupted the architecture of adjacent pulmonary parenchyma. Nematode adults, with (#5, 11, 16) or

without (#6, 8, 10, 37, 40) microfilaria were present in these inflammatory lung lesions in eight (20%) animals. In addition, mild, multifocal lymphoplasmacytic and variably eosinophilic bronchointerstitial pneumonia was present in dolphins of all age classes, both sexes and species. Pneumonia was also frequently accompanied by follicular lymphoid hyperplasia of bronchus associated lymphoid tissue (18 animals; 45%).

Clustered or scattered connective tissue nodules enclosing variably mineralized necrotic debris, mixed with eosinophils, lymphocytes and plasma cells and, in some cases sections of nematodes, occurred throughout the lung parenchyma (<4 cm diameter), often close to bronchioles (16 animals, 40%). Mild to moderate, multifocal, subacute lymphoplasmacytic and variably eosinophilic tracheobronchitis, with no apparent relationship to areas of bronchiectasis or parasites, was present in 12 (44%) *T. aduncus* calves and juveniles.

Small numbers of firm, white, pleural or subpleural plaques or nodules (<5 mm diameter), occasionally containing caseous material, were seen in 16 (40%) animals. These consisted histologically of chronic pleuritis characterized by variably thick fibrous connective tissue foci containing variably mineralized

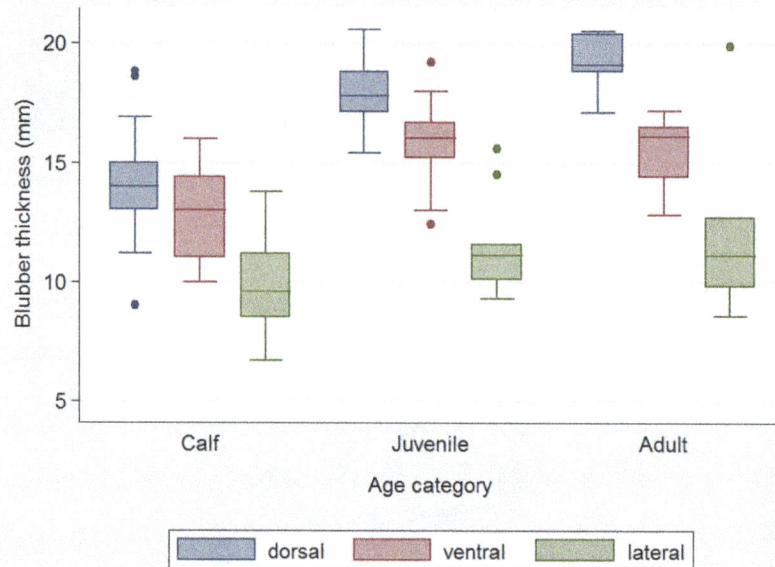

Figure 2. Blubber thickness (mm) of *T. aduncus* **in three age classes.** Box extends from 25th to 75th percentile, horizontal line represents the median, whiskers extend to the smallest and largest observations that are <1.5 times removed from the interquartile range (IQR), and dots represent outliers.

necrotic inflammatory and cellular debris with moderate lymphoid follicular hyperplasia and mild pleural and interstitial fibrosis in the adjacent tissue. Mild, multifocal lymphoplasmacytic and variably eosinophilic pleuritis that was not detected on gross examination was found in 12 calves and juveniles (30%) of both species. Pleural arterioles were prominent on the visceral pleura. One male *T. aduncus* (#14) had a large subpleural focus of bronchiectasis (8 cm diameter) lined by compressed lung tissue (2–3 mm thick) and bronchiolar epithelium which contained a few fine filamentous white helminths (<1 mm thick, 3–5 cm long). Thick white firmly attached adhesions between the parietal and visceral pleura and the diaphragm were present in two female *T. aduncus* (#1, 23). Histologically, these consisted of bands of mature fibrous connective tissue infiltrated with small foci of lymphocytes and plasma cells. The pleural surfaces of one juvenile and one adult male *S. plumbea* (#38, 40) were covered in small fibrovascular tags (<1 cm long) with variably plasmacytic and eosinophilic pleuritis and moderate pleural and interstitial fibrosis. In *T. aduncus* no association was found between pneumonia and pleuritis (p = 0.653).

Autolysis and freezing artefact precluded detailed assessment of lymphoid tissue, however, mild to moderate follicular and paracortical lymphoid hyperplasia were seen in ten animals with respiratory tract inflammation (#6, 16, 17, 19, 22–25, 34, 36, 38) and six with lung marginal lymph node serositis characterized by aggregates of small numbers of eosinophils, lymphocytes, macrophages and plasma cells in the lung marginal lymph node connective tissue capsule (# 9, 17, 19, 22, 36, 38). Inflammation also often extended to the connective tissue between the lung and the lung marginal lymph node. Lymphoid tissue appeared depleted in two female juvenile *T aduncus* (#27, 28). Mild, focal, neutrophilic and histiocytic, necrotising lung marginal lymph node lymphadenitis was seen in association with suspected fungal hyphae in a juvenile male *T. aduncus* (#25), although the lesion was not present on serial sections stained with GMS. While 12 lung sections contained small to large numbers of mixed bacteria

in blood vessels, interstitium and alveoli (H&E and Gram stains), these were not associated with necrosis or neutrophilic inflammation. A variety of bacteria were isolated on routine lung cultures, including *Pantoea agglomerans*, *Enterococcus solitarius*, *Enterobacter gergoviae*, *Shewanella algae* and *S. putrefaciens*, *Photobacterium damselae*, *Aeromonas media*, *Lactococcus garviae*, *Clostridium tertium*, *Streptococcus* from the *viridians* group, *Psychrobacter* sp, *Enterococcus* sp., *Micrococcus* sp., *Lactobacillus* sp., *Brevundimonas* sp., *Bacillus sp.*, *Acinetobacter* sp. *and Proteus* sp. Lung samples from 16 animals tested by immunohistochemistry contained no dolphin morbillivirus or *Toxoplasma* antigen.

Multiple variably mineralized deposits were common, occurring beneath or replacing the bronchial and bronchiolar mucosae. Unfortunately, details of the lesions in these animals were obscured by autolysis of the bronchiolar epithelium. Both affected and unaffected dolphins originated from both regions, were from all age classes, and of both sexes and species.

In *T. aduncus*, all three gastric compartments contained raised, firm tan nodules with central pores (<1 cm diameter); lesions were more common in the 3rd compartment (p = 0.004). Moderate to severe, multifocal, chronic lymphoplasmacytic and eosinophilic pyloric gastritis with variable calcification of the adjacent mucosa was associated with trematodes of the subfamily Brachycladiinae (Figure 4). Prevalence increased with age (p = 0.097), although this was not statistically significant in the multivariable model (p = 0.123). Eosinophilic and lymphoplasmacytic gastritis of variable severity and chronicity that was not detected on gross examination affected all three gastric compartments. The prevalence of this gastritis also increased with age (p = 0.034), as did the prevalence of similar enteritis (p = 0.002). Adult nematodes (*Anisakidae*) were found in gastro-intestinal tract of two *T. aduncus* (#26, 33). Lingual myocytes contained sarcocysts, without associated inflammation, in one *T. aduncus* calf (#13, Figure 5).

Although the livers were macroscopically unremarkable, eosinophilic and variably lymphoplasmacytic, and occasionally necro-

Table 1. Common pathology observed in Indian Ocean bottlenose (*Tursiops aduncus*) and Indo-Pacific humpback (*Sousa plumbea*) dolphins incidentally caught in shark nets, and bivariable association with species.

Lesion/abnormality	Total (%)	Species (n)		
		T. aduncus	S. plumbea	p*
Combined pneumonia	93	32/35	5/5	1.000
Bronchopneumonia	18	7/35	0/5	0.565
Interstitial pneumonia	63	22/35	3/5	1.000
Broncho-interstitial pneumonia	30	9/35	3/5	0.149
Pulmonary parasites	15	6/35	0/5	1.000
Pleuritis	30	10/35	2/5	0.627
Bronchiolar mucosal calcification	83	29/35	4/5	1.000
Pulmonary anthracosis	8	2/35	1/5	0.338
Gastritis all compartments	68	24/34	2/4	0.577
First and second compartment gastritis	63	23/34	1/4	0.132
Third compartment gastritis	65	14/21	1/2	1.000
Parasitic nodules all compartments	32	12/34	0/4	0.556
Parasitic nodules in the first and second gastric compartments	8	3/34	0/4	1.000
Parasitic nodules in the third gastric compartment	43	10/21	0/2	0.486
Pyloric mucosal calcification	26	5/21	1/2	0.462
Enteritis	68	25/35	2/5	0.307
Periportal hepatitis	54	21/35	0/4	**0.037**
Hepatic serositis	23	9/35	0/4	0.556
Periportal fibrosis	26	9/35	1/4	1.000
Hepatic trematode eggs	8	3/35	0/4	1.000
Bile ductular hyperplasia	44	15/35	2/4	1.000
Splenic filamentous peritonitis	45	17/35	1/5	0.355
Splenic serositis	28	11/35	0/5	0.298
Cervical lymph node serositis	26	10/34	0/5	0.302
Mesenteric lymphnode serositis	46	15/34	3/5	0.647
Marginal lymph node serositis	43	11/27	2/3	0.565
Marginal lymph node anthracosis	10	3/27	0/3	1.000
Endometritis	42	10/24	1/2	1.000
Metritis	23	5/24	1/2	0.415
Oophoritis	19	4/24	1/2	0.354
Mastitis	43	3/7	-	-
Mammary corpora amylacea	43	3/7	-	-
Testicular serositis	38	3/10	2/3	0.510
Endo-, myo- and epicarditis	51	20/35	0/4	**0.047**
Cardiac fibrosis	26	9/35	1/4	1.000
Meningoencephalitis	39	7/16	0/2	0.497
Myositis	19	6/32	1/5	1.000
Combined serositis	75	26/35	4/5	1.000
Abdominal serositis	60	20/35	4/5	0.631
Thoracic serositis	20	18/35	2/5	1.000

*Fisher's exact test; statistically significant results (p<0.100) in bold.

tizing, periportal hepatitis and cholangitis of variable severity and chronicity were present in 21 (60%) *T. aduncus*. Adults were more often affected than calves (p = 0.044), although the association was not significant on multivariable analysis (p = 0.112). Green-brown, triangular trematode eggs (Figure 6) were found in the portal triads of three *T. aduncus* (#13, 26, 32). Moderate to marked hyperplasia of the bile duct epithelium was present in a *T. aduncus* (#21) and two *S. plumbea* (#36, 38) with cholangitis. Significantly, although two *S. plumbea* had cholangitis, no animals of this species had hepatitis (p = 0.037). Mild to severe, multifocal to diffuse increases in periportal mature fibrous connective tissue was observed with age in *T. aduncus* (p = 0.020). The presence of

Table 2. Associations of age, sex and region with presence of various lesions in *T. aduncus*: results of multivariable exact logistic regression models.

Variable and level		Age class			Sex	Region
		Calf (<2 y)	Juvenile (2–12 y)	Adult (>12 y)	male vs. female	south vs. north
Pleuritis	OR[1]	1*	1.54	0.26	**6.50**	1.17
	95% C.I.[2]	–	0.19, 13.65	0.00, 2.53	**0.98, 59.17**	0.00, 11.33
	p*	–	0.952	0.270	**0.053**	1.000
Pulmonary pneumoconiosis	OR	1*	1.00	**9.52**	1.50	3.00
	95% C.I.	–	0.00, ∞	**0.72, ∞**	0.04, ∞	0.08, ∞
	p	–	–	**0.085**	0.800	0.500
Enteritis	OR	1*	**15.26***	**6.55**	0.33	0.17
	95% C.I.	–	**1.95, ∞**	**0.82, ∞**	0.02, 3.78	0.00, 2.48
	p	–	**0.006**	**0.080**	0.573	0.303
Gastritis	OR	1*	5.66	**6.21**	1.13	1.97
	95% C.I.	–	0.57, 291.4	**0.78, ∞**	0.14, 9.89	0.23, 26.05
	p	–	0.201	**0.090**	0.141	0.785
Gastritis	OR	1*	7.38	**7.02**	1.36	1.09
(compartments 1&2)	95% C.I.	–	0.76, 376.8	**0.89, ∞**	0.17, 10.91	0.12, 10.17
	p	–	0.104	**0.066**	1.000	1.000
Periportal fibrosis	OR	1*	3.02	**12.64**	1.21	1.71
	95% C.I.	–	0.30, 41.55	**1.17, 223.9**	0.13, 9.89	0.18, 15.94
	p	–	0.482	**0.033**	1.000	0.884
Splenic tags	OR	1*	2.20	4.33	2.01	**7.75**
	95% C.I.	–	0.29, 18.27	0.42, 67.24	0.33, 14.21	**1.10, 99.82**
	p	–	0.607	0.300	0.621	**0.037**
Splenic serositis	OR	1*	2.81	5.41	**11.07**	3.3
	95% C.I.	–	0.27, 40.96	0.37, 117.1	**1.51, 152.0**	0.36, 45.61
	p	–	0.553	0.307	**0.012**	0.408
Cervical lymph node serositis	OR	1*	1.42	4.77	**7.42**	3.90
	95% C.I.	–	0.13, 15.20	0.36, 90.18	**1.04, 95.28**	0.45, 54.29
	p	–	1.000	0.327	**0.045**	0.297
Mesenteric lymph node serositis	OR	1*	2.85	**16.82**	3.56	0.94
	95% C.I.	–	0.42, 23.36	**1.92, ∞**	0.53, 29.82	0.10, 7.54
	p	–	0.377	**0.009**	0.247	1.000
Endometritis	OR	1*	0.92	**8.10**	–	0.92
	95% C.I.	–	0.07, 9.14	**0.87, ∞**	–	0.07, 9.14
	p	–	1.000	**0.067**	–	1.000
Cardiac fibrosis	OR	1*	**13.97**	**51.63**	4.29	0.71
	95% C.I.	–	**1.54, ∞**	**5.35, ∞**	0.26, 280.2	0.04, 13.09
	p	–	**0.017**	**0.001**	0.498	1.000
Myositis	OR	1*	5.73	**14.31**	0.26	0.33
	95% C.I.	–	0.20, 470.3	**1.31, ∞**	0.00, 2.31	0.00, 3.55
	p	–	0.473	**0.029**	0.246	0.381
Abdominal serositis	OR	1*	4.05	**11.18**	3.00	0.73
	95% C.I.	–	0.63, 35.27	**1.37, ∞**	0.44, 26.62	0.08, 5.42
	p	–	0.177	**0.022**	0.362	1.000

[1]OR = Odds ratio.
[2]95% C.I. = 95% confidence interval.
*statistically significant results (p<0.100) in bold.

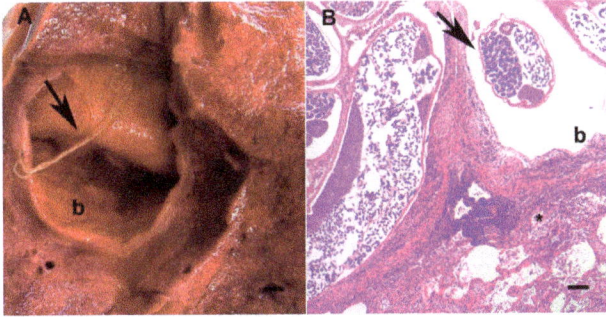

Figure 3. Parasitic pneumonia. A: Ectatic bronchus (b) containing thin (1–2 mm diameter), long, white helminths identified as *Halocercus* sp. (arrow). Bar = 5 mm. B: Pulmonary helminths (arrow) in an ectatic bronchiole (b) with eosinophilic and lymphoplasmacytic interstitial pneumonia (*) and an adjacent follicle of mildly hyperplastic bronchiolar-associated lymphoid tissue (HE, bar = 250 μm).

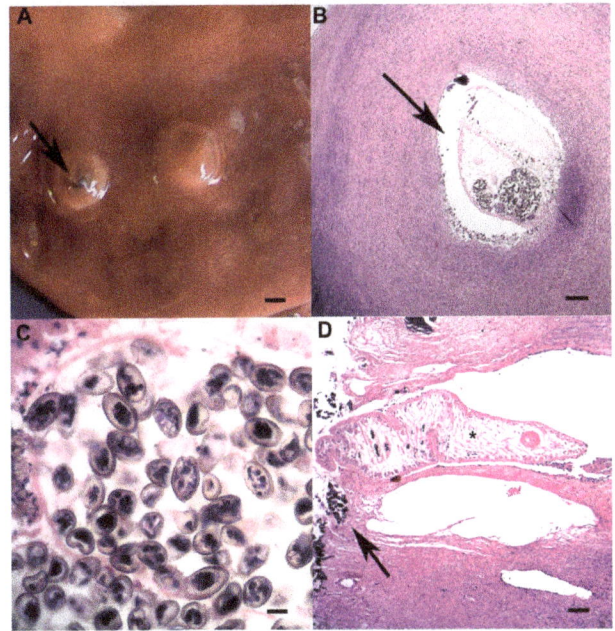

Figure 4. Gastric trematode associated lesions. A: Firm, round parasitic nodules (<1 cm diameter) with a small pore opening to the gastric lumen (arrow). Bar = 0.4 cm. B: Adult trematode (arrow) in the center of a focus of extensive fibrosis (HE, bar = 0.5 mm). C: Embryonated trematode eggs (280×160 μm, HE, bar = 150 μm). D: Parasitic nodule with adult trematode blocking the pore and irregular mineralized foci (arrow) in the adjacent superficial gastric epithelium (HE, bar = 500 μm).

increased portal connective tissues was positively associated with the presence of trematode eggs (p = 0.013). Mildly to moderately increased numbers of small bile ductules in the portal triads and under the hepatic capsule were interpreted as mild to moderate bile ductular hyperplasia in 17 (42.5%) animals of all ages and both sexes. Portal connective tissue was positively associated with bile ductular hyperplasia (p = 0.009) but not with portal hepatitis (p = 0.468).

Subjectively, increased numbers of eosinophilic cell lines were present in the rib bone marrow in 22 animals of both species (75% of *T. aduncus* and 33% of *S. plumbea*) and from both regions (80% north and 63% south). Mild to moderate, multifocal, variably eosinophilic and lymphoplasmacytic oophoritis that was not detected on gross examination was found in 21% of *T. aduncus* females and one *S. plumbea* female (#37). Endometritis was more common in adults (100%) than in calves (31%) and juveniles (29%), (p = 0.044) and consisted of small clusters of lymphocytes, plasma cells and variable numbers of eosinophils and neutrophils in the endometrium. A single adult *T. aduncus* (#32) had a trematode egg associated with the endometritis. Mild to moderate, multifocal, variably eosinophilic and lymphoplasmacytic metritis that was not detected on gross examination, was found in five *T. aduncus* (#13, 30, 31, 35, 36) and a single *S. plumbea* (#37). A positive association with age was found (p = 0.019), although this association was not significant on multivariable analysis (p = 0.107). Stamps stain for *Brucella* bacteria was negative in 12 females and all five males tested.

Mild to moderate, focal to multifocal, lymphoplasmacytic epicarditis, endocarditis and myocarditis (Figure 7A), that were not detected on gross examination, were seen in *T. aduncus* (20; 51%) but not in *S. plumbea* (p = 0.047). The highest prevalence was in juveniles (80%) (p = 0.060), although this was not significant in the multivariable model (p = 0.451). Immunohistochemistry of affected histologic sections did not demonstrate *T. gondii* antigen. Mild, focal to multifocal myocardial fibrosis (Figure 7B) was found in ten (51%) animals of both species for which heart was examined (#21, 24, 26, 29–34, 38). Prevalence increased with age (p = 0.001) and was positively associated with adrenal cortical hyperplasia (p = 0.043) but not correlated with epi-, endo-, or myocarditis (p = 0.393).

Mild, multifocal, lymphocytic meningoencephalitis was found in only seven (39%) *T. aduncus* (#7, 9, 18, 21, 25, 27, 29). Stamps and Gram histologic stains and immunohistochemistry of affected

sections did not demonstrate *Brucella*, other bacteria, *T. gondii* or dolphin morbillivirus antigen.

Multiple slightly raised, firm, white serosal nodules (<1 cm diameter) were present on various abdominal organs, mainly in *T. aduncus*. Animals from both regions and all age classes were affected (Figure 8). Histologically, these corresponded to mild, variably eosinophilic lymphoplasmacytic and necrotizing serositis

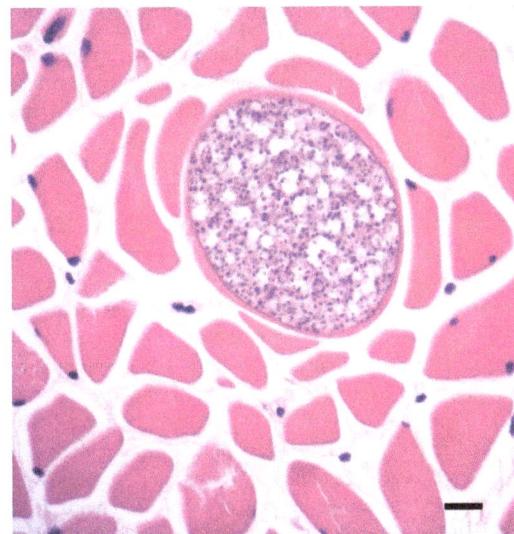

Figure 5. Lingual Sarcocystis. Sarcocyst containing myriad metrocytes in a muscle fiber of the tongue (HE, bar = 150 μm).

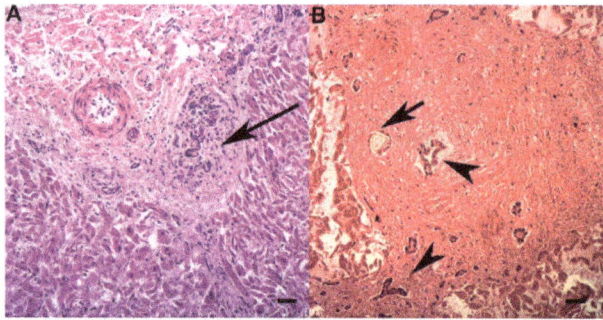

Figure 6. Hepatic lesions in *T. aduncus*. A: Mild proliferation (hyperplasia) of small portal bile ductules (arrow) (HE, bar = 100 μm): B: Severe hepatic periportal fibrosis associated with a trematode egg (arrow, 100 μm diameter). Note the bile ductules with hyperplastic epithelium (arrowheads, HE, bar = 100 μm).

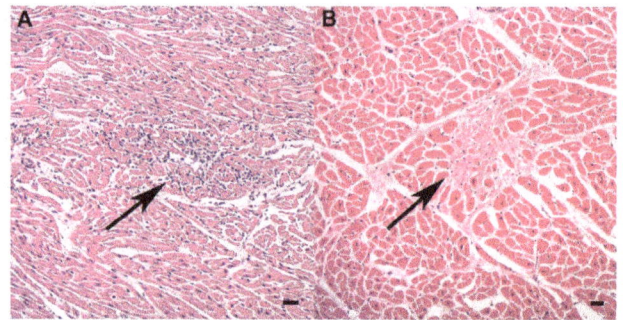

Figure 7. Myocardial lesions. A: Mild focal lymphoplasmacytic myocarditis (arrow) (HE, bar = 50 μm). B: Mild focal myocardial fibrosis (arrows) (HE, bar = 120 μm).
doi:10.1371/journal.pone.0107038.g007

affecting the fibrous capsule of the mesenteric lymph node (#2, 9, 13, 15, 17, 18, 20, 21, 23–26, 29, 30, 34, 35, 37, 38, 40), spleen (#9, 13, 17, 18, 23–26, 29, 32, 33, 35), liver (#4, 9, 15, 17, 21, 24, 25, 33), testis (#18, 24, 33, 38, 39), kidney (#5, 32, 36, 39), diaphragm (#7, 26, 30), and epididymis (#40), as well as adipose tissue adjacent to the mesenteric lymph node (#7, 8). Multifocal to diffuse, lymphoplasmacytic and eosinophilic inflammation was present in the mesenteric lymph node in five animals (#24, 30, 31, 34, 35), the testis in a *T. aduncus* calf (#18) and the spermatic cord in a juvenile *T. aduncus* (#38). Nematode larvae were associated with the mesenteric lymph node serositis in two juvenile male *T. aduncus* (#25, 29). These lesions were variably associated with mesenteric lymph node lymphoid hyperplasia (Table S1). The prevalence of the mesenteric lymph node serositis increased significantly with age in *T. aduncus* (p = 0.009). Male *T. aduncus* were more often affected with splenic serositis than females (p = 0.015). Renal serositis was not associated with the mild, multifocal, mainly lymphoplasmacytic, renal interstitial nephritis seen in 11 animals (Table S1).

Long, slender, splenic tags occurred in a higher proportion of *T. aduncus* (49%) than *S. plumbea* (20%) (Figure 9). Histologically, these filamentous projections of the splenic capsule consisted of fibrovascular connective tissue with minimal or mild, multifocal, lymphoplasmacytic and eosinophilic inflammation. Splenic tags were significantly more common in dolphins from the southern coast (80%) than the northern coast (36%) (p = 0.027) and were associated with splenic serositis (p = 0.034).

Mild, multifocal, lymphoplasmacytic interstitial skeletal myositis was present in ten (27%) dolphins of both species and sexes from the northern region (#9, 21, 22, 23, 30, 31, 32, 33, 27, 38). Prevalence increased with age (p = 0.007). Immunohistochemistry of affected histologic sections did not demonstrate *T. gondii*. Multiple raised, pale pink cystic lesions (<1 cm diameter) containing adult *Crassicauda* sp. were associated with moderate, locally extensive, chronic, eosinophilic myositis in the musculature next to the mammary gland in one *T. aduncus* adult female (#22), which also had round basophilic crystalline structures with variable mineralized cores (interpreted as *corpora amylacea*) in the adjacent otherwise unremarkable mammary gland. Mild, multifocal, interstitial mammary gland inflammation with pleocellular infiltrates was present in two *T. aduncus* calves (#3, 6) and one juvenile (#21). Sarcocysts, without associated inflammation, were found in neck and intercostal muscle of one *T. aduncus* calf (#13).

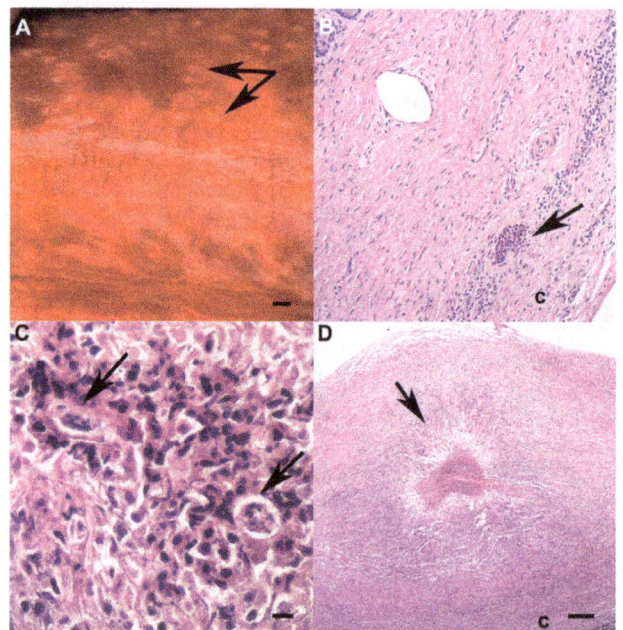

Figure 8. Abdominal serositis. A: Peritoneum overlying the testis contains multiple, slightly raised, firm white nodules, some of which contain depressed red centers (arrows). Bar = 5 mm. B: Eosinophil aggregate (arrow) and lymphoplasmacytic serositis in the testicular capsule (c). Note the seminiferous tubule in the upper left corner (HE, bar = 100 μm). C: Mesenteric lymph node serositis with intra-lesional nematode larvae in the capsule (60 μm diameter, arrows) (HE, bar = 30 μm). D: Severe focal granulomatous testicular serositis in the testicular capsule (c) with a central area of necrosis (arrow) resembling a helminth migration tract. Note seminiferous tubules at bottom right (HE, bar = 500 μm).

All animals had superficial cutaneous linear abrasions (net marks), particularly over the thorax, flippers, flukes and head, associated with subcutaneous congestion or haemorrhage in some cases (#1, 9, 20, 30, 32). An adult male *S. plumbea* (#40) had two flat, pale-tan, lobular, cutaneous soft masses below the dorsal fin (10 mm diameter) with a light brown exudate on the cut surface. Histologically, large numbers of large foamy macrophages and rare multinucleate giant cells infiltrated the skin and subcutis with a large number of intra-lesional round yeasts (7–10 μm diameter) that stained positive on both GMS and PAS, consistent with

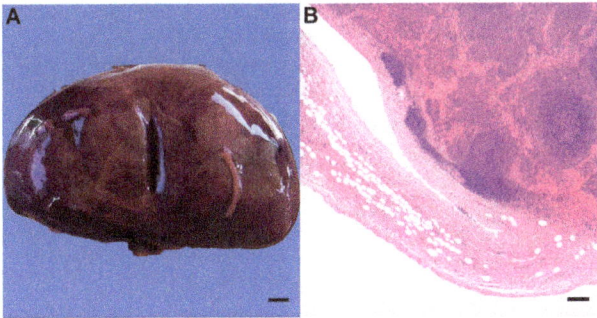

Figure 9. Splenic filamentous peritonitis. A: Fine long filamentous tags (1×2×30 mm) on the splenic capsule. Bar = 10 mm. B: Splenic tag consisting of mature fibrovascular connective tissue (HE, bar = 500 μm).

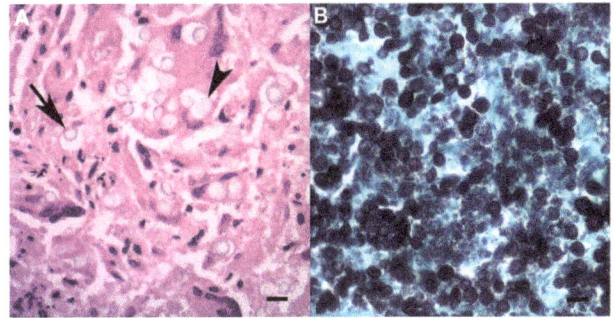

Figure 10. Cutaneous lobomycosis. A: Moderate numbers of round to oval refractile yeasts occur free in the subcutis (arrow) or within multinucleate giant cells (arrowhead, HE, bar = 10 μm). B: Large numbers of deep blue-black staining yeasts (GMS, bar = 10 μm).

lobomycosis (Figure 10). Small aggregates of lymphocytes, plasma cells, neutrophils or eosinophils occurred in the mammary gland interstitium of two *T. aduncus* calves (#3, 6) and one *T. aduncus* juvenile (# 8).

Mild, focal, lymphoplasmacytic and eosinophilic steatitis affecting the adipose tissue around the cervical lymph node was present in four animals (#11, 15, 20, 26). Mild, multifocal, lymphoplasmacytic and histiocytic inflammation of the capsule of the cervical lymph node and or surrounding adipose tissue was present in nine animals (#13, 17, 18, 22, 24, 26, 33, 34, 35), affecting more males (55%) than females (17%) (p = 0.045). This finding had no association with mild to moderate follicular and paracortical lymphoid hyperplasia present in this lymph node in 17 animals (Table S1).

Discussion

This study is the first reported systematic health assessment of incidentally caught dolphins in the Southern Hemisphere. This valuable information on the current prevalence of disease in the coastal dolphin populations of South Africa can be used as a baseline for future monitoring projects.

The degree of autolysis and freezing artefact varied between animals and organs, and likely masked subtle histological features such as necrosis and tissue and inflammatory cellular detail. The presence and patterns of inflammation and parasites could, however, be confidently diagnosed, as has been documented in harbour porpoises (*Phocoena phocoena*) [38,40,41] and fur seals (*Arctocephalus forsteri*) [42].

Correct interpretation of tissue changes as pathological was hampered by the small sample size and the lack of standardized descriptions of tissue anatomy in dolphins. Also, in contrast to regularly dewormed domestic species, establishing normal tissue parameters is complex in free-ranging mammals which often harbour large numbers of internal parasites that may vary with age, geographical location and season. Focal (#16, 19, 25, 26, 32), multifocal (#33, 39) or diffuse (#6, 35) increases in the amounts of mature connective tissue spatially unrelated to pneumonia were compared to pulmonary connective tissue amounts in the remaining animals and subjectively diagnosed as pulmonary fibrosis. Similarly, increased amounts of periportal mature connective tissue was positively associated with age, but not with the presence of periportal hepatitis (Table S3). This may therefore be an age-related change in *T. aduncus*, although it is not clear whether this is related to trematode infections which are more numerous in older animals (Table S2). Increased numbers of small bile ducts in the portal triads and under the hepatic capsule were

noted in 17 animals (Table S1); this change was subjectively associated with increased amounts of mature connective tissue, based on comparison between livers in other animals in this series and on our knowledge of similar lesions in terrestrial mammals. Too few animals were examined to assess whether the number of small bile ducts in the hepatic portal zone is variable in these species or is related to inflammatory changes. Documentation of the amount of connective tissue in well preserved tissues from newborn animals and any age-related increases, in the absence of pathological changes, would facilitate correct interpretation of the amount of pulmonary and hepatic connective tissue in these species.

Widespread tissue congestion and pulmonary emphysema and oedema are described in other net-captured cetaceans and are likely due to terminal heart failure and or drowning [37–39]. Clear, round vacuoles in various tissues and air emboli in blood vessels in a wide range of tissues were possibly a result of either drowning or supersaturation [84,88]; however without more detailed studies regarding the pathophysiology of drowning in cetaceans the distinction between these two possibilities is uncertain. The histological location, absence of nuclei, and variable size of tissue and intravascular vacuoles excluded adipocytes; the absence of bacteria associated with the vacuoles (HE and Gram) make gas produced by saprophytic bacteria unlikely. However, only bubble content analysis would confirm supersaturation [89].

As expected, most of the lesions noted in these incidentally caught dolphins were mild to moderate and severe lesions were mostly focal (Table S1) and the dolphins were judged to be healthy. Blubber thickness measurements were within previously published ranges for *T. aduncus* from the KwaZulu-Natal coast [29]. No reference ranges or prior data are available for blubber thickness in *S. plumbea* from the KwaZulu-Natal coast. None of the animals with the thinnest blubber had major or multiple significant lesions and no statistical association between thinner blubber and pathology could be demonstrated. Therefore, we concluded that all animals were at least in fair nutritional condition. Although parasite levels in free-ranging animals generally have little effect on the host, factors such as stress, altered nutrition, anthropogenic factors, pollutants or concurrent disease may compromise the host's immune system and increase the severity and prevalence of parasitic infections [40]. Parasite burdens may then be used as indicators for the overall health status of an individual [40]. This assumption should, however, be made with caution, as environmental factors such as pollution may also negatively affect parasite populations [43]. Pollutant analysis on stored tissues from these dolphins would be valuable.

However, myocardial inflammation and fibrosis as well as meningoencephalitis may affect organ function and therefore be significant for the individual dolphin. Fertility and therefore population dynamics could also be affected by oophoritis, endometritis and orchitis but since one pregnant female had mild metritis, this lesion alone may not impair fertility.

Although autolysis and freezing artefact likely obscured subtle lesions, visible lesions in the respiratory and gastro-intestinal tracts were largely parasitic, as expected in incidentally caught free-ranging animals. Lesions were generally mild compared to those described in other health investigations [38,39,41,44]. The presence of lungworms was less common (20%) than has been reported for stranded *T. truncatus* (77%) and *S. coeruleoalba* (76.5%) from the Northern Hemisphere [45,46]. Eosinophilic pneumonia, even in the absence of visible parasites, was likely parasitic [47,48].

Halocercus spp. are common in the lungs of many dolphin species, although the complete life cycle remains unknown [44,49]. They are generally considered to be of no clinical importance in *T. truncatus* from Florida [45]. Since parasites were recovered more often from calves than from juveniles, and no parasites recovered from adults, the infestation is likely established *in utero* or through milk ingestion [45,49]. Adult animals more often showed only chronic or resolving infections; however, heavily infested adults that died due to parasitism would have been missed in this survey. The variable lymphoplasmacytic inflammation and accompanying follicular lymphoid hyperplasia may indicate the presence of persistent foreign antigen and activation of the adaptive immune response despite clearance of the infestation in older animals [50]. As has been described previously [45], pulmonary interstitial fibrosis was significantly more common in older animals. Interstitial pulmonary fibrosis is a sequel to repetitive, persistent, or severe damage to the endothelial or epithelial cells, inflammation of the alveolar septa, or chronic pulmonary hypertension [51]. In dolphins it has commonly been reported in chronic morbillivirus [52–54] and parasitic infections [44,45]. However, no association between fibrosis and pneumonia or pulmonary verminosis could be demonstrated in this study.

Gastric parasitic nodules due to the trematode *Pholeter gastrophilus* infestation are a common incidental finding in dolphins [49,55,56]. As described previously, nodules were mainly in the pyloric compartment. Nematodes belonging to the family Anisakidae have an indirect life cycle, with animals ingesting infective larvae in infected fish and squid [49]. This likely explains the higher prevalence in juveniles and adults, since calves only become infected once they start consuming fish. Observed species differences in the prevalence of parasitic lesions in the liver, stomach, spleen, lung and lymph nodes may be a result of the small sample size of *S. plumbea*. Alternatively, the parasites that cause these lesions could be host specific due to consumption of different fish and squid species that act as intermediate or paratenic hosts [49]. Of the 94 prey species recorded in *T. aduncus* and 54 prey species in *S. plumbea*, only 25 species are eaten by both *T. aduncus* and *S. plumbea* [57,58]. Changing diet due to changes in prey population dynamics, climate change and or anthropogenic influences may affect parasite loads and is a key topic for future research.

Parasites, including the trematodes *Campula*, *Oschmarinella*, and *Brachycladium* (formerly *Zalophotrema*) which have been found in hepatic ducts, were the most likely cause of the hepatitis and periportal hepatitis in *T. aduncus* [44,49,56]. The life cycle of these brachycladiids is not known [49]. The eosinophilic

oophoritis, endometritis, metritis and orchitis were also probably caused by parasites, supported by the trematode egg present in one case. The positive association with age (up to 100% of adult animals) suggests an indirect life cycle. Small sample size, bias towards younger animals and autolysis precludes a definitive diagnosis of increased bone marrow eosinophilic myelopoiesis; however, a predominance of eosinophilic bone marrow cell lines could reflect the widespread parasitism in these dolphins. Sarcocysts have not previously been reported in dolphins from South African waters, although they have been reported in other cetacean populations [18,49,59–62]

Widespread serosal eosinophilic or fibrotic abdominal serosal lesions were reported to have increased in prevalence in 2009 (*pers. comm.* S. Plön). Similar lesions are described in in domestic horses with *Strongylus* spp migrations, and in domestic pigs due to chronic bacterial serositis. Most of the lesions were chronic with no definitive indication of aetiology. However, parasite larvae were found in the capsules of two mesenteric lymph nodes, and a necrotic tract suggestive of a migration tract was found in another mesenteric lymph node. Lack of association between serosal lesions and pulmonary verminosis, hepatic trematode eggs, or gastric trematodes may be due to the fact that these parasites were not the cause of the lesions, or perhaps due to temporal changes in lesion location and severity over the life cycle of the parasite. Changes in the ecology of food species acting as parasite intermediate hosts could explain the apparent changes in the prevalence of these lesions. Further research is needed on the identity of the parasite, its life cycle and the possible changes in host, environment and prey factors that may influence parasitic loads. Although the inflammatory nature of the splenic serositis resembles that in other abdominal organs, the aetiology of the splenic tags remains uncertain and further research is needed to determine their significance and explain why they are more common in *T. aduncus*, particularly from the southern region.

No histological or immunohistochemical evidence of dolphin morbillivirus infection, brucellosis or toxoplasmosis was found. However, cetacean morbillivirus antibodies were previously found in a *D. delphinus* that stranded approximately 350 km south of the study area [65]. Regrettably, no pathological information is available for this animal and paired serum samples could not be taken to confirm active infection. This population of dolphins may be less susceptible to these diseases than other populations. Alternately, the prevalence of these diseases may have been too low to detect in our study. However, the absence of histological or immunohistochemically stained antigen in the tissues from these dolphins may also have occurred due to poor tissue preservation or loss of antigen integrity due to formalin fixation. The antibody used to detect morbillivirus antigen was a pan-morbillivirus antibody and has been used with success in *Phoca vitulina* (harbour seal) [33], and *S. coeruleoalba* [38]. Commercially available immunohistochemical stains used in this study have been used effectively to detect *T. gondii* in dolphins [66]. The modified ZN (Stamps) stain is an accepted method of demonstrating *Brucella* spp. organisms in tissues [67,68]. This is a crucial area for future research, given the presence of inflammatory lesions compatible with these diseases and their worldwide distribution. Continued monitoring of these dolphin populations is needed as reliable detection of infectious agents present at low prevalence can only be accomplished by testing larger numbers of animals but access to live free-ranging coastal dolphins is limited [65,69]. Microbiological culture and biotyping of brain, spleen and reproductive tract isolates will be conducted in future. Serological and molecular diagnostic tests for *Brucella* spp. and *T. gondii* are also needed. If these dolphin populations are in fact naïve to these pathogens, their introduction could have devastating consequences, as has been documented previously in other populations elsewhere during morbillivirus epidemics [5,52,54,70–73].

No animals had lesions consistent with bacterial pneumonia and no primary bacterial pathogens were isolated from the lung. However, autolysis and freezing may have compromised culture success. Isolation of opportunistic bacteria such as *Aeromonas media* and *Photobacterium damselae* is consistent with previous reports [72]. *Shewanella algae* is commonly isolated from marine environments, and is an opportunistic human pathogen [74]. Remaining bacteria were considered contaminants or normal commensals.

Granulomatous dermatitis associated with fungi is consistent with the zoonotic disease lobomycosis [5,75,76]. This is, to our knowledge, the first confirmed report of lobomycosis in South African waters, although macroscopic lobomycosis-like disease has been documented in other Indian Ocean populations of *T. aduncus* [77]. Impaired adaptive immunity was found in endemically affected *T. truncatus* from the Indian River Lagoon, Florida [78]. The exact aetiology of the immunosuppression in dolphins has not yet been determined, but both environmental contaminants, such as mercury and polychlorinated biphenyls, and chronic stress as result of anthropogenic factors have been suggested [76,78]. No evidence of immunosuppression was found histologically in the dolphins in this study, although differential white cell counts, determination of lymphocyte subpopulations, phagocytic activity and lysozyme activity, amongst other tests [78], were not possible in incidentally caught animals.

While some variation in the width of the adrenal cortex and occasional cortical nodules were seen in the cortex or medulla in these animals, such variation could have been due to differing planes of section. Blood and faecal adrenocortical hormone assays, adrenal weights and objective measurement of adrenal cortico-medullary ratios by point-counting techniques [32] as well as systematic evaluation of the pituitary are needed to evaluate the possibility of stress in this dolphin population. Adrenal hyperplasia has been attributed to chronic stress from long-term debilitating disease or injury in *T. truncatus* in the Gulf of Mexico [32,79]; however, the animals in this study had relatively mild pathology. Environmental stressors, such as competition for resources, and anthropogenic factors, such as boat traffic, seismic or military activities warrant evaluation. Myocardial fibrosis is a non-specific indication of prior tissue damage due to inflammation or necrosis. Myocardial necrosis and fibrosis in stranded and incidentally caught *T. truncatus* and *S. coeruleoalba* from the Gulf of Mexico were attributed to the acute and chronic effects, respectively, of high catecholamine levels [79]. The association of cardiac fibrosis with age may indicate that the effects are cumulative. Cardiac fibrosis was not associated with myocarditis in *T. aduncus*; however, the small sample size precludes definitive conclusions on the aetiology of either lesion. Similarly, the small sample size, including only one adult *S plumbea*, may account for the absence of epicarditis, endocarditis or myocarditis seen in this species. Although mild cutaneous depigmentation (#1), lacerations (#6, 26, 31), and barnacles (#3) were documented, inter and intra-specific aggression could not be reliably distinguished from boat strike or other anthropogenic injury.

The higher numbers of *T. aduncus*, caught in the nets all along the coast likely reflects the relative population size and more widespread distribution of this species [21,22]. All five *S. plumbea* were caught on two adjacent beaches in the northern region (Figure 1), where they occur in higher numbers than in the south [80,81]. The fact that calves and juveniles are more inquisitive and inexperienced may explain why *T. aduncus* calves and juveniles

were caught more often than adults [82]. Females with calves also feed closer to shore, and therefore to the nets, which results in higher capture rates of adult females and calves [64,83].

Mineralization of the bronchiolar epithelium has previously been attributed to lungworm infection [85,86]. Bronchiolar mineralization is not a common feature of verminous pneumonia in cetaceans [38,44], but is occasionally seen in harbour porpoises from the North Sea (P. Wohlsein, *pers. comm.*). Foreign particles are thought to accumulate in the lung due to the inability of dolphins to cough. These particles become inspissated, undergo calcification and are later incorporated into the bronchial wall [85]. Additional investigations are underway to determine the distribution and exact location of the material. Small foci of mineralisation were present in 24 dolphins in a wide range of tissues, in addition to the airways (Table S1). In mammals, metastatic tissue mineralisation due to disturbed calcium and phosphorus metabolism typically occurs on the intercostal pleura, pulmonary and renal cortical basement membrane, and the middle and deep gastric mucosa [90]. Since these sites were not involved, and no indication of renal failure, neoplasia, or granulomatous inflammation that could result in secondary hyperparathyroidism were present, the mineralisation seen in these dolphins was assumed to be dystrophic changes due to minor tissue damage. However, since neither pituitary nor parathyroid glands were routinely sampled we cannot rule out the possibility of altered calcium homeostasis in these dolphins. We consider that nutritional hyperparathyroidism (due to altered calcium, phosphate or Vitamin D metabolism) is unlikely to be common in free-ranging animals; and cannot rule out the possibility of emerging secondary marine plant intoxication through ingestion of herbivorous fish.

Conclusion

In the first systematic health assessment of incidentally caught coastal dolphins in the Southern Hemisphere, we report the first confirmed cases of lobomycosis and sarcocystosis in dolphins from the South African coast. While optimum samples are not provided by frozen, incidentally caught animals, this study still yielded valuable information on the current prevalence of disease in the two dolphin populations, which can be used as a baseline for future monitoring projects, not only of the health status of the population, but also that of the environment. This may prove particularly important for *S. plumbea*, whose coastal habitat, restricted distribution range, and small population size make it prone to a number of threats, including anthropogenic impacts. These findings further highlight the importance of disease investigation in marine mammals.

Supporting Information

Table S1 Summary of mild, moderate and severe lesions and overall health status for each of 35 Indian Ocean bottlenose (T. aduncus) and five Indo-Pacific humpback (S. plumbea) dolphins incidentally caught in shark nets along the KwaZulu-Natal coast, South Africa, 2010-2012.

Table S2 Complete pathological findings for indicating occurrence (lesion/number of organ evaluated) and percentage per species, age group, and region (for both species combined).

Table S3 Common pathology observed in Tursiops aduncus and associations with sex, age and region.

Acknowledgments

The authors would also like to thank the staff and students of the Port Elizabeth Museum, in particular Dr. Greg Hofmeyr; staff of the National Zoological Gardens of South Africa; staff of the KwaZulu-Natal Sharks Board, in particular Geremy Cliff; laboratory staff of the Department of Pathology, Faculty of Veterinary Science, University of Pretoria, and the Department of Pathology, University of Veterinary Medicine, Hannover, Foundation, Germany. Particular thanks go to Dr. David Zimmerman for necropsy data and sampling in 2010; Dr. Maryke Henton (IDEXX South Africa) for culture and identification of the bacteria; Drs. Kerstin Junker (Agricultural Research Council – Onderstepoort Veterinary Institute, Pretoria, South Africa) and Kristina Lehnert (Institute of Terrestrial and Aquatic Wildlife, University of Veterinary Medicine, Hannover, Foundation, Germany) for parasite identification; and Dr. Ingrid de Wet for help during dissections.

Disclaimer

Any opinion, findings and conclusions or recommendations expressed in this material are those of the author(s) and therefore the NRF does not accept any liability in regard thereto.

Author Contributions

Conceived and designed the experiments: MdW EPL US PW PT SP. Performed the experiments: MdW EPL US PW SP. Analyzed the data: MdW EPL PW PT. Contributed reagents/materials/analysis tools: EPL PT SP. Wrote the paper: MdW EPL US PW PT SP.

References

1. Ryser-Degiorgis M (2013) Wildlife health investigations: Needs, challenges and recommendations. BMC Vet Res 9: 223–240.
2. Bossart GD (2006) Marine mammals as sentinel species for oceans and human health. Oceanography 19: 134–137.
3. Harvell CD, Mitchell CE, Ward JR, Altizer S, Dobson AP, et al. (2002) Climate warming and disease risks for terrestrial and marine biota. Science 296: 2158–2162.
4. Gulland FMD, Hall AJ (2007) Is marine mammal health deteriorating? Trends in the global reporting of marine mammal disease. Ecohealth 4: 135–150.
5. Van Bressem MF, Raga JA, Guardo G, Jepson PD, Duignan PJ, et al. (2009) Emerging infectious diseases in cetaceans worldwide and the possible role of environmental stressors. Dis Aquat Org 86: 143–157.
6. Raga JA, Banyard A, Domingo M, Corteyn M, Van Bressem MF, et al. (2008) Dolphin morbillivirus epizootic resurgence, Mediterranean Sea. Emerg Infect Dis 14: 471–473.
7. Harvell CD, Kim K, Burkholder JM, Colwell RR, Epstein PR, et al. (1999) Emerging marine diseases - climate links and anthropogenic factors. Science 285: 1505–1510.
8. Ward JR, Lafferty KD (2004) The elusive baseline of marine disease: Are diseases in ocean ecosystems increasing? PLoSBiol 2, e120 2: 542–546.
9. Epstein RP, Sherman D, Spanger-Siegfried E, Langston A, Prasad S (1998) Marine ecosystems: Emerging diseases as indicators of change. Boston: Harvard Medical School. MA. 85 p.
10. Geraci JR, Lounsbury VJ (2009) Health. In: Perrin WF, Würsig B, Thewissen JGM, editors. Encyclopedia of Marine Mammals (second edition). London: Academic Press. pp.546–553.
11. Bar K, Slooten E (1999) Effects of tourism on dusky dolphins at Kaikoura. Conservation Advisory Science Notes 229: 5–10.
12. Lafferty KD, Porter JW, Ford SE (2004) Are diseases increasing in the ocean? Annu Rev Ecol Evol Syst 35: 31–54.
13. Wells RS, Rhinehart HL, Hansen LJ, Sweeney JC, Townsend FI, et al. (2004) Bottlenose dolphins as marine ecosystem sentinels: Developing a health monitoring system. EcoHealth 1: 246–254.
14. Reddy ML, Dierauf LA, Gulland FMD (2001) Marine mammals as sentinels of ocean health. In: Dierauf LA, Gulland FMD, editors. CRC Handbook of marine mammal medicine. Boca Raton: CRC Press Inc. pp.3–13.
15. O'Shea TJ, Bossart GD, Fournier M, Vos JG (2003) Conclusions and perspectives for the future. In: Vos JG, Bossart GD, Fournier M, O'Shea TJ, editors. Toxicology of Marine Mammals. New York: Taylor & Francis. pp.595–613.
16. Riva GT, Johnson CK, Gulland FMD, Langlois GW, Heyning JE, et al. (2009) Association of an unusual marine mammal mortality event with Pseudo-nitzschia spp. blooms along the southern California coastline. J Wildl Dis 45: 109–121.
17. Flewelling LJ, Naar JP, Abbott JP, Baden GD, Barros NB, et al. (2005) Red tides and marine mammal mortalities. Nature 435: 755–756.
18. Dubey JP, Zarnke R, Thomas NJ, Wong SK, Bonn WV, et al. (2003) Toxoplasma gondii, Neospora caninum, Sarcocystis neurona, and Sarcocystis canis-like infections in marine mammals. Vet Parasitol 116: 275–296.
19. Miller MA, Gardner IA, Kreuder C, Paradies DM, Worcester KR, et al. (2002) Coastal freshwater runoff is a risk factor for Toxoplasma gondii infection of southern sea otters (Enhydra lutris nereis). Int J Parasitol 32: 997–1006.
20. Best PB (2007) Whales and dolphins of the Southern African subregion. Cape Town: Cambridge University Press. 352 p.
21. Cockcroft VG, Ross GJB (1990) Age, growth and reproduction in bottlenose dolphins (Tursiops truncates) from the east coast of Southern Africa. Fish B-NOAA 88: 289–302.
22. Cockcroft VG, Ross GJB, Peddemors VM (1990) Bottlenose dolphin Tursiops truncatus distribution in Natal's coastal waters. S Afr J Marine Sci 9: 1–10.
23. Karczmarski L, Cockcroft VG, Mclachlan A (2000) Habitat use and preferences of Indo-Pacific humpback dolphins Sousa chinensis in Algoa Bay, South Africa. Mar Mamm Sci 16: 65–79.
24. Cockcroft VG (1994) Is there common cause for dolphin captures? A review of dolphin catches in shark nets off Natal, South Africa. Report of the international whaling commission Special issue 15: 541–547.
25. KwaZulu-Natal Sharks Board (2009) Catch statistics. http://shark.co.za, last accessed 15 January 2013.
26. Norris KS (1961) Standardized methods for measuring and recording data on the smaller cetaceans. J Mammal 42: 471–476.
27. Geraci JR, Lounsbury VJ (2005) Specimen and data collection. In: Geraci JR, Lounsbury VJ, editors. Marine mammals ashore: A field guide for strandings (second edition). Baltimore, Maryland: National Aquarium in Baltimore. pp.167–251.
28. Jefferson TA, Hung SK, Robertson KM, Archer FI (2012) Life history of the Indo-Pacific humpback dolphin in the pearl river estuary, southern china. Mar Mamm Sci 28: 84–104. Adapted by Z. Nolte, Rhodes University.
29. Young DD (1998) Aspects of condition in captive and free-ranging dolphins. PhD thesis. Grahamstown, Rhodes University. 436 p.
30. De Wet M (2013) A systematic health assessment of two dolphin species by-caught in shark nets off the KwaZulu-Natal coast, South Africa. MSc Thesis. Pretoria: University of Pretoria. 152 pp.
31. Böck P, Romeis B (1989) Romeis mikroskopische technik. Munchen, Germany: Publisher Urban and Schwarzenberg. 697 p.
32. Clark LS, Cowan DF, Pfeiffer DC (2006) Morphological changes in the Atlantic bottlenose dolphin (Tursiops truncatus) adrenal gland associated with chronic stress. J Comp Pathol 135: 208–216.
33. Stimmer L, Siebert U, Wohlsein P, Fontaine JJ, Baumgärtner W, et al. (2010) Viral protein expression and phenotyping of inflammatory responses in the central nervous system of phocine distemper virus-infected harbour seals (Phoca vitulina). Vet Microbiol 145: 23–33.
34. Lehnert K, Raga JA, Siebert U (2007) Parasites in harbour seals (Phoca vitulina) from the German Wadden Sea between two phocine distemper virus epidemics. Helgol Mar Res 61: 239–245.
35. Peddemors VM (1999) Delphinids of Southern Africa: A review of their distribution, status and life history. J Cetac Res Manage 1: 157–165.
36. Natoli A, Peddemors VM, Hoelzel AR (2008) Population structure of bottlenose dolphins (Tursiops aduncus) impacted by by-catch along the east coast of South Africa. Conserv Genet 9: 627–636.
37. Kuiken T, Simpson VR, Allchin CR, Bennett PM, Codd GA, et al. (1994) Mass mortality of common dolphins (Delphinus delphis) in south-west England due to incidental capture in fishing gear. Vet Rec 134: 81–89.
38. Siebert U, Wünschmann A, Weiss R, Frank H, Benke H, et al. (2001) Post-mortem findings in harbour porpoises (Phocoena phocoena) from the German North and Baltic Seas. J Comp Pathol 124: 102–114.
39. Duignan PJ (2003) Disease investigations in stranded marine mammals, 1999–2002. Department of Conservation, Science Internal Series. 32 p.
40. Siebert U, Joiris C, Holsbeek L, Benke H, Failing K, et al. (1999) Potential relation between mercury concentrations and necropsy findings in cetaceans from Gerrman waters of the North and Baltic Seas. Mar Pollut Bull 38: 285–295.
41. Siebert U, Tolley K, Víkingsson GA, Ólafsdottir D, Lehnert K, et al. (2006) Pathological findings in harbour porpoises (Phocoena phocoena) from Norwegian and Icelandic waters. J Comp Pathol 134: 134–142.
42. Roe WD, Gartrell BD, Gartrell BD, Hunter SA (2012) Freezing and thawing of pinniped carcasses results in artefacts that resemble traumatic lesions. Vet J 194: 326–331.
43. Torchin ME, Lafferty KD, Kuris AM (2002) Parasites and marine invasions. Parasitology 124: S137–S151.
44. Jauniaux T, Petitjean D, Brenez C, Borrens M, Borrens L, et al. (2002) Postmortem findings and causes of death of harbour porpoises (Phocoena phocoena) stranded from 1990 to 2000 along the coastlines of Belgium and northern France. J Comp Pathol 126: 243–253.

45. Fauquier DA, Kinsel MJ, Dailey MD, Sutton GE, Stolen MK, et al. (2010) Prevalence and pathology of lungworm infection in bottlenose dolphins Tursiops truncatus from southwest Florida. Dis Aquat Org 88: 85–90.

46. Cornaglia E, Rebora L, Gili C, Guardo G (2000) Histopathological and immunohistochemical studies on cetaceans found stranded on the coast of Italy between 1990 and 1997. J Vet Med A 47: 129–142.

47. Van Dijk JE, Gruys E, Mauwen JMVM (2007) Colour atlas of veterinary pathology. Spain: Sauderns Elsevier. 200 p.

48. Bossart GD, Reidarson TH, Dierauf LA, Duffield DA (2001) Clinical pathology. In: Dierauf LA, Gulland FMD, editors. CRC Handbook of marine mammal medicine. Boca Raton: CRC Press Inc. pp.383–436.

49. Raga JA, Fernández M, Balbuena JA, Aznar FJ (2009) Parasites. In: Perrin WF, Würsig B, Thewissen JGM, editors. Encyclopedia of Marine Mammals (second edition). London: Academic Press. pp.821–830.

50. King DP, Aldridge BM, Kennedy-Stoskopf S, Scott JT (2001) Immunology. In: Dierauf LA, Gulland FMD, editors. CRC Handbook of marine mammal medicine. Boca Raton: CRC Press Inc. pp.237–252.

51. Caswell JL, Williams KJ (2007) Lungs. In: Maxie MG, editor. Jubb, Kennedy & Palmer's pathology of domestic animals (fifth edition). Edinburgh: Elsevier. pp.540–575.

52. Domingo M, Visa J, Pumarola M, Marco AJ, Ferrer L, et al. (1992) Pathologic and immunocytochemical studies of morbillivirus infection in striped dolphins (Stenella coeruleoalba). Vet Pathol 29: 1–10.

53. Kennedy S (1998) Morbillivirus infections in aquatic mammals. J Comp Pathol 119: 201–225.

54. Lipscomb TP, Kennedy S, Moffett D, Krafft A, Klaunberg BA, et al. (1996) Morbilliviral epizootic in bottlenose dolphins of the Gulf of Mexico. J Vet Diagn Invest 8: 283–290.

55. Aznar FJ, Fognani P, Balbuena JA, Pietrobelli M, Raga JA (2006) Distribution of Pholeter gastrophilus (digenea) within the stomach of four odontocete species: The role of the diet and digestive physiology of hosts. Parasitology 133: 369–380.

56. Geraci JR, St. Aubin DJ (1987) Effects of parasites on marine mammals. Int J Parasitol 17: 407–414.

57. Venter K (2009) Diet of humpback dolphins (Sousa plumbea) along the south-eastern coast of South Africa.

58. Kaiser SML (2012) Feeding ecology and dietary patterns of the Indo-Pacific bottlenose dolphin (Tursiops aduncus) incidentally caught in the shark nets off KwaZulu-Natal, South Africa. MSc thesis. Port Elizabeth: Nelson Mandela Metropolitan University. 68 p.

59. Daily M, Stroud R (1978) Parasites and associated pathology observed in cetaceans stranded along the Oregon coast. J Wildl Dis 14: 503–511.

60. Munday BL, Mason RW, Hartley WJ, Presidente PJ, Obendorf D (1978) Sarcocystis and related organisms in Australian wildlife: survey findings in mammals. J Wildl Dis 14: 417–433.

61. Resendes AR, Juan-Salls C, Almeria S, Maj N, Domingo M, et al. (2002) Hepatic sarcocystosis in a striped dolphin (Stenella coeruleoalba) from the Spanish Mediterranean coast. J Parasitol 88: 206–209.

62. Lehnert K, Seibel H, Hasselmeier IWP, Iversen M, Nielsen NH, et al. Change in parasite burden and associated pathology in harbour porpoises (Phocoena phocoena) in west Greenland. In Press.

63. Brown CC, Baker DC, Barker IK (2007) Peritoneum and retroperitoneum. In: Maxie MG, editor. Jubb, Kennedy & Palmer's pathology of domestic animals (fifth edition). Edinburgh: Elsevier. pp.279–296.

64. Cockcroft VG, Ross GJB (1990) Food and feeding of the Indian Ocean bottlenose dolphin off southern Natal, South Africa. In: Leatherwood S, Reeves RR, editors. The bottlenose dolphin. San Diego : Academic Press. pp.295–308.

65. Van Bressem MF, Waerebeek K, Jepson PD, Raga JA, Duignan PJ, et al. (2001) An insight into the epidemiology of dolphin morbillivirus worldwide. Vet Microbiol 81: 287–304.

66. Di Guardo G, Proietto U, Francesco CE, Marsilio F, Zaccaroni A, et al. (2010) Cerebral toxoplasmosis in striped dolphins (Stenella coeruleoalba) stranded along the Ligurian Sea coast of Italy. Vet Pathol 47: 245–253.

67. Alton GG, Jones LM, Pietz DE (1975) Laboratory techniques in brucellosis. Geneva: World Health Organization. 80 p.

68. Foster G, MacMillan AP, Godfroid J, Howie F, Ross HM, et al. (2002) A review of Brucella sp. infection of sea mammals with particular emphasis on isolates from Scotland. Vet Microbiol 90: 563–580.

69. Dubey JP, Fair PA, Bossart GD, Hill DFR, Sreekumar C, et al. (2005) A comparison of several serologic tests to detect antibodies to Toxoplasma gondii in naturally exposed bottlenose dolphins (Tursiops truncatus). J Parasitol 91: 1074–1081.

70. Calzada N, Lockyer C, Aguilar A (1994) Age and sex composition of the striped dolphin die-off in the western Mediterranean. Mar Mamm Sci 10: 299–310.

71. Di Guardo G, Marruchella G, Agrimi U, Kennedy S (2005) Morbillivirus infections in aquatic mammals: a brief overview. J Vet Med A 52: 88–93.

72. Keck N, Kwiatek O, Dhermain F, Dupraz F, Boulet H, et al. (2010) Resurgence of morbillivirus infection in Mediterranean dolphins off the French coast. Vet Rec 166: 654–655.

73. Lipscomb TP, Schulman FY, Moffett D, Kennedy S (1994) Morbilliviral disease in Atlantic bottlenose dolphins (Tursiops truncatus) from the 1987–1988 epizootic. J Wildl Dis 30: 567–571.

74. Tsai M, You H, Tang Y, Liu J (2008) Shewanella soft tissue infection: case report and literature review. Int J Infect Dis 12: e119–124.

75. Higgens R (2000) Bacteria and fungi of marine mammals: a review. Canadian Vet J 41: 105–116.

76. Reif JS, Mazzoil MS, McCulloch SD, Varela RA, Goldstein JD, et al. (2006) Lobomycosis in Atlantic bottlenose dolphins from the Indian River Lagoon, Florida. J Am Vet Med Assoc 228: 104–108.

77. Kiszka J, Van Bressem MF, Pusineri C (2009) Lobomycosis-like disease and other skin conditions in Indo-Pacific bottlenose dolphins, Tursiops aduncus, from the Indian Ocean. Dis Aquat Org 84: 151–157.

78. Reif JS, Peden-Adams M, Romano TA, Rice CD, Fair PA, et al. (2009) Immune dysfunction in Atlantic bottlenose dolphins (Tursiops truncatus) with lobomycosis. Med Mycol 47: 125–135.

79. Turnbull BS, Cowan DF (1998) Myocardial contraction band necrosis in stranded cetaceans. J Comp Pathol 118: 317–327.

80. Atkins S, Cliff G, Pillay N. (2013) Humpback dolphin by-catch in the shark nets in KwaZulu-Natal, South Africa. Biol Conserv 159: 442–449.

81. Durham B (1994) The distribution and abundance of humpback dolphin (Sousa chinensis) along the Natal coast, South Africa. MSc thesis. Durban: University of Natal. 32 p.

82. Peddemors VM (1995) The aetiology of bottlenose dolphin capture in shark nets off Natal, South Africa. PhD thesis. Port Elizabeth: University of Port Elizabeth. 83 p.

83. Cockcroft VG (1992) Incidental capture of bottlenose dolphins (Tursiops truncatus) in shark nets: an assessment of some possible causes. J Zool 226: 123–134.

84. Moore MJ, Bogomolni AL, Dennison SE, Early G, Garner MM, et al. (2009) Gas bubbles in seals, dolphins, and porpoises entangled and drowned at depth in gillnets. Vet Pathol 46: 536–547.

85. Woodard JC, Zam SG, Caldwell DK, Caldwell MC (1969) Some parasitic diseases of dolphins. Vet Pathol 6: 257–272.

86. Zappulli V, Mazzariol S, Cavicchioli L, Petterino C, Bargelloni L, et al. (2005) Fatal necrotizing fasciitis and myositis in a captive common bottlenose dolphin (Tursiops truncatus) associated with Streptococcus agalactiae. J Vet Diagn Invest 17: 617–622.

87. KwaZulu-Natal Sharks Board (2011) Shark nets, drumlins and safe swimming. Available: http://shark.co.za. Accessed 2013 Jan 15.

88. Bernaldo de Quirós Y, González-Díaz O, Arbelo M, Sierra E, Sacchini S, et al. (2012) Decompression vs. decomposition: distribution, amount, and gas composition of bubbles in stranded marine mammals. Frontiers in Physiology 3 (177): 1–19.

89. Bernaldo de Quirós Y, González-Díaz O, Saavedra P, Arbelo M, Sierra E, et al. (2011) Methodology for in situ gas sampling, transport and laboratory analysis of gases from stranded cetaceans. Scientific Reports 1 (193): 1–10.

90. Maxie MG, Newman SJ (2007) Urinary System. In: Maxie MG, editor. Jubb, Kennedy & Palmer's pathology of domestic animals (fifth edition). Edinburgh: Elsevier. pp.433–436.

7

Assessing the Underwater Acoustics of the World's Largest Vibration Hammer (OCTA-KONG) and Its Potential Effects on the Indo-Pacific Humpbacked Dolphin (*Sousa chinensis*)

Zhitao Wang[1,2¤], **Yuping Wu**[3], **Guoqin Duan**[4], **Hanjiang Cao**[4], **Jianchang Liu**[5], **Kexiong Wang**[1]*, **Ding Wang**[1]*

1 The Key Laboratory of Aquatic Biodiversity and Conservation of the Chinese Academy of Sciences, Institute of Hydrobiology, Chinese Academy of Sciences, Wuhan, P. R. China, 2 University of Chinese Academy of Sciences, Beijing, P. R. China, 3 School of Marine Sciences, Sun Yat-sen University, Guangzhou, P. R. China, 4 Hongkong-Zhuhai-Macao Bridge Authority, Guangzhou, P. R. China, 5 Transport Planning and Research Institute, Ministry of Transport, Beijing, P. R. China

Abstract

Anthropogenic noise in aquatic environments is a worldwide concern due to its potential adverse effects on the environment and aquatic life. The Hongkong-Zhuhai-Macao Bridge is currently under construction in the Pearl River Estuary, a hot spot for the Indo-Pacific humpbacked dolphin (*Sousa chinensis*) in China. The OCTA-KONG, the world's largest vibration hammer, is being used during this construction project to drive or extract steel shell piles 22 m in diameter. This activity poses a substantial threat to marine mammals, and an environmental assessment is critically needed. The underwater acoustic properties of the OCTA-KONG were analyzed, and the potential impacts of the underwater acoustic energy on *Sousa*, including auditory masking and physiological impacts, were assessed. The fundamental frequency of the OCTA-KONG vibration ranged from 15 Hz to 16 Hz, and the noise increments were below 20 kHz, with a dominant frequency and energy below 10 kHz. The resulting sounds are most likely detectable by *Sousa* over distances of up to 3.5 km from the source. Although *Sousa* clicks do not appear to be adversely affected, *Sousa* whistles are susceptible to auditory masking, which may negatively impact this species' social life. Therefore, a safety zone with a radius of 500 m is proposed. Although the zero-to-peak source level (SL) of the OCTA-KONG was lower than the physiological damage level, the maximum root-mean-square SL exceeded the cetacean safety exposure level on several occasions. Moreover, the majority of the unweighted cumulative source sound exposure levels (SSELs) and the cetacean auditory weighted cumulative SSELs exceeded the acoustic threshold levels for the onset of temporary threshold shift, a type of potentially recoverable auditory damage resulting from prolonged sound exposure. These findings may aid in the identification and design of appropriate mitigation methods, such as the use of air bubble curtains, "soft start" and "power down" techniques.

Editor: Michael L. Fine, Virginia Commonwealth Univ, United States of America

Funding: The research was supported by grants from the Ministry of Science and Technology of China (2011BAG07B05-3), the National Natural Science Foundation of China (31170501 and 31070347), the Knowledge Innovation Program of the Chinese Academy of Sciences (KSCX2-EW-Z-4), the State Oceanic Administration of China (201105011-3) and the Special Fund for Agro-scientific Research in the Public Interest of the Ministry of Agriculture of China (201203086). The funders had no role in designing the study, collecting data, analyzing data, preparing the manuscript or deciding where to publish the manuscript.

Competing Interests: The authors have declared that no competing interests exist.

* Email: wangk@ihb.ac.cn (KW); wangd@ihb.ac.cn (DW)

¤ Current address: Marine Mammal Research Program, Hawaii Institute of Marine Biology, University of Hawaii, Kaneohe, Hawaii, United States of America

Introduction

Over the past few decades, anthropogenic (human-generated) noise in aquatic environments has generated worldwide concern due to its potential adverse effects on the environment and aquatic life [1–5]. Of particular concern are the intense impulsive sounds from explosive detonations, seismic surveys and pile driving, common activities in the construction of renewable-energy marine wind farms, docks and bridges. The effects on marine mammals have been of particular interest [6–9]. This concern is partly due to the protected status of marine mammals under state laws and international conventions, such as the Convention on International Trade in Endangered Species of Wild Fauna and Flora, as well as their vulnerability to ambient noise. Cetaceans have a sophisticated acoustic sensory system with wideband hearing sensitivity [10], and they are heavily dependent on the acoustic environment for many life functions. They have evolved sophisticated vocalizations and multiple sound-reception pathways, and they rely on acoustic stimuli for social interaction, navigation and foraging in the marine environment [10].

The Greater Pearl River Delta is one of the most economically developed regions in China [11]. However, land transport between its western (such as the Zhuhai and Macao Special Administrative Regions) and eastern regions (such as the Hongkong Special Administrative Region) is limited by the Pearl

River Estuary (PRE). To increase the region's economic competitiveness and to facilitate economic collaboration, e.g., by reducing the costs involved in transporting people and goods between the regions, the Hongkong-Zhuhai-Macao Bridge (HZMB) is being constructed to connect these three cities. The HZMB Island Tunnel Project is a large-scale, cross-boundary sea crossing involving more than 300 supporting bridge piles, an underwater tunnel and two artificial islands (Fig. 1). Construction began on 15 December 2009 and is expected to continue into 2016 [12].

The PRE (22°16′S; 113°43′E) is a hot spot for the Indo-Pacific humpbacked dolphin (*Sousa chinensis*, locally called the Chinese white dolphin), which is distributed in shallow coastal waters from South Africa in the west to southern China in the east [13]. This species is currently assessed as Near Threatened; the *chinensis*-type geographic form (found from the east coast of India to China) is categorized as Vulnerable by the International Union for the Conservation of Nature Red List of Threatened Species [13] and as a Grade One National Key Protected Animal by China's Wild Animal Protection Law, issued in 1988. The population size of humpbacked dolphins in the PRE was estimated to be 2555 and 2517 during the wet and dry seasons, respectively [14], representing the largest known humpbacked dolphin population in China [15,16] and the world [14,17]. To better protect the dolphin population, the Pearl River Estuary Chinese White Dolphin National Nature Reserve (PRECWDNNR) (Fig. 1) was established in the PRE in 1999.

Unfortunately, the HZMB, which uses thousands of piles driven into the bottom of the estuary, is located across the PRECWDNNR. To minimize any adverse effects on protected species, the following strategies were adopted: (1) An underwater tunnel was designated to replace a bridge structure in the core area of the reserve (Fig. 1). (2) A marine mammal safety zone (an exclusion zone of 200 m radius [18]) was established in the vicinity of the bridge construction sites. Qualified marine mammal observers scan for the presence of marine mammals within the exclusion zone. If marine mammals are observed in the safety zone, operations halt until the animals have left the zone. (3) An acoustic deterrent device (Future Oceans 70 kHz Dolphin Pinger; Future Oceans, Queensland, Australia) that emits a 145 dB signal for 300 m every 4 s is used to warn any marine mammals away

from the safety zone both before and during construction. (4) A hydraulic vibration hammer is used for pile driving in addition to an impact hammer, which generates substantially louder impulse sounds.

However, due to a limited understanding of the sound produced by the construction activities, the safety zone was not established based on robust experimental or theoretical information. Efforts to protect animals are generally hampered when only limited data are available for establishing criteria for interim protection. Research on the characteristics of the underwater sound field produced during bridge construction is needed. In particular, pile driving, which produces loud underwater sounds, requires study to improve environmental impact analyses and aid in the identification and design of appropriate mitigation methods [19].

Compared with the conventionally used impact hammers, the vibratory hammer is a much more economical tool for construction companies [20]. In addition to its ability to extract piles, other advantages include (1) a lighter weight than conventional hammers, (2) faster operation at a lower noise level than conventional hammers and (3) lack of requirement of a temporary guide frame for driving free-standing piles [20]. Accordingly, it represents an alternative tool, or a complementary tool, to impact hammers from a conservation perspective.

In the waters of western Hong Kong, *S. chinensis* has been observed to travel at higher speeds during percussive pile driving. Moreover, the animals tend to partially and temporarily abandon the pile driving area [21]. Given that the peak pressure levels produced by a normal vibration hammer during the driving of normal, cast-in-steel-shell piles range from approximately 175 dB to 205 dB [19], the OCTA-KONG (American Piledriving Equipment Inc., Kent, WA, USA), which is the world's largest vibration hammer and is capable of driving and extracting piles, may impose a substantial threat to marine mammals. The use of this hammer further emphasizes the need for an assessment of underwater noise in and around the HZMB.

The present study had two main purposes. The first was to characterize the acoustic properties of the operating sounds of the OCTA-KONG, including pile driving and extraction. The second was to assess the potential impacts of this anthropogenic noise on *Sousa* with respect to three factors: *Sousa* sounds (whistles and

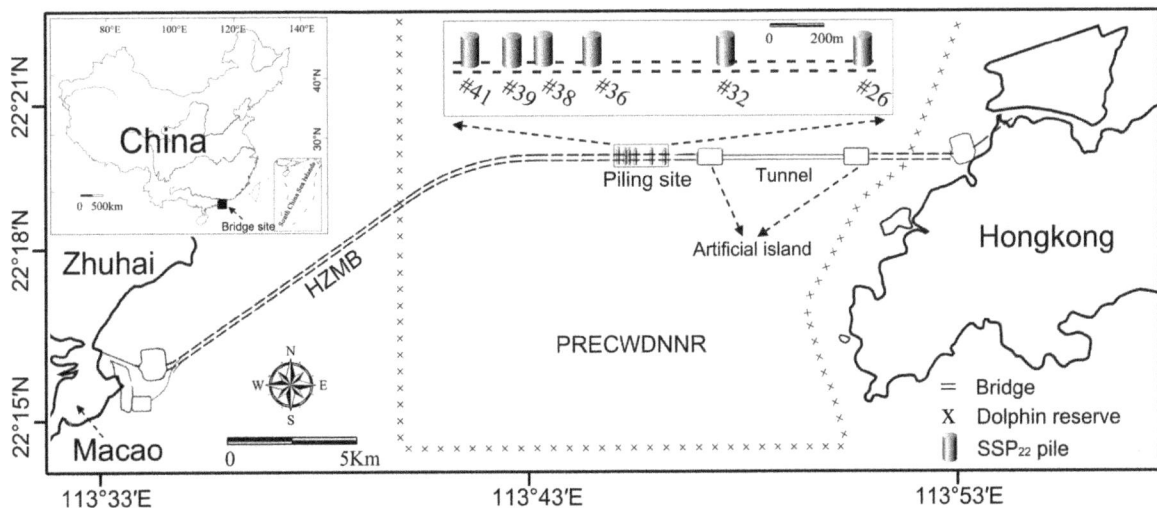

Figure 1. Map of the OCTA-KONG vibration monitoring area. HZMB: Hongkong-Zhuhai-Macao Bridge; PRECWDNNR: The Pearl River Estuary Chinese White Dolphin National Nature Reserve. The eastern boundary of the PRECWDNNR is also the boundary of the Zhuhai and Hongkong Special Administrative Regions. An exclusion zone of 200 m radius was established along the bridge.

Table 1. Specification of the OCTA-KONG hammer, power unit and SSP_{22} pile.

Hammer		Power unit		Pile	
Type	OCTA-KONG	Type	CAT32	Type	Steel shell pile
Total Drive Force	40000 000 N	Maximum power	882 600 W(1200 HP)	Diameter	22 m
Frequency	6.67 Hz–23.33 Hz (400–1400 vpm)	Operating speed	800 r/min to 2050 r/min(rpm)	Pile wall width	0.016 m
Pile clamp force	1176 000 N*	Maximum drive pressure	33 096 Pa	Height	39 m–60 m
Line pull for extraction	3131 000 N*	Clamp pressure	33 096 Pa	Weight	450 000 kg–600 000 kg

OCTA-KONG: a Multiple Linked Vibro System with 8× APE 600's connected in a tandem combination; vpm = vibrations per minute; rpm = revolutions per minute; hp = horsepower. * indicates the results for each APE 600 hammer. The rpm of the hammer was controlled by the vpm of the power unit and was approximately vpm/1.44.

clicks) recorded in the same district during a previous dolphin acoustic survey by the first author; *Sousa* audiograms [22,23]; the safety exposure level established by the U.S. National Marine Fisheries Service (NMFS) [24] and the marine mammal noise exposure criteria proposed by a panel of experts from a wide range of disciplines in acoustic research [7] and the National Oceanic and Atmospheric Administration (NOAA)[25].

Methods

Vibration piling

The OCTA-KONG is the world's largest hydraulic vibratory driver/extractor. It consists of a Multiple Linked Vibro System with 8 APE 600's connected in a tandem combination (Table 1, Fig. 2), with each APE 600 powered by a Model 1200 power unit (Fig. 2). The OCTA-KONG was used to drive 22 m diameter steel shell piles (SSP_{22}) during the construction of the main wall of the two artificial islands (Fig. 1) from 15 May 2011 to 25 December 2011, and more than 120 SSP_{22} were installed. It was subsequently used to drive and/or pull SSP_{22} at the construction sites of the bridge piers from #16 to #53 and from #60 to #89 beginning 15 October 2012; the estimated completion date is in June 2015 (Zeng TQ, personal communication).

Vibration piling components. The major components of the hammer are as follows: (1) the suppressor housing (bias-weight), with a rubber elastomer isolated suppressor; (2) the

Figure 2. OCTA-KONG vibration operation. During vibration, the pile and hammer are rigidly connected (A). The OCTA-KONG was a tandem combination of 8× APE 600 (B), with each APE600 composed of a suppressor housing, a vibrator gearbox and a clamping attachment (C).

vibrator gearbox, incorporating the phased high-amplitude eccentric weights; and (3) the clamping attachment (Fig. 2C).

Principles of vibration piling. During pile driving and extraction, the pile and hammer (except for the suppressor housing) are rigidly connected by the clamping attachment, forming a hammer-pile complex oscillating exciter (Fig. 2). Vibration is caused by the vertical movement produced by the centrifugal force that arises when the pairs of eccentrics are counter-rotated [20].The continuous pulses of energy transferred from the hammer-pile complex to the soil can temporarily change the stress-strain behavior, such as soil displacement (e.g., at the penetrating pile tip), and they can create excess pore water pressures or even complete fluidization of the soil. As a consequence, the frictional (i.e., both the internal friction of the soil and the pile-soil friction) and tip resistances are strongly reduced during vibratory driving, enabling the pile to penetrate under the low vertical thrust produced from the combined action of centrifugal force and the self-weight of the hammer-pile complex [20]. The detailed mechanics are not discussed further here.

Piling procedure. The SSP_{22} were 38 m–60 m long and weighed 450 000 kg–600 000 kg (Table 1). During the initial stage of pile installation (i.e., the pre-OCTA-KONG driving session), one of the SSP_{22} was rigidly connected to the clamping attachment of the OCTA-KONG, moved to the predesignated location by crane (Fig. 2) and sunk approximately 20 m by the self-weight of the hammer-pile complex. During the OCTA-KONG driving session, the hammer was used to further drive down the hammer-pile complex to the desired depth. The average sink depth during OCTA-KONG piling was 5 m (range 4 m to 6 m), depending on the substrate. During pile extraction, the OCTA-KONG was powered at the outset to reduce the pile-soil friction and to extract the pile using the line pull of the crane. At a certain point, the operation of the OCTA-KONG was stopped, and only the crane was used to extract the pile.

Ethical statement

Permission to conduct the study was granted by the Ministry of Science and Technology of the People's Republic of China. The research permit was issued to the Institute of Hydrobiology of the Chinese Academy of Sciences (Permit number: 2011BAG07B05).

Acoustic data recording system

Two sets of recording systems were adapted for underwater sound recording. The first was a boat-based system (hereafter referred to as BS) consisting of a Reson piezoelectric hydrophone (model TC-4013-1; Reson Inc., Slangerup, Denmark), a 1 MHz bandwidth EC6081 voltage pre-amplifier with a band-pass filter (model VP2000; Reson Inc.), a high-speed, 16 bit, multifunction data acquisition (DAQ) card (model NI USB-6251 BNC; National Instruments (NI), Austin, TX, USA),a laptop computer and LabVIEW 2011 SP1 (NI) software. Underwater signals were detected with a Reson hydrophone (sensitivity: −211 dB re 1 V/μPa at 1 m distance; frequency response: 1 Hz to 170 kHz +1/−7 dB) and conditioned by a VP2000 pre-amplifier. Further high-pass filtering at 10 Hz was conducted to reduce system and flow noise, and low-pass filtering at 250 kHz was conducted to prevent aliasing before inflow into the NI USB-6251 BNC DAQ card. The acoustic data were then stored directly on the hard drive of a computer in binary format with a sampling rate of 512 kHz, using LabVIEW software. The second recording system was a Song Meter Marine Recorder (hereafter referred to as SM2M), which included an HTI piezoelectric hydrophone (model HTI-96-MIN; High Tech, Inc., Long Beach, MS, USA) with a sensitivity of −

165 dB re 1 V/μPa at 1 m distance and a frequency response of 2 Hz–48 kHz +/−2 dB. It also included a programmable autonomous signal processing unit, integrated with a band-pass filter and a pre-amplifier, which can log data at a resolution of 16 bits and up to a 96 kHz sample rate, with a storage capacity of 512 GB (4×128GB SDXC cards). The signal processing unit was sealed inside a waterproof PVC housing and was submersible to a depth of 150 m. The Reson hydrophone and the SM2M system were calibrated prior to shipment from the factory. The remaining components of the BS system, including the amplifier, filter, DAQ card, LabVIEW software and laptops, were lab-calibrated prior to the field survey by inputting a calibration signal generated by an OKI underwater sound level meter (model SW1020; OKI Electric Industry Co., LTD., Tokyo, Japan). Signal transmission was also simultaneously monitored with an oscilloscope (model TDS1002C; Tektronix Inc., Beaverton, OR, USA).

Data collection

Acoustic recordings were made on 5 days between 21 October, 2013 and 4 January, 2014 at the construction site of the HZMB, China (21°16′–21°16′S; 113°33′–113°55′E) (Fig. 1, Table 2). Surveys were conducted from a 7.5 m recreational power boat with a 102 970 W (140 horsepower) outboard engine. Both stationary and floating recording methods were used during sound recording. For stationary recording, either peripheral static buoys were used to suspend the submersible SM2M or the research vessel was moored with an anchor to form a static platform for the boat-based BS recording system. For floating recordings, the vessel's engine was turned off after approaching a pile, allowing the boat to drift. The recording system was then deployed from the side of the boat. If the boat drifted too far from the pile, recording was stopped, and the boat was repositioned. During sound recording, the vessel's engine remained off. The hydrophones were deployed to 2 m depth using an attached weight to limit movement due to water flow. Furthermore, pile driving was performed primarily during the slack water period, when tidal influence on the water depth and currents were both minimal.

The distance to the construction site was measured using Nikon laser rangefinders (model Ruihao 1200S; Nikon Imaging (China) Sales Co., Ltd., Shanghai, China) with a performance range of 10 m to 1100 m and an accuracy of ±1 m. The locations for both stationary recording and floating recording were also logged using a GPS receiver (model GPSMAP 60CSx; Garmin Corporation, Sijhih, Taiwan). The water depth and quality, including temperature, salinity and pH, was measured with a Horiba Multi-parameter Water Quality Monitoring System (model W-22XD; Horiba, Ltd., Kyoto, Japan). Ambient noise was recorded before OCTA-KONG piling operations.

Acoustic data analysis

Acoustic signals, including OCTA-KONG vibration sounds and ambient noise, were continuously sliced into a time window segment of 1 s. Segments with obvious interference were deleted. Analysis was conducted with SpectraLAB 4.32.17 software (Sound Technology Inc., Campbell, CA, USA) and MATLAB 7.11.0 (The Mathworks, Natick, MA, USA) routines and custom programs.

Sound pressure levels (*SPLs*) and sound exposure levels (*SELs*). The measured parameters included sound pressure levels (SPLs) and sound exposure levels (SELs). *SPLs* were derived directly from the pressure metrics, including the zero-to-peak sound pressure (i.e., the maximum of the unweighted absolute instantaneous sound pressure in the measurement bandwidth (p_{max})) and the root-mean-square sound pressure (i.e., the average of the square of the unweighted instantaneous sound pressure $(p(t)$

Table 2. Descriptions of OCTA-KONG vibration sites, sound recording equipment and method.

	Type	Site	Longitude	Latitude	System	Recording type	Depth(m)	Duration(s)
10/21/2013	Piling	#32	22°16'59"	113°45'40"	SM2M	Fixed	8	137
12/4/2013	Piling	#39	22°16'60"	113°45'25"	BS	Float	7	150
12/13/2013	Piling	#38	22°16'61"	113°45'17"	BS	Fixed	8	142
12/23/2013	Piling	#41	22°16'62"	113°45'05"	BS	Fixed	7	156
1/4/2014	Piling	#36	22°16'63"	113°45'25"	BS	Fixed	7	139
12/23/2013	Extract	#26	22°16'64"	113°46'03"	BS, SM2M	Fixed and float	8	2218

Duration: the OCTA-KONG vibration duration.

in the measurement bandwidth integrated over the analyzed signal duration (T)). The zero-to-peak SPL (SPL_{zp}) is ten times the logarithm to the base 10 of the ratio of the square of the zero-to-peak sound pressure to the square of the reference sound pressure of 1 µPa (p_{ref1}). Similarly, the root-mean-square SPL (SPL_{rms}) is ten times the logarithm to the base 10 of the ratio of the square of the root-mean-square sound pressure to the square of the reference sound pressure of 1 µPa. The single SEL (SEL_{ss}) is ten times the logarithm to the base 10 of the ratio of the integral of the squared sound exposure of a signal of 1 s time window to the reference sound exposure of 1 µPa²s (p_{ref2}). Absolute pressure levels were derived by subtracting the sensitivity of the hydrophone and the gain due to the amplifier [10].

Spectrogram, power spectral density and 1/3 octave band frequency spectrum

The frequency composition of the signals was determined using spectrograms, which express a signal's amplitude, frequency and time, portraying amplitude as a graph plotted in a dark color on a two-dimensional time-frequency plane [10]. Power spectral density (PSD) level routines (dB re 1 µPa² Hz^{-1}), i.e., narrowband spectra in 1 Hz bands, which represent the averaged sound power in each 1 Hz band, were applied to investigate detailed tonal signatures [26]. The 1/3 octave band frequency spectrum, i.e., the sum of the squared pressure of all 1 Hz bands within a 1/3 octave, was investigated to assess impacts on mammalian hearing, as 1/3 of an octave approximates the effective filter bandwidth of cetaceans [5]. Both the spectrograms and narrowband spectra were obtained using the fast Fourier transform (FFT) method, combining a Hanning smoothing window function with an overlap of 85% for the averaging. For the BS (sample rate 512 kHz) and SM2M data (sample rate 96 kHz), the FFT size was 262 144 samples and 65 536 samples, respectively, resulting in a frequency grid resolution of 1.95 Hz and 1.46 Hz, respectively, and a temporal grid spacing of 76.80 ms and 102.40 ms, respectively. Narrowband spectra were further normalized to PSD by dividing by the frequency grid resolution.

Cetacean auditory weighted SEL (SEL_{ws}). As the damage risk criteria for marine mammals exposed to noise should incorporate the exposure frequency [27], cetacean auditory weighting (CA-weighting) [25] functions were used to incorporate the animals' auditory sensitivity to certain frequencies by emphasizing those frequencies where sensitivity to noise is high and de-emphasizing frequencies where sensitivity is low. The CA-weighting function ($W_{CA}(f)$) was merged with a marine mammal weighting function ($W_M(f)$, Equation 1) [7] and an equal-loudness weighting function curve ($W_{EQL}(f)$, Equation 2). Function $W_{EQL}(f)$ was derived from bottlenose dolphin (*Tursiops truncatus*) frequency-specific temporary threshold shift data [27,28] and equal-loudness contours [29]. Equal-loudness contours represent the SPLs of a sound that are perceived as equal in loudness magnitude in a testee as a function of sound frequency. They are considered to reveal the frequency characteristics of the testee's auditory system [30]. The contours are derived from subjective loudness experiments that ask candidates to judge the relative loudness of two tones of different frequencies [29,31]. At each frequency, the amplitude of the $W_{CA}(f)$ is defined using the larger value from the two component curves (Equation 3). The cetacean auditory-weighted SEL (SEL_{ws}) is ten times the logarithm to the base 10 of the ratio of the integral of the squared sound exposure of an CA-weighted signal of 1 s time window to the reference sound exposure of 1 µPa²s. This SEL_{ws} can be simplified (Equation 4), as the integral of the squared sound exposure of an CA-weighted signal is equal to the overall energy of the CA-weighted PSD contour ($PSD_W(f)$) multiplying its frequency resolution.

$$W_M(f) = K_1 + 20\log_{10}\left\{ (b_1^2 f^2)/[(a_1^2+f^2)(b_1^2+f^2)] \right\} \qquad (1)$$

$$W_{EQL}(f) = K_2 + 20\log_{10}\left\{ (b_2^2 f^2)/[(a_2^2+f^2)(b_2^2+f^2)] \right\} \qquad (2)$$

$$W_{CA}(f) = \text{maximum}\left\{ W_M(f), W_{EQL}(f) \right\} \qquad (3)$$

$$SEL_{ws} = 10\log_{10}\left\{ \int_1^{\frac{FFT}{2}} \left(10^{\frac{PSD_w(f)}{10}} \Big/ p_{ref2}^2 \right) df \right\}, \qquad (4)$$

where $W(f)$ is the weighting function amplitude (in dB) at frequency f (in Hz), a and b are constants related to the lower and upper hearing limits (the "roll off" and "cut off" frequencies), respectively, and K is a constant used to normalize the equation at a particular frequency [32]. For *Sousa*, which belongs to the mid-frequency cetacean functional hearing group, K_1, a_1 and b_1 are − 16.5, 150 and 160 000, respectively, and K_2, a_2 and b_2 are 1.4, 7829 and 95 520, respectively [25,32].

Source levels and source SELs. OCTA-KONG source levels (SLs), including the zero-to-peak SL (SL_{zp}, dB re 1 μPa), root-mean-square SL (SL_{rms}, dB re 1 μPa), and source SELs (SSELs), including unweighted SSEL ($SSEL_{ss}$, dB re 1 μPa^{2}s) and CA-weighted SSEL ($SSEL_{ws}$, dB re 1 μPa^{2}s), were obtained by combining measures of received level (RL) and transmission loss (TL) (Equation 5). TL was estimated from the distance from the source (r) as a result of the depth-dependent spreading loss plus frequency-dependent absorption (Equation 6) [33].

$$SL = RL + TL \qquad (5)$$

$$TL = A \times \log_{10}(r) + ar, \qquad (6)$$

where r is the range in meters; A is the spreading loss coefficient, which generally varies from 10 (cylindrical spreading) to 20 (spherical spreading); and a is the frequency-dependent absorption coefficient in dB/m. As the dominant frequency of vibration pile driving was below 10 kHz [19], the absorption term does not significantly contribute to transmission loss and can generally be ignored for those recordings with greatest measurement ranges of less than 1 km [34,35]. Therefore, Equation 6 can be simplified to Equation 7 for the estimates of SL_{zp}, SL_{rms}, $SSEL_{ss}$ and $SSEL_{ws}$. Sound propagation in shallow water environments (<200 m deep) is complex [33]. Attenuation may vary with depth depending on the sediment type, pressure and sediment porosity [36], and the frequency dependence of the acoustic response is sensitive to the details of the geoacoustic structure of the seabed [37]. Previous geophysical studies indicated that the surficial sediments of the bridge construction site were almost flat (TQ Zeng, personal communication). Cores taken in the vicinity of the bridge construction site indicated that the sediment was largely Quaternary sediment with approximately five layers. The top layer, deposited during the Holocene series, consists primarily of silt-clays. The second to the fourth layers, deposited during the Pleistocene series, are predominantly sand, gravel and clay. Mudstone occurs at a depth of approximately 70 m below the sea floor. The fifth layer consists of Yansanian granites (TQ Zeng, personal communication). The transmission loss equation was derived by fitting a least squares regression to the SPL_{zp}, SPL_{rms}, SEL_{ss} and SEL_{ws} measured at different distances during pile driving and extraction using the floating recording method. The derived equation was also used to estimate the source level of other piling sites where the stationary recording strategy was adopted.

$$TL = A \times \log_{10}(r). \qquad (7)$$

Unweighted and CA-weighted cumulative SSEL. Cumulative SSEL is ten times the logarithm to the base 10 of the ratio of the summation over a specified duration of sound exposures to the reference sound exposure of 1 μPa^{2}s. It can be simplified as the average $SSEL_{ss}$ for the unweighted cumulative SSEL ($SSEL_{cum}$, Equation 8) and as the average $SSEL_{ws}$ for the CA-weighted cumulative SSEL ($SSEL_{wcum}$, Equation 9) plus the log transformation of the duration of sound exposure divided by the duration of the 1 s reference time window (t_{ref}).

$$SSEL_{cum} = SSEL_{ss} + 10\log_{10}(duration\ of\ exposure/t_{ref}) \qquad (8)$$

$$SSEL_{wcum} = SSEL_{ws} + 10\log_{10}(duration\ of\ exposure/t_{ref}) \qquad (9)$$

Audibility range. Sound audibility is determined by both external conditions, such as the characteristics of received sound and background noise conditions, and internal conditions, such as the hearing capability of the receiving system (also called the hearing audiogram). As the lowest frequency of the available *Sousa* audiogram was 5.6 kHz [22,23], we were unable to analyze sound audibility for low-frequency sound by referencing the audiogram. Thus, the audible sound range was conservatively estimated as the range from which the sound source is attenuated by absorption and spreading loss with distance, measured at the point where the received sound is equal to the ambient noise level. The sound audible range was estimated using the transmission loss, Equation 9, to incorporate the deviation between the SL of the OCTA-KONG and the ambient noise level (the spreading loss coefficient derived above was adopted here).

Possible impacts on *Sousa*. The potential effects of anthropogenic noise on marine mammals include, but are not limited to, behavioral responses, auditory masking, and physiological effects [5]. Potential behavioral responses include exposure avoidance, behavioral disturbance or no response [7]. Auditory masking refers to the disruption of the reception of auditory signals by noise in the adjacent frequency bands (the so-called critical band) [7], resulting in partial or complete reduction in the audibility of the signals [7,38,39]. Physiological effects include temporary or (in extreme cases) permanent threshold shifts (TTS, PTS), a type of increase in the threshold of the audibility portion of an individual's hearing range or at a specified frequency above a previously established reference level producing states of temporary and recoverable shifts (TTS) or permanent, irreversible ones (PTS) [7]. As no *Sousa* were encountered during the recording period, documenting the behavioral responses was beyond the scope of the present study. Previous recordings of *Sousa* acoustics, including dolphin clicks and whistles recorded from within the same district, and *Sousa* audiograms [22,23] were used to analyze potential auditory masking. The *Sousa* whistles and clicks were recorded by following a focal group of dolphins (an aggregation of dolphins that were engaged in the same behavior and separated by less than 100 m) [40]. Using the vocalizations to determine the animals' location was difficult because only one hydrophone system was used. However, we can confirm that the recorded dolphin sound was from the focal group because no other groups of dolphins were present within approximately 1000 m. During the sound recording, the location of the dolphins was determined within a 50 m radius of our boat based on the successive sites at which they were observed to surface and breathe. The OCTA-KONG sound level was compared with both the cetacean safety exposure level and the proposed acoustic threshold levels for the onset of TTS and PTS for the analysis of potential physiological effects. The cetacean safety exposure level established by NMFS is 180 dB (SPL_{rms}) [24], and the proposed PTS and TTS acoustic threshold levels for *Sousa* (which are mid-frequency cetaceans) exposed to vibration driving noise (a non-impulsive sound source) are: (1) SL_{zp} of 230 dB and 224 dB re 1 µPa, respectively; (2) $SSEL_{cum}$ of 195 dB and 215 dB re 1 µPa²s, respectively; and (3) $SSEL_{wcum}$ of 178 dB and 198 dB re 1 µPa²s, respectively [25], using whichever level is first exceeded.

Statistical analysis

Statistical analyses were conducted using SPSS 16.0 (SPSS Inc., Chicago, IL, USA). Descriptive statistics of all measured SPLs (SPL_{zp} and SPL_{rms}) and SELs (SEL_{ss} and SEL_{ws}) were obtained, including means, standard deviations (SD) and ranges (minimum - maximum values). The mean SPLs and SELs were calculated in Pa and converted to dB. A Levene's test and a Kolmogorov-Smirnov test were used to analyze the homogeneity of the variances and data normality, respectively. Nonparametric methods [41] were adopted for parameters that were non-normally distributed (Kolmogorov-Smirnov test: $p < 0.05$). A Mann-Whitney U-test [41] was applied to analyze whether the SPLs and SELs of the OCTA-KONG varied significantly between pile driving and pile extraction (by comparing data recorded at the same distance to the pile and using the same system) and to test for differences in recorded noise level between the two recording systems. A Kruskal–Wallis test [41] was adopted to examine the overall ambient noise differences across different recording days. A Duncan's multiple comparison test [41] was used for post hoc comparisons of differences in ambient noise level among different recording days. Differences in the ambient noise level among different times within the same day were tested using a Mann-Whitney U-test. Differences were considered significant at $p < 0.05$.

Results

OCTA-KONG pile driving (Fig. 3) was recorded on 5 days at the sites between SSP_{22} #32 and #41, and pile extraction was monitored on SSP_{22} #26 (Table 2). One floating recording for both piling and extraction was obtained (Table 2). Water depths at the recording sites were shallow, ranging from 7 to 8 m (Table 2).

SPL_{zp}, SPL_{rms}, SEL_{ss} and SEL_{ws}

The acoustic signals were sliced into a time window of 1 s segments; therefore, SPL_{rms} is numerically equivalent to SEL_{ss}. Over all of the recording sessions, SPL_{zp} ranged from 146.99 dB to 164.49 dB and 140.83 dB to 164 dB for pile driving and extraction, respectively (Table 3). Both the SPL_{rms} and $SSEL_{ss}$ ranged from 137.77 dB to 153.11 dB and 128.83 dB to 154.58 dB for pile driving and extraction, respectively (Table 3).The OCTA-KONG vibration noise recorded by the BS system at a distance of 70 m during the driving of SSP_{22} #41 and the extraction of SSP_{22} #26 was not significantly different in SPL_{zp} (Mann-Whitney U-test; z = −1.21, df = 337, $p = 0.23$) but significantly different in SPL_{rms}, SEL_{ss} and SEL_{ws} (Mann-Whitney U-test: z = −9.03, df = 337, $p < 0.01$;Mann-Whitney U-test: z = −9.03, df = 337, $p < 0.01$ and Mann-Whitney U-test: z = −14.05, df = 337, $p < 0.01$; respectively) (Table 3). Ambient noise was inspected aurally and via spectrogram, and no bio-acoustic sound generation was observed. The ambient noise could have resulted primarily from wind-driven waves and sea-surface agitation [42]. No significant differences in SPL_{zp}, SPL_{rms}, SEL_{ss} and SEL_{ws} were observed between the ambient noise recorded by the BS and SM2M systems at SSP_{22} #26 (Mann-Whitney U-test: z = −0.30, df = 125, $p = 0.76$; Mann-Whitney U-test: z = −1.73, df = 125, $p = 0.08$; Mann-Whitney U-test: z = −1.73, df = 125, $p = 0.08$ and Mann-Whitney U-test: z = −1.35, df = 125, $p = 0.18$; two-tailed; respectively) (Table 4); therefore, we pooled the data from the two systems. Significant differences in ambient noise were observed among different recording days; i.e., in SPL_{zp}, SPL_{rms}, SEL_{ss} and SEL_{ws} (Kruskal-Wallis $\chi^2 = 27.18$, df = 4, $p < 0.01$; Kruskal-Wallis $\chi^2 = 41.21$, df = 4, $p < 0.01$; Kruskal-Wallis $\chi^2 = 41.21$, df = 4, $p < 0.01$ and Kruskal-Wallis $\chi^2 = 215.34$, df = 4, $p < 0.01$; respectively, Table 4). In particular, significant variation was observed in SPL_{zp} between SSP_{22} #38 and #41 vs #39 and #41 vs #36 (Duncan's multiple-comparison test; $p < 0.05$) (Table 4). Significant differences were observed in SPL_{rms}, SEL_{ss} and SEL_{ws} between SSP_{22} #39,#38 and #36 vs #32, between#39 and #38 vs #41 and #36 vs #41 (Duncan's multiple-comparison test; $p < 0.05$) (Table 4). Significant ambient noise differences were observed between the morning (before pile driving) and afternoon (before extraction)of the same day; i.e., differences in SPL_{zp}, SPL_{rms}, SEL_{ws} and SEL_{ws} (Mann-Whitney U-test: z = −3.97, df = 214, $p < 0.05$; Mann-Whitney U-test: z = −4.12, df = 214, $p < 0.05$; Mann-Whitney U-test: z = −4.12, df = 214, $p < 0.05$ and Mann-Whitney U-test: z = −4.66, df = 214, $p < 0.05$; two-tailed; respectively) (Table 4).

Spectrogram, PSD and 1/3 octave band spectrum

The recorded fundamental frequency of the OCTA-KONG vibration ranged from 15 Hz (Fig. 3) to 16 Hz (Fig. 4, 5). The noise increments were below 20 kHz, with the dominant frequency and most energy contained below approximately 10 kHz (Fig. 4, 5, 6).

SEL_{ws}

The recorded SEL_{ws} ranged from 112.74 dB to 128.86 dB and 111.92 dB to 138.07 dB for pile driving and pile extraction, respectively (Table 3).

SL_{zp}, SL_{rms}, $SSEL_{ss}$ and $SSEL_{ws}$

The best-fit sound propagation models for SPL_{zp}, SPL_{rms}, SEL_{ss} and SEL_{ws} for pile driving and pile extraction are shown in Figure 7A and 7B. The estimated mean SL_{zp} during pile driving and extraction ranged from 179.79 dB to 189.01 dB and

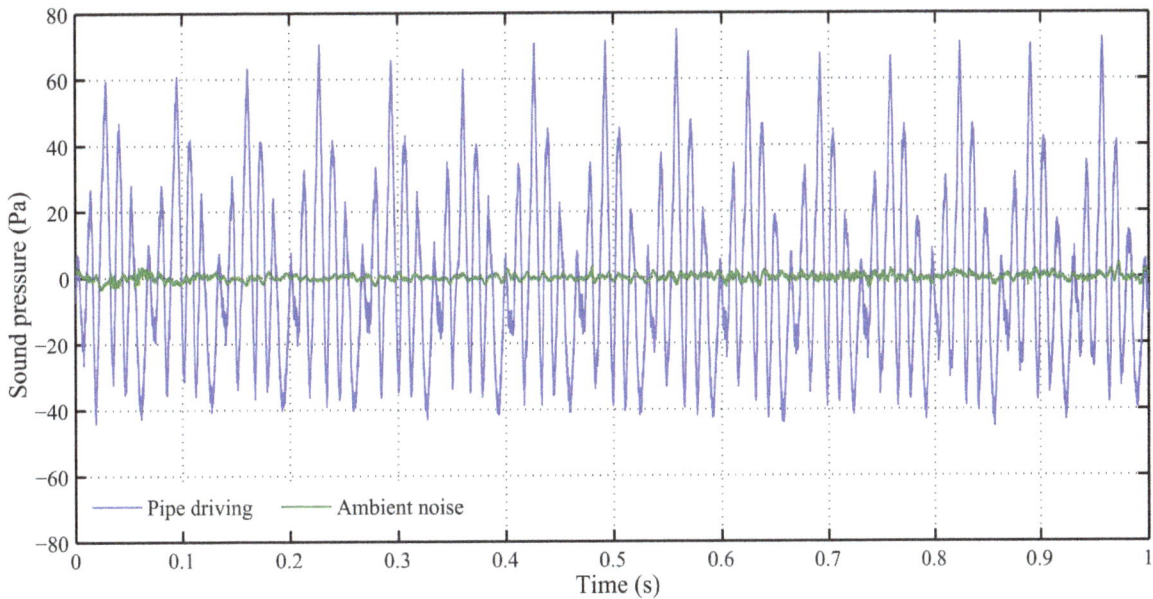

Figure 3. Wave form of the OCTA-KONG SSP$_{22}$ #32 vibration sound and ambient noise. The fundamental frequency of the vibration sound was 15 Hz.

185.70 dB to187.49 dB, respectively. The estimated mean SL_{rms} and $SSEL_{ss}$ during pile driving and extraction ranged from 168.90 dB to 179.96 dB and 173.00 dB to175.26 dB, respectively. The estimated mean $SSEL_{ws}$ during pile driving and extraction ranged from 142.95 dB to 157.20 dB and 157.00 dB to158.90 dB, respectively (Table 5).

Audibility range

The frequency-dependent sound absorption constant a was estimated at 0.0006 [35] for the specific pH of 8, a salinity of 33‰ and a water temperacure of 20°C (measured at the piling sites during the sound recording period) at a frequency of 10 kHz; the majority of OCTA-KONG vibration noise power is found below this frequency (Figs. 4, 5, 6). The transmission loss equations with correlation to the distance r for SL_{zp}, SL_{rms}, $SSEL_{ss}$ and $SSEL_{ws}$ were 15.1 $\log_{10}(r)$+0.0006r, 15.0 $\log_{10}(r)$+0.0006r, 15.0 $\log_{10}(r)$+0.0006r and 15.4 $\log_{10}(r)$+0.0006r, respectively, for pile driving and 19.1 $\log_{10}(r)$+0.0006r, 19.4 $\log_{10}(r)$+0.0006r, 19.4 $\log_{10}(r)$+0.0006r and 19.7 $\log_{10}(r)$+0.0006r, respectively, for pile extraction. The estimated audible range of SL_{zp} during pile driving and pile extraction ranged from 448 m to 1546 m and from 196 m to 236 m, respectively. The estimated audible range of SL_{rms} and $SSEL_{ss}$ during pile driving and pile extraction ranged from 818 m to 3489 m and from 192 m to 229 m, respectively. The estimated audible range of $SSEL_{ws}$ during pile driving and pile extraction ranged from 483 m to 2954 m and from 557 m to 765 m, respectively (Table 5).

Impact on *Sousa*

Auditory masking. The *Sousa* audiogram was revised from the two available audiograms [22,23], with the lowest threshold at each frequency defining the merged audiogram curve. Both the OCTA-KONG vibration sound and the ambient noise level recorded in this study were above the threshold of the *Sousa* audiogram (Fig. 6). The 1/3 octave band sound pressure level of the *Sousa* click sound at a distance of less than 50 m with a dominant frequency range of 20 kHz to 200 kHz would not be masked by the OCTA-KONG vibration sound recorded at a distance of 200 m. However, the 1/3 octave band sound pressure level of the *Sousa* whistle recorded at a distance of less than 50 m with a dominant frequency range from 3 kHz to 6 kHz would be masked by the vibration sound recorded at a distance of 200 m (Fig. 6).

Cetacean safety exposure level. The maximum SL_{rms} of SSP$_{22}$ driving (#32 and #36) and extraction (#26) exceeded the established cetacean safety exposure SPL_{rms} level of 180 dB (Table 5). However, the maximum SL_{rms} of SSP$_{22}$ pile driving of #38, #39 and #41 were lower than 180 dB (Table 5).

Physiological impact. All the calculated SL_{zp} values of OCTA-KONG pile driving and pile extraction (with maximums of 193.23 dB and 193.15 dB, respectively) (Table 5) were well below 224 dB, the proposed SL_{zp} threshold for the onset of TTS for mid-frequency cetaceans exposed to non-impulsive sound. However, the calculated $SSEL_{cum}$ values for pile driving of SSP$_{22}$ #32, #39 and #36 and pile extraction of SSP$_{22}$ #26 were 201.33 dB, 195.05 dB, 199.75 dB and 207.59 dB re 1 µPa^{2}s, respectively (Table 6), exceeding the proposed 195 dB threshold for the onset of TTS in mid-frequency cetaceans exposed to non-impulsive sound. In addition, the calculated $SSEL_{wcum}$ for the pile driving of SSP$_{22}$ #39 and pile extraction of SSP$_{22}$ #26 was 179.05 dB and 191.41 dB re 1 µPa^{2}s, respectively, greater than the proposed threshold of TTS onset at 178 dB. All the calculated SSELs were lower than the threshold of the onset PTS for mid-frequency cetaceans exposed to non-impulsive sound.

Discussion

Inshore marine mammals are highly susceptible to habitat loss, fragmentation, and degradation [5]. Marine mammals have a well-developed sense of hearing, and the importance of sound reception to these mammals makes them susceptible to the effects of anthropogenic noise [5].

The impacts of anthropogenic noise on marine life have been widely assessed [43]. The St. Lawrence River beluga (*Delphinapterus leticas*) may change its vocalization SPLs in direct response to

Table 3. Descriptive statistics of the SPL$_{zp}$, SPL$_{rms}$ and SEL$_{ws}$ values of the OCTA-KONG vibration.

Data			Vibration				
			SPL$_{zp}$ (dB re 1 μPa)	SPL$_{rms}$ (dB re 1 μPa)	SEL$_{ws}$ (dB re 1 μPa^{2}s)	N	Distance (m)
Piling	#32	Mean±SD	154.75±2.11	145.44±1.85	118.90±1.68	76	200
		Range	149.9-158.11	140.58-148.8	113.78-121.70		
	#39	Mean±SD	153.64±2.02	142.99±1.95	121.38±2.05	87	90-145
		Range	148.29-160.02	137.77-146.8	116.56-123.95		
	#38	Mean±SD	153.66±1.12	143.16±0.94	121.62±1.09	50	60
		Range	151.48-155.81	141.29-144.79	118.96-123.49		
	#41	Mean±SD	151.93±2.18	141.22±1.54	114.54±0.93	90	70
		Range	146.99-159.66	138.05-147.47	112.74-116.79		
	#36	Mean±SD	160.27±1.97	149.77±2.13	125.25±2.6	99	80
		Range	154.96-164.49	144.35-153.11	119.53-128.86		
Extract	#26a	Mean±SD	152.25±1.99	139.47±1.44	120.65±0.93	247	70
		Range	148.26-157.91	136.02-144.35	118.28-124.91		
	#26b	Mean±SD	151.27±4.98	137.7±5.6	123.17±5.47	1471	15-180
		Range	140.83-164	128.83-154.58	111.92-138.07		

Parameters are given as the mean ± standard deviation (SD), with the range denoting minimum and maximum values. SEL$_{ss}$ was identical to SPL$_{rms}$. N: sample size. Subscript 'a' denotes sound recorded by the BS recording system, and 'b' denotes sound recorded by the SM2M recording system.

Table 4. Descriptive statistics of the SPL_{zp}, SPL_{rms} and SEL_{ws} of the ambient noise.

			SPL_{zp} (dB re 1 µPa)	SPL_{rms}(dB re 1 µPa)	SEL_{ws} (dB re 1 µPa^{2}s)	N
Piling	#32	Mean±SD	140.41±4.23[a]	124.72±3.51[abc]	99.12±2.6[abc]	53
		Range	131.24–147.66	117.36–133.31	94.47–103.90	
	#39	Mean±SD	142.45±3.49[bc]	126.22±2.45[bd]	108.34±2.63[bd]	67
		Range	135.94–148.28	121.58–133.52	103.56–112.68	
	#38	Mean±SD	140.2±3.09[b]	125.64±2.08[ce]	107.37±1.69[ce]	45
		Range	135.43–146.58	123.18–134.09	104.28–113.58	
	#41	Mean±SD	139.37±5.21[cd]	123.74±4.63[def]	100.71±1.98[def]	89
		Range	130.32–153.48	114.75–137.99	98.01–106.18	
	#36	Mean±SD	140.4±3.36[d]	127.34±2.9[af]	103.33±2.08[af]	54
		Range	134.15–151.71	122.17–134.01	101.16–114.17	
Extract	#26$_a$	Mean±SD	142.07±3.39	129.3±3.09	102.57±1.58	94
		Range	134.42–153.1	121.77–139.16	99.95–106.94	
	#26$_b$	Mean±SD	141.85±3.27	128.6±3.46	101.63±1.93	31
		Range	134.46–150.7	120.64–138.36	99.20–106.79	
	#26$_c$	Mean±SD	142.12±3.31	129.1±3.14	102.34±1.62	125
		Range	134.42–153.1	120.64–139.16	99.20–106.94	

Parameters are given as the mean ±SD, with ranges denoting minimum and maximum values. SEL_{ss} was identical to SPL_{rms}, Means with different lowercase superscripts refer to post hoc Duncan's multiple-comparison tests that yielded significant results ($p<0.05$) for OCTA-KONG pile driving. Subscript 'a' denotes sound recorded by the BS recording system, 'b' denotes sound recorded by the SM2M recording system and 'c' denotes the combined results of the BS and SM2M recording systems.

changes in the noise field (Lombard effect) [44] or shift its frequency bands when exposed to vessel noise [45]. Killer whales (*Orcinus orca*) may adjust their vocal behavior, showing longer call durations [46], or exhibit a Lombard effect [47] to compensate for masking boat noise. Bowhead whales (*Balaena mysticetus*) may change migration routes and exhibit avoidance reactions when exposed to air gun noise [48] or travel at increased speeds in the presence of anthropogenic noise [5]. Harbor porpoises (*Phocoena phocoena*) tend to reduce their acoustic activity when exposed to the construction noise of pile driving [49] or reduce their buzzing activity when exposed to impulse noise from seismic surveys [50]. Bottlenose dolphins will significantly increase their whistle rate at the onset of an approach by a vessel [51], and Indo-Pacific bottlenose dolphins (*Tursiops aduncus*) tend to produce whistles

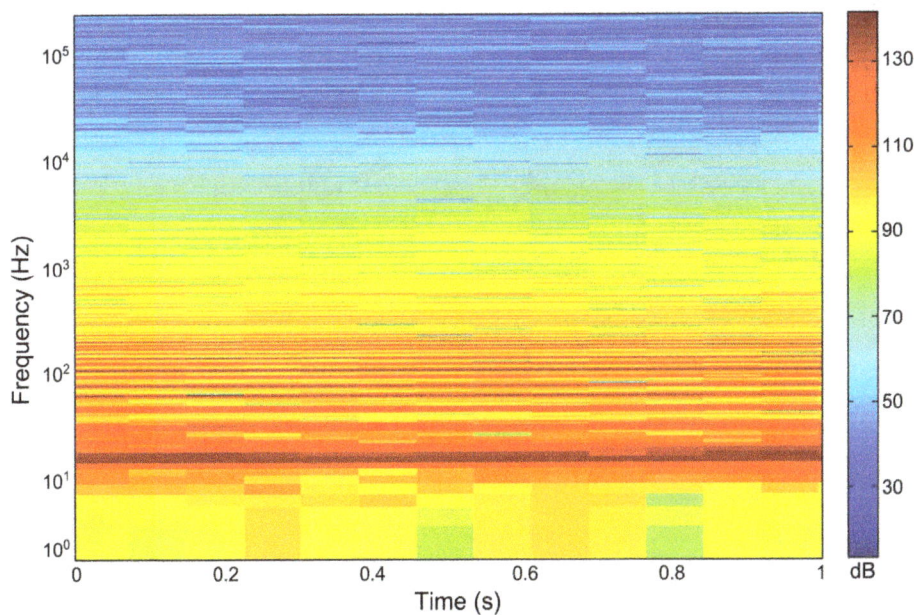

Figure 4. Spectrogram of the OCTA-KONG SSP$_{22}$ #36 driving sound. Spectrogram configuration: temporal grid resolution, 76.80 ms; overlap samples per frame, 85%; frequency grid spacing, 1.95 Hz; window size, 262 144; FFT size, 262 144; window type, Hanning. The fundamental frequency of the vibration sound was 16 Hz.

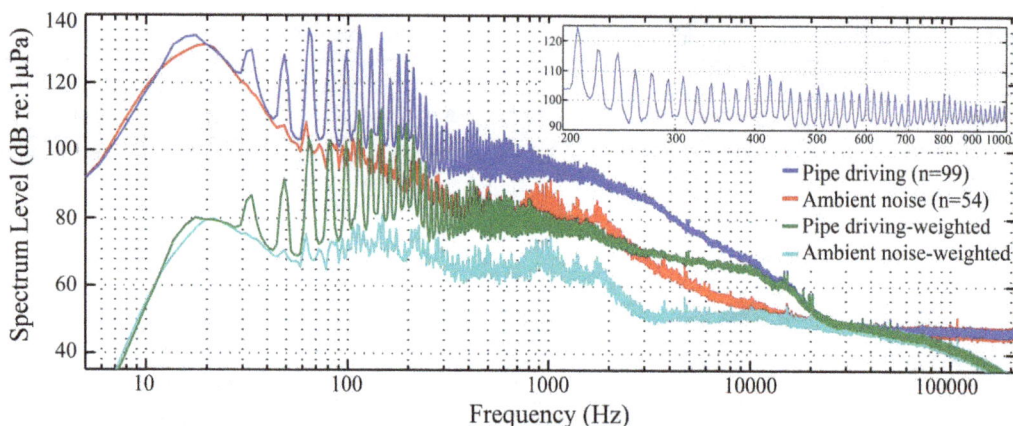

Figure 5. Power spectral density of the OCTA-KONG SSP₂₂ #36 driving sound and noise. Spectrum configuration: temporal grid resolution, 76.80 ms; overlap samples per frame, 85%; frequency grid spacing, 1.95 Hz and normalized to 1 Hz; window size, 262 144; FFT size, 262 144; window type, Hanning. The inset in the upper right corner shows a magnified frequency scale of the unweighted piling sound. The fundamental frequency of the vibration sound was 16 Hz. Pile driving sounds were recorded at a distance of 80 m from the vibration hammer.

with less frequency modulation at lower frequencies In habitats with greater ambient noise [52].

Acoustic impact models that estimate the effects of anthropogenic noise on the hearing and communication of fish and marine mammals by comparing noise spectra, audiograms and the vocalizations of the animal of interest have been widely applied, e.g., in research on the effects of ambient and boat noise on *Chromis chromis*, *Sciaena umbra* and *Gobius cruentatus* living in a marine protected area in Italy [53] and on the Lusitanian toadfish (*Halobatrachus didactylus*) in Portugal [54], on the impact of sounds resulting from construction and pipe-driving at an oil production island in Alaska on ringed seals (*Phoca hispida*) [34], on the potential effects of pile-driving at an offshore wind farm in

the Moray Firth, NE Scotland on marine mammals [6], on the possible sensitivity of bottlenose dolphins to pile-driving noise [55], on the potential effects of underwater noise produced by whale-watching boats on killer whales in southern British Columbia and northwestern Washington State [56] and on the effects of the high-speed hydrofoil ferry in West Hong Kong waters on the Chinese white dolphin [57].

Transmission loss is correlated with bathymetry, substrate type, and sound speed profile along the direction of transmission [36], and the fit obtained for site-specific transmission loss may not apply to transmission in other directions from the source if these conditions are different in those directions [33]. Because the bathymetry and substrate type in the studied construction site are

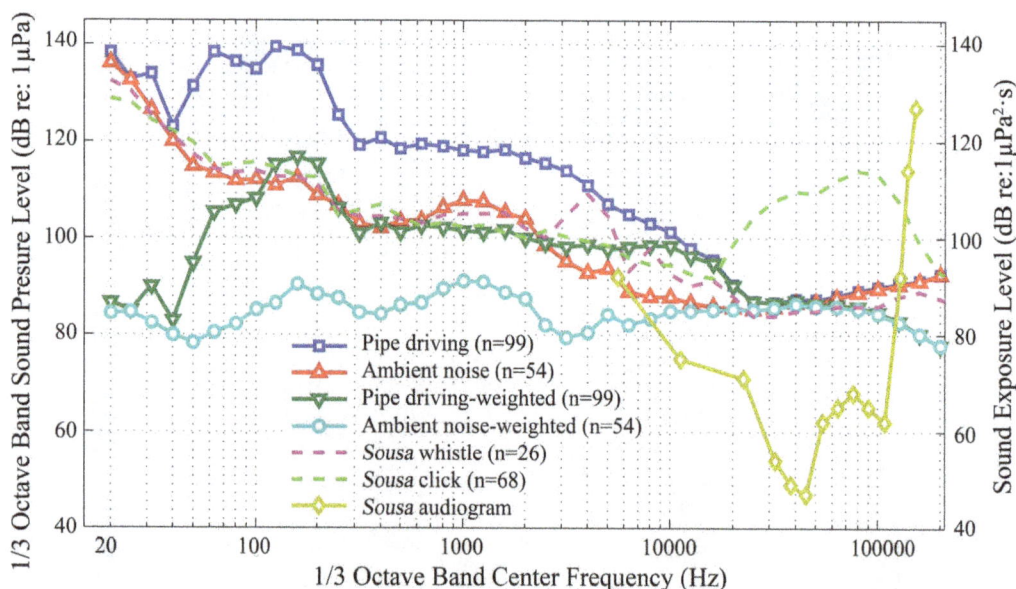

Figure 6. 1/3 octave band frequency spectrum and A-weighted sound exposure level of the SSP₂₂ #36 driving sound. Spectrum configuration: temporal grid resolution, 76.80 ms; overlap samples per frame, 85%; frequency grid spacing, 1.95 Hz; window size, 262 144; FFT size, 262 144; window type, Hanning. The *Sousa* audiogram was modified from previous sources [22,23], with the lowest threshold at each frequency defining the merged audiogram curve. n denotes the number of samples. Pile driving sounds were recorded at a distance of 80 m from the vibration hammer.

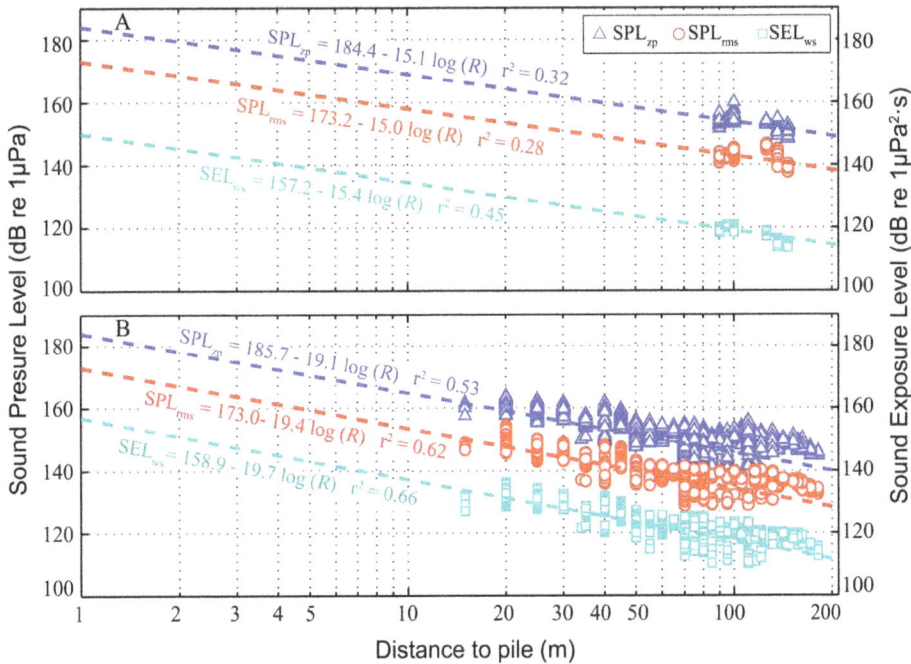

Figure 7. Broadband SPL_{zp}, SPL_{rms} **and** SEL_{ws} **of vibration sound as a function of distance from the noise source and the best-fit sound propagation model.** A: OCTA-KONG SSP$_{22}$ #39 driving; B: OCTA-KONGSSP$_{22}$ #26 extraction. The sound propagation equations that predicted the received SPLs and SELs based on distance were derived by applying a least squares regression to the measurements obtained via the floating recording method for pile driving and extraction, respectively.

Table 5. SL_{zp}, SL_{rms} and $SSEL_{ws}$ of the OCTA-KONG pile driving and pile extraction and the audible range.

			OCTA-KONG	Ambient noise	Sensation level(dB)	Audible range(m)
Piling	SL$_{zp}$	#32	189.5 (184.65–192.86)	140.41	49.09	1546
		#39	184.4	142.45	41.95	569
		#38	180.51 (178.33–182.66)	140.2	40.31	448
		#41	179.79 (174.85–187.52)	139.37	40.42	455
		#36	189.01 (183.7–193.23)	140.4	48.61	1456
	SL$_{rms}$	#32	179.96 (175.1–183.32)*	124.72	55.24	3489
		#39	173.2	126.22	46.98	1212
		#38	169.83 (167.96–171.46)	125.64	44.19	818
		#41	168.9 (165.73–175.15)	123.74	45.16	939
		#36	178.32 (172.9–181.66)*	127.34	50.98	2068
	SSEL$_{ws}$	#32	154.33(149.21–157.13)	99.12	55.21	2954
		#39	157.20	108.34	48.86	1324
		#38	149.00(146.34–150.87)	107.37	41.63	483
		#41	142.95(141.15–145.20)	100.70	42.25	527
		#36	154.55(148.83–158.16)	103.33	51.23	1802
Extract	SL$_{zp}$	#26$_a$	187.49 (183.5–193.15)	142	45.49	236
		#26$_b$	185.7	141.85	43.85	196
	SL$_{rms}$	#26$_a$	175.26 (171.81–180.14)*	129.30	45.96	229
		#26$_b$	173	128.6	44.40	192
	SSEL$_{ws}$	#26$_a$	157.00 (154.64–161.26)	102.57	54.43	557
		#26$_b$	158.90	101.63	57.27	765

$SSEL_{ss}$ was identical to SL_{rms}. The average levels of the OCTA-KONG and ambientnoise are provided. Numbers in parentheses indicate the range. Sensation level was derived by dividing the vibration sound by the ambient noise level. SLs and SSELs are re 1 µPa and 1 µPa^{2}s, respectively. Subscript 'a' denotes sound recorded by the BS recording system, 'b' denotes sound recorded by the SM2M recording system. * denotes results that exceeded the proposed cetacean safety exposure level of 180 dB (SPL_{rms}).

Table 6. $SSEL_{ss}$, $SSEL_{ws}$, $SSEL_{cum}$ and $SSEL_{wcum}$ of the OCTA-KONG vibration.

	Date	Sites	$SSEL_{ss}$ (dB re 1 µPa²s)	$SSEL_{ws}$ (dB re 1 µPa²s)	Duration(s)	$10\log(t)$	$SSEL_{cum}$ (dB re 1 Pa²s)	$SSEL_{wcum}$ (dB re 1 µPa²s)
Piling	10/21/2013	#32	179.96	154.33	137	21.37	201.33*	175.70
	12/4/2013	#39	173.2	157.20	153	21.85	195.05*	179.05*
	12/13/2013	#38	169.83	149.00	142	21.52	191.35	170.52
	12/23/2013	#41	168.9	142.95	156	21.93	190.83	164.88
	1/4/2014	#36	178.32	154.55	139	21.43	199.75*	175.98
Extract	12/23/2013	#26	174.13	157.95	2219	33.46	207.59*	191.41*

Average $SSEL_{ss}$ and $SSEL_{ws}$ of #26 obtained from the results of the BS and SM2M recording systems (Table 5). *and ** denote results that exceeded the proposed acoustic threshold levels for the onset of TTS (178 dB and 195 dB for $SSEL_{wcum}$ and $SSEL_{cum}$ respectively) and PTS (198 dB and 215 dB for $SSEL_{wcum}$ and $SSEL_{cum}$ respectively), respectively.

consistent, the site-specific empirical fit method that we used to determine the transmission loss can be applied to transmission in other locations.

Spectrogram, PSD and 1/3 octave band frequency spectrum

The spectrograms and PSD levels allowed us to evaluate the detailed frequency composition of the signal (Figs. 4, 5); however, they did not consider the critical band theory of the mammalian auditory system. Therefore, they offer little insight into either how these mammals perceive noise or the extent of the masking effect of the noise PSD levels [9]. The 1/3 octave band sound pressure level information provided us with a starting point for evaluating the frequency components of the construction sounds that are audible to the dolphins [34]. Although there is little noise energy above the ambient noise levels between 20 kHz and 120 kHz, the Chinese white dolphin shows the greatest sensitivity to sound in this range, as is normal for toothed whales [5], and the majority of the noise increments above the ambient noise levels of 5.6 kHz to 20 kHz were greater than 15 dB (Fig. 6). Both the OCTA-KONG vibration noise and the ambient noise level were above the threshold of the *Sousa* audiogram at frequency bands between 5.6 kHz and 128 kHz (Fig. 6), indicating that sound detection in these frequency bands was limited by the ambient noise rather than by the Sousa audiogram.

Impacts on *Sousa*

Sound masking. The dominant noise level of the OCTA-KONG operation was below 20 kHz, suggesting that *Sousa* clicks were not adversely affected (Fig. 6). This interpretation is further supported by the finding that the peak frequency of *Sousa* clicks ranges from 43.5 kHz to 142.1 kHz [23]. By contrast, the *Sousa* whistle, with a fundamental frequency ranging from 520 Hz to 33 kHz [15], was most susceptible to auditory masking and could be completely masked at a distance of 200 m (Fig. 6). As whistles play a significant role in dolphin communication, such auditory masking may disrupt activities such as feeding and sexual behavior [5]. The adopted safety zone of approximately 200 m radius, as suggested by NOAA [18], should be enlarged to a more conservative region of 500 m radius, as recommended by the Joint Nature Conservation Committee [58]. Beyond this distance, the audibility of certain OCTA-KONG vibrations to *Sousa* is negligible (Table 5).

Physiological impact. Although the SL_{zp} of the SSP_{22} vibration was lower than the proposed physiological damage level, 60% (3 out of 5 piles) of the $SSEL_{cum}$ values during SSP_{22} driving, the $SSEL_{wcum}$ values during SSP_{22} #39 driving and both the $SSEL_{cum}$ and $SSEL_{wcum}$ values during SSP_{22} extraction exceeded the acoustic threshold levels for the onset of TTS (Table 6). In general, the $SSEL_{cum}$ and/or $SSEL_{wcum}$ values could exceed the PTS or TTS threshold in a multitude of ways, depending on the exposure levels and durations [25]. The average $SSEL_{ss}$ values for all six SSP_{22} sites were lower than the cetacean safety exposure level (180 dB) (Table 6); therefore, the surpassed $SSEL_{cum}$ and $SSEL_{wcum}$ levels were due to the prolonged duration of the operation (as a function of $10\log(t)$). The average durations of OCTA-KONG vibration during pile driving and pile extraction are 3 min and 30 min, respectively, with a range of 2 min to 6 min and 20 min to 40 min, respectively (YP Wang, personal communication).

Mitigation method

As the $SSEL_{cum}$ and $SSEL_{wcum}$ values were exceeded due to the prolonged sound exposure periods, the PTS could potentially be avoided by alternating the OCTA-KONG vibration with periods of inactivity, e.g., operations on nonconsecutive days to reduce the sound exposure. In addition, as several of the maximum SL_{rms} values for OCTA-KONG vibration exceeded the cetacean safety exposure level, an air bubble curtain could be introduced. Such curtains can substantially reduce underwater noise at frequencies between 400 Hz and 6400 Hz [59]. During the present study, the power unit rotated primarily in the 1300 r/min–1500 r/min range; however, a maximum of 1700 r/min was used during the construction of the two artificial islands (YF Yang, personal communication),which may have introduced more intense operation noise. Moreover, in addition to the use of pings, "soft start" and "power down" techniques should be adopted [18]. Specifically, at the beginning of each pile installation or extraction, vibratory hammers should be activated at low power for 15 s, followed by a 1-min waiting period (i.e., at a duty cycle of 20%, repeated at least twice) before full power is achieved (i.e., a "soft start"). Additionally, if dolphins are observed within the exclusion zone during the in-situ vibration, operations should either cease or substantially reduce the vibration power (i.e., "power down"). Pile-driving operations should occur during periods when threatened or endangered species are less abundant, as suggested by NOAA [19].

Limitations

The present study had two limitations: First, dolphin behavioral responses during pile driving and extraction were not addressed. In view of the limited current knowledge of the noise dose-response relationship, we are unable to assess whether the noise generated by the OCTA-KONG may cause behavioral disruption. Second, although the adopted audiogram was derived from two Chinese white dolphins of different ages [22,23], there is individual variation in cetacean audiograms [60,61]. Therefore, the two audiograms used here should not be considered representative of the hearing sensitivity of this species. In addition, the *Sousa* audiogram data were sparse and did not extend below a lower frequency limit of 5.6 kHz, further limiting noise exposure assessment at lower frequencies. Audiograms covering a wider frequency range for *Sousa* are needed to quantitatively analyze the impact of the noise.

Conclusions and Future Research

The fundamental frequency of the OCTA-KONG vibration ranged from 15 Hz to 16 Hz, with noise increments below 20 kHz and a dominant frequency and energy below 10 kHz. The vibration zone detectable by *Sousa* extends beyond 3.5 km. *Sousa* clicks do not appear to be adversely affected, whereas *Sousa* whistles are susceptible to auditory masking; therefore, a safety zone of 500 m radius is proposed. Although the SL_{zp} value of the OCTA-KONG was lower than the physiological damage level, the maximum SL_{rms} value sometimes exceeded the cetacean safety exposure level, and the majority of $SSEL_{cum}$ and $SSEL_{wcum}$ values exceeded the acoustic threshold levels for the onset of TTS. Moreover, the TTS was due to the prolonged production of the vibration sound. These findings can help improve environmental impact analyses. Future research that evaluates the real-time noise conditions accompanying underwater construction and the associated behavioral responses of nearby dolphins is recommended to address, in a more direct and robust manner, the possible impacts of human-generated noise on these animals. An increased understanding of the dose effects of noise exposure will provide us with valuable information on how to mitigate possible impacts during the underwater project; this information is important for *Sousa* conservation. In addition, prey are a critical resource for cetaceans [62–64], but little is known about the effects of construction noise on fish [1,65,66]. Dolphins can identify and locate their prey through passive listening during the search phase of the foraging process [67,68]. Therefore, further research is needed to identify the potential adverse impacts on fish, including the masking of prey sounds by anthropogenic noise, particularly of those species that are important prey for marine mammals.

Acknowledgments

The sound recording program was provided courtesy of Hao Zhou (National Instruments Corporation), and the SM2M system was provided by the Public Technology Service Center, IHB, CAS. We gratefully acknowledge Tiequan Zeng and Yipeng Wang (CCCC First Harbor Engineering Company Ltd.), Hua Wen (Hong Kong-Zhuhai-Macao Bridge Authority), David J. White and Yunfu Yang (Shanghai ZhenLi Equipment Company) for information and assistance. Individual thanks are due to Wenjun Xu (Central China Normal University) for her statistical assistance. Special thanks are also extended to the academic editor and two anonymous reviewers for their helpful critique of an earlier version of this manuscript.

Author Contributions

Conceived and designed the experiments: ZW YW GD HC JL KW DW. Performed the experiments: ZW YW KW DW. Analyzed the data: ZW KW DW. Contributed reagents/materials/analysis tools: ZW KW DW. Wrote the paper: ZW YW GD HC JL KW DW.

References

1. Popper AN, Hawkins A (2012) The effects of noise on aquatic life. New York: Springer Science & Business Media. 695 p.
2. Wahlberg M, Westerberg H (2005) Hearing in fish and their reactions to sounds from offshore wind farms. Marine Ecology Progress Series 288: 295–309.
3. Popper AN, Hastings MC (2009) The effects of human-generated sound on fish. Integrative Zoology 4: 43–52.
4. Casper BM, Popper AN, Matthews F, Carlson TJ, Halvorsen MB (2012) Recovery of Barotrauma Injuries in Chinook Salmon, *Oncorhynchus tshawytscha* from Exposure to Pile Driving Sound. PLoS ONE 7: e39593.
5. Richardson WJ, Greene CRJ, Malme CI, Thompson DH (1995) Marine Mammals and Noise. San Diego: Academic Press. 576 p.
6. Bailey H, Senior B, Simmons D, Rusin J, Picken G, et al. (2010) Assessing underwater noise levels during pile-driving at an offshore windfarm and its potential effects on marine mammals. Marine Pollution Bulletin 60: 888–897.
7. Southall BL, Bowles AE, Ellison WT, Finneran JJ, Gentry RL, et al. (2007) Marine Mammal Noise Exposure Criteria:Initial Scientific Recommendations. Aquatic Mammals 33: 411–521.
8. Nowacek DP, Thorne LH, Johnston DW, Tyack PL (2007) Responses of cetaceans to anthropogenic noise. Mammal Review 37: 81–115.
9. Madsen PT, Wahlberg M, Tougaard J, Lucke K, Tyack PL (2006) Wind turbine underwater noise and marine mammals: implications of current knowledge and data needs. Marine Ecology Progress Series 309: 279–295.
10. Au WWL, Hastings MC (2008) Principles of marine bioacoustics. New York: Springer Science. 679 p.
11. Yeung YM, Shen JF (2008) The Pan-Pearl River Delta: An Emerging Regional Economy in a Globalizing China. Hongkong: the chinese university press. 581 p.
12. Cheung E, Chan AP (2009) Is BOT the best financing model to procure infrastructure projects?: A case study of the Hong Kong-Zhuhai-Macau Bridge. Journal of Property Investment & Finance 27: 290–302.
13. Reeves RR, Dalebout ML, Jefferson TA, Karczmarski L, Laidre K, et al. (2008) *Sousa chinensis*. IUCN Red List of Threatened Species (Version 2013-2). Available: http://www.iucnredlist.org. Accessed 25 January 2014.

14. Chen T, Hung SK, Qiu YS, Jia XP, Jefferson TA (2010) Distribution, abundance, and individual movements of Indo-Pacific humpback dolphins (*Sousa chinensis*) in the Pearl River Estuary, China. Mammalia 74: 117–125.

15. Wang ZT, Fang L, Shi WJ, Wang KX, Wang D (2013) Whistle characteristics of free-ranging Indo-Pacific humpback dolphins (*Sousa chinensis*) in Sanniang Bay, China. The Journal of the Acoustical Society of America 133: 2479–2489.

16. Chen BY, Zheng DM, Yang G, Xu XR, Zhou KY (2009) Distribution and conservation of the Indo-Pacific humpback dolphin in China. Integrative Zoology 4: 240–247.

17. Preen A (2004) Distribution, abundance and conservation status of dugongs and dolphins in the southern and western Arabian Gulf. Biological Conservation 118: 205–218.

18. NOAA (2013) Taking and importing marine mammals; taking marine mammals incidental to construction and operation of offshore oil and gas facilities in the Beaufort Sea. Federal Register 78: 75488–75510.

19. Reyff JA (2005) Underwater sound pressure levels associated with marine pile driving: Assessment of impacts and evaluation of control measures. Transportation Research Record: Journal of the Transportation Research Board 11: 481–490.

20. Jonker G (1987) Vibratory pile driving hammers for pile installations and soil improvement projects. the 19th Annual Offshore Technology Conference. Houston, Texas. pp. 549–560.

21. Würsig B, Greene Jr CR, Jefferson TA (2000) Development of an air bubble curtain to reduce underwater noise of percussive piling. Marine Environmental Research 49: 79–93.

22. Li S, Wang D, Wang K, Taylor EA, Cros E, et al. (2012) Evoked-potential audiogram of an Indo-Pacific humpback dolphin (*Sousa chinensis*). The Journal of Experimental Biology 215: 3055–3063.

23. Li S, Wang D, Wang K, Hoffmann-Kuhnt M, Fernando N, et al. (2013) Possible age-related hearing loss (presbycusis) and corresponding change in echolocation parameters in a stranded Indo-Pacific humpback dolphin. The Journal of Experimental Biology 216: 4144–4153.

24. NMFS (2000) Taking and importing marine mammals; taking marine mammals incidental to construction and operation of offshore oil and gas facilities in the Beaufort Sea. Federal Register 65: 34014–34032.

25. NOAA (2013) Draft guidance for assessing the effects of anthropogenic sound on marine mammals: Acoustic threshold levels for onset of permanent and temporary threshold shifts. 12-18-2013 ed. Silver Spring, Maryland: National Oceanic and Atmospheric Administration. Aailable: http://www.nmfs.noaa.gov/pr/acoustics/guidelines.htm. Accessed 20 February 2014.

26. Sims PQ, Vaughn R, Hung SK, Wursig B (2012) Sounds of Indo-Pacific humpback dolphins (*Sousa chinensis*) in West Hong Kong: A preliminary description. The Journal of the Acoustical Society of America 131: EL48–EL53.

27. Finneran JJ, Schlundt CE (2010) Frequency-dependent and longitudinal changes in noise-induced hearing loss in a bottlenose dolphin (*Tursiops truncatus*). The Journal of the Acoustical Society of America 128: 567–570.

28. Finneran JJ, Schlundt CE (2013) Effects of fatiguing tone frequency on temporary threshold shift in bottlenose dolphins (*Tursiops truncatus*). The Journal of the Acoustical Society of America 133: 1819–1826.

29. Finneran JJ, Schlundt CE (2011) Subjective loudness level measurements and equal loudness contours in a bottlenose dolphin (*Tursiops truncatus*). The Journal of the Acoustical Society of America 130: 3124–3136.

30. Suzuki Y, Takeshima H (2004) Equal-loudness-level contours for pure tones. The Journal of the Acoustical Society of America 116: 918–933.

31. Robinson DW, Dadson RS (1956) A re-determination of the equal-loudness relations for pure tones. British Journal of Applied Physics 7: 166–181.

32. Finneran J, Jenkins A (2012) Criteria and thresholds for US Navy acoustic and explosive effects analysis. San Diego, California: Space and Naval Warfare Systems Center Pacific. 60 p.

33. Urick RJ (1983) Principles of underwater sound. New York: McGraw-Hill.

34. Blackwell SB, Lawson JW, Williams MT (2004) Tolerance by ringed seals (*Phoca hispida*) to impact pipe-driving and construction sounds at an oil production island. The Journal of the Acoustical Society of America 115: 2346–2357.

35. Fisher FH, Simmons VP (1977) Sound absorption in sea water. The Journal of the Acoustical Society of America 62: 558–564.

36. Hamilton EL (1976) Sound attenuation as a function of depth in the sea floor. The Journal of the Acoustical Society of America 59: 528–535.

37. Knobles DP, Koch RA, Thompson LA, Focke KC, Eisman PE (2003) Broadband sound propagation in shallow water and geoacoustic inversion. The Journal of the Acoustical Society of America 113: 205–222.

38. Finneran JJ, Schlundt CE, Dear R, Carder DA, Ridgway SH (2002) Temporary shift in masked hearing thresholds in odontocetes after exposure to single underwater impulses from a seismic watergun. The Journal of the Acoustical Society of America 111: 2929–2940.

39. Finneran JJ, Schlundt CE, Carder DA, Ridgway SH (2002) Auditory filter shapes for the bottlenose dolphin (*Tursiops truncatus*) and the white whale (*Delphinapterus leucas*) derived with notched noise. The Journal of the Acoustical Society of America 112: 322.

40. Hawkins ER, Gartside DF (2010) Whistle emissions of Indo-Pacific bottlenose dolphins (*Tursiops aduncus*) differ with group composition and surface behaviors. The Journal of the Acoustical Society of America 127: 2652–2663

41. Zar JH (1999) Biostatistical analysis. Upper Saddle River, NJ: Prentice-Hall. 929 p.

42. Hildebrand JA (2009) Anthropogenic and natural sources of ambient noise in the ocean. Marine Ecology Progress Series 395: 5–20.

43. Richardson WJ, Würsig B (1997) Influences of man-made noise and other human actions on cetacean behaviour. Marine and Freshwater Behaviour and Physiology 29: 183–209.

44. Scheifele PM, Andrew S, Cooper RA, Darre M, Musiek FE, et al. (2005) Indication of a Lombard vocal response in the St. Lawrence River beluga. The Journal of the Acoustical Society of America 117: 1486–1492.

45. Lesage V, Barrette C, Kingsley MCS, Sjare B (1999) The effect of vessel noise on the vocal behavior of belugas in the St. Lawrence river estuary,Canada. Marine Mammal Science 15: 65–84.

46. Foote AD, Osborne RW, Hoelzel AR (2004) Environment: Whale-call response to masking boat noise. Nature 428: 910–910.

47. Holt MM, Noren DP, Veirs V, Emmons CK, Veirs S (2009) Speaking up: Killer whales (*Orcinus orca*) increase their call amplitude in response to vessel noise. The Journal of the Acoustical Society of America 125: EL27–EL32.

48. Richardson WJ, Würsig B, Greene CR (1986) Reactions of bowhead whales, *Balaenamysticetus*, to seismic exploration in the Canadian Beaufort Sea. The Journal of the Acoustical Society of America 79: 1117–1128.

49. Brandt MJ, Diederichs A, Betke K, Nehls G (2011) Responses of harbour porpoises to pile driving at the Horns Rev II offshore wind farm in the Danish North Sea. Marine Ecology Progress Series 421: 205–216.

50. Pirotta E, Brookes KL, Graham IM, Thompson PM (2014) Variation in harbour porpoise activity in response to seismic survey noise. Biology Letters. doi: 10.1098/rsbl.2013.1090.

51. Buckstaff KC (2004) Effects of watercraft noise on the acoustic behavior of bottlenose dolphins, *Tursiops truncatus*, in Sarasota bay, Florida. Marine Mammal Science 20: 709–725.

52. Morisaka T, Shinohara M, Nakahara F, Akamatsu T (2005) Effects of ambient noise on the whistles of Indo-Pacific bottlenose dolphin populations. Journal of Mammalogy 86: 541–546.

53. Codarin A, Wysocki LE, Ladich F, Picciulin M (2009) Effects of ambient and boat noise on hearing and communication in three fish species living in a marine protected area (Miramare, Italy). Marine Pollution Bulletin 58: 1880–1887.

54. Vasconcelos RO, Amorim MCP, Ladich F (2007) Effects of ship noise on the detectability of communication signals in the Lusitanian toadfish. Journal of Experimental Biology 210: 2104–2112.

55. David JA (2006) Likely sensitivity of bottlenose dolphins to pile-driving noise. Water and Environment Journal 20: 48–54.

56. Erbe C (2002) Underwater noise of whale-watching boats and potential effects on killer whales (*Orcinus orca*), based on an acoustic impact model. Marine Mammal Science 18: 394–418.

57. Sims PQ, Hung SK, Wursig B (2012) High-Speed vessel noises in west Hong kong waters and their contributions relative to Indo-Pacific humpback dolphins (*Sousa chinensis*). Journal of Marine Biology.doi:10.1155/2012/169103.

58. JNCC (2008) Draft guidelines for minimising acoustic disturbance to marine mammals from seismic surveys. Aberdeen: Joint Nature Conservation Committee. Available:http://www.jncc.gov.uk/marine. Accessed 3 January 2014.

59. Würsig B, Greene Jr C, Jefferson T (2000) Development of an air bubble curtain to reduce underwater noise of percussive piling. Marine Environmental Research 49: 79–93.

60. Popov VV, Supin AY, Pletenko MG, Tarakanov MB, Klishin VO, et al. (2007) Audiogram variability in normal bottlenose dolphins (*Tursiops truncatus*). Aquatic Mammals 33: 24–33.

61. Houser DS, Gomez-Rubio A, Finneran JJ (2008) Evoked potential audiometry of 13 Pacific bottlenose dolphins (*Tursiops truncatus gilli*). Marine Mammal Science 24: 28–41.

62. Wang ZT, Akamatsu T, Wang KX, Wang D (2014) The Diel Rhythms of Biosonar Behavior in the Yangtze Finless Porpoise (*Neophocaena asiaeorientalis asiaeorientalis*) in the Port of the Yangtze River: The Correlation between Prey Availability and Boat Traffic. PLoS ONE 9: e97907.

63. Au WWL, Giorli G, Chen J, Copeland A, Lammers M, et al. (2013) Nighttime foraging by deep diving echolocating odontocetes off the Hawaiian islands of Kauai and Ni'ihau as determined by passive acoustic monitors. The Journal of the Acoustical Society of America 133: 3119–3127.

64. Wang ZT, Akamatsu T, Mei ZG, Dong LJ, Imaizumi T, et al. (2014) Frequent and prolonged nocturnal occupation of port areas by Yangtze finless porpoises (*Neophocaena asiaeorientalis*): forced choice for feeding?. Integrative Zoology. doi: 10.1111/1749-4877.12102.

65. Casper BM, Halvorsen MB, Matthews F, Carlson TJ, Popper AN (2013) Recovery of Barotrauma Injuries Resulting from Exposure to Pile Driving Sound in Two Sizes of Hybrid Striped Bass. PLoS ONE 8: e73844.

66. Halvorsen MB, Casper BM, Woodley CM, Carlson TJ, Popper AN (2012) Threshold for Onset of Injury in Chinook Salmon from Exposure to Impulsive Pile Driving Sounds. PLoS ONE 7: e38968.

67. Burros NB, Myrberg AA (1987) Prey detection by means of passive listening in bottlenose dolphins (*Tursiops truncatus*). The Journal of the Acoustical Society of America 82: S65–S65.

68. Gannon DP, Barros NB, Nowacek DP, Read AJ, Waples DM, et al. (2005) Prey detection by bottlenose dolphins, *Tursiops truncatus*: an experimental test of the passive listening hypothesis. Animal Behaviour 69: 709–720.

8

Using Models of Social Transmission to Examine the Spread of Longline Depredation Behavior among Sperm Whales in the Gulf of Alaska

Zachary A. Schakner[1]*, Chris Lunsford[2], Janice Straley[3], Tomoharu Eguchi[4], Sarah L. Mesnick[4]*

1 Department of Ecology and Evolutionary Biology, University of California Los Angeles, Los Angeles, CA, United States of America, **2** Alaska Fisheries Science Center, National Marine Fisheries Service, NOAA, Auke Bay Laboratories, Juneau, AK, United States of America, **3** University of Alaska Southeast, Sitka, AK, United States of America, **4** Southwest Fisheries Science Center, National Marine Fisheries Service, NOAA, La Jolla, CA, United States of America

Abstract

Fishing, farming and ranching provide opportunities for predators to prey on resources concentrated by humans, a behavior termed depredation. In the Gulf of Alaska, observations of sperm whales depredating on fish caught on demersal longline gear dates back to the 1970s, with reported incidents increasing in the mid-1990s. Sperm whale depredation provides an opportunity to study the spread of a novel foraging behavior within a population. Data were collected during National Marine Fisheries Service longline surveys using demersal longline gear in waters off Alaska from 1998 to 2010. We evaluated whether observations of depredation fit predictions of social transmission by fitting the temporal and spatial spread of new observations of depredation to the Wave of Advance model. We found a significant, positive relationship between time and the distance of new observations from the diffusion center ($r^2 = 0.55$, p-value $= 0.003$). The data provide circumstantial evidence for social transmission of depredation. We discuss how changes in human activities in the region (fishing methods and regulations) have created a situation in which there is spatial-temporal overlap with foraging sperm whales, likely influencing when and how the behavior spread among the population.

Editor: Garet P. Lahvis, Oregon Health and Science University, United States of America

Funding: Z.A.S. is supported by a National Science Foundation Graduate Research Fellowship. The funder had no role in study design, data collection and analysis, decision to publish, or preparation of the manuscript.

Competing Interests: The authors have declared that no competing interests exist.

* Email: zschakner@ucla.edu (ZS); sarah.mesnick@noaa.gov (SM)

Introduction

Fishing, farming and ranching provide opportunities for predators to prey on resources concentrated by humans, a behavior termed depredation. In the oceans, increasing global fishing effort provides a multitude of opportunities for marine mammals to exploit prey caught on lines, in nets or aquaculture pens. The concentrated prey provides strong energetic incentives for predators presumably because of reduced costs of foraging. Attention to depredation by marine mammals is growing [1,2]. It is observed in many fisheries, is economically costly and can cause injurious or lethal entanglement in gear [1,2]. The global scope of depredation, along with observations of depredation rapidly spreading through some marine mammal populations raise questions about how new foraging behaviors arise and are transmitted and also have led to suggestion that the behavior may be socially transmitted [3,4]. Quantifying the mechanisms by which behavioral traits spread in any wild population, however, is challenging because field studies rarely allow for reliably distinguishing between asocial and social mechanisms [5]. As stated in [6], these problems are exacerbated by the logistical challenges of accessibility and visibility inherent in studying marine mammals at sea.

Depredation of demersal longline catches is observed in high latitude feeding grounds frequented by adult male sperm whales (*Physeter macrocephalus*) in the eastern North Pacific, the North Atlantic and the Southern Ocean [7]. Early observations of sperm whale depredation of sablefish, (*Anoplopoma fimbria*) in the Gulf of Alaska are not well documented, but anecdotal accounts date back to mid-1970s [V. O'Connell 2012 pers. comm.]. In 1996, reports of depredation began to substantially increase in the region coincident with changes in fishery management that lengthened the fishing season [8,9]. Longitudinal observations of sperm whale depredation provide an opportunity to examine the mechanisms underlying the diffusion of a new and complex behavior through a wild population. Studies of social learning in humans and terrestrial animals commonly utilize the assumption that socially transmitted behaviors are expected to show accelerated diffusion [10,11], while traits acquired independently are not expected to arise in a spatially linked pattern nor to spread as quickly. Yet, while social transmission may increase the likelihood of an accelerating increase over time, the pattern can be generated by entirely asocial processes, and the shape of diffusion curves alone cannot be reliably interpreted as an indicator of social learning [12]. Newer methods, such as network based diffusion analysis, that take into account individual level information are more reliable [6,13] yet are beyond the scope of many studies of wild animals. Using data collected during National Marine Fisheries Service (NMFS) assessment surveys for sablefish, we plot the occurrence of depredation over time to graphically illustrate the rate of behavioral transmission. We then investigate the spatial and temporal distribution of depredation using an indicator of social transmission, the Wave-of-Advance model.

The Wave-of-Advance model quantifies the relationship between first observation of a behavior and its spread over time [14,15]. If a novel behavior is socially transmitted through a population, it is expected to have a diffusion center or geographical location where the innovation first arises, with new occurrences of the behavior progressively radiating outward through time. The Wave-of-Advance uses simple linear regression to test for a positive correlation between the distance a behavior has spread and time. This pattern was observed during the presumed social transmission of agriculture in Neolithic Europe [15,16]. Quantitatively, social transmission is evidenced by a linear increase in the distances of new observations of the behavior from the diffusion center over time. Alternatively, if a behavior arises independently among multiple innovators, there would be scattered pattern with wide variation between space and time.

Material and Methods

(a) Data Collection

Survey background. Data were collected during annual sablefish longline surveys conducted by the National Marine Fisheries Service from 1998 to 2010. These standardized, fishery-independent surveys cover the upper continental slope and selected gullies of the eastern Bering Sea, Aleutians Islands, and Gulf of Alaska. The surveys cover nearly all areas where adult sablefish are found and overlap commercial demersal longline fishing regions within the U.S. Exclusive Economic Zone. Sampling occurs annually in the summer and lasts three months. The survey follows a systematic design by placing stations that are 30–60 km apart at depths of 150 to 1000 m (Figure 1; Table S1 in File S1). Each year, approximately 90 stations were sampled. Beginning in 1998, observers began collecting data on depredation. Depredation was defined to occur when sperm whales were found near the vessel and adjacent to the longline during haulback (typically 100 m or less from the vessel) and when damaged sablefish were retrieved [16]. Characteristics of damaged sablefish include missing body parts, shredded tissue or lips remaining on hooks. At each station, the presence/absence of depredation was recorded. During a depredation event, the number of sperm whales present and the number of damaged sablefish were also recorded, but individual identification of the whales was not recorded.

(b) Statistical analysis

We plotted the occurrence of depredation over time to illustrate the rate of behavioral transmission. The rate of behavioral transmission was modelled with the cumulative number of stations observed with depredation (y) as functions of time (x). Four functions were fit to the data: a function previously shown to indicate a constant rate [(linear ($y = a_1 + b_1*x$), decelerating rate [logarithmic ($y = a_2 + b_2* \log x$)], and two functions indicating an accelerative rate of transmission [exponential ($y = a_3*e^{(b_3*x)}$) and logistic ($y = k/(1 + e^{(a_4 - r*x)})$). We used an iterative least squares curve fitting procedure implemented by SigmaPlot (v.11). To compare fit of these models, we used Akaike's Information Criterion (AICc) for small sample sizes [17].

We mapped geographic distribution of depredation through time to construct the Wave-of-Advance model. We estimated that the center of diffusion occurred in the location where depredation was observed in the first year of our study. Since depredation occurred at four stations in the Central Gulf (Figure 2a), we measured the midpoint of those stations for the diffusion center).

With each successive survey year, the distance of the farthest new observation of depredation from the diffusion center was measured. These data were plotted with the distance from the center of diffusion as ordinate and survey year as abscissa and analyzed with a linear regression model in R [18] to determine if the relationship fit a linear spatial advance. We tested the null hypothesis that the slope is equal to zero. Under the null hypothesis, there is no relationship between the timing of new observations of depredation and their distance from the diffusion center, which would suggest the independent origin of the behavior. Statistical significance was tested at the type I error rate of 0.05.

Results

Over the study period, depredation was observed on 126 occasions (Table S2 in File S1). The number of stations with depredation in a given year ranged from 4 (1998) to 22 (2008). The number of individuals present at stations with depredation ranged from 1–7 individuals (mean of 3.0). Depredation occurred only in the Gulf of Alaska and not along the Aleutian Islands or in the Bering Sea. The cumulative number of stations to observe depredation over time is presented in Figure 2. Over the study period, observations of depredation showed an accelerative pattern of increase over time. All four functions explained the variability in data (i.e., high r^2 values), but the logistic function was the best model based on AICc (Table 1, Figure 2; see Figure S1 in File S1 for all functions).

Depredation was first observed at four stations in 1998 and radiated from the West Yakutat region to the southeast and southwest during the 12 years (Figure 3). The Wave-of-Advance model shows a significant, positive relationship between time and the distance of new observations of depredation from the diffusion center ($r^2 = 0.55$, p-value $= 0.003$), thus we can reject the null hypothesis (slope = 0). The speed at which the behavior radiated is implied from the slope of the solid blue line, 81.3±47.9 km/year (95% confidence interval).

Discussion

Exploration of the spatiotemporal spread of depredation behavior using both the rate of transmission and Wave-of-Advance analyses provides insights that neither method can alone. The observed spatiotemporal radiation of depredation in the Gulf of Alaska provides evidence, albeit circumstantial, for social transmission and provides context for further investigations as to why and how it evolved.

We acknowledge the limitations to both models which use stations as proxies for individuals yet enable us to provide standardized data throughout the study area. We are also unable to control for factors such as individual sperm whale movement or changes in commercial fishing effort that may provide alternative explanations for the observed spread of depredation. Male sperm whales are capable of travelling long distances [19,20], are known to follow fishing vessels and some are likely to be repeat offenders among stations and years [20]. If the original innovator were a single whale, or a few individuals, that had learned to depredate commercial boats at the diffusion center and then expanded their range outward, this might explain the observed spatial pattern. This scenario, however, is unlikely because depredation continued to be observed at all intervening stations, including the diffusion center, during the course of all subsequent survey seasons. If there were changes in the distribution of sablefish, or of commercial fishing for sablefish in the Gulf of Alaska this might also influence the observed spatial radiation. If, for example, the distribution of

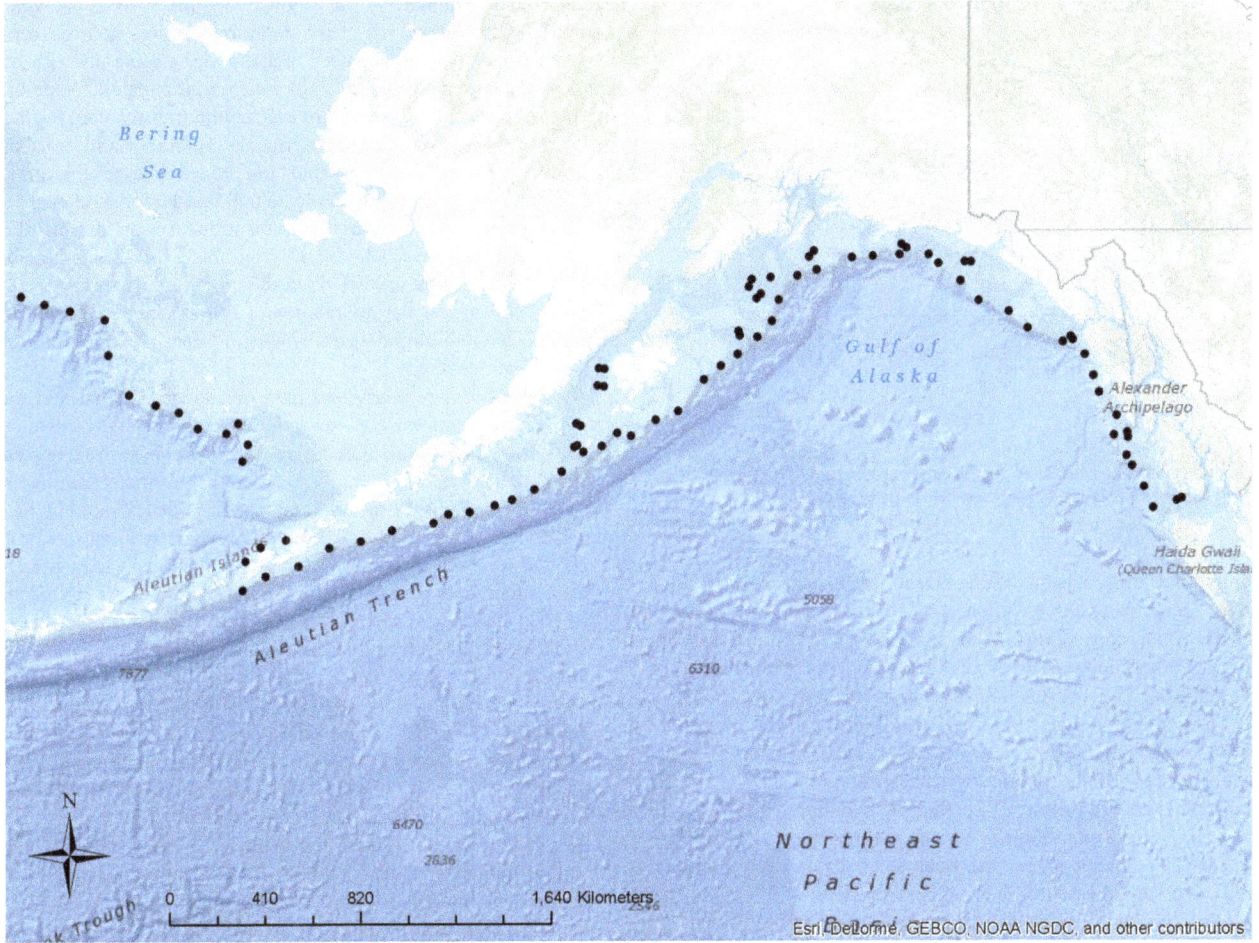

Figure 1. Survey area for sablefish assessment survey.

fish or fishing effort increased outside the diffusion center in Central Gulf, there could potentially be more opportunities for individuals to acquire the behavior, which would be reflected in our survey as new stations with depredation. In addition, it is important to note that despite commercial fishing for sablefish throughout the study area, this study only observed depredation at research stations in the Gulf of Alaska. If the behavior was independently acquired, we would expect to see occurrences arise randomly anywhere in the study area, including in the Aleutians and Bering Sea. Further studies with the commercial fishing sector could help to illuminate these factors. In addition, further studies which relate the foraging preferences of sperm whales to fish availability may shed light on ecological factors influencing the

behavior. For example, historic whaling records found that fish occurred in the stomachs of sperm whales more often in the Gulf of Alaska than along the Aleutians or in the Bering Sea [21] which may explain depredation being limited to the Gulf of Alaska.

The plot of the number of stations to experience depredation through time, and results of the function fitting exercise suggest an accelerated rate of increase in the occurrence of depredation among Gulf of Alaska sperm whales. We are cautious to infer social transmission from the transmission rate analysis alone, however, because of potential asocial explanations [12]. For instance, a decrease in neophobia (the avoidance response to novel stimuli), can influence transmission rates [22]. If the whales become less neophobic toward fishing gear over time, the rate at

Table 1. Rate of transmission analyses using Akaike Information Criterion (*best fit).

Model	# of Parameters	Parameter estimates	Uncorrected residual	r^2	ΔAIC_c	Σw_i
Logistic*	3	k = 182 r = 10.2, a_4 = 3.5	124.7	.99	0	.995
Linear	2	a_1 = 11.19 b_1 = −10.04	382	.97	11.04	.003
Exponential	2	a_3 = 15.4 b_3 = .17	476.5	.97	14	.001
Logarithmic	2	a_2 = −21.4, b_2 = 46.4	3561	.79	40.14	.000

Cumulative number of depredated stations

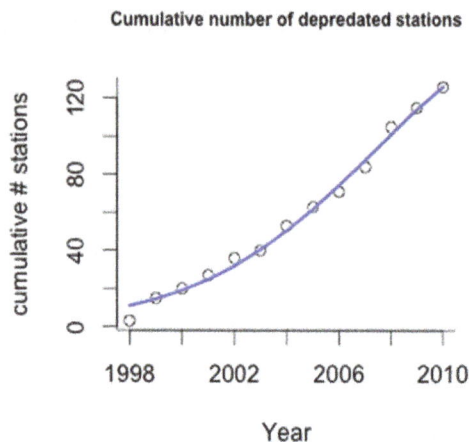

Figure 2. The cumulative number of stations with depredation over time with best fitting function (logistic).

which they approach the fishing vessels might increase – resulting in an acceleratory curve without social transmission. In the case of sperm whale depredation, there is much room for the development of other methods, especially those that incorporate individual identification for investigating social transmission.

The conditions that gave rise to the original innovation are not well known, but the presence of longline fishing provides opportunities for individuals to learn to acquire prey off of the lines. Anecdotal accounts of longline depredation extend back to the 1970s, suggesting that this time period may represent a period of early innovation before the behavior became more common and widespread in the late 1990s. Beginning in 1984, the previously year-round sablefish season in the Gulf of Alaska was shortened, as short as 10 days in some years. In 1995,

management shifted to an individual quota system, which extended the season to eight months. Fishing is now open March-November, overlapping spatially and temporally with the time when sperm whales naturally forage in the region. While our data are suggestive of a period of accelerated transmission, the lack of pre-1998 survey data is a major limitation and prevents examination of the events that may have caused the initial association or its expansion in any detail. In addition to changes in fishing season duration, other potential influences on the spread of depredation may be at play such as on-board fish processing and whether there is fishery offal (discard) in the water that may attract whales, changes in the abundance of fish stocks over time, and changes in the abundance of whales after the cessation of commercial whaling.

Despite a lack of behavioral data on individual whales at this time, the mechanisms (i.e., imitation, emulation, local enhancement) by which individuals might acquire behaviors consistent with depredation from conspecifics deserves further examination with other methods. While adult male sperm whales found on high latitude feeding grounds are generally thought to forage solitarily [19], our results also show that groups are more likely to be present during depredation events than solitary animals [Table S1 in File S1]. This observation is consistent with other studies in the region which have found that individually-identified males exhibit differing levels of association with vessels [23]. Depredation in groups was also more widespread in the later years of the study [Table S1 in File S1]. Aggregating around prey resources may facilitate social learning and may reveal an underlying male sociality, evidence of which is also observed in the existence of "bachelor schools" (loose aggregations of subadult males [19].) On-going studies to observe the behavior of depredating whales through satellite tagging and photo-identification studies are documenting male movements and may shed light on interactions among adult males. In addition, acoustic studies suggest individuals may eavesdrop for information on food aggregations [24].

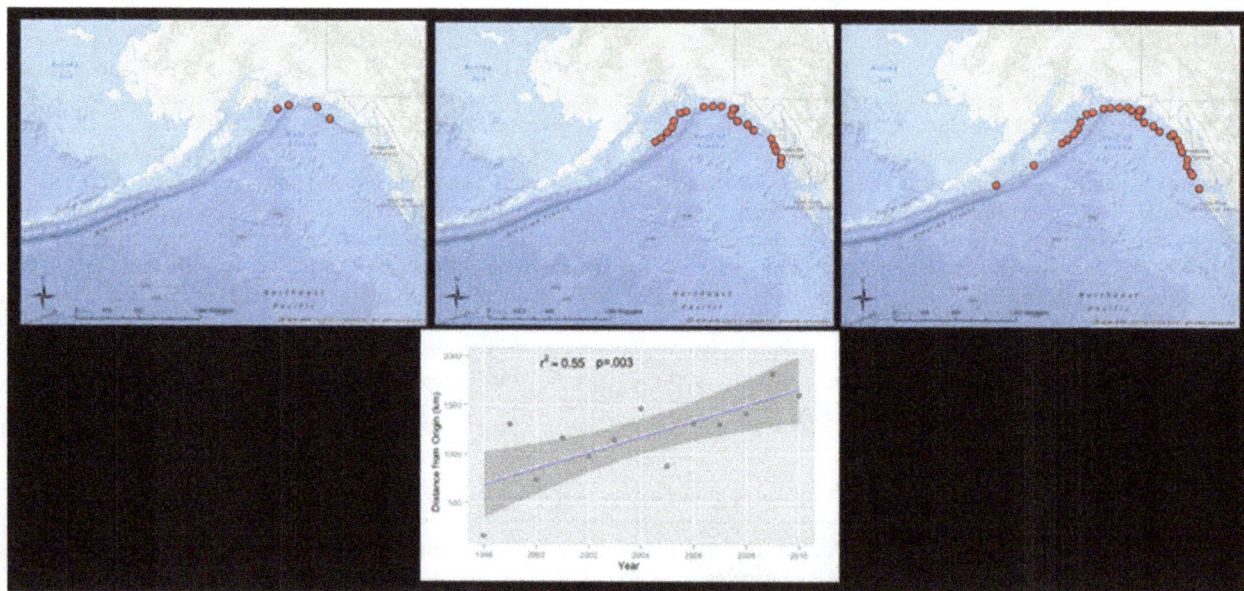

Figure 3. Spatial radiation of depredation and Wave-of-Advance model shows a positive correlation between time and the distance of new observations of depredation from origin ($r^2 = .55$, p-value = 0.003). Red dots represent stations to observe depredation. The panels, from left to right represent 1998, 1998–2003, and 1998–2010. The speed at which the behavior radiated is the slope of solid blue line, or 81.3±42.6 km/year (95% confidence interval). The solid blue line represents regression line and the gray shaded area is the 95% confidence interval for the line of best fit.

Novel vocalization patterns produced during depredation [25] may convey cues used by nearby individuals and increase the probability of interaction through local or stimulus enhancement.

Human hunting, fishing and farming provides foraging opportunities for wildlife, resulting in the spread of novel foraging behaviors. The underlying transmission mechanisms have consequences for the rate and geographical spread of depredation. Social transmission can function as a multiplier in that new traits can spread more quickly and completely through a population than by individual acquisition. Knowledge of these mechanisms is important for management, which may find mitigation more tractable early on, before the behavior spreads through the population.

Supporting Information

File S1 Figure S1, Cumulative number of stations fit with linear, exponential, logarithmic, and sigmoid functions. **Table S1,** Survey Coordinates. **Table S2,** Stations with depreda-

tion, 1998–2010. Station numbers are given along with the number of individuals observed at those stations (in parenthesis) for the years in which these data were recorded.

Acknowledgments

We thank Karin Forney, Robert Brownell, Jr., Peter Fashing, members of the Blumstein Lab and two anonymous for their valuable comments; the crew and scientists on the NMFS longline survey for data collection; and SEASWAP (Sperm Whale and Longline Fisheries Interactions in the Gulf of Alaska) team for their expertise on sperm whale depredation in the region.

Author Contributions

Conceived and designed the experiments: ZS SM. Performed the experiments: ZS SM TE. Analyzed the data: ZS SM TE. Contributed reagents/materials/analysis tools: CL JS. Wrote the paper: ZS SM.

References

1. Read AJ (2008) The looming crisis: Interactions between marine mammals and fisheries. J Mammal 89: 541–548.
2. Hamer DJ, Childerhouse SJ, Gales NJ (2012) Odontocete bycatch and depredation in longline fisheries: A review of available literature and of potential solutions. Mar Mammal Sci 28: E345–E374. (doi:10.1111/j.1748-7692.2011.00544.x).
3. Powell JR, Wells RS (2011) Recreational fishing depredation and associated behaviors involving common bottlenose dolphins (Tursiops truncatus) in Sarasota Bay, Florida. Mar Mammal Sci 27: 111–129. doi:10.1111/j.1748-7692.2010.00401.x.
4. Donaldson R, Finn H, Bejder L, Lusseau D, Calver M (2012) The social side of human–wildlife interaction: wildlife can learn harmful behaviours from each other. Anim Conserv: 1–9. doi:10.1111/j.1469-1795.2012.00548.x.
5. Kendal RL, Galef BG, Schaik CPV (2010) Social learning research outside the laboratory: How and why? Learning & Behavior 38: 187–194. doi:10.3758/LB.38.3.187.
6. Allen J, Weinrich M, Hoppitt W, Rendell L (2013) Network-based diffusion analysis reveals cultural transmission of lobtail feeding in humpback whales. Science 340: 485.
7. Arangio R (2012) Minimising whale depredation on longline fishing. Nuffield Australia Farming Scholars Report, Project No.1201. Unpublished report. Available: http://nuffield.com.au.
8. Hill PS, Laake JL, Mitchell E (1999) Results of a pilot program to document interactions between sperm whales and longline vessels in Alaska waters. U.S. Dep. Commer., NOAA Tech.Memo.NMFS-AFSC-108, 42 p.
9. Hanselman DH, Lunsford CR, Rodgveller CJ (2011) Assessment of the sablefish stock In Alaska Stock Assessment and Fishery Evaluation Report for the Groundfish Resources of the Gulf of Alaska, North Pacific Fishery Management Council, 329–468.
10. Lefebvre L (1995) The opening of milk bottles by birds – evidence for accelerating learning rates, but against the wave-of-advance model of cultural transmission. Behav Process 34: 43–53.
11. Lefebvre L (1995) Culturally transmitted feeding behavior in primates: Evidence for accelerating learning rates. Primates 36: 227–239.
12. Reader S (2004) Distinguishing social and asocial learning using diffusion dynamics. Learn Behav 32: 90–104.
13. Franz M, Nunn CL (2009) Network-based diffusion analysis: a new method for detecting social learning. Proc R Soc B 276: 1829–1836. doi:10.1098/rspb.2008.1824.13.
14. Ammerman AJ, Cavalli-Sforza LL (1984) The neolithic transition and the genetics of populations in Europe. Princeton University Press. 176 p.
15. Ammerman AJ, Cavalli-Sforza LL (1971) Measuring the rate of spread of early farming in Europe. Man 6: 674–688. doi:10.2307/2799190.
16. Sigler MF, Lunsford CR, Straley JM, Liddle JB (2008) Sperm whale depredation of sablefish longline gear in the northeast Pacific Ocean. Mar Mammal Sci 24: 16–27.
17. Burnham KP, Anderson DR (2002) Model selection and multi-model inference: a practical information-theoretic approach. New York, Springer Verlag.
18. R Core Team (2014) R: A language and environment for statistical computing. R Foundation for Statistical Computing, Vienna, Austria. Available: http://www.R-project.org/.
19. Whitehead H (2003) Sperm whales: social evolution in the ocean. Chicago: University of Chicago Press.
20. Straley JM, Schorr GS, Thode AM, Calambokidis J, Lunsford CR, et al. (2014) Depredating sperm whales in the Gulf of Alaska: local habitat use and long distance movements across putative population boundaries. End. Spec. Res.
21. Kawakami T (1980) A review of sperm whale food. Sci Rep Whales Res Inst.
22. Hoppitt W, Kandler A, Kendal JR, Laland KN (2010) The effect of task structure on diffusion dynamics: Implications for diffusion curve and network-based analyses. Learning & Behavior 38: 2438: 243ehavio.3758/LB.38.3.243.
23. Straley J, O'Connell T, Mesnick SL, Behnken L, Liddle J (2005) Sperm Whale and Longline Fisheries Interactions in the Gulf of Alaska. North Pacific Research Board Final Report R0309.
24. Madsen PT, Wahlberg M, Møhl B (2002) Male sperm whale acoustics in a high-latitude habitat: implications for echolocation and communication. Behav Ecol Sociobiol 53: 31–41.
25. Mathias D, Thode AM, Straley J, Calambokidis J, Schorr GS, et al. (2012) Acoustic and diving behavior of sperm whales (Physeter macrocephalus) during natural and depredation foraging in the Gulf of Alaska. J Acoust Soc Am 132: 518–532.

9

Temporal Dynamics of Top Predators Interactions in the Barents Sea

Joël M. Durant[1]*, Mette Skern-Mauritzen[2], Yuri V. Krasnov[3], Natalia G. Nikolaeva[4], Ulf Lindstrøm[5], Andrey Dolgov[6]

1 Centre for Ecological and Evolutionary Synthesis (CEES), Department of Biosciences, University of Oslo, Oslo, Norway, 2 Institute of Marine Research, Bergen, Norway, 3 Murmansk Marine Biological Institute, Murmansk, Russian Federation, 4 White Sea Biological Station, Department of Biology, Lomonosov Moscow State University, Moscow, Russian Federation, 5 Institute of Marine Research, Tromsø, Norway, 6 Knipovich Polar Research Institute of Marine Fisheries and Oceanography (PINRO), Murmansk, Russian Federation

Abstract

The Barents Sea system is often depicted as a simple food web in terms of number of dominant feeding links. The most conspicuous feeding link is between the Northeast Arctic cod *Gadus morhua*, the world's largest cod stock which is presently at a historical high level, and capelin *Mallotus villosus*. The system also holds diverse seabird and marine mammal communities. Previous diet studies may suggest that these top predators (cod, bird and sea mammals) compete for food particularly with respect to pelagic fish such as capelin and juvenile herring (*Clupea harengus*), and krill. In this paper we explored the diet of some Barents Sea top predators (cod, Black-legged kittiwake *Rissa tridactyla*, Common guillemot *Uria aalge*, and Minke whale *Balaenoptera acutorostrata*). We developed a GAM modelling approach to analyse the temporal variation diet composition within and between predators, to explore intra- and inter-specific interactions. The GAM models demonstrated that the seabird diet is temperature dependent while the diet of Minke whale and cod is prey dependent; Minke whale and cod diets depend on the abundance of herring and capelin, respectively. There was significant diet overlap between cod and Minke whale, and between kittiwake and guillemot. In general, the diet overlap between predators increased with changes in herring and krill abundances. The diet overlap models developed in this study may help to identify inter-specific interactions and their dynamics that potentially affect the stocks targeted by fisheries.

Editor: Brian R. MacKenzie, Technical University of Denmark, Denmark

Funding: This work was supported by the Research Council of Norway (http://www.forskningsradet.no/en/Home_page/1177315753906) through the ADMAR project (grant no. 200497/130). The funders had no role in study design, data collection and analysis, decision to publish, or preparation of the manuscript.

Competing Interests: The authors have declared that no competing interests exist.

* Email: joel.durant@ibv.uio.no

Introduction

The Barents Sea is an open Arcto-boreal shelf-sea with an average depth of about 230 m. This ecosystem is both of large applied interest due to the large commercial fisheries, and also an interesting biological system showing clear bottom-up effects [1,2], top-down effects [3–5] and climate effects [1,6]. The climate appears to have a strong effect on the trophic control in the Barents Sea in that both climate and trophic control change with a decadal periodicity [7]. Understanding linkages between climate and trophic interactions is important for understanding the changes in the Barents Sea biodiversity expected to follow climate and harvesting changes.

Fairly simple pelagic Arctic ecosystems such as the Barents Sea [8] may be more vulnerable to changes in the abundance of the few key species [9] compared to more diverse system in terms of link strengths [10]. For instance, the collapse of the Barents Sea capelin *Mallotus villosus* stock in the 1980s significantly affected several trophic levels including the capelin prey, zooplankton [11], capelin predators such as the Northeast Arctic (NEA) cod *Gadus morhua* [12] and the harp seal *Pagophilus groenlandicus* [13], and alternative prey of capelin predators such as shrimp [14,15].

In recent years, understanding and predicting food web dynamics in the Barents Sea have become a priority with the aim at improving the management of marine resources. As a result, there has been an increased focus on ecosystem or multispecies models [16–18]. Indeed, on an ecological time scale, top-predators can affect the abundance of other species through predation or competition. The top-predator (species at the top of their food chain) community in the Barents Sea consists of about 33 seabird species [19] and 21 mammal species [16], in addition to large demersal fish, of which cod is the most abundant species. Interactions between several of these species have previously been identified, such as between the Black-legged kittiwake *Rissa tridactyla* and the Common guillemot *Uria aalge* [20], between the Minke whale *Balaenoptera acutorostrata* and the NEA cod [21], and between the harp seal and the NEA cod [22]. Such interactions should be included when investigating population dynamics in an ecosystem context, i.e., taking into account species interactions.

The sensitivity of each top predator species to changes in prey availability depends on the availability of, and ability to use, alternative prey species. The sensitivity of the top predator community, being the sum of each predator sensitivity, depends on the response diversity within the community [23]. If the majority of the predators of the community responds to a stressor in the same way then the predator community will show a sensitivity to the change of this particular stressor. On the contrary, if the predator responses are diverse, the community is more robust. Based on the strong effects across species of the two first capelin collapses we may expect a low response diversity and hence a sensitive community. However, the top predators are typically generalists, foraging on different prey species depending on spatiotemporal overlap between predator and prey distributions, and abundance of the different prey species in the system [24,25]. It is therefore likely that the response diversity of these top predators may depend on the availability of alternative prey in the system.

In this study we explore the interactions between some of the major top predators of the Barents Sea ecosystem, by investigating the diet overlap among them. These species may only interact (e.g., compete for food) if they share a certain amount of the prey resources. The prey availability in the system is varying both due to natural cycles [e.g., capelin, 26] and to anthropogenic pressures [e.g., fishing, 27]. We therefore expect that the top predator diets are varying through time. If the predators demonstrate species specific responses to changing prey abundances the number of response types may be high and lead to a year-to-year changes in both trophic and competitive interactions. To address these topics we have conducted a diet overlap analysis based on stomach content over the years and run generalised additive model to try to explain the temporal changes observed. We expect that the diet changes may be due to changes in prey abundance and distribution as well as in climate that can affect those.

Methods

Data

Following a simple food web description of the Barents Sea [8], the main predators in terms of total consumption [16] are the NEA cod (thereafter cod), the Minke whale, the harp seal and seabirds. The latter group includes black-legged kittiwake and the common guillemot, thereafter kittiwake and guillemot respectively. The diet data of four of these species is displayed in Table 1 and Fig. S1 in File S1; unfortunately the harp seal was omitted from the analysis due to sparse data.

Collection of data, spatial and temporal extent of data

The origin of the diet data used is summarized in the Table 1. Data were transformed to annual average percentages by mass. To make inter-specific comparisons we merged some prey categories (see supplementary material, Table S1 and S2 in File S1). Seabird diets were obtained during the breeding period at Kharlov Island on the coast of the Barents Sea [28,29]. We calculated the whale diet for the entire Barents Sea as well as for a subset of the data restricted to the southern Barents Sea (Minke whale sampled south of the 75°N, Fig. S1 in File S1). The complete data are used to analyse the change in the whale diet over time and to compare with the diet of the cod. The subset data are used to make inter-specific comparisons with seabirds because they are central place foragers and limited to the southern Barents Sea during breeding (the period when the seabird data were collected). Data on cod diet were taken from the joint Russian-Norwegian PINRO-IMR data base [30,31], diet for the entire Barents Sea as well as for a subset of it (<72°N) to compare with seabirds were calculated.

We used two types of environmental variables, climate indices and prey abundances, as predictors in the statistical analyses (given in Table 2). As climate indices we considered the average Barents Sea surface temperature (ST, annual), an index of the areal coverage of cold, Arctic water in the Barents Sea and the winter North Atlantic Oscillation index (wNAO). The rationale for analysing the effect of climate indices on the diet changes is that these variables may influence the spatial distribution of both the predators and the prey [25,32,33], which is unknown in our study. Temperature influences zooplankton productivity [34] and also acts as a proxy for various direct and indirect effects [35]. In particular, high temperature has been associated with inflow of warm, and potentially zooplankton-rich, waters from the Norwegian Sea [35]. The North Atlantic Oscillation index measures large-scale climate effects, is positively correlated with inflow and temperature, and was found to be the best climatic predictor of zooplankton biomass in the Barents Sea in spring and summer [e.g., for plankton 5] and thus linked to the productivity of the system.

Capelin, euphausiids (krill) and, to some extent also, juvenile Norwegian Spring Spawning herring *Clupea harengus* (thereafter herring) were found to be the major prey species in all predators (see Fig. S1 in File S1). These prey species are also considered major players in the trophic dynamics of the Barents Sea [e.g., 26, 35–39]. Thus the abundances of these prey species were used as predictors in our models (Table 2).

Table 1. Species studied.

Species	Description and source	Years
Black-legged kittiwake *Rissa tridactyla*	Regurgitation of 653 adults on the breeding colony Kharlov Island on the coast of the Barents Sea (BS) during the breeding season (April-May).	1982–1999 (lacking data for 1984 and 1985)
Common guillemot *Uria aalge*	Observation of 1951 fish deliveries at the breeding colony Kharlov Island on the coast of the BS during the breeding season (April-May).	1984–1999 (lacking data for 1985)
NEA cod *Gadus morhua*	Stomach content [a,b]. To compare with the seabirds we used data for 68–72°N and 20–40°E only (Mar-July) and to compare with the whale data for 70–80°N and 5–40°E (July-Sept.). [c]	1984–2009
Minke whale *Balaenoptera acutorostrata*	Stomach content of 345 whales caught in the BS between May-Sept. To compare with the seabirds a subset for the area <75°N was used. [d]	1992–2004

[a]Report of the ICES Arctic Fisheries Working Group [61], Table 1.3 p 55.
[b]The Russian-Norwegian data base on cod diet, further details see Mehl and Yaragina [31], and Dolgov et al. [30]
[c]the subsets from this base.
[d]Further details on the capture and the stomach sampling is given in Haug et al. [62]

Table 2. Explanatory variables used for the GAM analyses. Subscript *t* refers to year.

Variable	Description and source
$ST_{(t)}$	Mean Barents Sea (BS) temperature in °C for and January$_t$ to December$_t$ at 0–200 m depth in Atlantic water parts of the Kola section (70.5–72.5°N, 33.5°E) over 1921–2009[a].
$NAO_{(t)}$	Principal component based winter (December$_{t-1}$ – March$_t$) North Atlantic Oscillation (NAO) index[b]
$Cap_{(t)}$	Biomass of capelin in the BS in 10^3 t[c].
$Krill_{(t)}$	Euphausiids, abundance indices covering 1984 to 2004 from the Polar Research Institute of Marine Fisheries and Oceanography (PINRO)[d]. Data are for southern (Krill.S) and the northwestern (Krill.NW) BS. Krill is the sum of both area.
$Herr_{(t)}$	Biomass of immature Norwegian Spring Spawning herring (1–2 years of age) in the BS in 10^3 t[e].

[a]Tereschenko [63, http://www.pinro.ru/], [b]Hurrell [64, https://climatedataguide.ucar.edu/sites/default/files/climate_index_files/nao_station_djfm.txt], [c]Report of the ICES Arctic Fisheries Working Group [65], Table 9.5 p 498, [d]Zhukova et al., 2009 data used with permission, and [e]Report of the ICES Arctic Fisheries Working Group AFWG Table 9.6 p 499.

Diet overlap Index and Niche breadth

Schoener's [40] index of niche overlap, which is the most commonly used diet overlap index [41,42], was used to calculate the diet overlap among predators:

$$O_{jk} = 1 - 0.5 \times \sum |p_{ij} - p_{ik}|$$

where O_{jk} is the overlap between the species j and the species k, p_{ij} is the proportion of species j feeding on prey species/group i and p_{ik} is the proportion of species k feeding on prey species/group i. O_{jk} values range from 0 to 1. Overlap in diet between species j and k is complete when $O_{jk} = 1$ and is absent when $O_{jk} = 0$ [20,41]. Values exceeding 0.6 are considered to represent "biologically significant" overlap in diet composition [42]. However, we considered that when mean $O_{jk} > 2 \cdot SD$ the diet overlap between species j and species k is significant [20,41].

Using original (non-merged) diet data, we have calculated the Schoener's index O for consecutive years (overlap of diet between years) for each predator species (kittiwake, guillemot, Minke whale, and cod). To do this we adapted the equation above to calculate diet overlap between years, replacing P_{ij} with $P_{i,j,t}$ and $P_{i,k}$ with $P_{i,j,t+1}$, where j denotes a predator species and t year. This way we obtained a diet overlap O_t between year t and t-1 and ultimately a time series of O indices of temporal trends in diet overlap between years within the species. Furthermore, year to year changes in O were then related to environmental descriptors using Generalized Additive Model (GAM, see below).

We have also calculated O for each pair of predators over common period of time, and O for each pair of predators from year to year.

To assess the complexity of the diet for each species, we used the Shannon-Wiener niche breadth index D [43]. The D index has the advantage of not being greatly affected by sample size. D was calculated as follows:

$$D = - \sum p_i \times \ln (p_i)$$

where p_i is the proportion of the species considered feeding on prey species/group i.

Statistical analyses

The temporal variability in diet overlap (O_t ranging from 0 to 1) was analysed with respect to prey abundance and climate variables (both regional and large-scale climate indices) using Generalized Additive Models (GAM) with a logit link function in the formulation (family quasi-binomial) using the mgcv library in R 2.14.1 [44,45]. Note that the quasi-binomial distribution takes into account overdispersion and underdispersion of the data.

We then modelled the observations Ot as coming from a quasi-binomial distribution with an expected value equal to logit($\alpha + \Sigma_i s_i (X_{i,t})$) where $s_i(\cdot)$ is a nonparametric smoothing function of covariate X_i on the dependent variable O. Note that the GAM analysis was conducted only for the predator pairs where the diet overlap O was considered significant (i.e., pairs where the diets overlapped over the whole studied period and not for some particular years only; see above).

The GAM procedure automatically selects the degree of smoothing based on the Generalized Cross Validation (GCV) score. GCV is a proxy for the model's predictive performance analogous to the Akaike's Information Criterion. However, to avoid spurious and ecologically implausible relationships, we constrained the model to be at maximum a quadratic relationship implying that we set the maximum degrees of freedom for each smooth term to 2 (i.e., k = 3 in the GAM formulation). The maximum number of explanatory variables on the starting models was depending on the time series length (number of variables should be ≤ to n/4, see Table 3). These explanatory variables were selected on two criteria that were availability and biological meaning.

We wanted a parsimonious model which described the response well but was as simple as possible. We entered every candidate predictor in a GAM model and conducted a shrinkage model selection by using thin plate regression spline smoother with "shrinkage" for each term of the model [46]. Unimportant terms were shrank to zero, i.e., effectively removing the term, by the fitting procedure, and thus selecting a reasonably optimal model in one step (i.e. the model that includes all of the terms that were not shrunk to zero). There was no temporal autocorrelation (using autocorrelation function ACF) in the residuals of the models.

Results

Capelin was the overall most important prey for the selected top predator species, ranging from an average of 27.5% in Minke whale diet to 34.9% in guillemot diet (Fig. S2 in File S1). However, sandeel was the most important prey for the guillemots (ca 49%) and krill was the most important prey for the Minke whale (ca 40%), when considering the whole Barents Sea. In these two cases capelin was the second most eaten prey. Herring was also an abundant prey in the diet of the predators (13–24%, apart for the cod where it represented only 2.7%) as was the krill (10–40%, apart for the guillemots which are not foraging on krill).

Table 3. Results of the generalized additive models selected by shrinkage method of the relationship between diet overlap and different explanatory variables.

Overlap	Species	Variable 1		Variable 2		Variable 3		Variable 4		n	R²
Intra-specific	Kittiwake	ST(t)*	1.87	ln(Cap)(t)	1.59	Year(t)	0.00			14	0.66
	Guillemot	ST(t)**	1.98	Cap(t)	0.40	Year(t)**	1.00			14	0.78
	Minke whale	Herr(t)*	1.72	wNAO(t)*	1.01	Cap(t)	0.00			12	0.52
	Cod	ln(Cap)(t)*	1.92	Year(t)***	1.06	ST(t)	0.00	Krill_North(t)	0.00	23	0.55
Inter-specific	Minke whale vs cod	ln(Herr)(t)*	0.85	Krill(t)*	1.83	Year(t)	0.00			13	0.66
	Kittiwake vs guillemot	ln(Herr)(t)·	0.58	wNAO(t)*	1.79	Year(t)·	0.68			14	0.70

Models are written $O_t = \alpha + s_1(X_{1t}) + s_2(X_{2t}) + s_3(X_{3t}) + ... + \varepsilon_t$ with s_s a nonparametric smoothing function specifying the effect of the covariates X_i on the dependent variable O for year t; α, intercept; and ε, stochastic noise term. The estimated degrees of freedom (edf) for each explanatory variable is indicated as is the significance (** p.Value <0.01, * <0.05 and · <0.10). Variables with edf=0.00 were shrank by the fitting procedure and thus effectively removed from the formulation. See Fig 2–3 for the model fit and for the confidence intervals of the retained variables.

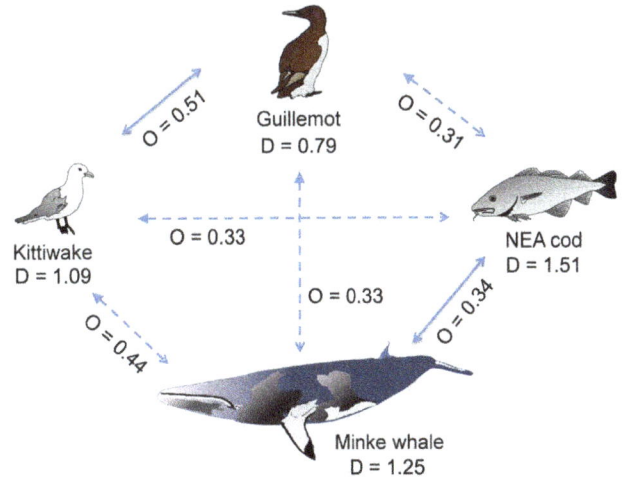

Guillemot D = 0.79

O = 0.51 O = 0.31

Kittiwake D = 1.09   O = 0.33   NEA cod D = 1.51

O = 0.44   O = 0.33   O = 0.34

Minke whale D = 1.25

Figure 1. Trophic relationships between the main components of the food web in the Barents Sea ecosystem. Average Schoeners' diet overlap index O for the five predator pairs studied and their respective Shannon-Wiener niche breadth D (see Table S1 in File S1). The significant relationship are given in plain arrows (Fig. 3, Table 3) The shape of the arrow head indicates the interpretation on how one species may affect another based on biomass [55]. Different arrow heads indicate unbalanced biomass between a predator pair (filled head indicates a potential stronger effect than open head).

Diet breadth results show that the two central placed foragers (i.e., the two seabird species) had a narrower diet than the two other species (Minke whale and cod, Table S1 in File S1). Figure 1 gives the niche breadth for each predator. The cod had the broadest diet, followed by the minke whale, the kittiwake and the guillemot.

Intraspecific year to year variation in diet

Table 3 shows the best models selected by shrinkage technique explaining the year to year change in diet for each predator.

The kittiwake diet varied over time, with O ranging from 16% to 88% of overlap with the previous year (60 ± 23(SD), Fig. 2). The diet changes can be explained by the generally positive relationship with the sea temperature (ST, over 3.9°C) and with the capelin biomass (log transformed, until ca $2.4\ 10^6$ t) (Table 3, Fig. 2). With increasing ST and increase of capelin abundance the diet became more similar.

The guillemot diet also varied over time (16–92%, 60 ± 23(SD)). The diet overlap decreased with time (Fig. 2). This trend taken into account, the year-to-year change in diet can be explained by the changes in the ST (Fig. 2). With increasing ST the diet became more similar until ca 4.1°C, when the relationship was reversed, i.e., the diet was more variable in extreme temperatures. With the increase of capelin abundance the diet became to some extent more variable.

The Minke whale diet was relatively stable (42–79%, 65 ± 13(SD)). The change in diet can be explained by the changing herring abundances combined to the changes in the winter NAO index (wNAO, Fig. 2). The more abundant the herring (up to an abundance of ca $1.43\ 10^6$ t) and lower the wNAO the smaller was the diet overlap (Fig. 2). Hence, the diet varied more in years with high herring abundances, and in years of positive wNAO.

The cod diet was remarkably constant with very small variation compared to the other species during the studied period (56–92%, 80 ± 10(SD)) and showed a clear positive time trend. The year-to-year fluctuations in diet can be explained by the annual variation

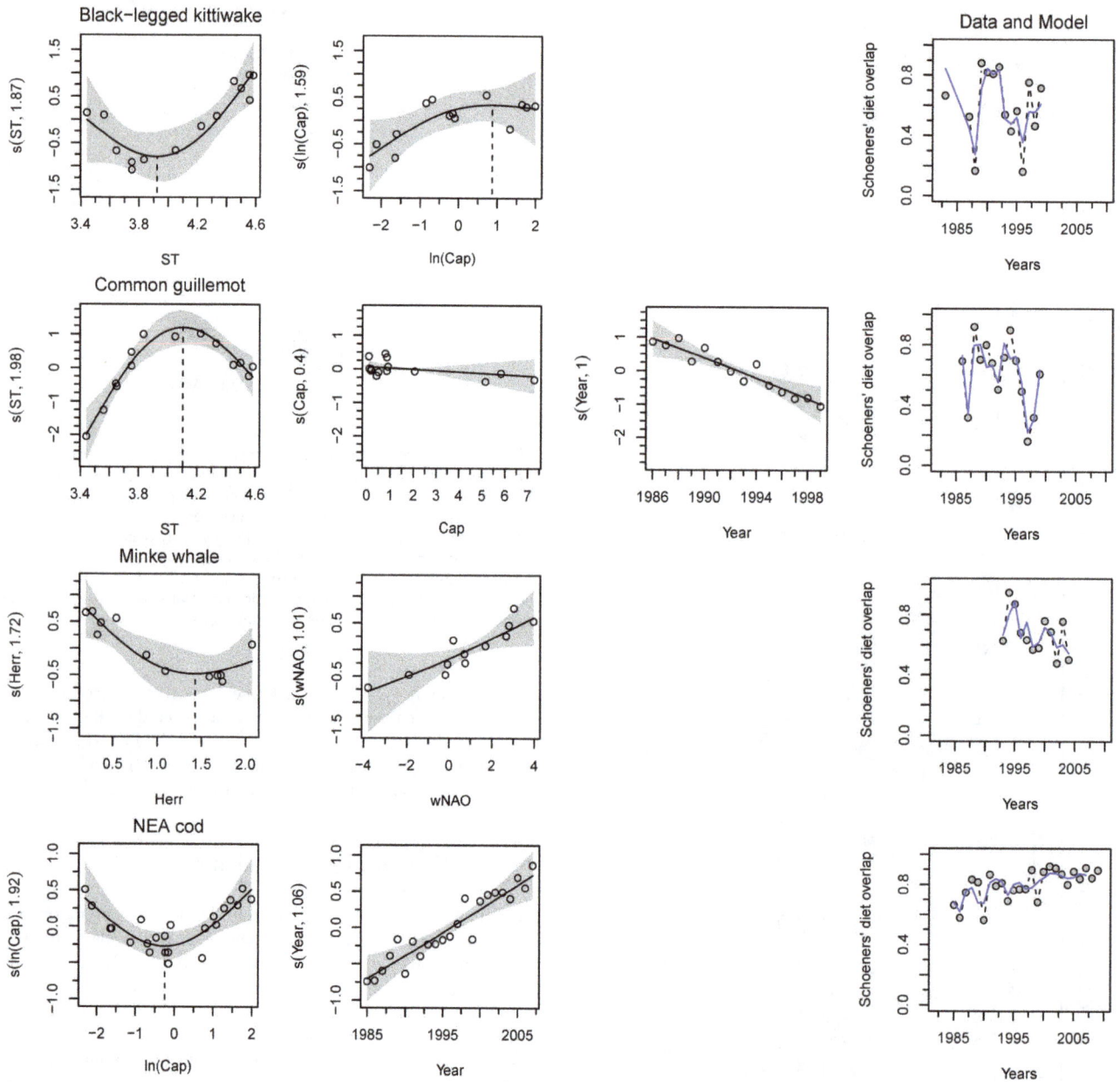

Figure 2. Intraspecific diet dynamics of the main predator species in the Barents Sea. The generalized additive models (GAMs) are presented for each predator. For each plot, the x-axes show the covariate and the y-axes the partial effect that each covariate has on the response variable. The line is the smooth term effect of the considered covariate on the elasticity with the pointwise 95% confidence interval around the mean prediction (grey-shaded area). The dots are the partial residuals calculated by adding to the effect of the concerned covariate to the residuals, the model prediction at any given point is given by the sum of all partial effects plus a constant. When it applies, the dotted line locates the inflection point. Abbreviation are explained in Table 2 and the models in Table 3. Superimposed on the overlap data (grey filled dots) in the last column is the corresponding GAM prediction (plain line).

in capelin abundance (log transformed, Fig. 2). With increasing capelin abundance the diet became less similar until ca 0.78 10^6 t when the relationship was reversed (Fig. 2).

Interspecific year to year variation in diet overlap

Among all pairwise comparison of predators, only two exhibited significant diet overlap, but it all cases diet overlap varied annually.

The Minke whale/cod pair had a diet overlap ranging from 22–61% (34±10(SD), Fig. 3). The change in the diet overlap between the two predators can be explained by the positive effects of

herring abundance (log transformed) and of the krill abundance (Fig. 3).

The kittiwake/guillemot pair had a diet overlap ranging from 14–80% (51±20(SD), Fig. 3). The diet overlap exhibited a slight positive time trend (Fig. 3). The change in the diet overlap between the two predators can be explained by the positive effects of herring abundance (log transformed) and of the wNAO (Fig. 3). With increasing wNAO and herring abundance the diets became more similar.

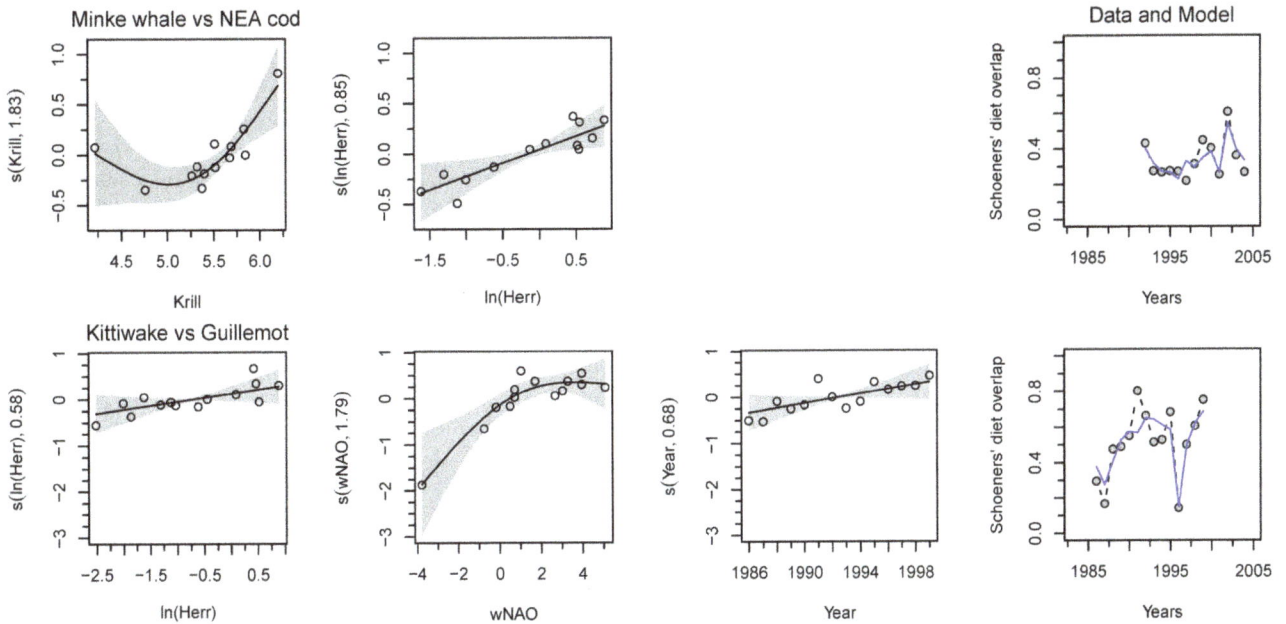

Figure 3. Interspecific diet overlap for the main predator species in the Barents Sea. The generalized additive models (GAMs) are presented for each pair or predator. For each plot, the x-axes show the covariate and the y-axes the partial effect that each covariate has on the response variable. The line is the smooth term effect of the considered covariate on the elasticity with the pointwise 95% confidence interval around the mean prediction (grey-shaded area). The dots are the partial residuals calculated by adding to the effect of the concerned covariate to the residuals, the model prediction at any given point is given by the sum of all partial effects plus a constant. When it applies, the dotted line locates the inflection point. Abbreviations are explained in Table 2 and the models in Table 3. Superimposed on the overlap data (grey filled dots) in the last column is the corresponding GAM prediction (plain line).

The cod and the kittiwake pair (6–81%, 33±21(SD)), the Minke whale and the guillemot pair (5–77%, 33±27(SD)), the Minke whale and the kittiwake pair (11–72%, 44±24(SD)), the cod and the guillemot pair (5–54%, 31±17(SD)) had no significant diet overlap. Note that the two last pairs had near significant diet overlap.

Discussion

Diet overlaps were obtained by using the commonly accepted Schoener's Index for niche overlap computed on stomach contents. Linton et al. [47] showed that the Schoener's index gives a more accurate representation of true overlap when the overlap is ranging between 7–90% as is the case in our study when compared to other often used indices [48]. It results that our models displayed the trend in diet overlap fairly well; their relative stiffness being likely due to the small amount of covariates used.

However, there is limitation to the diet overlap techniques. The first is the availability of the data that requires heavy logistics to obtain. This is well illustrated by the harp seal case where the data series available to us were too short for our analysis. On the same level is the spatial coverage of the data. For instance while still possible, the calculation of diet overlap index has meaning only if the two predators compared feed in the same area at the same time. This is why we have restricted the NEA cod and the Minke whale data to the southern Barents Sea when comparing with the seabirds data. To some extent there is also a similar problem with the season when the data are collected, explaining why we have also restricted the seasonal extent of the NEA cod data (Table 1). Optimally, data should be collected for all predators studied at the same geographical area, during the same season and over a sufficient amount of consecutive years.

1. Annual change in the diet of the main predators in the Barents Sea

Our results revealed that cod and Minke whale, the predators with a large niche breadth due to predation on a wide variety of prey species (Fig. 1 and Table S2 in File S1), had more stable diets across time than the seabirds foraging on fewer prey species (Fig. 2). In cod and Minke whale, fluctuations in prey abundances resulted in relatively small changes in use of many alternative prey species, compared to the larger changes in use of fewer prey species in the seabirds (Fig. 2). Being central place foragers during reproduction, the seabirds choice of prey is limited to the vicinity of the breeding site. It is then the local distribution of prey that explains the variation in the seabirds diet much more than the prey abundance. In this respect, the diets of cod and Minke whale appear to be more robust to fluctuations in the prey base. Indeed, wider distributions, and no spatial limitation to areas neighbouring a central place (colony) during foraging is likely important factors increasing the dietary flexibility and robustness of the cod and the Minke whale in comparison to the seabirds (e.g., large impact of prey availability on survival explaining seabird population decline such as the one observed for common guillemot in 1986–1987 [49,50]). Nevertheless, cod, Minke whale and seabirds were negatively impacted by past fluctuations in prey abundance [36,51,52].

Capelin abundance is the major driver causing changes in cod and seabird diets. Nevertheless, the dietary response (i.e., change of O-index) to changing capelin abundance differed from U-shaped, positive and negative for cod, kittiwake and guillemot, respectively (Fig. 2). While the major prey eaten by cod is the capelin, which has highly variable abundance and distribution [33], the diet of the cod remains remarkably constant. However, the small changes observed between years are explained by the variation in capelin abundance. The U-shaped dietary response relative to capelin abundance indicates that the cod diet is similar in periods with either low or high capelin abundance, but varies during transitions between high and low capelin abundances. We suggest that the U-shape of the relationship is due to the particular dynamics of the capelin in the Barents Sea with regular periods with low abundance [26]. In such years, the cod shifts to juvenile cod and haddock as alternative prey [12,53] or other prey with high abundance, and back to capelin when capelin stock recovers. However, it seems that juvenile cod was an important prey for adult cod only during the mid 90's capelin collapse when there was strong recruiting year classes of cod [53], but not so much during the capelin collapses in mid 80s or 2000 (Fig. S1 in File S1). Note that in the recent years, cod appears to respond to the warming by expanding its distribution range [25].

The two seabird species show remarkably mirror responses to changing capelin abundance and sea temperature. Since it was shown that the kittiwake is a competitor to the common guillemot [20], the mirror response may reflect that what is good for kittiwake is bad for the guillemot hence the remarkably similar inflection point in sea temperature at ca 4°C for both species. The change in their diet is explained by climatic variables such as sea temperature, that may be a proxy of the local condition in term of prey availability spectrum. Note that the overlap of diet between the two seabird species is stronger when winter NAO index is high which corresponds to high temperatures in the Barents Sea. This may indicate that high winter NAO index is stabilizing the prey availability around the breeding site by for example favouring one prey species over the others. A study on the spatial distribution showed that in the Barents Sea the seabird distribution at sea was relatively stable over the years (Fauchald pers. comm.).

Similarly to the cod, the Minke whale displays a relatively constant diet. The small changes in its diet are explained by variation in the abundance of the juvenile herring in the Barents Sea and not of the capelin. However, the Minke whale's body condition was found to be poorer in years when both capelin and herring was at a low abundance level [36] indicating a dependence on capelin availability. The diet was also more similar in years with low herring abundance. Years with little herring in the diet coincided with periods with little capelin but increased krill in the diet (Fig. S1 in File S1); krill is an important alternative prey for these whales when pelagic fish abundances are low [36]. However, during the recent years, the Minke whale distribution was relatively constant and independent of prey distribution [54], similarly to the seabirds (Fauchald pers. comm.). The past decade showed an increasing abundances of krill and shrimp associated with large stocks of demersal and pelagic fish in the Barents Sea [7]. During this recent period the whale condition might have remain good despite the period of low capelin abundance thanks to alternate prey (e.g. krill). Unfortunately our data on Minke whale do not cover the recent years; i.e., they stop in 2004. However, our model may have caught this effect through the positive effect of winter NAO index on the diet similarity; positive NAO being the signature of the later years (Fig S2 in File S1).

2. Annual change in diet overlap of the major predators in the Barents Sea

When exploring the trophic interaction between predators we should always keep in mind that the species are not representing the same biomass in carbon. For instance in the Barents Sea, the cod biomass is some hundred mg C m^{-2}, the whales ca 100 mg C m^{-2}, while the seabirds all together are only up to 2.5 mg C m^{-2} [55]. This difference in biomass must be taken in to account when comparing interspecific diets. For instance, a diet overlap between

cod and kittiwake may indicate a potential competition of cod on kittiwake but not the reverse (or very locally). On the other hand, Minke whale and cod populations or the two seabird populations having similarly scaled biomass may engage in a direct two-way competition [20]. Another factor to consider is that digestion rate can be different between species – from 3–10 hours needed for full digestion of fish (sandeel and whiting) in seabirds [56] and up to 1–3 days in cod (capelin, herring, shrimp and other prey) (see refs in [16]). More important is that the digestion rates ratio between prey type (e.g., digestion rate for crustacean/digestion rate for fish prey…) is similar for the predators compared. If not, some prey species may be overly represented in the diet of some predators and not others. Unfortunately such information is not available. Among the pairs tested only the Minke whale/NEA cod and kittiwake/common guillemot pairs display significant diet overlap (i.e., have a regular diet overlap over the years, Fig. 3). Food competition may thus occur between Minke whale and cod, but the implications for interspecific competition need to be mathematically tested [20]. Taking into account the difference in biomass and well known effect of cod predation on capelin [26] we may also find food competition of cod on seabirds, notably with kittiwake where the overlap is nearly significant.

Changes in pairs of diet overlap are explained by a positive effect of herring and/or krill abundance. Herring is an essential food source, e.g., during chick raising at Kharlov (Krasnov pers. com.), that may explain why the two seabird populations have a more similar diet when the juvenile herring are abundant in the Barents Sea. The same is true for the Minke whale and the cod, however, previous works have shown that both Minke whale [36] and cod [57] switch to krill and amphipods as prey in periods with low herring and capelin abundance. It seems that there is two alternative states where cod and Minke whale have high similarity in their diet; one with high abundance of capelin/herring (the two stocks show a similar dynamic with time, see Fig S2 in File S1) and one with low abundance of these pelagic prey but high abundance of krill (note that krill and capelin populations tend to have inverse temporal dynamic, Fig. S2 in File S1).

3. Conclusions

The Barents Sea predators demonstrated a diversity both in their diets, and in change in diets within and between species. Also the responses to possible drivers of diet change, such as abundances of key prey species and ocean climate were diverse, both within species and between pairs of species. The potential for interspecific competition could perhaps be strongest if top predator diets became more similar when prey abundances were low, i.e. that the top predators were switching to the same alternative prey species. However, the dietary response diversity observed in this study indicate that the top predator community could be relatively robust to changes in the ecosystem. As with the diversity of species that contribute to the same ecosystem function is regarded as an important property for ecosystem resilience [23,58,59], the diversity of responses [23] to environmental changes within functional groups will increase the probability of compensation for one species by the others and thereby secure the continuation of an ecosystem function [60].

Supporting Information

File S1 Figure S1. Diet of the main predator species in the Barents Sea over time. Note that for the black-legged kittiwakes and common guillemots the amphipods and krill prey species where not dissociated and are assembled in one category "krill". There are two minke whale diet plots: for the whole

Barents Sea (left) and restricted to the Southern Barents Sea part (70–74°N and 20–40°E). The first data are used to analyse the change in the minke whale diet over time and to compare with the diet of the NEA cod. The second data are used to compare with the diet of the seabirds that are central place foragers and limited to the Southern Barents Sea during reproduction (period when the seabird diet data were collected). There are three NEA cod plots: for the ICES data (1984–2009) used for the intraspecific analysis and for restricted area of the Barents Sea to compare with the seabirds' diet (March to July, 68–72°N and 20–40°E) and with the minke whale's diet (July to September, 70–80°N and 5–40°E).

Figure S2. Time series used as explanatory variables in the study. Data for the winter NAO come from https://climatedataguide.ucar.edu/sites/default/files/climate_index_files/nao_station_djfm.txt. Data for the sea temperature come from PINRO. They are yearly average sea temperature measured monthly at 0–200 m depth on the Russian Kola meridian transect (33° 30′ E, 70° 30′ N to 72° 30′ N). Data for capelin and herring biomass come from ICES report (Table 9.5 p 498 in ICES 2012).

Figure S3. Interspecific diet overlap for the main predator species in the Barents Sea. Change of diet from one year to another is presented by a Schoeners' diet overlap index (grey filled dots). Higher is the index higher is the overlap. **Table S1.** Diet of the different predators. **Table S2.** Prey species and categories used for the calculation of the Schoeners' index.

Acknowledgments

This work is collaborative work between CEES and IMR under the ADMAR project (grant no. 200497/130).

Author Contributions

Conceived and designed the experiments: JMD. Performed the experiments: JMD. Analyzed the data: JMD. Contributed reagents/materials/analysis tools: JMD AD UL YVK. Wrote the paper: JMD MSM YVK NGN UL AD.

References

1. Hjermann DØ, Bogstad B, Eikeset AM, Ottersen G, Gjosaeter H, et al. (2007) Food web dynamics affect Northeast Arctic cod recruitment. Proc R Soc Lond B 274: 661–669.
2. Yaragina NA, Marshall CT (2000) Trophic influences on interannual and seasonal variation in the liver condition index of Northeast Arctic cod (Gadus morhua). ICES J Mar Sci 57: 42–55.
3. Gjøsæter H, Bogstad B (1998) Effects of the presence of herring (Clupea harengus) on the stock-recruitment relationship of Barents Sea capelin (Mallotus villosus). Fish Res 38: 57–71.
4. Bogstad B, Haug T, Mehl S (2000) Who eats whom in the Barents Sea? NAMMCO Sci Publ 2: 98–119.
5. Stige LC, Lajus DL, Chan KS, Dalpadado P, Basedow SL, et al. (2009) Climatic forcing of zooplankton dynamics is stronger during low densities of planktivorous fish. Limnol Oceanogr 54: 1025–1036.
6. Ottersen G, Hjermann DO, Stenseth NC (2006) Changes in spawning stock structure strengthen the link between climate and recruitment in a heavily fished cod (Gadus morhua) stock. Fish Oceanogr 15: 230–243.
7. Johannesen E, Ingvaldsen RB, Bogstad B, Dalpadado P, Eriksen E, et al. (2012) Changes in Barents Sea ecosystem state, 1970–2009: climate fluctuations, human impact, and trophic interactions. ICES Journal of Marine Science: Journal du Conseil 69: 880–889.
8. Link JS, Bogstad B, Sparholt H, Lilly GR (2009) Trophic role of Atlantic cod in the ecosystem. Fish Fish 10: 58–87.
9. Paine RT (1980) Food Webs: Linkage, Interaction Strength and Community Infrastructure. J Anim Ecol 49: 667–685.
10. Frank KT, Petrie B, Shackell NL (2007) The ups and downs of trophic control in continental shelf ecosystems. Trends Ecol Evol 22: 236–242.
11. Dalpadado P, Borkner N, Bogstad B, Mehl S (2001) Distribution of Themisto (Amphipoda) spp in the Barents Sea and predator-prey interactions. ICES J Mar Sci 58: 876–895.
12. Hjermann DØ, Stenseth NC, Ottersen G (2004) The population dynamics of Northeast Arctic cod (Gadus morhua) through two decades: an analysis based on survey data. Can J Fish Aquat Sci 61: 1747–1755.
13. Haug T, Nilssen KT (1994) Ecological implications of harp seal Phoca groenlandica invasions in northern Norway. In: A . S . Blix, L . Walloe and O. . Ulltang, editors. International Symposium on the Biology of Marine Mammals in the North-East Atlantic. Tromso, Norway. pp. 545–556.
14. Worm B, Myers RA (2003) Meta-analysis of cod-shrimp interactions reveals top-down control in oceanic food webs. Ecology 84: 162–173.
15. Berenboim BI, Dolgov AV, Korzhev VA, Yaragina NA (2000) The impact of cod on the dynamics of Barents Sea shrimp (Pandalus borealis) as determined by mutlispecies models. J Northw Atl Fish Sci 27: 69–75.
16. Jakobsen T, Ozhigin VK (2011) The Barents Sea: Ecosystem, Resources, Management. Half a century of Russian - Norwegian cooperation. Trondheim, Norway: Tapir Academic Press. pp. 825.
17. Lindstrøm U, Smout S, Howell D, Bogstad B (2009) Modelling multi-species interactions in the Barents Sea ecosystem with special emphasis on minke whales and their interactions with cod, herring and capelin. Deep-Sea Res Part II 56: 2068–2079.
18. Blanchard JL, Pinnegar JK, Mackinson S (2002) Exploring marine mammal-fishery interactions using 'Ecopath with Ecosim': modelling the Barents Sea ecosystem. Science Series Technical Report 117: 52pp.
19. Anker-Nilssen T, Bakken V, Strøm H, Golovkin AN, Bianki VV, et al. (2000) The status of Marine Birds Breeding in the Barents Sea Region. Norwegian Polar Institute, Tromsø.
20. Durant J, Krasnov Y, Nikolaeva N, Stenseth N (2012) Within and between species competition in a seabird community: statistical exploration and modeling of time-series data. Oecologia 169: 685–694.
21. Sivertsen SP, Pedersen T, Lindstrøm U, Haug T (2006) Prey partitioning between cod (Gadus morhua) and minke whale (Balaenoptera acutorostrata) in the Barents Sea. Mar Biol Res 2: 89–99.
22. Nilssen KT, Pedersen OP, Folkow LP, Haug T (2000) Food consumption estimates of Barents Sea harp seals. NAMMCO Sci Publ 2: 9–27.
23. Elmqvist T, Folke C, Nyström M, Peterson G, Bengtsson J, et al. (2003) Response diversity, ecosystem change, and resilience. Front Ecol Environ 1: 488–494.
24. Haug T, Skern-Mauritzen M, Lindstrøm U (2011) Predation by marine mammals. In: T. Jakobsen and V. K. Ozhigin, editors. The Barents Sea: Ecosystem, Resources, Management Half a century of Russian - Norwegian cooperation. Trondheim, Norway: Tapir Academic Press. pp. 485–494.
25. Johannesen E, Lindstrøm U, Michalsen K, Skern-Mauritzen M, Fauchald P, et al. (2012) Feeding in a heterogeneous environment: spatial dynamics in summer foraging Barents Sea cod. Mar Ecol Prog Ser 458: 181–197.
26. Hjermann DØ, Bogstad B, Dingsør GE, Gjøsæter H, Ottersen G, et al. (2010) Trophic interactions affecting a key ecosystem component: a multi-stage analysis of the recruitment of the Barents Sea capelin. Can J Fish Aquat Sci 67: 1363–1375.
27. Ottersen G (2008) Pronounced long-term juvenation in the spawning stock of Arcto-Norwegian cod (Gadus morhua) and possible consequences for recruitement. Can J Fish Aquat Sci 65: 523–534.
28. Barrett RT, Bakken V, Krasnov JV (1997) The diets of common and Brünnich's guillemots Uria aalge and U. lomvia in the Barents Sea region. Polar Res 16: 73–84.
29. Barrett RT, Krasnov YV (1996) Recent responses to changes in stocks of prey species by seabirds breeding in the southern Barents Sea. ICES J Mar Sci 53: 713–722.
30. Dolgov AV, Yaragina NA, Orlova EL, Bogstad B, Johannesen E, et al. (2008) 20th anniversary of the PINRO-IMR cooperation in the investigations of fish feeding in the Barents Sea – results and perspectives. In: T. Haug, O. A. Misund, H. Gjøsæter and I. Røttingen, editors. Long term bilateral Russian-Norwegian scientific co-operation as a basis for sustainable management of living marine resources in the Barents Sea. Proceedings of the 12th Norwegian-Russian Symposium Tromsø. pp. 44–78.
31. Mehl S, Yaragina NA (1992) Methods and results in the joint PINRO-IMR stomach sampling program. In: B. Bogstad and S. Tjelmeland, editors. Interrelations between fish populations in the Barents Sea. Bergen, Norway: Proceedings of the fifth PINRO-IMR Symposium, Murmansk, 12-16 August 1991. Institute of Marine Research. pp. 5–16.
32. Zatsepin VJ, Petrova NS (1939) Feeding of cod in the south part of the Barents Sea (by observations in 1934–1938). Trudy PINRO: 5-170 (in Russian).
33. Fauchald P, Mauritzen M, Gjøsæter H (2006) Density-dependent migratory waves in the marine pelagic ecosystem. Ecology 87(11): 2915–2924.
34. Ellingsen I, Dalpadado P, Slagstad D, Loeng H (2008) Impact of climatic change on the biological production in the Barents Sea. Clim Chang 87: 155–175.
35. Dalpadado P, Ingvaldsen RB, Stige LC, Bogstad B, Knutsen T, et al. (2012) Climate effects on Barents Sea ecosystem dynamics. ICES Journal of Marine Science: Journal du Conseil.
36. Haug T, Lindstrøm U, Nilssen KT (2002) Variations in Minke Whale (Balaenoptera acutorostrata) Diet and Body Condition in Response to Ecosystem Changes in the Barents Sea. Sarsia 87: 409–422.

37. Barrett RT (2002) Atlantic puffin *Fratercula arctica* and common guillemot *Uria aalge* chick diet and growth as indicators of fish stocks in the Barents Sea. Mar Ecol Prog Ser 230: 275–287.

38. Krasnov YV, Nikolaeva NG, Goryarev YI, Ezhov AV (2007) Current status and population trends in the Kittiwake (*Rissa tridactyla*), Common (*Uria aalge*) and Brunnich's (*U. lomvia*) guillemots at Kola Peninsula, European Russia. Ornithologia 34: 65–75. Moscow State University, Moscow. In Russian.

39. Zhukova NG, Nesterova VN, Prokopchuk IP, Rudneva GB (2009) Winter distribution of euphausiids (Euphausiacea) in the Barents Sea (2000–2005). Deep-Sea Res Part II 56: 1959–1967.

40. Schoener TW (1968) Anolis lizards of Bimini - Resource partitioning in a complex fauna. Ecology 49: 704–726.

41. Mysterud A (2000) Diet overlap among ruminants in Fennoscandia. Oecologia 124: 130–137.

42. Wallace RK (1981) An Assessment of Diet-Overlap Indexes. Trans Am Fish Soc 110: 72–76.

43. Spellerberg IF (2008) Shannon–Wiener Index. In: S. E. Jørgensen and B. D. Fath, editors. Encyclopedia of Ecology. Oxford: Academic Press. pp. 3249–3252.

44. R Development Core Team (2013) R: A language and environment for statistical computing. R Foundation for Statistical Computing, Vienna, Austria: URL http://www.R-project.org.

45. Wood SN, Augustin NH (2002) GAMs with integrated model selection using penalized regression splines and applications to environmental modelling. Ecol Model 157: 157–177.

46. Wood SN (2006) Generalized Additive Models: An Introduction with R. Boca Raton: Chapman & Hall/CRC. 391 p.

47. Linton LR, Davies RW, Wrona FJ (1981) Resource Utilization Indices: An Assessment. J Anim Ecol 50: 283–292.

48. Pianka ER (1974) Niche Overlap and Diffuse Competition. Proceedings of the National Academy of Sciences 71: 2141–2145.

49. Vader W, Barrett RT, Erikstad KE, Strann K-B (1990) Differential responses of Common and Thick-Billed Murres to a crash in the capelin stock in the southern Barents Sea. Stud Avian Biol 14: 175–180.

50. Krasnov YV, Barrett RT (1995) Large-scale interactions among seabirds, their prey and humans in the southern Barents Sea. In: H. R. Skjoldal, C. Hopkins, K. E. Erikstad and H. P. Leinaas, editors. Ecology of fjords and coastal waters. Amstredam, The Netherlands: Elsevier Science B.V. pp. 443–470.

51. Erikstad KE, Reiertsen TK, Barrett RT, Vikebo F and Sandvik H (2013) Seabird-fish interactions: the fall and rise of a common guillemot *Uria aalge* population. Mar Ecol Prog Ser 475: 267–276.

52. Gjøsæter H, Bogstad B, Tjelmeland S (2009) Ecosystem effects of the three capelin stock collapses in the Barents Sea. Mar Biol Res 5: 40–53.

53. Dolgov AV (1999) Impact of predation on recruitment dynamics of the Barents Sea cod. In: V. N. Shleinik, editor editors. Biology and regulation of fisheries of demersal fish in the Barents Sea and the North Atlantic. Murmansk, PINRO Press. pp. 5–118 (in Russian).

54. Skern-Mauritzen M, Johannesen E, Bjørge A, Øien N (2011) Baleen whale distributions and prey associations in the Barents Sea. Mar Ecol Prog Ser 426.

55. Sakshaug E, Johnsen G, Kovacs KM (2009) Ecosystem Barents Sea. Trondheim: Tapir Academic Press.

56. Hilton GM, Ruxton GD, Furness RW, Houston DC (2000) Optimal digestion strategies in seabirds: A modelling approach. Evol Ecol Res 2: 207–230.

57. Prozorkevich DV, Ushakov NG (2010) Capelin of the Barents Sea and adjacent waters. In: M. S. Shevelv, editor editors. Development of Russian fisheries on the North Basin after 200 miles zones establishment. Murmansk, Russia: PINRO Press. pp. 248–255 (in Russian).

58. Pimm SL (1982) Food webs. London, UK: Chapman and Hall. 219 p.

59. Folke C, Carpenter S, Walker B, Scheffer M, Elmqvist T, et al. (2004) Regime shifts, resilience and biodiversity in ecosystem management. Annu Rev Ecol Evol Syst 35: 557–581.

60. Naeem S (2002) Ecosystem consequences of biodiversity loss: the evolution of a paradigm. Ecology 83: 1537–1552.

61. ICES (2010) Report of the Arctic Fisheries Working Group (AFWG), 22–28 April 2010, Lisbon, Portugal/Bergen, Norway. ICES CM 2010/ACOM 05: 664 pp.

62. Haug T, Gjøsæter H, Lindstrøm U, Nilssen KT (1995) Diet and food availability for north-east Atlantic minke whales (*Balaenoptera acutorostrata*), during the summer of 1992. ICES J Mar Sci 52: 77–86.

63. Tereschenko VV (1996) Seasonal and year-to-year variations of temperature and salinity along the Kola meridian transect. ICES CM 1996/C:11.

64. Hurrell JW (1995) Decadal trends in the North Atlantic Oscillation: Regional temperatures and precipitation. Science 269: 676–679.

65. ICES (2012) Report of the Arctic Fisheries Working Group 2012 (AFWG), 20–26 April 2012, ICES Headquarters, Copenhagen. ICES CM 2012/ACOM 05: 633 pp.

10

Platelet-Rich Plasma and Adipose-Derived Mesenchymal Stem Cells for Regenerative Medicine-Associated Treatments in Bottlenose Dolphins (*Tursiops truncatus*)

Richard J. Griffeth[1], Daniel García-Párraga[2], Maravillas Mellado-López[1,5], Jose Luis Crespo-Picazo[2], Mario Soriano-Navarro[3], Alicia Martinez-Romero[4], Victoria Moreno-Manzano[1,5]*

1 Centro de Investigación Príncipe Felipe, Tissue and Neuronal Regeneration Lab, Valencia, Spain, 2 Oceanogràfic (grupo Parques Reunidos), Valencia, Spain, 3 Centro de Investigación Príncipe Felipe, Electron Microscopy Unit, Valencia, Spain, 4 Centro de Investigación Príncipe Felipe, Cytomics Unit, Valencia, Spain, 5 FactorStem, Ltd. Valencia, Spain

Abstract

Dolphins exhibit an extraordinary capacity to heal deep soft tissue injuries. Nevertheless, accelerated wound healing in wild or captive dolphins would minimize infection and other side effects associated with open wounds in marine animals. Here, we propose the use of a biological-based therapy for wound healing in dolphins by the application of platelet-rich plasma (PRP). Blood samples were collected from 9 different dolphins and a specific and simple protocol which concentrates platelets greater than two times that of whole blood was developed. As opposed to a commonly employed human protocol for PRP preparation, a single centrifugation for 3 minutes at 900 rpm resulted in the best condition for the concentration of dolphin platelets. By FACS analysis, dolphin platelets showed reactivity to platelet cell-surface marker CD41. Analysis by electron microscopy revealed that dolphin platelets were larger in size than human platelets. These findings may explain the need to reduce the duration and speed of centrifugation of whole blood from dolphins to obtain a 2-fold increase and maintain proper morphology of the platelets. For the first time, levels of several growth factors from activated dolphin platelets were quantified. Compared to humans, concentrations of PDGF-BB were not different, while TGFβ and VEGF-A were significantly lower in dolphins. Additionally, adipose tissue was obtained from cadaveric dolphins found along the Spanish Mediterranean coast, and adipose-derived mesenchymal stem cells (ASCs) were successfully isolated, amplified, and characterized. When dolphin ASCs were treated with 2.5 or 5% dolphin PRP they exhibited significant increased proliferation and improved phagocytotic activity, indicating that in culture, PRP may improve the regenerative capacity of ASCs. Taken together, we show an effective and well-defined protocol for efficient PRP isolation. This protocol alone or in combination with ASCs, may constitute the basis of a biological treatment for wound-healing and tissue regeneration in dolphins.

Editor: Eva Mezey, National Institutes of Health, United States of America

Funding: This study was supported by the Spanish Ion Channel Initiative (CSD2008-00005): VMM, and the Instituto de Salud Carlos III (PI10/01683 and PI13/00319): VMM. The funders had no role in study design, data collection and analysis, decision to publish, or preparation of the manuscript.

* Email: vmorenom@cipf.es

Introduction

Dolphins exhibit an extraordinary capacity to heal deep soft-tissue injuries, such as those following shark bites [1]. Dolphins in captivity often experience external soft tissue injuries as a result of repetitive exercises and movements, such as open wounds on the underside of their lower mandible which slowly develop and worsen during training regimens. Often times, dolphins which were injured in the open ocean are rescued and rehabilitated in captivity [1]. Nevertheless, accelerated wound healing in wild or captive dolphins may help minimize infection and other side effects associated with open wounds in marine animals.

Platelet-rich plasma (PRP) is a fraction of plasma with a higher number of platelets compared to whole blood, thereby containing increased concentrations of growth factors [2]. The alpha granules in platelets are the source of multiple growth factors including vascular endothelial growth factor (VEGF), platelet-derived growth factor (PDGF), and transforming growth factor beta (TGFβ) among others [3]. These growth factors play an essential role in the complex processes of wound healing and tissue regeneration [4]. PRP stimulates type 1 collagen, matrix metalloproteinase 1, and increases the expression of regulators of cell cycle progression to accelerate wound healing [5,6], and has been widely used in many species, including humans, for regenerative medicine in an increasing variety of surgical fields. Successful clinical applications have been reported using PRP for wound repair, soft tissue healing [2,7], cosmetic surgery [8–10], burns [11], nervous tissue [12,13], chronic skin ulcers [14], maxillofacial and long bone defects as well as in the treatment of joints in various mammals [3,15–18]. However, some studies have suggested that PRP had little or no benefit, which most likely was the result of poor quality PRP [16,19,20]. Nevertheless, PRP has already been used in a wide variety of applications for regenerative medicine purposes. Moreover, at this point there is no agreed upon gold standard protocol for PRP generation and little characterization has been performed on the obtained products. Often times protocols vary across and within species, including the use of protocols defined for certain species being used for others without any additional characterization [21]. Well-defined simple procedures will result in very useful therapeutic tools, especially for veterinary medicine. The optimization of centrifugation conditions is fundamental to obtaining high quality PRP with minimal manipulation. Quantification and identification of platelets and lymphocytes as well provides a proper characterization of the PRP concentration procedure. Additionally, maintaining platelet integrity and quality without damaging or lysing them allows them to fully secrete growth factors upon controlled activation. Furthermore, PRP treatment enhances angiogenesis [22] and stimulates stem cell proliferation and cell differentiation for tissue regeneration [9]. Undifferentiated stem cells migrate to the site of growth factors delivered from PRP applications and trigger proliferation of the stem cells at the site [15].

Mesenchymal stem cells are an attractive cell population for regeneration of musculoskeletal tissues and wound healing [23–25]. Multiple sources of mesenchymal stem cells have been described including bone marrow, ligaments, lung, umbilical cord, and adipose tissue [26]. Adipose-derived stem cells (ASC) in particular are an appealing source because of their abundant availability and excellent ability to expand and proliferate in culture. In humans, ASC have been used successfully to treat soft tissue defects, scars, and burn injuries and to regenerate various damaged tissues [10]. Recently, ASCs from dolphins have been isolated, cultured, and differentiated into adipogenic, chondrogenic and osteogenic cell lineages [27], thereby demonstrating that dolphin ASCs may have similar regenerative potential as other already documented mammals. Here we have defined a simple and well-characterized protocol for efficient isolation of both ASCs and PRP in dolphins. The use of PRP separately or in combination with ASCs has the potential to provide a safe and efficient treatment for soft tissue injuries and regeneration not previously described in this species.

Materials and Methods

Animals and blood collection

Blood samples were collected from the tail vein plexus from 9 different dolphins at a local aquarium (Oceanografic; http://www. cac.es/oceanografic) for routine hematological and biochemical testing. To prevent clotting, whole blood was collected into tubes containing sodium citrate and the excess blood (~10 ml) from each dolphin was used for these studies. In accordance with the European Parliament and Council normative 2010/63/UE (22nd September 2010) on the protection of animals used for scientific purposes and with the Real Decreto 53/2013 (1st February 2013), under the standards for the protection of animals used for experimental and other scientific purposes including teaching, "The non-experimental clinical veterinary practice" (RD 53/2013 Article 2, section 5) is excluded from the scope of the legislation and therefore approval from the corresponding ethical committee was not required. The Centro de Investigación Príncipe Felipe (CIPF) and the Oceanografic have a signed collaborative agreement for research purposes. Under this agreement CIPF obtained consent, specifically for the use of surplus dolphin blood collected for routine general exams. As part of the preventative

medical care program, blood samples are taken every two months from these dolphins. The excess blood collected was used for this study. Whole blood from adult male and female bottlenose dolphins (*Tursiops truncatus*) between the ages of 7–25 years were utilized. Blood was transported to the adjacent laboratory at 4°C immediately and processed within 30 min after collection. To compare samples of dolphin blood with those of human blood, surplus blood samples from healthy anonymous donors from a local blood donation program (Unidad de Transfusiones de la Comunidad Valenciana; http://centro-transfusion.san.gva.es) were collected and processed separately but identically to that of dolphin blood. Human blood samples had been donated for use in transfusions, however after a certain time in storage, these samples were no longer recommended for transfusion and this blood was to be discarded and destroyed. Therefore, we were able to take advantage of this surplus blood for use in research and accordingly informed consent for the use of this blood for research purposes was not required.

Centrifugation of blood samples and PRP isolation

Equal volume (1 ml) of whole blood samples in tubes containing sodium citrate were gently inverted multiple times before centrifugation. Following centrifugation, the plasma fraction was divided into two parts. The upper half was considered platelet-poor plasma and removed while the lower half was considered platelet-rich plasma and used for further analysis (Figure 1A). Centrifugation was performed at room temperature in an Eppendorf 5810R centrifuge with a swing-bucket rotor (A-4-62, Eppendorf) at the following centrifugation speeds (and equivalent forces) and durations: 1) one spin at 900 rpm (equivalent to 106× g) for 3 min, 2) one spin at 900 rpm (equivalent to 106× g) for 6 min, 3) one spin at 1380 rpm (250× g) for 3 min, 4) two spins at 1380 rpm (250× g) for 3 min each (this consisted of a first spin at 1380 rpm (250× g). The bottom half of the plasma fraction was collected and then spun again at 1380 rpm (250× g) for an additional 3 min and the bottom half of this fraction was considered PRP and used for analysis), 5) one spin at 1870 rpm (460× g) for 8 min, 6) one spin at 2700 rpm (958× g) for 3 min, and 7) one spin at 4000 rpm (2102× g) for 3 min.

FACS analysis

To determine the concentration of platelets in whole blood and in PRP fractions, the human cell platelet marker CD41 was utilized (BD Bioscience, USA). For absolute numbers of platelets, BD Trucount Tubes (BD Bioscience, USA) were used and the fold change of platelet concentration in each PRP fraction was compared to whole blood. Fifty μl of whole blood or 50 μl of each PRP fraction were incubated with 10 μl CD41 antibody conjugated with PC5 (BD Bioscience, USA) for 30 minutes at room temperature in the dark. Subsequently, 20 μl of this mix was then transferred to the BD Trucount Tubes and diluted with 1 ml of PBS and used for FACS analysis. Flow cytometry acquisition and analysis was performed on a FC500 flow cytometer (Beckman Cultek, USA). For absolute platelet number the following formula was utilized:

(Number of events in region containing cells/Number of events in absolute count bead region)

× (Number of beads per test ∗/test volume) =

absolute count of cell

*Number of beads per test: 52250

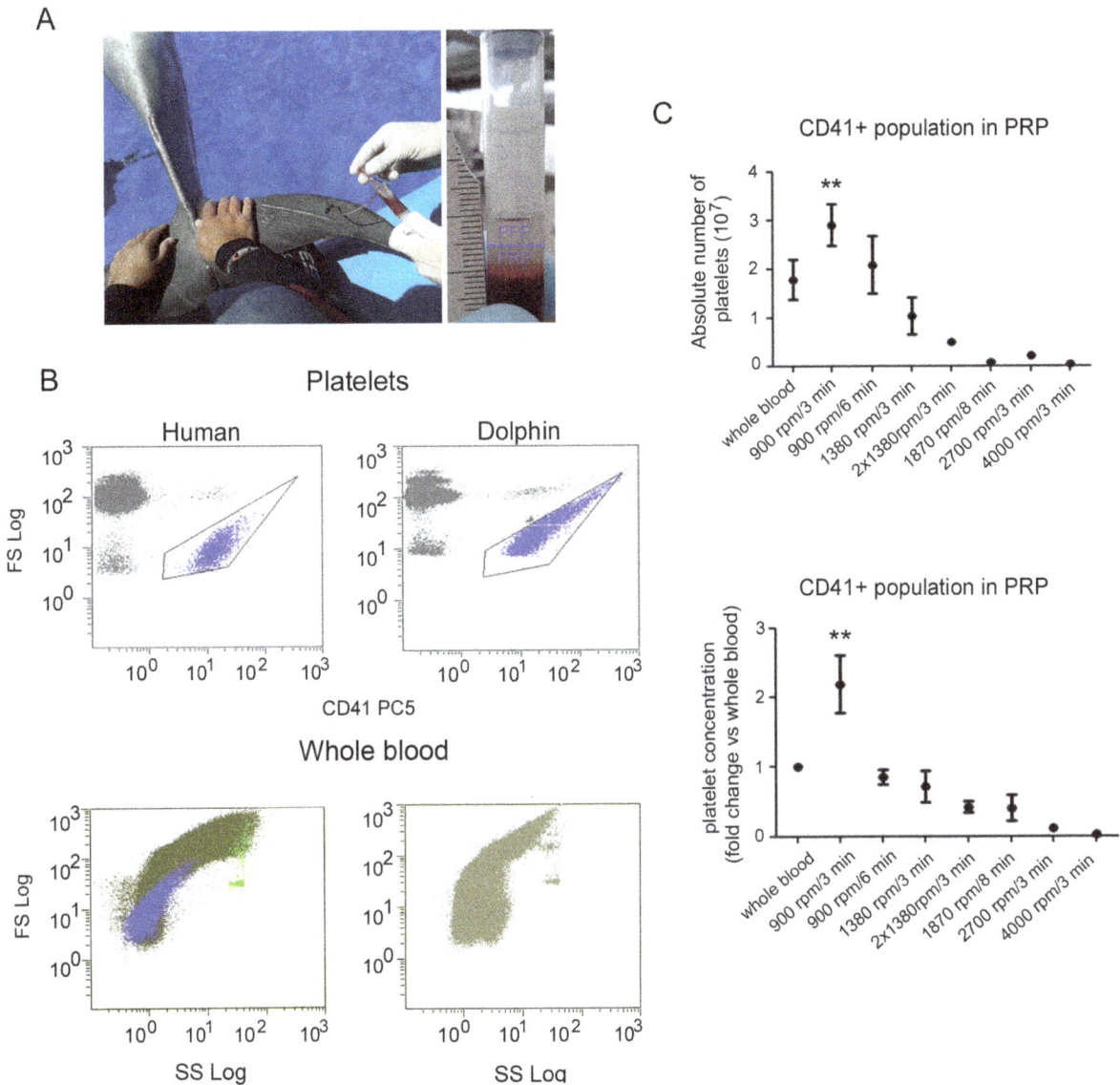

Figure 1. Efficient dolphin platelet-rich plasma concentration protocol. (A) Adult male and female bottlenose dolphins (*Tursiops truncatus*) between the ages of 7–25 years were utilized. Blood samples were collected from the tail vein plexus from 9 different dolphins at a local aquarium and placed into tubes containing sodium citrate. After centrifugation the upper half of the plasma was considered platelet-poor plasma (PPP) and discarded while the lower half was considered platelet-rich plasma (PRP) and used for subsequent experiments. (B) Representative images of FACS analysis utilizing a human CD41 antibody which recognized human and dolphin platelets (upper panels); Acquisition profile of FS versus SC of both, human and dolphin whole blood are shown in lower panels (in blue is represented the CD41 positive population in human sample). (C) Whole blood samples were subjected to multiple centrifugation protocols to determine which was the most efficient in concentrating platelets in a small volume of plasma. Significant increases in absolute number of platelets and platelet concentration as determined by fold change compared to whole blood were observed when whole blood samples were centrifuged at 900 rpm for 3 min. Asterisks denote a significant difference compared to whole blood; ** $P < 0.01$.

For ASC characterization, cell suspension after passage 2 was assayed for cell surface protein expression of CD90-PE, CD44-PE-Cy7, CD105-PE, CD34-PE-Cy5 and CD45-FITC (BD Pharmigen, USA). Cells were trypsinized and pelleted, resuspended in PBS at a concentration of 10^5 cells/100 μl, and incubated at a 1:100 dilution for each antibody or alone for background controls. Cells were incubated in the dark for 45 min at room temperature and then washed three times with PBS and resuspended in 0.5 ml of PBS for FACS analysis. The mean ± SD of the 2 different tested samples were determined for each condition, ASCs cultured in the presence of autologous serum (10% dolphin serum) or cultured with 10% fetal bovine serum (FBS).

Transmission electron microscopy

PRP fractions were fixed in 2.5% glutaraldehyde in 0.1M phosphate buffer (PB) for 1 hr. Then, the cells were washed with 0.1M PB three times and a single drop of 1.5% agar was added. Sections were post-fixed with 2% osmium, rinsed, dehydrated and embedded in Durcupan resin (Fluka, Sigma-Aldrich, St. Louis, USA). Semithin sections (1.5 μm) were cut with an Ultracut UC-6 (Leica, Heidelberg, Germany) and stained lightly with 1%

toluidine blue. Finally, ultra-thin sections (0.08 μm) were cut with a diamond knife, stained with lead citrate (Reynolds solution) and examined under a transmission electron microscope FEI Tecnai G2 Spirit (FEI Europe, Eindhoven, Netherlands) attached to a digital camera Morada (Olympus Soft Image Solutions GmbH, Münster, Germany). To quantify the area, diameter, and absolute number of alpha granules Image J software was utilized. The mean ± SD of 3 different tested samples were determined for dolphin and human PRP.

Quantification of growth factors

Activation of platelets is required to release the growth factors and was performed by adding 14.3units/ml thrombin and 1.4 mg/ml CaCl$_2$ to the PRP samples (human n = 5, dolphin n = 7), followed by incubation at 37°C for 1 hr. Non-activated PRP was included as a negative control, in this case no reactivity was detectable for any tested growth factor. The samples were then centrifuged at 4000× g for 10 min at room temperature and the supernatant was collected and stored at −80°C until growth factor quantification by enzyme-linked immunosorbent assay (ELISA), which was performed using Luminex xMAP Technology. Affymetrix kits (eBioscience) containing antibodies against human PDGF-BB, VEGF-A, and TGFβ were utilized according to the manufacturer's instructions and detection of growth factors was performed on a Luminex 200 system and analyzed using Exponent 3.1 software by extrapolating the absolute value from the standard curve for each growth factor.

Adipose tissue extraction, ASC isolation and cell culture

The Oceanografic aquarium is part of the Stranding Network through an agreement between the "Ciudad de las Artes y las Ciencias" and the "Conselleria de Infraestructuras, Territorio y Medio Ambiente". Through this agreement both institutions have transferred to the Oceanografic the rights for veterinary assistance in cases of stranded sea turtles and cetaceans. This agreement includes the rights to euthanize animals when required and perform autopsies in collaboration with the University of Valencia. The agreement also allows for the use of samples from the cadaveric tissue for research purposes. Once the health and condition of the animal(s) are evaluated, if euthanasia is required, the guidelines for euthanasia of non-domestic animals are followed [28]. Additionally, in accordance with the European Parliament and Council normative 2010/63/UE (22nd September 2010) and the Real Decreto 53/2013 (1st February 2013) in post-mortem tissue collection for research purposes, approval from the corresponding ethical committee is not required. In this study adipose tissue was not collected from live dolphins, therefore approval from the corresponding ethical committee was not required for the development of research-related studies from post-mortem animals.

On two separate occasions in 2013 the veterinary team at the Oceanografic in Valencia, Spain were notified about a stranded wild striped dolphin (Stenella coeruleoalba) found along the eastern Spanish Mediterranean coast. The first dolphin was already dead when encountered but the second dolphin was euthanized by intravenous administration of a lethal dose of pentobarbital [28,29] by the veterinarians. This dolphin was not euthanized specifically for use in this study. Collection of adipose tissue from both dolphins was completely opportunistic. Stranded dolphins are occasionally found along the eastern Spanish Mediterranean coast and are assessed by veterinarians from the Oceanografic. In the case of the euthanized dolphin, the official clinical evaluation by the veterinarians indicated that rehabilitation was not possible. This dolphin was unable to swim or keep normal flotation and

exhibited severe neurological-related abnormalities including tremors, convulsive episodes and loss of reflexes. Therefore, this dolphin was euthanized. Adipose tissue was opportunistically collected after confirmation of death by the official veterinarian and the local authorities.

In both instances, the recently postmortem (0±0.5 days) cadavers were transported to the Oceanografic and adipose tissue was obtained from the postnuchal fat pad, placed into a solution containing PBS plus antibiotic and transported to the adjacent laboratory. The adipose tissue was washed multiple times in PBS plus antibiotics to clean the tissue and remove residual blood. In a petri dish, 10 g of adipose tissue were added to a solution containing PBS, 100 units/ml penicillin and 100 μg/ml streptomycin (Gibco 15140) and collagenase type IA (0.07%, Sigma C9891 CA, USA) and the tissue was manually cut into small pieces using sterile surgical scissors in a laminar flow hood and digested overnight at 37°C, 20% O$_2$, 5% CO$_2$. The following day the digested adipose tissue was collected and washed multiple times with PBS plus antibiotic by centrifugation. The pellet was then resuspended in growth medium (DMEM medium containing 10% dolphin serum or 10% heat-inactivated FBS, 2 mM L-glutamine, 30% L-glucose, 100 units/ml penicillin and 100 μg/ml streptomycin), plated in petri dishes, and incubated overnight. The following day the medium was removed and replaced with fresh medium and attached cells were allowed to grow until nearly confluent then subsequently passaged three times and subjected to viability/proliferation assays, FACS analysis and phagocytosis assays.

ASC directed-differentiation. Once the ASC expanded in vitro and equivalently to previous procedures [27], passages after 4, were distributed to induce adipogenesis, osteogenesis and chondrogenesis differentiation process. All directed-differentiation mediums were obtained from Lonza catolog. *Adipogenesis*: ASC were seeded at a cell density of 10000 cells/cm^2 and when ASC have became>90% confluence the growth medium is substitute for differentiation medium containing, among others, insulin, Dexamethasone, IBMX (3-isobutyl-methyl-xantine) and indomethacin (Adipose Derived stem cell Basal Medium; Lonza Group Ltd). The cells were then incubated for 10–12 days. The adipogenic differentiation was evaluated by Oil Red staining of the lipid vacuoles in formalin fixed cultures; *Osteogenesis*: ASC were seeded at a cell density of 10000 cells/cm^2 in collagen I (Sigma; 10 mM) coated plates in medium containing among others 0.1 μM dexamethasone, 50 μM Asc2P and 10 mM μ-glycerophosphate (Osteogenic Basal Medium; Lonza Group Ltd) with 10% of fetal bovine serum (FBS). ASC cultures were maintained in this medium for 4 weeks (with medium changes every 3 days). For detection of extracellular calcium deposits the Alizarin Red staining was used in formalin fixed cultures; *Chondrogenesis*: The ASC culture was performed from cell "Micromass" starting form with a high concentration of cells in a minimal volume (1×10^5 cels/100 μl) in the presence of TGF-β 1 and 3 10 ng/ml, Asc 2P (50 μM) and insulin (6.25 μg/ml) (Chondro BulletKit; Lonza Group Ltd) for four weeks with medium changes every 3 days. Alcian blue was used to detect the presence of enrichment of sulfated proteoglycans in the extracellular matrix. Before staining, the micromass cultures were fixed in formalin, included in paraffin and sectioned into 10 μm. All samples were carried out in parallel with or without additional 2.5% PRP in the corresponding differentiation mediums.

ASC viability and proliferation

Briefly, 10^4 dolphin ASCs at passage 3–4 were seeded in 96-well plates and allowed to grow for 24 hr in growth medium containing

10% FBS. Serum deprived growth medium was then supplemented with 50 U/ml heparin and dolphin ASCs were treated with 0, 1, 2.5 or 5% dolphin PRP or the same concentration of FBS as a positive control. All groups were then subjected to the cell viability test, CellTiter 96 AQueous Non-Radioactive Cell Proliferation Assay (MTS assay; Promega, CA, USA). Every condition was assayed in quadruplicate in three different experiments for both lines of dolphin ASCs. The viability of cells at each assayed condition was expressed as the percentage ratio of the mean ± SD of colorimetric signal from treated cells in the presence of 1, 2.5, or 5% PRP compared to the absence of PRP.

For phagocytosis assays, 10^5 dolphin ASC at passage 3–4 were seeded into 35 mm petri dishes and allowed to grow for 24 hr in growth medium containing 10% FBS. Serum deprived growth medium was then supplemented with 50 U/ml heparin and dolphin ASCs were treated with 0 or 5% dolphin PRP in the presence of 2 μm diameter red fluorescent microspheres (Invitrogen F8826). After 24 hr incubation, the cells were fixed with 4% PFA, washed with PBS, and images were taken immediately. For Giemsa staining and morphological assessment, ASCs treated with microspheres were incubated for 24 h, then fixed in cold 100% methanol for 20 min, and then stained with Giemsa (Fluka, UK) for 1 hr. Giemsa was then removed and the ASCs were washed with tap water, allowed to air-dry, and images were taken immediately.

Statistics

Statistical comparisons were assessed by Student's t-test. All P values were derived from a two-tailed statistical test using the Graphpad Prism 5 Software. A P-value<0.05 was considered statistically significant.

Results

Low centrifugation speed and short duration yield the highest quality PRP in bottlenose dolphins

Low centrifugation speed and short duration yielded the highest quality PRP preparation from dolphin whole blood (Figure 1). Following blood collection and centrifugation, the top half of the separated plasma was considered platelet-poor plasma (PPP) and removed, while the bottom half was considered PRP and utilized for the experiments in this study (Figure 1A). A human antibody against the platelet cell surface marker CD41 showed reactivity for dolphin platelets by FACS analysis, and a similar cell profile for human and dolphin whole blood was found by CD41 immunoreactivity (Figure 1B). An additional population of non-gated cells also showed positive labeling for CD41 but at a higher size (FS) corresponding to a small population of activated platelets that bind to a fraction of leukocytes. For absolute numbers of platelets, BD Trucount Tubes (BD Bioscience, USA) were used and the fold change of platelet concentration in each PRP fraction was compared to whole blood. Quantification of the dolphin CD41+ cell population revealed that the centrifugation protocol with the most enriched PRP was 900 rpm (equivalent to 106× g) for 3 min (Figure 1C, upper panel). There was a significant increase in the absolute number of platelets and a significant, more than a 2-fold increase, in the concentration of platelets in this fraction of PRP (Figure 1C, lower panel). Using the same centrifugal force but increasing the duration to 6 min caused a slight but insignificant decrease in the platelet concentration compared to whole blood while the absolute number of platelets was slightly but insignificantly increased compared to whole blood (Figure 1C). Increasing the centrifugation speed to 1380 rpm (250× g) with duration of 3 min resulted in a slight but insignificant decrease in platelet

concentration and absolute number of platelets (Figure 1C). However, two sequential centrifugations at 1380 rpm for 3 min resulted in a significant decline in both absolute number of platelets and fold change compared to whole blood (Figure 1C). Three other conditions were evaluated, each with increasing centrifugal forces (1870 rpm = 460× g, 2700 rpm = 958× g, and 4000 rpm = 2102× g) and each resulted in significant decreases in absolute number of platelets and fold change compared to whole blood (Figure 1C).

Dolphin platelets are larger than human platelets

Transmission electron microscopy studies demonstrated that dolphin platelets have a significantly larger area compared to human platelets (Figure 2A–C). The mean area ± SEM of dolphin platelets was 4.53 μm^2±1.8 while that of human platelets was 3.26 μm^2±0.8. Likewise, linear measurements taken from sections cut through the major axis of nearly rounded elliptical platelets, were longer in dolphin platelets compared to human platelets (Figure 2A,C). However, similar numbers of alpha granules were found in cross sections of human platelets compared to dolphin platelets (Figure 2A,D).

Quantification of TGFβ, PDGF, and VEGF in dolphin PRP

The concentrations of TGFβ and VEGF-A in dolphin PRP were significantly lower than that in human samples, while there were no differences in the concentration of PDGF-BB (Figure 2E). Mean concentrations ± SEM for five humans and seven dolphins were: TGFβ; human 1016±177 pg/ml, dolphin 331±37 pg/ml, PDGF-BB; human 174±13 pg/ml, dolphin 163±30 pg/ml, and VEGF-A; human 46±8 pg/ml, dolphin 14±2 pg/ml.

Isolated Dolphin ASCs are plastic adherent, express mesenchymal-specific surface antigens and have the capacity for tri-lineage mesenchymal differentiation

The adipose tissue from the postnuchal fat pad dissected from recently postmortem (0±0.5 days) cadaveric dolphins (Figure 3A) was subjected to ASC isolation and characterization.

The putative dolphin ASCs were characterized according to the criteria put forth by the International Society for Cellular Therapy [30], and met all the criteria for status as mesenchymal stem cells. The ASCs were adherent to plastic culture dishes either growth in Dolphin autologus serum or in Fetal Bovine Serum containing medium (Figure 3B). FACS analysis of mesenchymal cell surface antigen-specific markers CD90 and CD44 was>98%, almost half of this population also showed CD105 reactivity, and a lack of expression of the hematopoietic antigens CD34 and CD45 (Figure 3C). Finally, the ASCs were induced to differentiate to adipocytes, osteocytes and chondrocytes under standard *in vitro* differentiation protocols (Figure 3D).

Dolphin PRP stimulates proliferation and activates phagocytosis in dolphin ASCs

Dolphin ASCs treated with 2.5 or 5% dolphin PRP exhibited significant increases in cell proliferation as assessed by cell viability assays (MTS assay), while treatment with 1% PRP did not lead to significant changes in proliferation compared to controls (Figure 3E). FBS, used as a positive control, showed a similar pattern of proliferation as that seen with PRP (Figure 3E). Morphologically, the photomicrographs clearly illustrate an increase in the number and density of ASCs in the presence of 5% PRP compared to non-treated ASCs (Figure 3E).

PRP also activates additional ASC properties, such as phagocytic activity. Dolphin ASCs cultured in the presence of 5% PRP

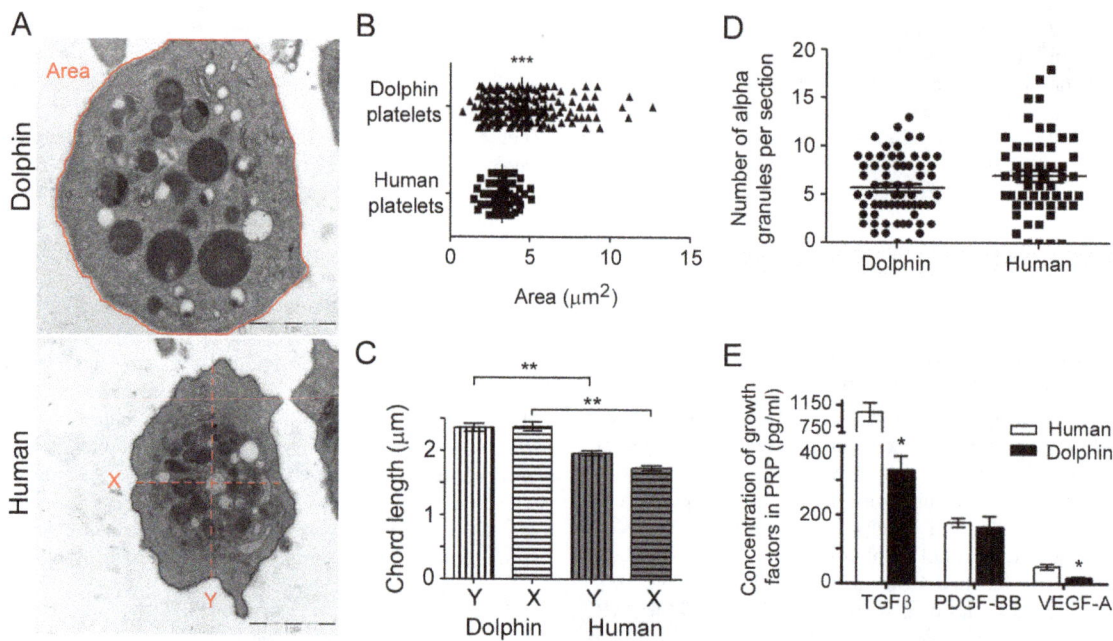

Figure 2. Characterization of dolphin platelets. (A) Representative images of the morphology of dolphin and human platelets by transmission electron microscopy revealed the larger size of dolphin platelets compared to human platelets. (B) Utilizing Image J software, the circumference of platelets that were sectioned through the major axis was traced and the area was calculated (see A upper panel). The area of dolphin platelets was significantly larger than those of human platelets. (C) Chords were drawn through the center of the platelet sections (approximate diameters) in the X and Y axes and these lengths were measured (see A lower panel). Chords from both axes were significantly longer in dolphin platelets than human platelets. (D) There were no significant differences in the number of alpha granules per section between dolphin and human platelets. Results presented in A–D are mean ± SEM of 3 human samples and 3 dolphin samples. A minimum of 50 measurements were taken from each sample. (E) The concentrations of TGFβ and VEGF-A were significantly reduced in dolphin PRP compared to human PRP, while there was no difference in the concentration of PDGF-BB. Results are mean ± SEM of 5 human samples and 7 dolphin samples. Asterisks denote significant differences; * $P<0.05$, ** $P<0.01$, *** $P<0.001$.

exhibited enhanced ability to phagocytose red fluorescent 2 μm microspheres which were added to the culture system (Figure 3F, upper panels). The enhanced phagocytic activity of ASCs induced by PRP is also clearly demonstrated by Giemsa staining. There is increased proliferation of ASCs treated with PRP and several microspheres have been phagocytosed and are unmistakably visible within the ASCs (Figure 3F, lower panels).

Discussion

The regenerative capability of multiple growth factors found in platelets has been harvested and used in the form of PRP for regenerative purposes in multiple species for several years. However, to our knowledge, no studies have utilized dolphin PRP. ASCs in dolphins were only recently identified [27], thus there is an obvious paucity of information regarding their characterization. In this study, we have developed a simple and reproducible centrifugation protocol that yields high quality PRP which is able to induce proliferation of dolphin ASCs *in vitro*. For the first time we have identified dolphin platelets and characterized them by transmission electron microscopy and measured the levels of three major growth factors in dolphin PRP. Furthermore, we derived and characterized dolphin ASCs and demonstrated that dolphin PRP is able to induce proliferation and activate phagocytotic activity of ASCs *in vitro*.

Especially in aquariums and zoos throughout the world, the need for simple, effective, and standardized procedures to treat non-experimental animals is a necessity. Injuries caused by enclosures or repetitive movements to animals in captivity require

immediate attention to avoid prolonged and chronic damage to the tissue. PRP has been linked with improvements in wound regeneration, such as reduced healing time, in multiple tissues and in several species [3,15–18]. However, detailed procedures for collection and isolation of PRP had never before been documented in dolphins. Therefore separation conditions by single centrifugation of sodium citrate collected whole blood were initially based on those commonly employed for use in other mammals, specifically humans and dogs, which have been more extensively described [7]. A commonly used condition for PRP collection is 1870 rpm ($460\times$ g) for 8 min [3], therefore a series of preliminary studies were performed using this condition. To identify platelets by FACS analysis the platelet glycoprotein CD41 was employed. CD41 constitutes the alpha subunit of a highly expressed platelet surface integrin protein and appears on the platelet surface before activation thereby rendering it a reliable marker of platelets [31]. Evolutionary analyses demonstrate that dolphins share common gene and protein expression patterns with humans and that dolphin physiology may be a reliable model for studying human disease [32,33]. Therefore, although commercial dolphin antibodies are not currently available, a CD41 antibody of human origin was utilized and indeed it did recognize dolphin platelets (Figure 1B). A similar profile for antibody binding and cellular size was found for the CD41+ population in both dolphins and humans. Of note, in the FACS analysis of both human and dolphin whole blood, there was a population of CD41+ cells found within the leukocyte-expected size spectrum that most likely aggregated with larger cells such as leukocytes during the antibody incubation process. The results of platelet concentration at

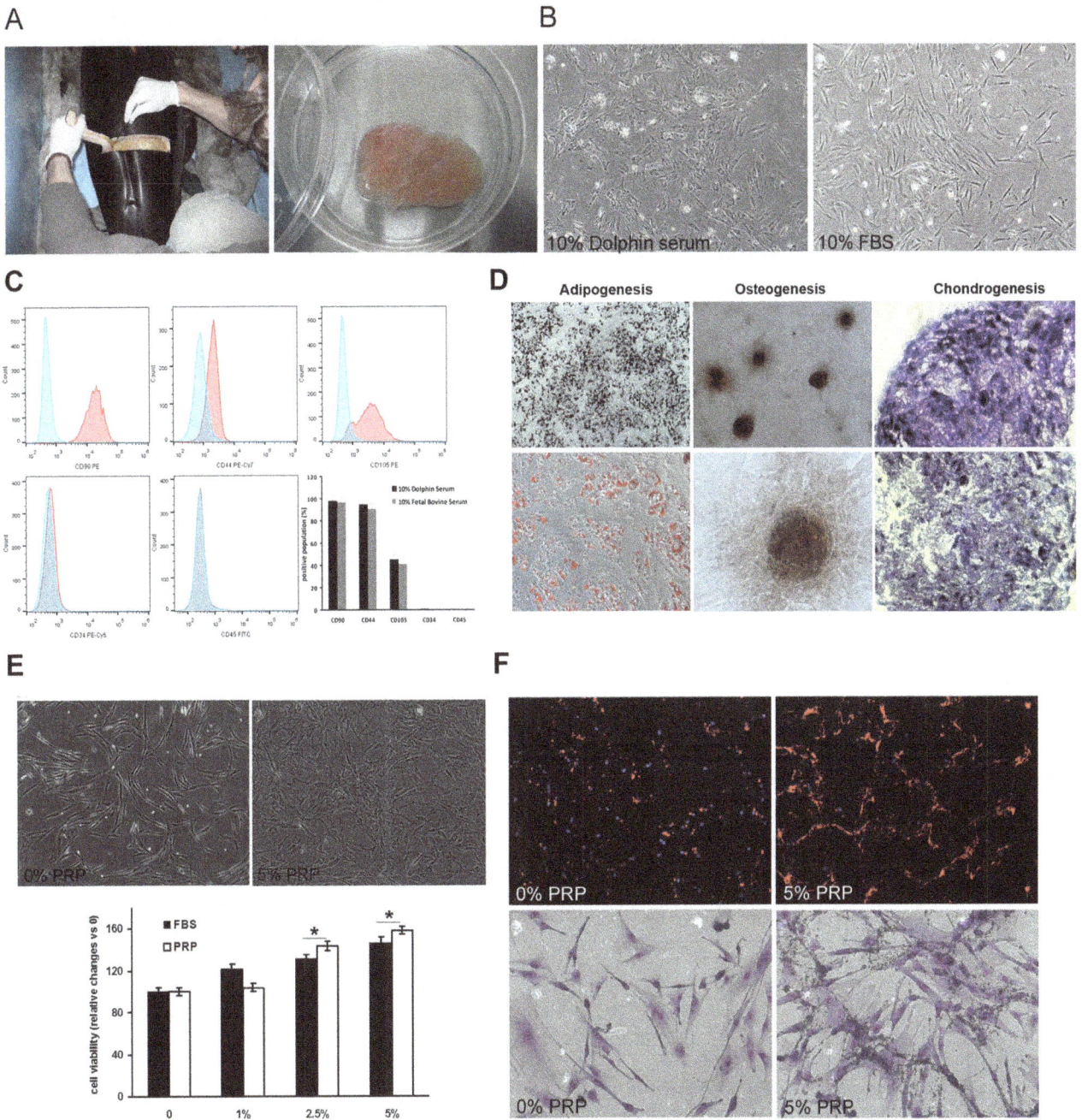

Figure 3. Dolphin PRP induces proliferation and phagocytic activity of dolphin ASCs. (A) Adipose tissue was collected from the postnuchal fat pad from recent postmortem wild striped dolphins (*Stenella coeruleoalba*) (n = 2) and dolphin ASCs were derived and characterized. (B) Dolphin ASCs are plastic adherent and are able to be cultured in the presence of both 10% FBS and 10% dolphin serum. The morphology of ASCs treated with 10% dolphin serum appeared less elongated and senescent compared to those cultured with 10% FBS. (C) Dolphin ASCs were positive for mesenchymal cell markers CD90, CD44, and CD105 and were negative for hematopoietic cell markers CD34 and CD45. The histograms for CD90, CD44, and CD105 show the shift in the positive population in pink versus the non-stained sample in blue. CD34 and CD45 did not show positive reactivity, thereby confirming that the putative ASCs are indeed of mesenchymal origin. (D) Dolphin ASCs were capable of tri-lineage mesenchymal differentiation. ASCs were differentiated under standard *in vitro* conditions to adipocytes (Oil Red O staining), osteocytes (Alizarin Red staining) and chondrocytes (Alcian blue staining). (E) Dolphin ASCs treated *in vitro* with 2.5 or 5% dolphin PRP exhibited significantly increased proliferation, while those treated with 1% PRP were not different than controls. Proliferation rates in ASCs treated with the same concentrations of FBS were similar but significantly lower at 2.5 and 5% compared to PRP. Morphologically there was an increase in the number and density of ASCs cultured with 2.5 or 5% PRP compared to controls. Representative images of dolphin ASCs treated with 0 or 5% dolphin PRP are shown. Results of MTS assays are mean ± SD of colorimetric signal from treated cells in the presence of 1, 2.5, or 5% PRP compared to the absence of PRP. Every condition was assayed in quadruplicate in three different experiments for both lines of dolphin ASCs. (F) In addition to inducing proliferation of ASCs, treatment with 5% PRP stimulates phagocytic activity in dolphin ASCs. Red fluorescent microspheres were highly phagocytosed by ASCs in the presence of PRP compared to those without PRP (upper panels). Similarly, when fixed and stained with Giemsa, there were clearly more ASCs indicating increased proliferation. Also visible are the increased number of microspheres which have been phagocytosed within the ASCs treated with 5% PRP compared to fewer microspheres inside untreated ASCs. Asterisks denote significant difference compared to controls (0% or PRP or FBS) * $P<0.05$.

1870 rpm (460× g) for 8 min were unexpectedly low as confirmed by FACS acquisition of CD41+ cells (preliminary data not shown, however this condition is shown in Figure 1C). Therefore, a range of centrifugal conditions were examined in order to optimize conditions for PRP isolation, ranging from 900–4000 rpm (106–2102× g) for 3–8 min. The initial evaluation was performed by qualitative analysis via optical microscopy of blood smears (data not shown) and subsequently quantified by FACS analysis. The most effective centrifugation condition, which yielded the highest absolute number of platelets and the highest platelet concentration compared to whole blood, was the lowest speed and duration, i.e., 900 rpm (106× g) for 3 min (Figure 1C). Increasing centrifugal force had an inverse relationship with absolute platelet number. At 900 rpm (106× g) for 3 min, the platelet concentration doubled compared to whole blood, however using the same force but doubling the duration of centrifugation (6 min) lead to a slight decline in platelet concentration compared to whole blood (Figure 1C). An alternative explanation is that in dolphins, platelet integrity could be compromised with increasing centrifugal forces and increasing time of centrifugation. Likewise, in humans, the concentration of sP-selectin, which is a marker of platelet activation and growth factor release [34], increases with elevated centrifugal forces (800–1600× g) [35], indicating compromised integrity of the platelets. Thus, platelet integrity is critical for proper growth factor release and responsiveness and for PRP to be an effective treatment in any species, platelets need to be intact, non-activated, and be able to secrete growth factors upon controlled activation. Therefore, high centrifugal forces applied to dolphin whole blood may compromise the integrity of platelets and render them inadequate for high quality PRP isolation due to loss of platelet integrity. This error in the technical step of PRP preparation and isolation may help to explain some reports indicating non-beneficial effects of PRP treatment [36]. Furthermore, our initial studies utilizing fewer revolutions (<900 rpm) were unable to concentrate the platelets to levels different than that of whole blood, i.e., lower centrifugal forces did not produce PRP in dolphin blood samples. Likewise, centrifugation at 900 rpm for a shorter duration (1 or 2 min) was unsuccessful in yielding sufficient plasma to collect and process. Therefore, to effectively isolate platelets for use in PRP-associated treatments with dolphin blood it is important to maintain the proper balance between speed and the duration of centrifugation.

To our knowledge, there are no data describing the morphology of dolphin platelets. Therefore, to characterize their morphology and further investigate the activity of platelets *in vitro* we have for the first time, described the ultra-structure of dolphin platelets via transmission electron microscopy and compared them to human platelets (Figure 2). Dolphin platelets have a larger area than human platelets (Figure 2A,B), likewise measurements across the x- and y-axes of platelets demonstrate that the distance of both of these measurements is longer in dolphin platelets than human platelets (Figure 2A,C). The larger sized platelets found in dolphin whole blood may explain the need for a reduction in centrifugation force and duration to obtain PRP compared to the optimal centrifugation conditions for human whole blood. Although dolphin platelets are larger in size, they contain similar numbers of alpha granules as human platelets (Figure 2D), although the alpha granules in dolphin platelets are also larger than those of human platelets. These observations may indicate that platelets from dolphins are extremely similar to those of humans but that components are proportionally larger. Further platelet analysis and research is needed to make decisive conclusions.

Multiple growth factors secreted by platelets are important for numerous functions including tissue regeneration, reducing inflammation, and wound healing. To our knowledge, no previous study has quantified growth factors in dolphin platelets, therefore it was necessary to evaluate essential growth factors that are associated with the improved regenerative ability of PRP and to determine if these factors are present and active in putative dolphin PRP. Therefore, three growth factors in particular were selected and measured by ELISA; platelet-derived growth factor (PDGF-BB), transforming growth factor beta (TGFβ), and vascular endothelial growth factor (VEGF-A). PDGF is known to induce proliferation of undifferentiated mesenchymal cells and some progenitor populations [37]. The tissue repair mechanisms induced by PDGF-BB appear to involve fibroblast proliferation, collagen production, and neovessel formation [38]. Several phase III human clinical trials have demonstrated the efficacy of PDGF [39], and topical application is safe, well-tolerated, and improves healing of chronic diabetic foot ulcers [40–42]. Dolphin PRP contained similar concentrations of PDGF-BB as in humans, hence PRP treatment in dolphins may also provide similar regenerative effects for soft tissue injuries associated with acute or chronic wounds which often occur in this species in captivity. In addition, wound repair requires the reestablishment of a functional vascular network. One of the most potent pro-angiogenic agents is VEGF, which binds the VEGF receptor on vascular endothelial cells [43], and initiates the MAPK signaling pathway which induces angiogenesis [44,45]. TGFβ is vital for cutaneous regeneration after injury [46], and induces fibroblast proliferation and migration into the site of injury [47]. TGFβ also triggers the production of a collagen-rich matrix, which induces differentiation of fibroblasts into myofibroblasts which promote wound closure by acquiring contractility and expressing α smooth muscle actin [47,48]. Thus, TGFβ is a critical component in the regenerative action of PRP. Interestingly in dolphins, concentrations of VEGF-A and TGFβ in platelets were significantly reduced compared to humans.

A recent study successfully established the use of ultrasound-guided liposuction to obtain ASCs from the postnuchal fad pad of bottlenose dolphins in captivity [27]. To circumvent surgical procedures and potential injuries to captive dolphins, adipose tissue was obtained from the postnuchal fat pad of two recent (0±0.5day) postmortem striped dolphins found along the eastern Spanish Mediterranean coast and two separate ASC lines were produced. The International Society for Cellular Therapy [30] recommends that a minimum of three criteria must be met to effectively characterize multipotent mesenchymal stromal cells, also known as mesenchymal stem cells. Even though these criteria were designed as a guide for the characterization of human tissue, we have followed these basic guidelines to characterize dolphin ASCs and also added further levels of characterization confirming the successful derivation of dolphin ASCs. The first criterion to define ASCs is that they are adherent to plastic. Our culture system has clearly shown this to be true about the dolphin ASCs that were isolated as shown in Figure 3B. In addition to culturing the putative dolphin ASCs in 10% FBS, we found that 10% dolphin serum was also effective in maintaining dolphin ASC cultures. The second major characterization criterion is that these cells express specific surface antigens as measured by flow cytometry. The putative dolphin ASCs exhibited greater than 95% positivity for both CD90 and CD44 (Figure 3C). It is equally important to exclude the possibility of heterogeneous cell populations within the putative ASC population by identifying the lack of expression, or negative markers, of mesenchymal stem cells. CD45, a pan-leukocyte marker, and CD34, a primitive hematopoietic progenitor and endothelial cell marker, are the cells most likely to be found in mesenchymal stem cell cultures [30,49].

As illustrated in Figure 3C, the dolphin ASCs lacked CD34 and CD45 expression. The third criterion and the most unique property of mesenchymal stem cells is that they must have the capacity for tri-lineage mesenchymal differentiation, that is, they must be able to differentiate *in vitro* to osteocytes, adipocytes, and chondrocytes. As shown in Figure 3D, we have demonstrated this under standard *in vitro* differentiation conditions by staining with Alizarin red (osteocytes), Oil Red O (adipocytes), and Alcian Blue (chondrocytes). Accordingly, taken together the data unequivocally demonstrate that the mesenchymal stromal cells obtained from the postnuchal fat pad of dolphins are indeed adipose-derived mesenchymal stem cells.

Cells are usually cultured *in vitro* using a serum-based component such as FBS. In addition to culturing the dolphin ASCs in 10% FBS, we found that 10% dolphin serum was also effective in maintaining the cells in a multipotent state. An interestingly observation was that cells treated with 10% dolphin serum were morphologically distinct to those treated with 10% FBS, appearing almost senescent.(Figure 3B). A possible explanation to why these cells appear different when cultured in serum from different sources is that in dolphins the coagulation cascade is markedly prolonged compared to humans. Dolphins, as well as killer whales, lack factor XII which is important in blood clotting [50,51]. Multiple additional physiological adjustments are part of the dive response in marine mammals, and while such changes favor blood coagulation in terrestrial mammals, these adaptations in dolphins allow them to thrive and hunt at depth and high pressure. Likewise, in the laboratory it was difficult to effectively separate the serum fraction from the erythrocyte fraction after centrifugation of dolphin whole blood, whereas human blood samples collected in the same coagulation tubes yielded two distinct and easily separable fractions, the upper serum and the lower erythrocytes. Therefore, there is the possibility that the 10% dolphin serum which was utilized for cell culture actually contained a lower percentage of serum and consequently reduced concentrations of the normal complement of growth factors and hormones found in fetal bovine serum. This may help explain the change in the morphology of the dolphin ASCs cultured with 10% dolphin serum compared to the normal appearance of those cultured with 10% FBS.

An array of *in vitro* functional experiments utilizing the dolphin ASCs were performed to determine if dolphin PRP was able to stimulate proliferation of dolphin ASCs. Similar results from both cell lines confirm proper ASC function in culture and impart the potential of a combinatorial treatment for improved wound healing applications. Quantification of cell viability assays (MTS assay) demonstrated that there was a dose-response increase in cell proliferation of dolphin ASCs when treated with dolphin PRP (Figure 3E). A significant increase in cell viability was observed when dolphin ASCs were treated with 2.5 or 5% dolphin PRP. Consistent with the quantified significant increase in cell proliferation from the MTS assay, visual morphological inspection of the cells treated with 5% PRP revealed an increase in cell number compared to untreated controls (Figure 3E, upper panels). Thus, in the presence of dolphin PRP, dolphin ASCs proliferate at an increased rate. These data support the fact that even a small percentage of high quality dolphin PRP is able to stimulate proliferation of dolphin ASCs. A similar response in cell viability was seen between ASCs treated with PRP or FBS. These data are encouraging for non-xenogenic and autologous tissue and/or cell transplant applications and regenerative medicine interventions in dolphins.

Furthermore, the viability of ASCs treated with 10% FBS was not different than cells treated with 5% PRP, although the morphology was slightly different (Figure 3B, E). ASCs when cultured with dolphin PRP show a morphology that appears to be pre-adipocytic. It may be thought that the change in ASC morphology as seen in ASCs treated with 5% dolphin PRP might indicate that the ASCs could be differentiating down a specific lineage pathway and potentially losing multi-potent regenerative capacity. However, this was not the case because PRP did not condition the differentiation potential in any of the three directed differentiation lineages (adipocytes, osteocytes, or chondrocytes). No differences were detected in any of the differentiation processes regardless of the presence or absence of PRP (data not shown).

Thus, for the first time, we have shown that dolphin PRP induces proliferation of dolphin ASCs. This demonstrates that dolphin PRP contains the same or similar active growth factors with analogous proliferative ability as PRP from other mammals including humans. The regenerative capacity of the growth factors found in PRP is an excellent source for assisting in shortening the recovery time of open wounds and various tissue injuries in several mammals. Dolphins in particular are quite remarkable in their ability to recover from deep tissue wounds. Dolphin blubber contains organohalogens which exhibit antimicrobial properties and antibiotic activity [52,53], likewise, isovaleric acid, another antimicrobial compound found in dolphin blubber, may help control microbial growth within and around damaged tissues [1]. Moreover, PRP has antimicrobial properties both *in vitro* and *in vivo*; seven antimicrobial peptides have been isolated from human platelets [54,55]. Therefore, both ASCs and PRP from dolphins may contain innate antibacterial properties which favor and accelerate the recovery of damaged tissue.

In addition, PRP exerts anti-inflammatory properties thereby aiding in the reduction of pain associated with tissue injuries [56]. Mesenchymal stem cells have the capacity to modulate the immune system via a plethora of mechanisms (reviewed in [57]), and phagocytosis is the first step in triggering host defense and inflammation. Dolphin PRP appears to activate the phagocytic activity in dolphin ASCs as evidenced by their enhanced ability to phagocytose red fluorescent microspheres (Figure 3F, upper panels). The enhanced phagocytic activity of ASCs induced by PRP was also revealed by Giemsa staining, where there was increased proliferation and higher density of ASCs and increased phagocytosis of microspheres (Figure 3F, lower panels). Taken together, *in vitro* manipulation of dolphin ASCs with dolphin PRP may provide an exciting combination therapy for regenerative medicine in this species. Further studies are essential for improvements in basic understanding of dolphin ASCs and PRP and should aid in future veterinary interventions in aquatic medicine.

In summary, the findings presented in this study demonstrate that PRP collection and isolation containing high quality, intact, non-activated platelets from dolphin whole blood requires low centrifugal force and duration. Morphological measurements show that dolphin platelets are larger than human platelets and contain similar numbers of growth factor-containing alpha granules. In addition, dolphin ASCs were derived and characterized from adipose tissue obtained from the postnuchal fat pad of recently deceased wild striped dolphins. These ASCs are plastic adherent, show positive cell-surface antigen expression of CD90 and CD44 and lack expression of CD45 and CD34, and are also capable of tri-lineage mesenchymal differentiation to osteocytes, adipocytes, and chondroctyes *in vitro*. Moreover, dolphin PRP is able to induce proliferation of dolphin ASCs *in vitro*, demonstrating that dolphin PRP contains active growth factors. Potential treatments using dolphin PRP alone may have the capacity to treat injuries such as soft tissue wounds, however a combination therapy of dolphin ASCs and dolphin PRP either applied at the same time or ASCs treated with PRP *in vitro* and then transplanted to the injury site, may have incredible potential to treat injuries of mesenchymal origin, such as soft tissue, bone, cartilage, or tendon in dolphins. Furthermore, these findings most likely will be able to be extrapolated and applicable to other Cetaceans and marine mammals.

Author Contributions

Conceived and designed the experiments: VMM RJG DGP. Performed the experiments: RJG VMM MML AMR MSN DGP JLC. Analyzed the data: VMM RJG DGP AMR. Contributed reagents/materials/analysis tools: VMM. Wrote the paper: RJG VMM.

References

1. Zasloff M (2011) Observations on the remarkable (and mysterious) wound-healing process of the bottlenose dolphin. J Invest Dermatol 131: 2503–2505.
2. Borrione P, Gianfrancesco AD, Pereira MT, Pigozzi F (2010) Platelet-rich plasma in muscle healing. Am J Phys Med Rehabil 89: 854–861.
3. Anitua E, Andia I, Ardanza B, Nurden P, Nurden AT (2004) Autologous platelets as a source of proteins for healing and tissue regeneration. Thromb Haemost 91: 4–15.
4. De La Mata J (2013) Platelet rich plasma. A new treatment tool for the rheumatologist? Reumatol Clin 9: 166–171.
5. Cho JW, Kim SA, Lee KS (2012) Platelet-rich plasma induces increased expression of G1 cell cycle regulators, type I collagen, and matrix metalloproteinase-1 in human skin fibroblasts. Int J Mol Med 29: 32–36.
6. Kim DH, Je YJ, Kim CD, Lee YH, Seo YJ, et al. (2011) Can Platelet-rich Plasma Be Used for Skin Rejuvenation? Evaluation of Effects of Platelet-rich Plasma on Human Dermal Fibroblast. Ann Dermatol 23: 424–431.
7. de Vos RJ, van Veldhoven PL, Moen MH, Weir A, Tol JL, et al. (2010) Autologous growth factor injections in chronic tendinopathy: a systematic review. Br Med Bull 95: 63–77.
8. Sommeling CE, Heyneman A, Hoeksema H, Verbelen J, Stillaert FB, et al. (2013) The use of platelet-rich plasma in plastic surgery: a systematic review. J Plast Reconstr Aesthet Surg 66: 301–311.
9. Hausman GJ, Richardson RL (2004) Adipose tissue angiogenesis. J Anim Sci 82: 925–934.
10. Cervelli V, Gentile P, Scioli MG, Grimaldi M, Casciani CU, et al. (2009) Application of platelet-rich plasma in plastic surgery: clinical and in vitro evaluation. Tissue Eng Part C Methods 15: 625–634.
11. Pallua N, Wolter T, Markowicz M (2010) Platelet-rich plasma in burns. Burns 36: 4–8.
12. Wu CC, Wu YN, Ho HO, Chen KC, Sheu MT, et al. (2012) The neuroprotective effect of platelet-rich plasma on erectile function in bilateral cavernous nerve injury rat model. J Sex Med 9: 2838–2848.
13. Shen YX, Fan ZH, Zhao JG, Zhang P (2009) The application of platelet-rich plasma may be a novel treatment for central nervous system diseases. Med Hypotheses 73: 1038–1040.
14. Villela DL, Santos VL (2010) Evidence on the use of platelet-rich plasma for diabetic ulcer: a systematic review. Growth Factors 28: 111–116.
15. Choi J, Minn KW, Chang H (2012) The efficacy and safety of platelet-rich plasma and adipose-derived stem cells: an update. Arch Plast Surg 39: 585–592.
16. Paoloni J, De Vos RJ, Hamilton B, Murrell GA, Orchard J (2011) Platelet-rich plasma treatment for ligament and tendon injuries. Clin J Sport Med 21: 37–45.
17. Taylor DW, Petrera M, Hendry M, Theodoropoulos JS (2011) A systematic review of the use of platelet-rich plasma in sports medicine as a new treatment for tendon and ligament injuries. Clin J Sport Med 21: 344–352.
18. Nikolidakis D, Jansen JA (2008) The biology of platelet-rich plasma and its application in oral surgery: literature review. Tissue Eng Part B Rev 14: 249–258.
19. Marx RE (2004) Platelet-rich plasma: evidence to support its use. J Oral Maxillofac Surg 62: 489–496.
20. Eppley BL, Pietrzak WS, Blanton M (2006) Platelet-rich plasma: a review of biology and applications in plastic surgery. Plast Reconstr Surg 118: 147e–159e.
21. Zimmermann R, Arnold D, Strasser E, Ringwald J, Schlegel A, et al. (2003) Sample preparation technique and white cell content influence the detectable levels of growth factors in platelet concentrates. Vox Sang 85: 283–289.
22. Hu Z, Peel SA, Ho SK, Sandor GK, Clokie CM (2009) Platelet-rich plasma induces mRNA expression of VEGF and PDGF in rat bone marrow stromal cell differentiation. Oral Surg Oral Med Oral Pathol Oral Radiol Endod 107: 43–48.
23. de Almeida AM, Demange MK, Sobrado MF, Rodrigues MB, Pedrinelli A, et al. (2012) Patellar tendon healing with platelet-rich plasma: a prospective randomized controlled trial. Am J Sports Med 40: 1282–1288.
24. Eskan MA, Greenwell H, Hill M, Morton D, Vidal R, et al. (2014) Platelet-rich plasma-assisted guided bone regeneration for ridge augmentation: a randomized, controlled clinical trial. J Periodontol 85: 661–668.
25. Silva A, Sampaio R (2009) Anatomic ACL reconstruction: does the platelet-rich plasma accelerate tendon healing? Knee Surg Sports Traumatol Arthrosc 17: 676–682.
26. Niemeyer P, Fechner K, Milz S, Richter W, Suedkamp NP, et al. (2010) Comparison of mesenchymal stem cells from bone marrow and adipose tissue for bone regeneration in a critical size defect of the sheep tibia and the influence of platelet-rich plasma. Biomaterials 31: 3572–3579.

27. Johnson SP, Catania JM, Harman RJ, Jensen ED (2012) Adipose-derived stem cell collection and characterization in bottlenose dolphins (Tursiops truncatus). Stem Cells Dev 21: 2949–2957.
28. United States Department of Agriculture National Agricultural Library (1993) 1993 Report of the AVMA Panel on Euthanasia. J Am Vet Med Assoc 202: 229–249.
29. Close B, Banister K, Baumans V, Bernoth EM, Bromage N, et al. (1996) Recommendations for euthanasia of experimental animals: Part 1. DGXI of the European Commission. Lab Anim 30: 293–316.
30. Dominici M, Le Blanc K, Mueller I, Slaper-Cortenbach I, Marini F, et al. (2006) Minimal criteria for defining multipotent mesenchymal stromal cells. The International Society for Cellular Therapy position statement. Cytotherapy 8: 315–317.
31. Ferkowicz MJ, Starr M, Xie X, Li W, Johnson SA, et al. (2003) CD41 expression defines the onset of primitive and definitive hematopoiesis in the murine embryo. Development 130: 4393–4403.
32. McGowen MR, Grossman LI, Wildman DE (2012) Dolphin genome provides evidence for adaptive evolution of nervous system genes and a molecular rate slowdown. Proc Biol Sci 279: 3643–3651.
33. Venn-Watson S, Carlin K, Ridgway S (2011) Dolphins as animal models for type 2 diabetes: sustained, post-prandial hyperglycemia and hyperinsulinemia. Gen Comp Endocrinol 170: 193–199.
34. Kostelijk EH, Fijnheer R, Nieuwenhuis HK, Gouwerok CW, de Korte D (1996) Soluble P-selectin as parameter for platelet activation during storage. Thromb Haemost 76: 1086–1089.
35. Perez AGM LJ, Rodrigues AA, Luzo ACM, Belangero WD, Santana MHA (2014) Relevant Aspects of Centrifugation Step in the Preparation of Platelet-Rich Plasma. ISRN Hematology 2014: 8.
36. Arenaz-Bua J, Luaces-Rey R, Sironvalle-Soliva S, Otero-Rico A, Charro-Huerga E, et al. (2010) A comparative study of platelet-rich plasma, hydroxyapatite, demineralized bone matrix and autologous bone to promote bone regeneration after mandibular impacted third molar extraction. Med Oral Patol Oral Cir Bucal 15: e483–489.
37. Andrae J, Gallini R, Betsholtz C (2008) Role of platelet-derived growth factors in physiology and medicine. Genes Dev 22: 1276–1312.
38. Pierce GF, Tarpley JE, Allman RM, Goode PS, Serdar CM, et al. (1994) Tissue repair processes in healing chronic pressure ulcers treated with recombinant platelet-derived growth factor BB. Am J Pathol 145: 1399–1410.
39. Smiell JM, Wieman TJ, Steed DL, Perry BH, Sampson AR, et al. (1999) Efficacy and safety of becaplermin (recombinant human platelet-derived growth factor-BB) in patients with nonhealing, lower extremity diabetic ulcers: a combined analysis of four randomized studies. Wound Repair Regen 7: 335–346.
40. Edmonds M, Bates M, Doxford M, Gough A, Foster A (2000) New treatments in ulcer healing and wound infection. Diabetes Metab Res Rev 16 Suppl 1: S51–54.
41. Perry BH, Sampson AR, Schwab BH, Karim MR, Smiell JM (2002) A meta-analytic approach to an integrated summary of efficacy: a case study of becaplermin gel. Control Clin Trials 23: 389–408.
42. Steed DL (2006) Clinical evaluation of recombinant human platelet-derived growth factor for the treatment of lower extremity ulcers. Plast Reconstr Surg 117: 143S–149S; discussion 150S–151S.
43. Bao P, Kodra A, Tomic-Canic M, Golinko MS, Ehrlich HP, et al. (2009) The role of vascular endothelial growth factor in wound healing. J Surg Res 153: 347–358.
44. Breen EC (2007) VEGF in biological control. J Cell Biochem 102: 1358–1367.
45. Swift ME, Kleinman HK, DiPietro LA (1999) Impaired wound repair and delayed angiogenesis in aged mice. Lab Invest 79: 1479–1487.
46. Hynes RO (2009) The extracellular matrix: not just pretty fibrils. Science 326: 1216–1219.
47. Hinz B (2007) Formation and function of the myofibroblast during tissue repair. J Invest Dermatol 127: 526–537.
48. Desmouliere A, Chaponnier C, Gabbiani G (2005) Tissue repair, contraction, and the myofibroblast. Wound Repair Regen 13: 7–12.
49. Lin CS, Xin ZC, Dai J, Lue TF (2013) Commonly used mesenchymal stem cell markers and tracking labels: Limitations and challenges. Histol Histopathol 28: 1109–1116.
50. Robinson AJ, Kropatkin M, Aggeler PM (1969) Hageman factor (factor XII) deficiency in marine mammals. Science 166: 1420–1422.
51. Tibbs RF TEM, Tran LT, Van Bonn W, Romano T, Cowan DF (2005) Characterization of the coagulation system in healthy dolphins: the coagulation

factors, natural anticoagulants, and fibrinolytic products. Comp Clin Path 14: 95–98.

52. Janssens JC, Steenackers H, Robijns S, Gellens E, Levin J, et al. (2008) Brominated furanones inhibit biofilm formation by Salmonella enterica serovar Typhimurium. Appl Environ Microbiol 74: 6639–6648.

53. Ezaki N, Koyama M, Kodama Y, Shomura T, Tashiro K, et al. (1983) Pyrrolomycins F1, F2a, F2b and F3, new metabolites produced by the addition of bromide to the fermentation. J Antibiot (Tokyo) 36: 1431–1438.

54. Tang YQ, Yeaman MR, Selsted ME (2002) Antimicrobial peptides from human platelets. Infect Immun 70: 6524–6533.

55. Li H, Hamza T, Tidwell JE, Clovis N, Li B (2013) Unique antimicrobial effects of platelet-rich plasma and its efficacy as a prophylaxis to prevent implant-associated spinal infection. Adv Healthc Mater 2: 1277–1284.

56. Zhang J, Middleton KK, Fu FH, Im HJ, Wang JH (2013) HGF mediates the anti-inflammatory effects of PRP on injured tendons. PLoS One 8: e67303.

57. Eggenhofer E, Hoogduijn MJ (2012) Mesenchymal stem cell-educated macrophages. Transplant Res 1: 12.

11

Environmental Predictors of Ice Seal Presence in the Bering Sea

Jennifer L. Miksis-Olds*, Laura E. Madden[†]

Applied Research Laboratory, The Pennsylvania State University, State College, Pennsylvania, United States of America

Abstract

Ice seals overwintering in the Bering Sea are challenged with foraging, finding mates, and maintaining breathing holes in a dark and ice covered environment. Due to the difficulty of studying these species in their natural environment, very little is known about how the seals navigate under ice. Here we identify specific environmental parameters, including components of the ambient background sound, that are predictive of ice seal presence in the Bering Sea. Multi-year mooring deployments provided synoptic time series of acoustic and oceanographic parameters from which environmental parameters predictive of species presence were identified through a series of mixed models. Ice cover and 10 kHz sound level were significant predictors of seal presence, with 40 kHz sound and prey presence (combined with ice cover) as potential predictors as well. Ice seal presence showed a strong positive correlation with ice cover and a negative association with 10 kHz environmental sound. On average, there was a 20–30 dB difference between sound levels during solid ice conditions compared to open water or melting conditions, providing a salient acoustic gradient between open water and solid ice conditions by which ice seals could orient. By constantly assessing the acoustic environment associated with the seasonal ice movement in the Bering Sea, it is possible that ice seals could utilize aspects of the soundscape to gauge their safe distance to open water or the ice edge by orienting in the direction of higher sound levels indicative of open water, especially in the frequency range above 1 kHz. In rapidly changing Arctic and sub-Arctic environments, the seasonal ice conditions and soundscapes are likely to change which may impact the ability of animals using ice presence and cues to successfully function during the winter breeding season.

Editor: Craig A. Radford, University of Auckland, New Zealand

Funding: Funding was provided by the Office of Naval Research Award N000140810391. The funders had no role in study design, data collection and analysis, decision to publish, or preparation of the manuscript.

Competing Interests: The authors have declared that no competing interests exist.

* Email: jlm91@arl.psu.edu

† Deceased.

Introduction

Ribbon (*Histriophoca fasciata*) and bearded seal (*Erignathus barbatus*) vocalizations are salient vocalizations recorded seasonally in the Bering Sea from January-June when sea ice is present [1–2]. These calls are most likely produced by males as a display to attract females and establish territory during the mating season [3–8]. During the winter breeding season, the Bering Sea is cold, dark, and ice covered; consequently, these aquatically mating species must locate potential mates while also maintaining positions within the ice sheets where breathing holes can be maintained or where they have access to open water at the ice edge or within polynyas. How they navigate in the low visibility conditions of the dynamic ice sheets to locate potential mates and maintain access to open water for breathing and mating is not fully known. Artificially introduced acoustic cues were shown to be extremely important for a blindfolded spotted seal (*Phoca largha*) in navigating under ice to locate breathing holes [9]; therefore, it is not unreasonable to hypothesize that ribbon and bearded seals

may also be using soundscape cues to orient under the ice. The goal of this work was to determine the strongest environmental predictors of ice seal vocal presence in the Bering Sea. Multiple environmental (ice and prey) and acoustic variables were considered in predictive models of ribbon and bearded seal presence during the winter in the Bering Sea, and it was the modeling results that provided insight as to the potential role the soundscape may play in under-ice navigation of these ice seals.

It is widely known that animals use sound to navigate through the environment. Bats and dolphins actively probe the environment with echolocation [10], and non-visual communication signals from conspecifics and heterospecifics guide animals in acquiring mates, foraging, and defense [11]. Over the past decade, there have been an increasing number of studies that explore how animals use information from the overall environmental "soundscape", the combination of biologic (biophony), abiotic (i.e. wind, rain, and other geologic sounds referred to as geophony), and man-made (anthrophony) sounds, gained via passive listening for orientation and navigation [12–15]. The concept of using ambient

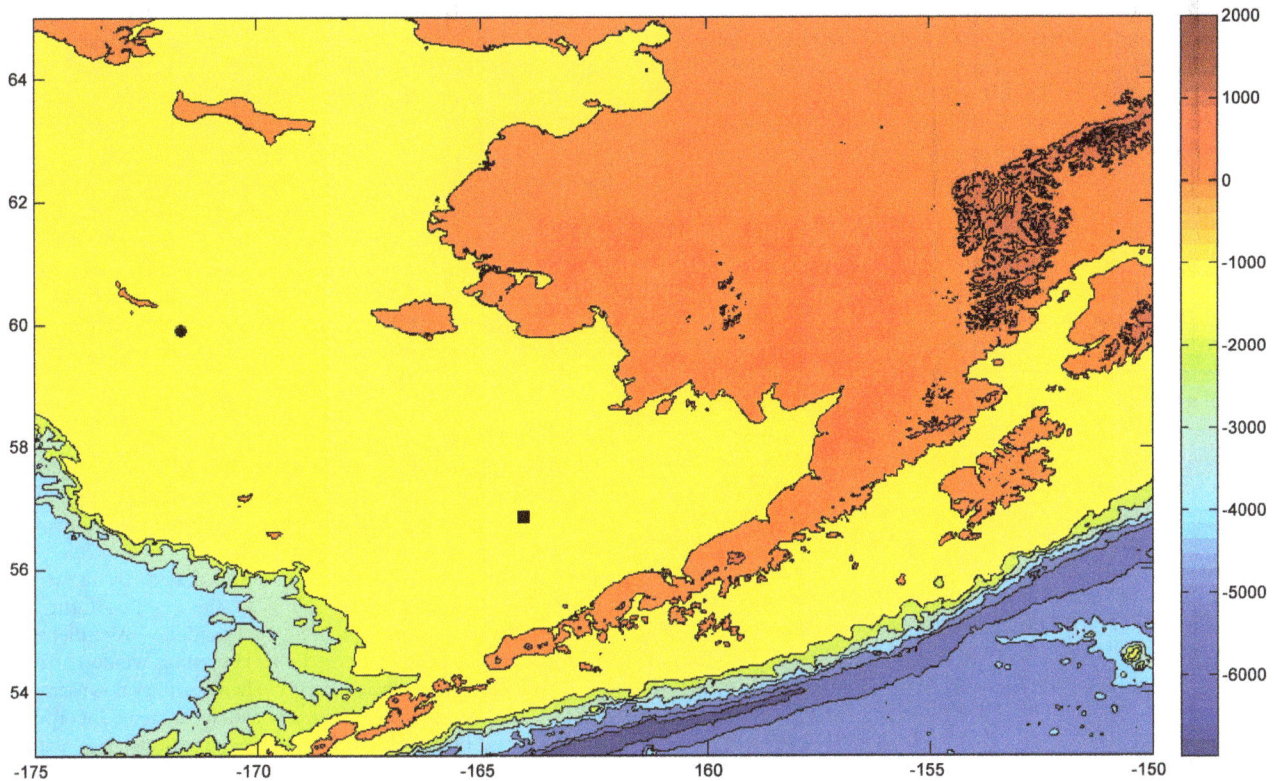

Figure 1. Mooring site locations in the Bering Sea.

or reflected sounds (as opposed to specific communication signals) to direct movement or identify appropriate habitats has recently been identified as a new field of study referred to as soundscape orientation, and the concept is also included within the broader field of soundscape ecology in the scientific literature [14,16–17].

In the marine environment where visual signals do not propagate very far, animals rely on sound as their primary means of obtaining information over any significant distance. It has been speculated that large baleen whales use ambient acoustic cues or acoustic landmarks to guide their migration [18–22]. Laboratory and field studies have demonstrated that both invertebrates (oyster and crab) and fish use soundscape cues for orientation and localization of appropriate settlement habitat. The commonality between all the soundscape orientation experiments conducted in the marine environment, and select terrestrial habitats, is that the soundscapes identified as having an impact on animal behavior originated from areas of high species diversity in association with reefs or rainforests [12,15,23–27]. Habitats with greater biodiversity are associated with richer acoustic soundscapes compared to low diversity habitats, which in itself may be an important cue for animal orientation [16,28–30]. This author is only aware of two studies that identify specific acoustic characteristics of the soundscape that are predictors of behavioral response. Stanley et al. (2011) [31] measured the sound intensity level required to elicit settlement and metamorphosis in several species of crab larvae, and Simpson et al. (2008) [32] discovered that coral reef fish responded more strongly to the higher frequency components (>570 Hz) of the reef soundscape.

It has been impossible to truly assess specific predictors, acoustic or otherwise, of behavioral response for ice seals living in conditions unhospitable to direct observation. However, advances in remote sensing technology and capabilities have provided the

means to begin identifying potentially important parameters that are deserving of more in-depth study. This study used multiple, synoptically sampled time series from remotely deployed sensors to gain a better understanding of the environmental parameters most likely influencing ice seal behavior during the breeding season in the Bering Sea.

Methods

Moorings

Active and passive acoustic sensors were incorporated into subsurface acoustic moorings deployed at two locations on the 70-m isobath of the eastern Bering Sea shelf at sites M2 (56° 51.570′N, 164° 3.801′W) and M5 (59° 54.285′N, 171° 42.285′W) in 2009 and 2007, respectively. Moorings at these locations have been deployed and maintained as part of the NOAA Ecosystems and Fisheries-Oceanography Coordinated Investigations (Eco-FOCI) Program (http://www.ecofoci.noaa.gov) since 1995 and 2004, respectively for the M2 and M5 moorings [33] (Figure 1). The acoustic sensors were integrated into the NOAA-deployed, observational moorings under a NOAA Request for Blanket Scientific Research Permit and did not require a specific permit for remote sensing. Moorings were deployed subsurface to prevent entanglement in seasonal sea ice and were serviced in Spring (April/May) and Fall (September/October) each year depending on weather conditions. The acoustic sensors were deployed on a separate, short mooring in conjunction with oceanographic moorings at each location. The oceanographic and acoustic moorings were separated by a distance of approximately 1 km to minimize noise produced by the oceanographic mooring hardware and sensors in the acoustic recordings. The data used in this study comes from acoustic data acquired 27 Sep 2009–19 May 2011 at location M2 and 26 Sep 2008–20 May 2011 at location M5.

The acoustic mooring consisted of a series of active and passive sensors including a 300 kHz RDI ADCP, three-frequency (125 kHz, 200 kHz, and 460 kHz) scientific echosounder system of Acoustic Water Column Profilers (AWCPs: ASL Environmental Sciences, Inc, Sidney, BC), Passive Aquatic Listener (PAL) recorder, and an AURAL (Autonomous Underwater Recorder for Acoustic Listening) (Multi-Électronique (MTE) Inc., Québec, Canada). The mooring was constructed in the following order from top (approximately 60–62 m) to bottom (approximately 70 m): 36″ floatation, 300 kHz RDI ADCP, 30″ floatation, AURAL, AWCP system, PAL, and acoustic release. The AWCP system was mounted in an upward-looking direction 15° off vertical to eliminate interference from flotation and instruments in the mooring line directly above the active acoustic system.

This study utilized data from the PAL and AWCP system. AWCPs record acoustic backscatter to monitor the presence and location of acoustic scatterers such as zooplankton and fish within the water column [34–35]. The transducers of the three different frequencies were positioned in the mooring cage so that the beam patterns were aligned to sample the same parcel of water nearly simultaneously. The echosounders sampled the water column for 5 min. each half hour. During each 5 min. sampling period, acoustic backscatter measurements were recorded every 2 s with 20 cm range bins from approximately 0.75 m above the transducer face to the water surface. Zooplankton net tows were conducted during mooring maintenance activities and on separate research cruises in the area using either a 25-cm diameter CalVET system (CalCOFI Vertical Egg Tow; [36]) having 0.15 mm mesh nets or double-oblique tows of paired bongo frames (60-cm frame with 0.333 mm mesh and 20-cm frame with 0.150 mm mesh) [2]. Data from the net tows provided information on dominant species, species composition, and numerical density to aid in defining size classes and interpretation of the acoustic data.

The PAL is an event detector, or adaptive sub-sampling acoustic recorder, with a temporal sampling strategy designed to allow the instrument to record data for up to one year [37–40]. The default sampling strategy was to record a 4.5 sec acoustic time series, or soundbite, at a sampling rate of 100 kHz every 5 minutes corresponding to a 1.5% duty cycle. When sampling in the default mode, onboard processing algorithms sub-sampled the 4.5 sec soundbite eight times and generated a power spectrum for each sub-sample. A preliminary detection algorithm identified signals of interest when a temporal feature of the sub-sampled power spectra in a soundbite exceeded one of three threshold criteria: 1) the matching of spectrum characteristics to known spectra, 2) exceeding a 12 dB threshold level between sequential samples indicating a transient source, or 3) the matching of predefined peaks (e.g. 300 Hz–3 kHz) indicating possible tonal or click vocalizations from marine mammals. If no signals of interest were detected, the spectra were averaged, and a single spectrum was saved to the hard disk. The soundbite time series was discarded in the default sampling mode. During periods of increased acoustic activity where signals of interest triggered a modified sampling protocol, the sampling interval was decreased to 2 minute intervals corresponding to a 4% duty cycle. In the modified sampling mode, individual spectra and the soundbites were saved to the hard disk. The PAL continued to operate in the modified sampling mode until no signals of interest were detected. The PAL then returned to the default sampling mode. Details on the adaptive sampling algorithms of the PAL are found in Miksis-Olds et al. (2010) [39].

Field data was collected under Observational Institutional Animal Care and Use Committee (IACUC) #36003 "Character-

izing Biological Scatter and Its Implications for Marine Mammals in the Bering Sea" from The Pennsylvania State University. There was no direct interaction with any vertebrates in this study, as all data from marine mammals were obtained remotely through passive acoustic listening.

Ice Data

Daily mean ice cover (or percent cover in this specific region) and ice thickness data were obtained from the images produced by the NOAA Ice Desk at the National Weather Service in Anchorage, Alaska. The images are posted on http://pafc.arh. noaa.gov/ice.php. Ice conditions surrounding the mooring locations were estimated within an approximate 20 km² around the mooring.

Data Processing

Ribbon and bearded seal presence was determined from the PAL soundbites with the understanding that detection of vocalizations indicates seal presence, and the lack of acoustic detection does not imply animal absence. Soundbites were reviewed by a human classifier and verified by a second independent human classifier blind to the results of the first reviewer. Sound sources present in the soundbites were identified from spectrograms (1024 point FFT, Hamming window, 87.5% overlap) made from the original 100 kHz recordings downsampled to 48 kHz using Adobe Audition 3.0 (Adobe Systems Incorporated). These settings provided a bandwidth of 61 Hz, with a frequency resolution of 47 Hz, and a time resolution of 2.7 ms. Marine mammal vocalizations were classified aurally and visually from the spectrograms by species (bowhead (*Balaena mysticetus*), gray whale (*Eschrichtius robustus*), killer whale (*Orcinus orca*), beluga whale (*Delphinapterus leucas*), walrus (*Odobenus rosmarus*), ribbon, and bearded seals). Ribbon seal grunts, roars, and downsweeps were used to indicate presence [1,41–42]. Bearded seal vocal presence was determined by the identification of trills, the most salient of the bearded seal vocalizations [6–7]. The adaptive sampling protocol and low sampling duty cycle of the PAL prevented calculations of the daily detection rate or overall number of seal vocalizations per day.

Analysis of PAL spectra included examination of spectral shape and levels. Temporal clusters of similarly distinctive sound spectra lasting tens of minutes to hours were manually identified and classified. Sound levels were computed from the time series of spectra. Each spectrum was computed from 1024 point samples of the 4.5 s time series. This resulted in a 513 point power spectral density with each of the bins covering 97 Hz of the 50 kHz usable bandwidth. The spectra were then reduced from 513 points to 64 points by averaging spectra levels over two bins below 3 kHz and over ten bins from 3 to 50 kHz. The resulting power spectral density, relative to $\mu 1Pa^2/Hz$, represents energy from the complete 50 kHz bandwidth with variable frequency resolution. To compute the sound level from these spectra, the values were converted to linear power spectral density and multiplied by the frequency resolution of the bins and then summed. The unit of the full bandwidth average is a sound pressure level, re 1 μPa. Processing of power spectral density was conducted for five frequencies over seven octaves (500 Hz, 2 kHz, 10 kHz, 20 kHz, 40 kHz) with units of dB re 1 $\mu Pa^2/Hz$.

To assess prey parameters related to zooplankton/fish abundance and community composition, the AWCP data were processed in 5 m vertical depth bins. Daily mean volume backscatter coefficient (mean S_v in units m^2/m^3) was calculated from 24 hour integrations over each 5 m depth layer using EchoView software (Myriax, Tasmania). Targets within each

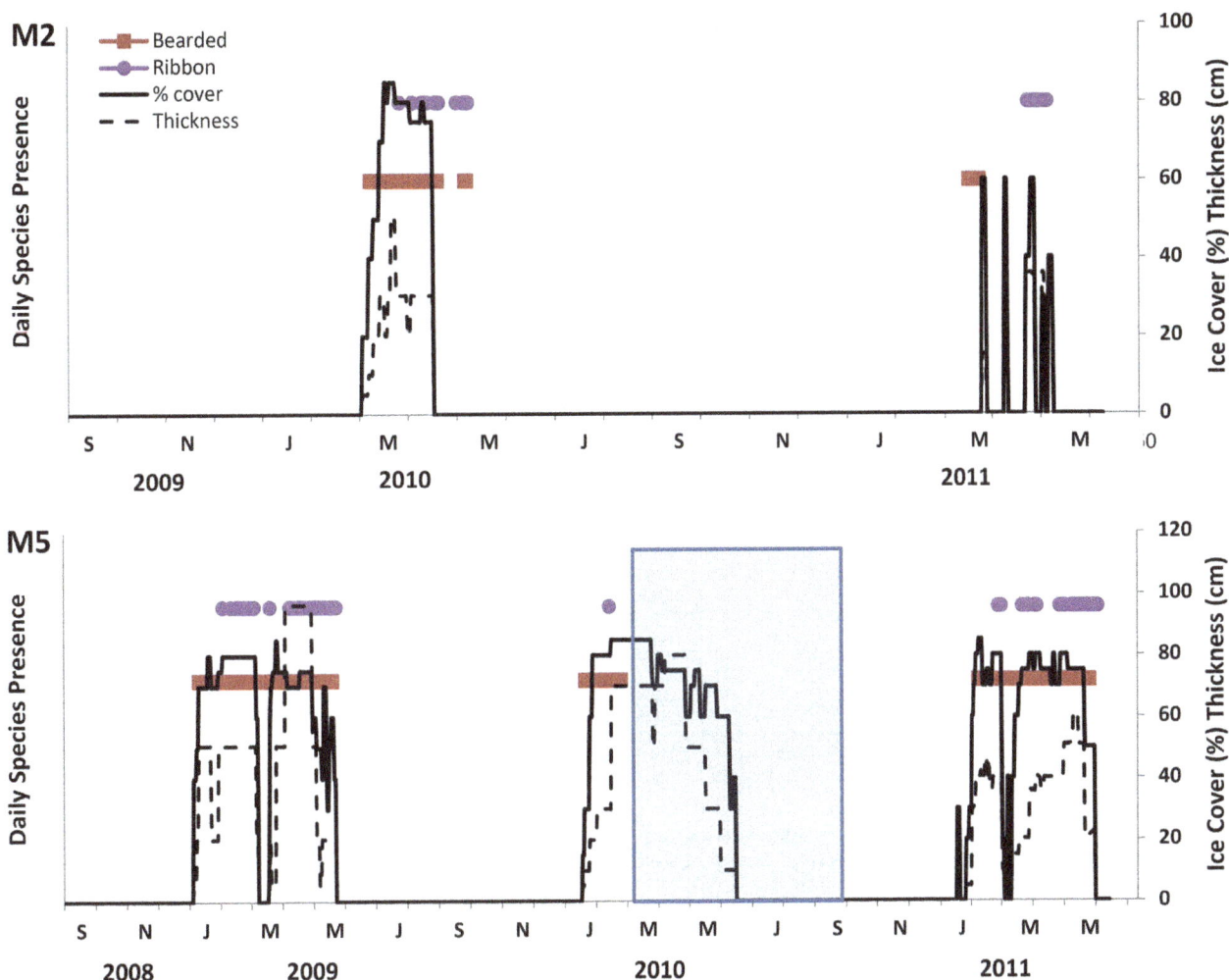

Figure 2. Time series of daily ice and ice seal presence over the M2 and M5 moorings where data exist from 2008–2011. Acoustic presence of species does not correspond to a numerical value on the y axis. The species-specific symbols reflect daily acoustic presence and are separated spatially for easy visualization. The blue box indicates a period of time where no acoustic data were available from the PAL.

depth and time bin were classified as to the likely source of the scattering based on differences in scattering amplitude between the three frequencies. Analyses using this dB-difference approach [2,43–45] are typically groundtruthed with information from net tows or video observations. However, given the low level of direct sampling of the water column in this study, a different approach was used and was consistent with Miksis-Olds et al. (2013) [2] summarized here. If scattering assemblages were monospecific, then the dB-difference for a single scatterer type and an aggregation of scatterers of this type would be identical, although the volume backscattering at each frequency would be different. Theoretical scattering curves for four different types of individual scatterers were generated and dB-differences between the three AWCP frequencies were calculated. Scattering amplitudes (and the subsequent dB differences) were generated using a Stochastic Distorted Wave Born Approximation model [46] for the following scatterers: 1) small scatterers such as neritic copepods (lengths of 1–5 mm) (*Pseudocalanus* spp., *Acartia longiremis, Oithona* spp. and *Calanus*), 2) medium scatterers (lengths of 5–15 mm) which includes juvenile krill, chaetognaths, and amphipods, 3) large scatterers such as adult euphausiids (lengths of 15–30 mm), 4) resonant scatterers which represents an organism with a gas-

inclusion such as a swim-bladdered fish or siphonophore, and 5) unknown. The acoustic system was not able to detect the weak scattering strengths of scatterers less than approximately 5 mm in length unless they were present in extremely dense aggregations. Aggregations were classified as belonging to one of the five categories (small, medium, or large scatterer; resonant; or unknown) by determining the shortest geometric distance between the three dB differences calculated for the aggregation and that of the theoretical scatterers. If the closest geometric distance was more than 12 dB (an arbitrarily chosen value), then the aggregation was classified as unknown.

Modeling. Daily presence-absence data for ribbon and bearded seals identified in the passive acoustic recordings was the response variable in the generalized linear and generalized additive models (GLM and GAM) designed to identify predictor variables of ice seal presence (Table S1). There is a high degree of temporal overlap between ribbon and bearded seal detections in the Bering Sea [1–2], so daily presence-absence data for the two species was combined into a single ice seal response variable to increase statistical power. Initial models included the following predictor variables: ice thickness, % ice cover, 200 kHz Sv, % prey composition (small, medium, large, and resonant scatterers), and

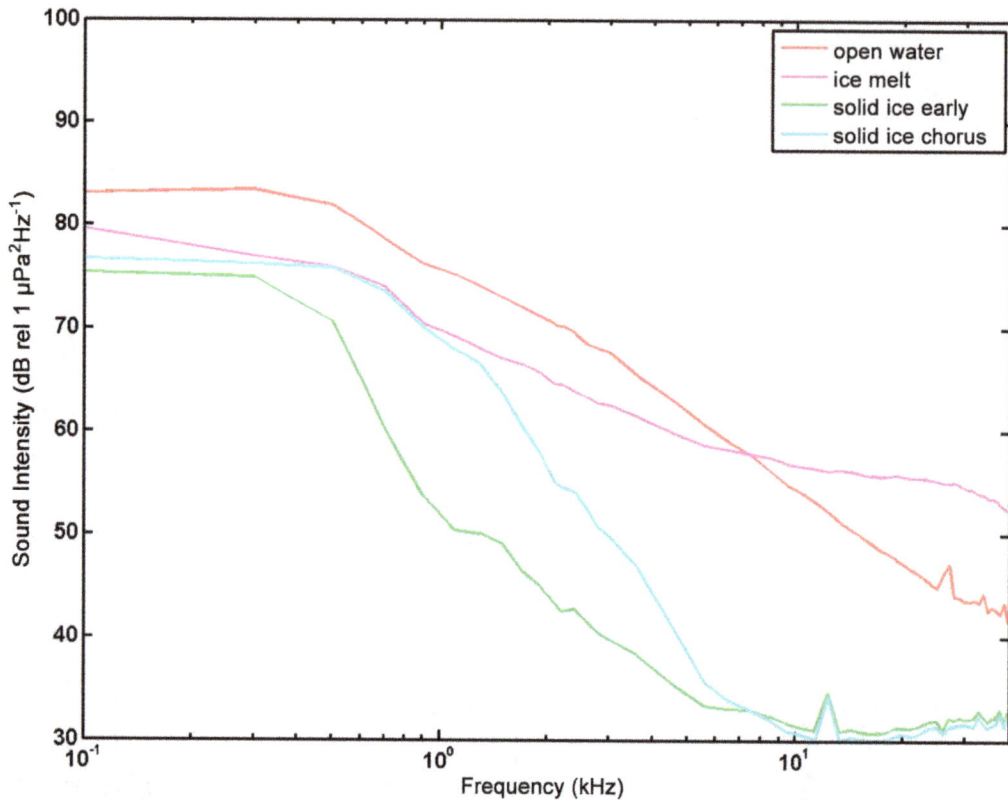

Figure 3. Representative spectra from the Bering Sea under different surface conditions. The Solid Ice Early spectrum represents the acoustic environment prior to the seasonal arrival of chorusing ice seals. The Solid Ice Chorus spectrum captures the acoustic environment when ice seals were observed to be chorusing in the acoustic record.

mean daily sound level (500 Hz, 2 kHz, 10 kHz, 20 kHz, 40 kHz) (Table S1 and Table S2). Data was first explored to identify potential outliers and evaluate distribution and collinearity among predictor variables and also with marine mammal presence using functions of the AED package in R [47–48]. Explanatory variables

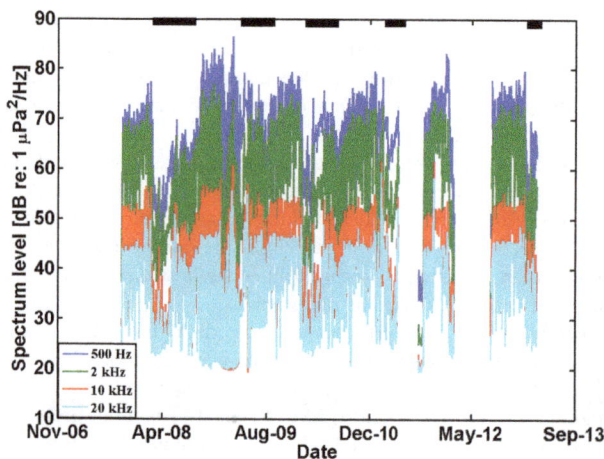

Figure 4. Sound pressure levels at four frequencies from the M5 location in the Bering Sea over a 4 year time period. Gaps in the data are periods when no data was available from the PAL. The black bars across the top of the figure indicate presence of regional ice cover.

were centered to allow better model convergence and interpretation, with the exception of ice cover and thickness. Ice cover and ice thickness showed a zero-inflated distribution and transformations failed to sufficiently address the skewed distributions. Zero values are meaningful for these measurements and thus these variables were not truncated or transformed. High collinearity was found between environmental sound level variables with the correlation highest between close frequencies. Ice cover and ice thickness were also highly collinear, although one or both of these variables were removed from the models during the selection process so this collinearity did not pose a problem. Final models including multiple noise variables were checked for collinearity using the *corvif* function from the R package *AED* [47–48]. All variables included in final models had VIFs well below 10 (the maximum VIF was 2.32), indicating sufficiently low collinearity [49].

Generalized linear models (GLMs) and generalized additive models (GAMs) allow model fitting to describe relationships between variables without constraints of linear regression models [50]. Generalized additive mixed models (GAMMs) and generalized linear mixed models (GLMMs) extend GAMs and GLMs to include random effects and correlation structures to deal with violations of independence that are often present in observational and time series data and are becoming popular in the analysis of ecological data (for example: Friedlaender et al., 2006 [51]; Wagner & Sweka, 2011 [52]). GLMs and GLMMs with a binomial distribution and logit link function were fit using a backward stepwise approach. Variables were selected for removal using the *drop1* command from the basic *stats* package in R (version 2.14.1; [53]) to apply an analysis of deviance test following a Chi-square distribution. Variables were removed based on a significance criteria of p<0.01 until all variables in the model were considered significant. Significance tests and p-values for analysis of deviance are approximate, thus a selection criteria below the standard 95% significance level was used to avoid inclusion of unnecessary terms. GLMMs were fit using the *glmmPQL* function from the *MASS* package in R [54]. This approach allowed the inclusion of a random effect for site to allow inference beyond the two stations sampled and a temporal correlation structure to address the lack of independence due to repeated sampling at each site. Convergence problems were frequently encountered when a temporal correlation structure for date grouped by site was included. Auto-correlation in the model residuals was examined to determine whether the temporal correlation structure was needed, as including a random effect for site imposes an implicit compound symmetry correlation structure that assumes a constant correlation within data points from the same site. GAMs and GAMMs with a binomial distribution and logit link function were fit using the same procedure described above with the *gamm* function from the *mgcv* package in R [55–56] to explore potential non-linear relationships.

GLMM and GAMM techniques are on the "frontier of statistical research" and as such model selection and validation for generalized models on absence-presence response data is difficult [47]. Standardized residuals were extracted and plotted against predictor variables and fitted values to look for patterns. Greater variation in residuals at zero ice coverage was discovered, likely due to the large number of zero values for ice cover. A new

data set, zero-truncated for ice cover, was then used to fit the final models to explore the potential for zero values to interfere with model function and selection.

Results

Seasonal ice was present at both mooring locations in the Bering Sea during each winter of the study (Figure 2). The ice cover over M5 on the central shelf was thicker and present longer compared to M2 on the southeastern shelf. Bearded seals were detected on 340 days, and ribbon seals were detected on 161 days over the study period from Sep 2008-May 2011. Seals were detected on fewer days at the southern mooring (M2) most likely due to the less persistent and shorter duration of ice cover compared to M5, but the proportion of daily detections for each species was similar at both mooring locations (39 days (63%) bearded and 23 days (37%) ribbon detected at M2; 301 days (68%) bearded and 138 days (32%) ribbon at M5) (Figure 2).

The grouping of PAL spectra identified four general sea surface conditions (open water, freeze up, solid ice, seasonal melting) (Figure 3). Validation of sea ice conditions from the passive acoustic data was inferred from the satellite ice thickness and mean ice cover calculations, seasonality, and recorded soundbites of physical processes. Overall sound levels during open water conditions were generally greater than when ice was present for frequencies of 1–10 kHz, which was consistent with previous studies (Figures 3 and 4) [2]. Above 10 kHz, melting conditions produced the greatest sound intensity. For frequencies less than 1 kHz, open water and initial freeze-up conditions had the greatest sound intensity. When solid ice was present above the moorings,

Table 1. GLM and GLMM final model results.

Variable	Parameter Estimate	Std. Error	DF	p-value
GLM				
Intercept	−4.353	0.380	902	<0.001
Ice cover	8.253	0.865	902	<0.001
c 10 kHz sound	−0.146	0.039	902	<0.001
c 40 kHz sound	0.130	0.045	902	0.004
c Large crustacean	0.036	0.017	902	0.031
Ice cover: c Large crustacean	−0.104	0.034	902	0.002
GLMM				
Random effect: site				
Intercept	−4.353	0.332	904	<0.001
Ice cover	8.254	0.756	904	<0.001
c 10 kHz noise	−0.146	0.034	904	<0.001
c 40 kHz noise	0.130	0.039	904	0.001
c Large crustacean	0.036	0.015	904	0.014
Ice cover: c Large crustacean	−0.104	0.030	904	<0.001
GLMM				
Random effect: site				
Correlation: CAR1				
Intercept	−3.700	0.314	1315	<0.001
Ice cover	5.947	0.482	1315	<0.001
c 10 kHz noise	−0.071	0.016	1315	<0.001

The letter *c* denotes centered variables. Ice cover is given as a fraction of cover from 0 to 1, large crustacean represents a percent composition, and both 10 kHz and 40 kHz are given in dB re 1 µPa²/Hz. The random intercept for site in the GLMM has a standard error of 0.0001 and residual standard error of 0.871. The explanatory variable large crustacean does not meet significance selection criteria (p<0.01), however is included due to the significance of its interaction term with ice cover.

sound levels were observed to be the lowest across the full frequency spectrum, with extremely low level intensity (<40 dB re 1 $\mu Pa^2/Hz$) above 5 kHz. The internal noise floor of the recorder was likely a limiting factor for sound levels below approximately 32 dB.

The initial models to determine predictors of ice seal presence included ice thickness, % ice cover, 200 kHz Sv, four size categories of % prey composition (small, medium, large, and resonant scatterers), and five sound levels (500 Hz, 2 kHz, 10 kHz, 20 kHz, 40 kHz). The final GLM and GLMM model both included ice cover, 10 kHz sound, 40 kHz sound, and an interaction between ice cover and large crustaceans as significant predictors of ice seal presence (Table 1). Ice seal presence showed a strong positive correlation with ice cover and a negative association with 10 kHz environmental sound levels (Table 1). Prey alone was not a good predictor of seal presence and an interaction between ice cover and large crustaceans indicates a negative relationship with seal presence, likely due to the somewhat non-linear relationship between ice seals and ice cover at M5 (discussed below with GAMM models).

Parameters estimated by both the GLM and GLMM are nearly identical (Table 1), suggesting little difference between M2 and M5. However, the inclusion of a random site effect in the GLMM was highly effective in addressing residual autocorrelation (Figure 5B). The inclusion of a temporal correlation structure in the GLMM reduced numerical stability (increased non-convergence problems) and captured less of the residual auto-correlation in the final model (Figure 5C). The GLMM with a random site effect and no temporal correlation structure (Figure 5B) was selected as the optimal model.

GAMM models showed primarily linear relationships between seal presence and ice cover and 10 kHz sound, with the exception of ice cover at M5 (Figure 6B, Table 2). The plateau in the seal-ice cover smoother seen at M5 may explain the negative slope of the ice cover: large crustacean interaction terms in Table 1. The final GAMM model included smooth terms for ice cover and 10 kHz sound with a random smoother for ice cover (Table 2). Including random smoothers for ice cover and 10 kHz sound resulted in neither 10 kHz smoother being significant. All GAMM models with 40 kHz sound failed to converge.

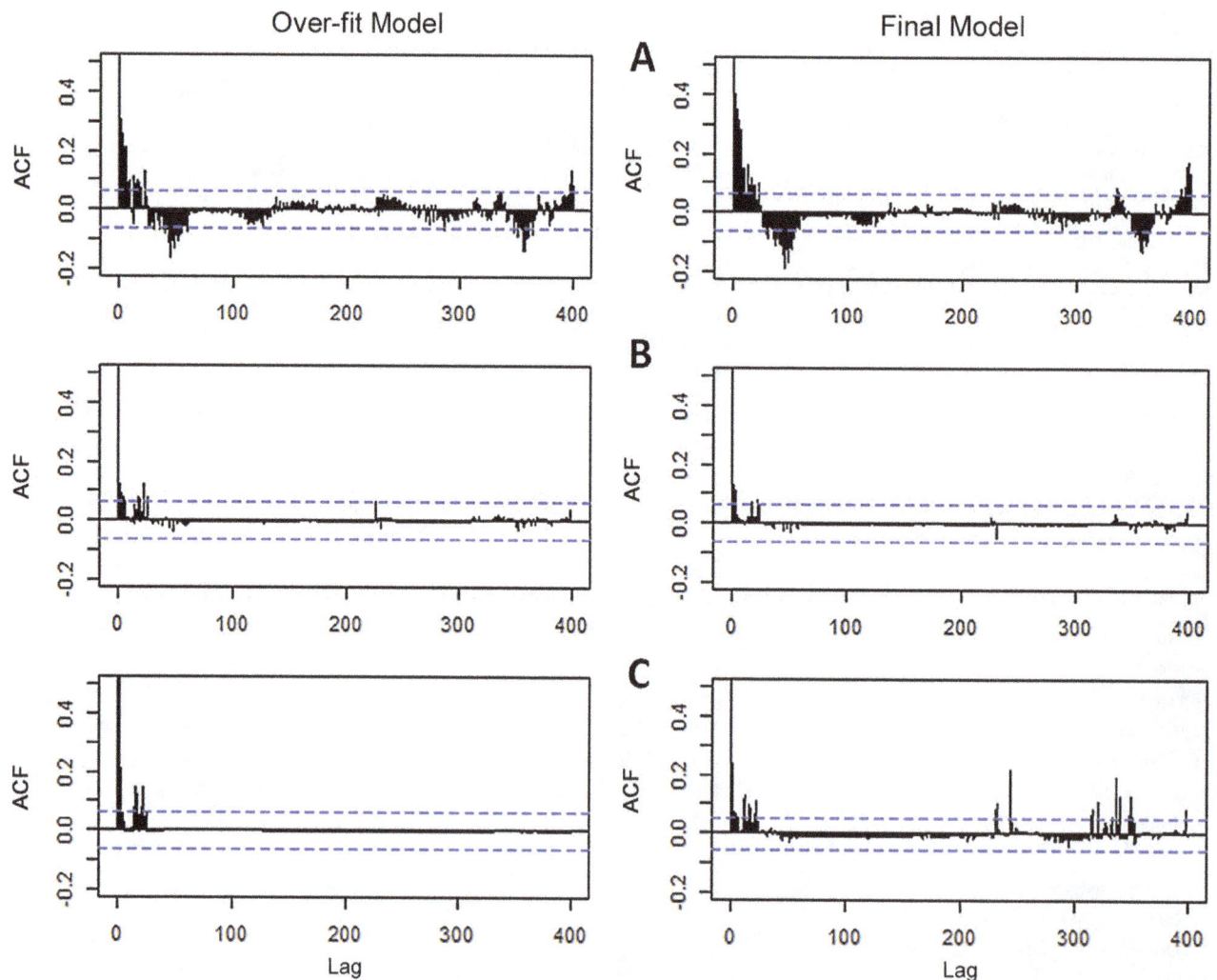

Figure 5. Auto-correlation function (ACF) of model residuals. Plots show auto-correlation of model residuals to 400 lags (400 days) for A) GLM with no random effects or temporal correlation structure, B) GLMM with random site effect and no temporal correlation structure and C) GLMM with random site effect and continuous AR-1 correlation structure. Over-fit models include all explanatory variables and interactions under consideration. Final models include only significant predictor variables after model selection.

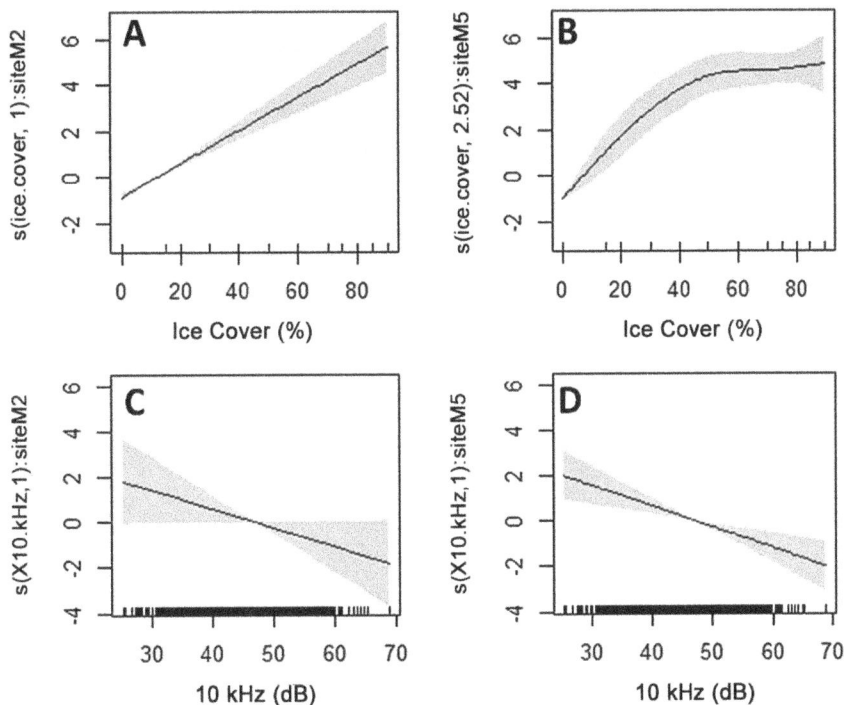

Figure 6. GAMM comparison of smooth functions by site for (A–B) percent ice cover and (C–D) 10 kHz sound. Shaded areas denote 95% confidence intervals. The smooth for M2 on 10 kHz sound was not significant (C). Increase in ice seal presence slows beyond 50% ice cover at M5 (B), although the relationship is still generally linear.

Figure 6B suggests a non-linear effect of ice cover on seal presence as the increasing trend in the smoother levels near 50 percent cover. However, 2.52 degrees of freedom alone is not strong evidence against a GLMM [47]. This relationship must also be regarded with caution as non-convergence issues disallowed inclusion of all predictor variables of interest in the fitting of this model. As a result, the GLMM with a random site effect was selected as the optimal model for this data. Ice cover and 10 kHz sound level appear to be significant predictors of seal presence, with 40 kHz sound and prey presence (combined with ice cover) as potential predictors as well.

Discussion

Sea ice and 10 kHz sound levels were the strongest predictors of ice seal vocal presence during the winter breeding season in the Bering Sea. The results indicate that as 10 kHz (and to a lesser extend 40 kHz) sound levels increased, the detection of ice seal vocalizations decreased. Neither ribbon seals nor bearded seals have a significant amount of energy in their vocalizations above 10 kHz [1,6], but if the underwater hearing capabilities of ribbon and bearded seals are comparable to other phylogenetically related, ice-dependent species (e.g. spotted seal (*Phoca largha*), ring seal (*Pusa hispida*), harbor seal (*Phoca vitulina*) [57]) then they are capable of hearing sound above 70 kHz [58–61]. The 10 kHz frequency falls directly within the frequency range of best hearing for related phocid species [58–61], so it is appropriate to conclude that ribbon and bearded seals can both detect and respond to sound signals in the 10–40 kHz range. Although it is known that ice seals hear and vocalize underwater, there is little direct evidence about how they use or rely on sound to direct their movements and behavior.

The modeling results directed a more detailed examination of the acoustic environment that ice seals encounter on an annual basis. Open water conditions are the loudest up to approximately 8 kHz (Figure 3). Above 8 kHz, conditions associated with the

Table 2. Final GAMM model parameters including a random smoother by site for ice cover.

Variable	edf	Std. Error	p-value
GAMM			
Intercept	–3.394	0.225	<0.001
s(ice cover): M2	1.000	–	0.003
s(ice cover): M5	2.522	–	<0.001
s(10 kHz)	1.000	–	<0.001

The estimated degrees of freedom (edf) indicate the "curviness" of the smooth terms with 1.00 representing a straight line. A linear relationship is indicated for 10 kHz sound (both sites) and ice cover at M2.

process of ice melting were loudest. Conversely, acoustic conditions associated with solid ice cover were the quietest over the entire spectrum from <5 Hz to 50 kHz. On average, there was a 20–30 dB difference between sound levels during solid ice conditions compared to open water or melting conditions. This difference provides a salient acoustic gradient between open water and solid ice conditions by which ice seals could orient to maintain their horizontal position within the ice sheet or proximity to the ice edge so that access to open water for breathing is preserved. By constantly assessing the acoustic environment to navigate along with the seasonal ice movement in the Bering Sea, it is possible that ice seals can gauge their safe distance to open water or the ice edge through the soundscape in dark, ice covered surroundings by orienting in the direction of higher sound levels, especially in the frequency range above 1 kHz.

This observational study was not able to establish a cause-effect relationship or identify a specific threshold or optimal sound level range that ribbon and bearded seals may employ to navigate under ice. Long-term tagging studies with acoustic dosimeters and GPS location capabilities will be needed to confirm this theory and provide direct evidence of the mechanisms of under-ice navigation in ice seals. It will also be useful to investigate this relationship across locations and regions in both the Arctic and Antarctic to assess whether this concept can be generalized to all ice-dependent species required to navigate under ice. This work presents a particularly timely observation, as the sea ice and acoustic conditions of the oceans, the strongest predictors of ice seal vocal presence during the winter breeding season, are changing due to climate change and industrialization related to shipping and energy exploration/production [62–63]. In order to fully access the risk this poses to ice seals, it is critical to gain a better understanding of how the seals use and rely on ice presence and its associated sound to survive in their extreme environments.

Supporting Information

Table S1   Response and predictor variables used in the GLM and GAM modeling at the M2 (A) and M5 (B) locations. Bearded and ribbon seal acoustic presence is a binary response: 0 for absent, 1 for present. The 200 kHz Sv are daily mean values, and the scatterer percent composition is reflective of the daily mean values within each size category. Ice cover % and ice thickness are daily assessment values.

Table S2   Daily mean sound levels at M2 (A) and M5 (B) used in the GLM and GAM modeling. Sound level units are dB re 1 $\mu Pa^2/Hz$.

Acknowledgments

Thanks are extended to the captains and crews of the R/V Miller Freeman and R/V Oscar Dyson for their efforts in deploying and recovering the mooring instruments. Support from Phyllis Stabeno, Bill Floering and Carol Dewitt (NOAA PMEL) made it possible to include the passive acoustic recorders into the moorings. Jeffrey Nystuen (APL UW) graciously provided the soundscape spectra image.

Author Contributions

Conceived and designed the experiments: JLMO LEM. Performed the experiments: JLMO. Analyzed the data: JLMO LEM. Contributed reagents/materials/analysis tools: JLMO LEM. Contributed to the writing of the manuscript: JLMO LEM.

References

1. Miksis-Olds JL, Parks SE (2011) Seasonal trends in acoustic detection of ribbon seals (*Histriophoca fasciata*) in the Bering Sea. Aquatic Mammals 37: 464–471.
2. Miksis-Olds JL, Stabeno PJ, Napp JM, Pinchuk AI, Nystuen JA, et al. (2013) Ecosystem response to a temporary sea ice retreat in the Bering Sea. Prog Oceanography 111: 38–51.
3. Poulter TC (1968) Marine mammals. p. 405–465. *in* Seboek T, ed. Animal Communication. Indiana University Press.
4. Ray C, Watkins WA, Burns J (1969) The underwater song of *Erignathus* (bearded seal). Zoologica 54: 79–83.
5. Burns JJ (1981) Bearded seal, *Erignathus barbatus* (Erxleben, 1777). P. 145–170. *in* Ridgway SS, Harrison RJ, eds. Handbook of Marine Mammals. Academic Press.
6. Stirling I, Calvert W, Cleator H (1983) Underwater vocalizations as a tool for studying the distribution and relative abundance of wintering pinnipeds in the High Arctic. Arctic 36: 262–274.
7. Cleator H, Stirling I, Smith TG (1989) Underwater vocalizations of the bearded seal (*Erignathus barbatus*). Can J Zool 67: 1900–1910.
8. Cleator HJ, Stirling I (1990) Winter distribution of bearded seals (*Erignathus barbatus*) in the Penny Strait Area, Northwest Territories, as determined by underwater vocalisations. Can J Fish Aquat Sci 47: 1071–1109.
9. Sonafrank N, Elsner R, Wartzok D (1983) Under-ice navigation by the spotted seal, *Phoca largha*. Abstract. Fifth Biennial Conf. on the Biol. of Mar. Mammals, Boston, November 1983.
10. Madsen PT, Surlykke A (2013) Functional convergence in bat and toothed whale biosonars. Physiology 28: 276–283.
11. Bradbury JW, Vehrencamp SL (1998) Principles of Animal Communication. Sinauer Associates, Inc. 882 p.
12. Simpson SD, Meekan M, Montgomery J, McCauley R, Jeffs A (2005) Homeward sound. Science 308: 221.
13. Slabbekoorn H, Bouton N (2008) Soundscape orientation: a new field in need of sound investigation. Anim Behav 76: e5–e8.
14. van Opzeeland I, Slabbekoorn H (2012) Importance of underwater sounds for migration of fish and aquatic mammals. 357–359. *in* Popper AN, Hawkins A, eds. Effects of Noise on Aquatic Life. Springer Science+Business Media, LLC.
15. Lillis A, Eggleston DB, Bohnenstiehl DR (2013) Oyster larvae settle in response to habitat-associated underwater sounds. PLoS ONE 8: e79337.
16. Pijanowski BC, Villanueva-Rivera LJ, Dumyahn SL, Farina A, Krause BL, et al. (2011) Soundscape ecology the science of sound in the landscape. BioScience 61 (3): 203–216.
17. Bormpoudakis D, Sueur J, Pantis JD (2013) Spatial heterogeneity of ambient sound at the habitat level: ecological implications and applications. Landscape Ecol 28: 495–506.
18. Norris KS (1967) Some observations on the migration and orientation of marine mammals. 101–102. *in* Storm RM, ed. Animal Orientation and Navigation. Proceedings of the 27th Annual Biology Colloquium. Oregon State University Press. 134p.
19. Able KP (1980) Mechanisms of orientation, navigation, and homing. 283–373. *in* Gauthreaux Jr SA, ed. Animal Migration, Orientation, and Navigation. Academic Press.
20. Kenney RD, Mayo CA, Winn HE (2001) Migration and foraging strategies at varying spatial scales in western North Atlantic right whales: a review of hypothesis. J Cet Res Manag (Special Issue) 2: 251–260.
21. Gordon JCD, Tyack P (2002) Acoustic techniques for studying cetaceans. 293–324. *in* Evans PGH, Raga JA, eds. Marine Mammals: Biology and Conservation. Kluwer Academic.
22. Mate BR, Urban-Ramirez J (2003) A note on the route and speed of a gray whale in its northern migration from Mexico to central California, tracked by satellite-monitored radio tag. J Cet Res Manag 5(2): 155–157.
23. Jeffs A, Tolimieri N, Montomery JC (2003) Crabs on cue for the coast: the use of underwater sound for orientation by pelagic crab stages. Mar Freshwat Res 54: 841–845.
24. Tolimieri N, Haine O, Jeffs A, McCauley R, Montgomery J (2004) Directional orientation of pomacentrid larvae to ambient reef sound. Coral Reefs 23: 184–191.
25. Radford CA, Stanley JA, Simpson SD, Jeffs AG (2011) Juvenile coral reef fish use sound to locate habitats. Coral Reefs 30: 295–305.
26. Stanley JA, Radford CA, Jeffs AG (2012) Location, location, location: finding a suitable home among the noise. Proc R Soc B 270: 3622–3631.
27. Rodriguez A, Gasc A, Pavoine S, Grandcolas P, Gaucher P, et al. (2014) Temporal and spatial variability of animal sounds within a Neotropical forest. Ecological Informatics 21: 133–143.
28. Sueur J, Pavoine S, Hamerlynck O, Duvail S (2008). Rapid acoustic survey for biodiversity appraisal. PLoS ONE 3: e4065.
29. Radford CA, Stanley JA, Tindle CT, Montgomery JC, Jeffs AG (2010) Localised coastal habitats have distinct underwater sound signatures. Mar Ecol Prog Ser 401: 21–29.
30. McWilliam JN, Hawkins AD (2013) A comparison of inshore marine soundscapes. J Exp Mar Biol Ecol 446: 166–176.

31. Stanley JA, Radford CA, Jeffs AG (2011) Behavioural response thresholds in New Zealand crab megalopae to ambient underwater sound. PLoS ONE 6: e28572.

32. Simpson SD, Meekan MG, Jeffs A, Montgomery JC, McCauley RD (2008) Settlement-stage coral reef fish prefer the higher-frequency invertebrate-generated audible component of reef noise. Anim Behav 75: 1861–1868.

33. Stabeno PJ, Napp J, Mordy C, Whitledge T (2010) Factors influencing physical structure and lower trophic levels of the eastern Bering Sea shelf in 2005: Sea ice, tides and winds. Prog Oceanography 85(3–4): 180–196.

34. Brierley AS, Saunders RA, Bone DG, Murphy EJ, Enderlein P, et al. (2006) Use of moored acoustic instruments to measure short-term variability in abundance of Antarctic krill. Limnology and Oceanography: Methods 4: 18–29.

35. Kunze E, Dower JF, Beveridge I, Dewey R, Bartlett KP (2006) Observations of Biologically Generated Turbulence in a Coastal Inlet. Science 22: 1768–1770.

36. Smith PE, Flerx W, Hewitt RP (1985) The CalCOFI vertical egg tow (CalVET) net. p. 23–33. in Lasker R, ed. An Egg Production Method for Estimating Spawning Biomass of Pelagic Fish: Application to the northern anchovy Engraulis mordax. NOAA Technical Report NMFS 36, U.S. Department of Commerce.

37. Nystuen JA (1998) Temporal sampling requirements for Autonomous Rain Gauges, J. Atmos. Ocean. Technol. 15: 1254–1261.

38. Nystuen JA, Amitai E, Anagnostou EN, Anagnostou MN (2008) Spatial averaging of oceanic rainfall variability using underwater sound: Ionian Sea Rainfall Experiment 2004, J Acoust Soc Am 123: 1952–1962.

39. Miksis-Olds JL, Nystuen JA, Parks SE (2010) Detecting marine mammals with an adaptive sub-sampling recorder in the Bering Sea. J App Acoust 71: 1087–1092.

40. Sousa-Lima RS, Norris TF, Oswald JN, Fernandes DP (2013) A review and inventory of fixed autonomous recorders for passive acoustic monitoring of marine mammals. Aquatic Mammals 39: 23–53.

41. Watkins WA, Ray GC (1977) Underwater sounds from ribbon seal, Phoca (Histriophoca) fasciata. Fish Bulletin 75: 450–453.

42. Jones JM, Thayre BJ, Roth EH, Mahoney M, Sia I, et al. (2014) Ringed, bearded, and ribbon seal vocalizations north of Barrow, Alaska: Seasonal presence and relationship with sea ice. Arctic 67: 203–222.

43. Watkins JL, Brierley AS (2002) Verification of acoustic techniques used to identify and size Antarctic krill. ICES J Mar Sci 59: 1326–1336.

44. Reiss CR, Cossio AM, Loeb VL, Demer DA (2008) Variations in the biomass of Antarctic krill (Euphausia superba) around the South Shetland Islands, 1996–2006. ICES J Mar Sci 65 (4): 497–508.

45. De Robertis A, McKelvey DR, Ressler PH (2010) Development and application of an empirical multifrequency method for backscatter classification. Can J Fish Aquat Sci 67: 1459–1474.

46. Demer DA, Conti SG (2003) Validation of the stochastic distorted-wave Born approximation model with broad bandwidth total target strength measurements of Antarctic krill. ICES J Mar Sci 60: 625–635.

47. Zuur AF, Ieno EN, Walker NJ, Saveliev AA, Smith GM (2009) Mixed effects models and extensions in ecology with R: Springer Science+Buisness Media.

48. Zuur A (2010) AED: Data files used in Mixed effects models and extensions in ecology with R (in Zuur et al. 2009). R package version 1.0.

49. Chatterjee S, Hadi AS (2006) Regression Analysis by Example. John Wiley and Sons.

50. McCullagh P, Nelder JA (1989) Generalized Linear Models, Chapman & Hall/CRC.

51. Friedlaender AS, Halpin PN, Qian SS, Lawson GL, Wiebe PH, et al. (2006) Whale distribution in relation to prey abundance and oceanographic processes in shelf waters of the Western Antarctic Peninsula. Mar Ecol Prog Ser 317: 297–310.

52. Wagner T, Sweka JA (2011) Evaluation of Hypotheses for Describing Temporal Trends in Atlantic Salmon Parr Densities in Northeast US Rivers. N Am J Fish Manag 31(2): 340–351.

53. R Development Core Team (2011) R: A Language and Environment for Statistical Computing. R Foundation for Statistical Computing, Vienna, Austria. Available: http://www.R-project.org/.

54. Venables WN, Ripley BD (2002) Modern Applied Statistics with S, Springer Verlag.

55. Wood SN (2006) Generalized Additive Models: An Introduction with R, CRC Press.

56. Wood SN (2011) Fast stable restricted maximum likelihood and marginal likelihood estimation of semiparametric generalized linear models. J Roy Stat Soc: Ser B (Statistical Methodology).

57. Berta A, Churchill M (2012) Pinniped taxonomy: review of currently recognized species and subspecies, and evidence used for their description. Mamm Rev 42: 207–234.

58. Terhune JM, Ronald K (1975) Underwater hearing sensitivity of two ringed seals (Pusa hispida). Can J Zool 53: 227–231.

59. Terhune JM (1988) Detection thresholds of a harbour seal to repeated underwater high-frequency, short-duration sinusoidal pulses. Can J Zool 66: 1578–1582.

60. Reichmuth C, Holt MM, Mulsow J, Sills JM, Southall BL (2013) Comparative assessment of amphibious hearing in pinnipeds. Journal of Comparative Physiology A 199: 491–507.

61. Sills JM, Southall BL, Reichmuth C (2014) Amphibious hearing in spotted seals (Phoca largha): underwater audiograms, aerial audiograms, and critical ratio measurements. J Exp Biol 217: 726–734.

62. Joseph JE, Chiu CS (2010) A computational assessment of the sensitivity of ambient noise level to ocean acidification. J Acoust Soc Am 128(3): EL144–EL149.

63. Boyd IL, Frisk G, Urban E, Tyack P, Ausubel J, et al. (2011) An International Quiet Ocean Experiment. Oceanography 24(2): 174–181.

12

Satellite Tagging and Biopsy Sampling of Killer Whales at Subantarctic Marion Island: Effectiveness, Immediate Reactions and Long-Term Responses

Ryan R. Reisinger[1]*, W. Chris Oosthuizen[1], Guillaume Péron[2], Dawn Cory Toussaint[1], Russel D. Andrews[3,4], P. J. Nico de Bruyn[1]

[1] Mammal Research Institute, Department of Zoology and Entomology, University of Pretoria, Pretoria, South Africa, [2] Centre for Statistics in Ecology, Environment and Conservation, Department of Statistical Sciences, University of Cape Town, Cape Town, South Africa, [3] School of Fisheries and Ocean Sciences, University of Alaska Fairbanks, Fairbanks, Alaska, United States of America, [4] Alaska SeaLife Center, Seward, Alaska, United States of America

Abstract

Remote tissue biopsy sampling and satellite tagging are becoming widely used in large marine vertebrate studies because they allow the collection of a diverse suite of otherwise difficult-to-obtain data which are critical in understanding the ecology of these species and to their conservation and management. Researchers must carefully consider their methods not only from an animal welfare perspective, but also to ensure the scientific rigour and validity of their results. We report methods for shore-based, remote biopsy sampling and satellite tagging of killer whales *Orcinus orca* at Subantarctic Marion Island. The performance of these methods is critically assessed using 1) the attachment duration of low-impact minimally percutaneous satellite tags; 2) the immediate behavioural reactions of animals to biopsy sampling and satellite tagging; 3) the effect of researcher experience on biopsy sampling and satellite tagging; and 4) the mid- (1 month) and long- (24 month) term behavioural consequences. To study mid- and long-term behavioural changes we used multievent capture-recapture models that accommodate imperfect detection and individual heterogeneity. We made 72 biopsy sampling attempts (resulting in 32 tissue samples) and 37 satellite tagging attempts (deploying 19 tags). Biopsy sampling success rates were low (43%), but tagging rates were high with improved tag designs (86%). The improved tags remained attached for 26±14 days (mean ± SD). Individuals most often showed no reaction when attempts missed (66%) and a slight reaction–defined as a slight flinch, slight shake, short acceleration, or immediate dive–when hit (54%). Severe immediate reactions were never observed. Hit or miss and age-sex class were important predictors of the reaction, but the method (tag or biopsy) was unimportant. Multievent trap-dependence modelling revealed considerable variation in individual sighting patterns; however, there were no significant mid- or long-term changes following biopsy sampling or tagging.

Editor: Jean-Benoit Charrassin, Musee National d'Histoire Naturelle, France

Funding: Funding was provided by the National Research Foundation's (NRF) Thuthuka and South African National Antarctic programmes, the South African Department of Science and Technology through the NRF, the Mohamed bin Zayed Species Conservation Fund (Project number: 10251290) and the International Whaling Commission's Southern Ocean Research Partnership. The funders had no role in study design, data collection and analysis, decision to publish, or preparation of the manuscript.

Competing Interests: The authors have declared that no competing interests exist.

* Email: rrreisinger@zoology.up.ac.za

Introduction

Cetaceans spend the vast majority of their lives under water and are highly mobile and often wide-ranging, which makes them a challenging taxon to study. Two field methods – tissue biopsy sampling and satellite-linked telemetry (or satellite tagging) – are becoming widely used in cetacean studies because they allow the collection of data which are difficult or impossible to obtain by other means. Tissues obtained by biopsy sampling can be used for a range of analyses including genetics, stable isotopes, fatty acids, contaminants, hormones and trace elements (see [1] for a review) and can so address aspects such as population structure, diet and animal health (e.g., [2–5]). Satellite tagging can elucidate the movement, distribution, behaviour and habitat use of cetaceans in relation to their physical environment (e.g., [6–8]). Such data are

critical to understanding the ecology of a species and its environmental role and, consequently, are vital to conservation or management efforts (e.g., [9,10]). The need for such information is particularly acute given the anthropogenic pressures many such populations and species face [10–12].

However, researchers must carefully consider their methods not only from an animal welfare perspective, but also to ensure the scientific rigour and validity of their results. The latter point is critical where methods may affect the subsequent behaviour or performance of individuals, thereby biasing the results obtained (e.g., [13–15]). From an ethical perspective researchers have an onus to assess the tradeoffs between the 'importance' of research, its likely benefit and its effect on animals before conducting work [16,17]; from a scientific perspective the responsibility is to design robust and valid studies [18]. Researchers should further evaluate animal effects and research methods *post-hoc*, refine these where needed and, importantly, publish such results [19,20].

Small cetaceans may be captured and restrained for satellite tagging and biopsy sampling (e.g., [21,22]) but this is impractical for most species and therefore remote techniques, which employ pole-mounted or projectile systems (typically fired from pneumatic rifles or crossbows) to biopsy sample or tag unrestrained animals, are most common. Remote biopsy sampling is an effective, mostly benign method of collecting fresh tissue samples from free-ranging cetaceans [1]. While cetaceans usually show some behavioural reaction to biopsy sampling, the reactions are typically mild and short-term (0.5–3 min) and the wounds made by the biopsy dart or punch heal quickly with no apparent adverse effects. Few studies, however, report on the behavioural and physiological impacts of remote biopsy sampling; this is important as different species and populations may react differently. No studies have shown long-term effects of biopsy sampling such as avoidance of the sampling area (e.g., [23]) or negative effects on reproduction and calf survival [24]; however, such effects are likely difficult to examine and only a small number of studies have attempted to do so [1].

Satellite tags are attached to animals using some form of sub-dermal retaining dart (e.g., [7,25]). As with biopsy sampling, relatively few remote satellite (and earlier radio) tagging studies describe the behavioural reactions of animals to tagging – if they do it is largely qualitative – and mid- to long-term follow up studies are rare. The majority of immediate reactions to tagging seem to be unnoticeable or mild and short-term [25–29]. Best and Mate [30] found no major effect of satellite tagging on the reproductive success of adult female southern right whales *Eubalaena australis* or the survival of their calves. Tagging also does not appear to affect the survival or reproductive success of humpback whales *Megaptera novaeangliae* [29,31].

One of the main challenges in remote satellite tagging systems is maximising the attachment durations of tags while minimising their invasiveness. Attachment durations have improved greatly (often hundreds of days currently compared to only a few days for the first attempts, see [25]) and tags have become smaller due to technological advances, but attachment duration remains highly variable. Remote satellite tagging studies were previously limited to large cetacean species, but the development of tags such as the 'Low Impact Minimally Percutaneous External-electronics Trans-mitter' configuration (LIMPET, [7]) has allowed tagging of smaller species such as killer whales *Orcinus orca*, Blainville's beaked whales *Mesoplodon densirostris*, false killer whales *Pseudorca crassidens* and pygmy killer whales *Feresa attenuata* [7,8,32–34].

Marion Island killer whales

Marion Island (46° 54′ S, 37° 45′ E), which lies in the Polar Frontal Zone in the Indian sector of the Southern Ocean, has a population of 58 identified killer whales which may occur at the island year round, but are most abundant between September and December [35,36]. This population has been observed preying on southern elephant seals *Mirounga leonina*, sub-Antarctic fur seals *Arctocephalus tropicalis* and three penguin species, and the peak killer whale abundance coincides with the breeding seasons of these seals and penguins [35]. It is entirely unknown what proportion of the whales' diet each species comprises and whether or not other prey (e.g., fishes, cephalopods) are taken, particularly when the whales are not observed at the island. Killer whales in the region depredate Patagonian toothfish *Dissostichus eleginoides* from longline fishing vessels [37], but it is unknown whether these individuals are from the Marion Island population or if toothfish are natural prey. When animals are not observed at the island their whereabouts and movements are unknown, although eight

individuals have been photographically identified at both Marion Island and the Crozet Islands, located approximately 950 km east of Marion Island [36,38]. The role of killer whales as drivers of seal and penguin population dynamics at Marion Island is important, but quantitatively uncertain [39]. The remoteness of Marion Island makes geographically wide-scale observations to elucidate diet and movement unfeasible and thus satellite tagging and biopsy sampling are vital methods to investigate the ecology of this population of killer whales.

Aims

In this paper we, firstly, report our methods for shore-based, remote biopsy sampling and satellite tagging of killer whales, the success of these methods and particularly the attachment duration and performance of LIMPET satellite tags. Secondly, we describe the immediate behavioural reactions of animals to biopsy sampling and satellite tagging and test for differences in the reactions to each. Thirdly, we test whether researcher experience influences biopsy sampling and satellite tagging. Lastly, using multievent capture-recapture analysis, we evaluate whether biopsy sampling and satellite tagging changed the behaviour of individuals, altering mid- (1 month) and long-term (<24 months) sighting patterns.

Methods

Ethics statement

Biopsy sampling and tagging was approved by the University of Pretoria's Animal Use and Care Committee (EC023-10) and the Prince Edward Islands Management Committee research and collection permits: 17/12; 1/2013; 1/2014.

Field methods

All killer whale studies at Marion Island are shore-based as boat-based work is not logistically possible or permitted [40]. Shore-based photographic identification (photo ID) has been successful at Marion Island as killer whales frequently approach within a few metres of the shore (Figure 1; [41]). This also allows work in weather conditions unsuitable for boat-based operations and importantly, in this study, allowed us to assess the reactions of animals to biopsy sampling and satellite tagging without any confounding reactions to boats.

We use 'sampled' and 'sampling' to refer to both biopsy sampling and satellite tagging; biopsy sampling is distinguished. We biopsy sampled and satellite tagged killer whales at two locations (Rockhopper Bay and Transvaal Cove) on the island's leeward east coast, near (<1.0 km) a long-term observation/photo ID site [41]. Both locations are low rock ledges, 1.0–2.0 m above the water surface. Sampling attempts were made primarily during 'dedicated observation sessions', in which the marksman would wait for killer whales for a predetermined length of time (typically 3–10 hours). We used a 68 kg draw weight recurve crossbow (Barnett Panzer V; Barnett Outdoors, LLC, Tarpon Springs, Florida, United States of America) equipped with a red dot sight for biopsy sampling and satellite tagging. Bolts were tethered with line and a fishing reel mounted on the crossbow (Methods S1, [42]). Biopsy and tagging attempts were made by two arbalesters during the study and reactions – described in Table 1– were scored by the arbalester. After October 2011 the arbalester usually wore a high-definition video camera (GoPro HD Hero and GoPro HD Hero 2; Woodman Labs, Inc., Half Moon Bay, California, United States of America) to record biopsy and tagging attempts (Figure 1).

Biopsy sampling. We obtained tissue samples using stainless steel biopsy tips (25 mm×7 mm) attached to the bolts; a steel

Figure 1. Satellite tagging of an adult male killer whale. Still frame from a point of view video showing satellite tagging of an adult male killer whale (M007) at Marion Island. The tag can be seen in the dorsal fin.

flange prevented penetration beyond 25 mm. Tips were sterilized before use and stored in clean plastic bags (Methods S1, [42]). The tissue samples obtained were stored for genetic, stable isotope and fatty acid analyses (Methods S1).

Satellite tagging. We deployed three models of satellite-linked telemetry devices: Sirtrack Kiwisat 202 (Sirtrack Ltd., Havelock North, New Zealand), Wildlife Computers SPOT5 and Wildlife Computers Mk10-A (Wildlife Computers, Redmond, Washington, United States of America). All three tag models allow estimation of geographic position via satellite using the Argos System (Collecte Localisation Satellites, Toulouse, France); the Mk10-A tag additionally includes a pressure (depth) sensor and a fast-response thermistor. Position estimates are classed by Collecte Localisation Satellites based on the estimated accuracy of the position, as follows: Class A and B – no estimate; 0– >1 500 m; 1– 500-1 500 m; 2–250-500 m; 3– <250 m (Table S2) [43]. To extend tag battery life while maintaining biologically sensible data capture, tags were programmed with various transmission schedules or 'duty cycles' (Table S2).

The tags were all in the LIMPET configuration where the tag is externally attached to the animal by sub-dermal darts which typically do not penetrate past the blubber layer (Figure 2; [7]).

Penetration deeper than the length of the darts is prevented by the tag itself. This is in contrast to a typical 'fully implantable' tag where the transmitter is largely sub-dermal and the attachment darts (or anchors) may often penetrate through the blubber into the muscle (e.g., [25,29]).

Kiwisat 202 tags were attached using 65 mm medical-grade stainless steel darts designed by RRR following [7]. Following an initial deployment with two darts (PTT 67764 in Table S2) we had difficulty attaching the tags and changed to a single dart design for these tags. SPOT5 and Mk10-A tags were attached using two 65 mm titanium darts designed and manufactured by RDA and Wildlife Computers (described in [7]). Tags (including darts) weighed 114 g (Kiwisat 202), 59 g (SPOT5) and 75 g (Mk10-A).

For deployment, tags were held on the crossbow bolt using urethane cups which fitted over the tag body (Figure 2). On impact with the animal, the sudden deceleration causes the tag to separate from the tag cup and bolt, which are retrieved using the tether (Figure 1; as for biopsy sampling). To prevent losing the tag if a shot was missed, Kiwisat 202 tags were additionally secured using two small screws which sheared the tag cup on impact with the animal and Wildlife Computer tags were secured using water

Table 1. Description of scores used to assess the immediate reactions of killer whales to biopsy sampling or tagging.

Score	Name	Description
0	None	No visible reaction
1	Slight	Slight flinch, slight shake, short acceleration, immediate submerge
2	Moderate	Pronounced flinch, pronounced shake, acceleration, prolonged dive
3	Strong	Prolonged dive and flight
4	Extreme	Breaching, tail slapping and flight (not observed in this study)

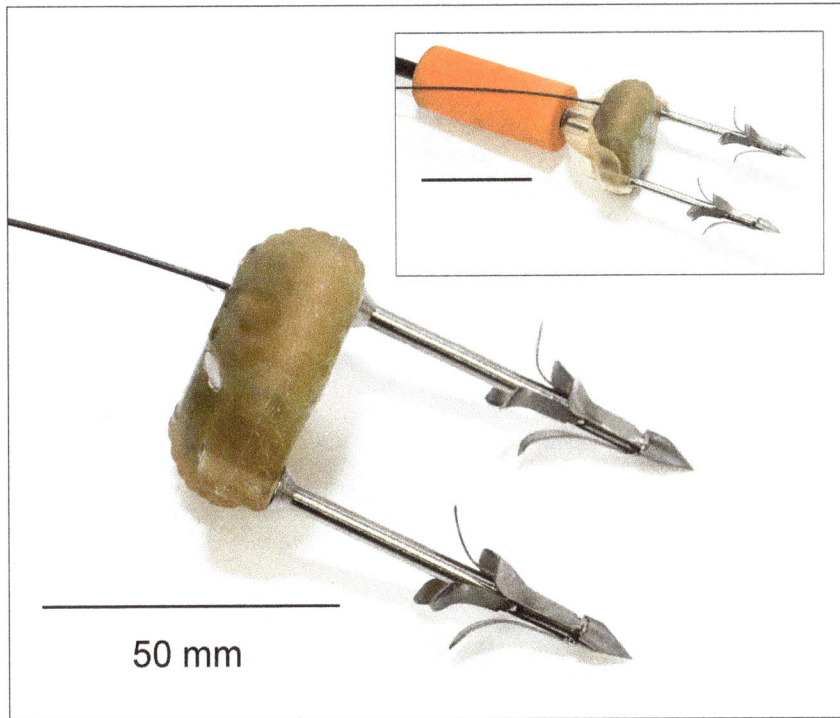

Figure 2. Wildlife Computers SPOT5 satellite-linked tag with attachment darts. The inset shows the tag in a deployment cup, attached to a crossbow bolt with float.

soluble tape (which tore or dissolved) and monofilament tethers (which broke) on impact.

Reactions to biopsy sampling and satellite tagging

We evaluated behavioural responses to tagging and biopsy by fitting generalized linear mixed models (GLMMs) using package lme4 in R [44,45]. We treated reactions as binomial; i.e., no response vs. response. The reaction observations (n = 103) were not independent because we resampled some individuals and we therefore included *individual* as a random effect. Our candidate models included combinations of three variables which potentially affected response: *biopsy/tag* (whether a biopsy sampling or tagging attempt), *hit/miss* (whether the tag or biopsy arrow hit or missed the animal), and *class* (adult male, adult female or juvenile) (Table 2). Interactions between explanatory variables were not considered. Models were compared using Akaike's Information Criterion corrected for small sample sizes (AIC$_c$). The model with the lowest AIC$_c$ is the most parsimonious model in the model set [46].

To test the validity of using binomial reactions rather than the reaction scores as defined in Table 1, we also compared the reaction scores using Kruskal-Wallis rank sum tests (kruskal.test in R) followed by multiple comparison tests where applicable (kruskalmc in package pgirmess in R; [47]).

Effect of arbalester experience

To test whether the experience of an arbalester influenced the probability of hitting the target individual in a sampling event (*hit/miss*, as above), we fitted generalized linear models (GLMs) with a binomial error distribution in R. Both arbalesters were proficient marksmen and underwent training before fieldwork; however, neither had field experience of remote biopsy sampling or satellite tagging prior to this study. We therefore used the cumulative number of sampling attempts by the arbalester as a proxy for their experience level at each sampling attempt. Candidate models included all combinations of the following predictor variables: *experience*, *biopsy/tag* (as above), *arbalester* (the identity of the arbalester) and *range* (estimated range of the shot, in meters) (Table 3). As for the GLMMs, interactions between variables were not considered and AIC$_c$ was used to compare models.

Sighting patterns

We used two approaches to detect changes in the sighting patterns of individuals after sampling using photographic identification sighting histories from 2006/04–2013/05 (sighting proportion) and 2008/05–2013/05 (mark-recapture). Briefly, dorsal fin photographs were taken during opportunistic (2006–2013) and dedicated (2008–2013) survey sightings and individuals were identified based on characteristic features such as scarring, mutilation and pigmentation. We stringently scored photographs based on their quality and used only good quality photographs to create a sighting history for each individual. All individuals were considered equally identifiable from good quality photographs, irrespective of the uniqueness of their characteristic features. Thus, individual variation in 'recognisability' should not affect the detection process (see [41] for methods). Sighting histories were restricted to sightings near (<1.0 km) the biopsy/tagging sites.

Sighting proportion. Firstly, following [23], we compared an individual's 'sighting proportion' before and after sampling. For a given period, the sighting proportion was simply the number of photographic sightings of a given individual in that period divided by the number of photographic sightings of all individuals in that period. Sighting proportions were calculated for all sampled individuals before and after each sampling attempt and compared with a Wilcoxon paired Rank Sum Test (wilcox.test in R).

Table 2. Model selection for the generalized linear mixed effects models (GLMMs) used to describe the reaction of killer whales to biopsy sampling and tagging.

Model	Np[a]	AIC$_c$[b]	ΔAIC$_c$[c]	ω_l[d]
class + hit/miss	5	136.90	0.00	0.65
class + hit/miss + biopsy/tag	6	139.10	2.24	0.21
hit/miss	3	140.60	3.69	0.10
hit/miss + biopsy/tag	4	142.40	5.51	0.04
class	4	151.30	14.45	0.00
NULL	2	151.50	14.57	0.00
class + biopsy/tag	5	153.40	16.53	0.00
biopsy/tag	3	153.60	16.68	0.00

The full model was *reaction ~class + hit/miss + biopsy/tag + (1|individual)*, where *reaction* was the response variable and *(1|individual)* denoted a random effect. All models included the random effect; only the predictor variables included in each model are shown.
Notes: [a]number of parameters; [b]small sample corrected Akaike Information Criterion; [c]difference between the AIC$_c$ score of the model in question and the best model; [d]Akaike weight: relative likelihood of model in question divided by the sum of relative likelihoods for all models.

Mark-recapture analysis. Secondly, we used multievent mark-recapture models [48] to determine whether sampling reduced future detection probabilities. Typically, when individuals are physically captured, they may seek (trap-happy) or avoid (trap-shy) the sampling area (the 'trap') on future occasions [49]. We considered two possible responses to sampling. Firstly, sampling may result in temporary avoidance of sampling area, affecting detection only at the time-step following the one when the animal was sampled ('trap-dependence' in capture-recapture parlance [49,50]). Alternatively, sampling may permanently alter individuals' behaviour, resulting in a permanent state change with reduced detection following sampling, i.e., long-term trap-dependence. In this long-term trap-dependence model, instead of

automatically returning to their initial state one time interval after being sampled [49,50], individuals permanently remained in a 'sampled' state. For the purpose of our study, 'normal' trap-dependence corresponded to the mid-term (1 month) effect of sampling (Data S1), while long-term trap-dependence corresponded to the long-term (up to 24 months) effect of sampling (Data S2). Thus, in the model where response to sampling was temporary, animals reverted back to the naïve state after one month. Where sampling was assumed to permanently influence behaviour, the state change was permanent.

Before trying to estimate the effect of sampling on individuals' behaviour, we had to account for intrinsic individual heterogeneity in detection, as failure to do so may lead to flawed inference [51].

Table 3. Model selection for the generalized linear models (GL Ms) used to describe factors influencing the probability of hitting the target animal (*hit/miss*) during a sampling attempt.

Model	np[a]	AIC$_c$[b]	ΔAIC$_c$[c]	ω_l[d]
range	2	140.50	0.00	0.24
experience + range	3	141.76	1.26	0.13
biopsy/tag + range	3	142.35	1.85	0.10
arbalester + range	3	142.60	2.10	0.09
NULL	1	142.99	2.50	0.07
experience + arbalester + range	4	143.47	2.97	0.06
biopsy/tag	2	143.52	3.02	0.05
experience + biopsy/tag + range	4	143.56	3.06	0.05
experience	2	144.29	3.79	0.04
biopsy/tag + arbalester + range	4	144.49	4.00	0.03
experience + biopsy/tag + range	3	144.62	4.12	0.03
arbalester	2	144.81	4.31	0.03
experience + arbalester	3	145.25	4.75	0.02
experience + biopsy/tag + arbalester + range	5	145.25	4.75	0.02
biopsy/tag + arbalester	3	145.44	4.95	0.02
experience + biopsy/tag + arbalester	4	145.63	5.13	0.02

The full model was *hit/miss ~experience + biopsy/tag + range + arbalester*. Only the predictor variables included in each model are shown.
Notes: [a]number of parameters; [b]small sample corrected Akaike Information Criterion; [c]difference between the AIC$_c$ score of the model in question and the best model; [d]Akaike weight: relative likelihood of model in question divided by the sum of relative likelihoods for all models.

One-sided directional test statistics (the signed square roots of the χ^2-statistics) for Test3.SR (a test for transience) and Test2.CT (a test for trap-dependence) in U-CARE [52] suggested significant heterogeneity in detection (Table S3, [53] and references therein). We used capture-recapture mixture models [54,55] that model heterogeneity using discrete 'classes' of individuals with low or high detection probability. Transience was accommodated by separately estimating the survival probability over the interval immediately following the first observation of the individual at Marion Island and survival during following intervals [56].

Mixture models specified the existence of two hidden states, representing individuals with distinct probabilities of detection. Our specification of two classes of individuals should not strictly be interpreted as evidence of the existence of two such classes; rather, these classes introduce heterogeneity in detection, improving model selection and reducing bias in parameter estimates [54].

Individual capture histories (n = 48) were based on photographic resightings between 2008 and 2013 (Data S1, Data S2). The full set of resightings for each individual was reduced to monthly 'capture occasions' (i.e., an individual was considered resighted or 'captured' in a month if it was photographed at least once in the month). At each occasion resighted individuals were known with certainty to be 'sampled' or 'not sampled'. We thus defined three events: 'not observed', 'resighted; not sampled' and 'resighted; sampled'. Depending on which of the above-described model structures we used, we defined up to nine states (Figure S2). Individuals moved in a Markovian way between the states. In the most complex model the states were thus: 'Seen$_{t-1}$; sampled', 'Not seen$_{t-1}$; sampled', 'Seen$_{t-1}$; not sampled' and 'Not seen$_{t-1}$; not sampled'. Assigning the four states to two hidden groups with different detectability increased the number of states to eight. Finally, 'death' was explicitly included as a state. Transitions between states were decomposed as: 1) survival, 2) detection conditional on survival, and 3) sampling, given survival and detection (Figure S2). Models were fitted using program E-SURGE 1.9.0 [57].

Seasonality was introduced by separating the peak in killer whale abundance (September – December) from the rest of the year. Two periods of varying observer intensity (2008–2011 and 2011–2013) were also considered. Sampling was only possible when animals were seen, and sampling probabilities were constrained to the sampling period (2011–2013).

For both mid-term and long-term response to sampling, the same four initial candidate models were ranked using QAIC$_c$ (sample size corrected, quasi-likelihood Akaike's Information Criterion [46]). This initial set of four models was designed to help us decide on the best model structure for seasonality (winter/summer) among the following four options: 1) no seasonality; 2) same seasonality effect for all individuals; 3) seasonality applying to all individuals but in different strength for two hidden groups (suggesting variation in seasonal attendance between individuals); 4) seasonality applying only to one of the hidden groups (suggesting 'residents' and 'migrants'). All models included two age classes for survival (transience model) and two periods of different field effort. They all included the effect of sampling (either long-term or mid-term). Having selected a seasonal model based on QAIC$_c$, we removed the sampling effect from the model and evaluated the change in QAIC$_c$.

Results

Overall, 109 biopsy and satellite tagging attempts were made, resulting in 71 hits (Table 4; Data S3). Of these, 101 attempts were made in 236 'dedicated observation sessions' (on 231 days)

totalling 1,645 hours – therefore an attempt was made every 16 h 17 m, overall. Biopsy hit rate was lower than tagging hit rates and biopsy sampling rate was low (43%). Tagging rate for Kiwisat 202 tags was very low (30%), reflecting–together with the short attachment durations (below and Figure 3)–the greater size and weight of these tags and the unsuccessful design of the attachment darts used with the tags. Tagging rate for the SPOT5 and Mk10-A tags was high (86%). Biopsy attempts were made at ranges from 3–20 m (average 8 m) and tagging attempts were made at ranges from 3–9 m (average 6 m).

Satellite tags

We deployed 19 tags (Table S2). One Kiwisat 202 tag and 1 Mk10-A tag never transmitted. Both animals were resighted without tags 5 days later. Excluding these two instances, attachment duration was 0.6–3.9 days (Kiwisat 202), 0.3–53.2 days (SPOT5) and 12.5–23.0 days (Mk10-A) (Figure 3). Mean attachment duration (± SD) was 1.8±1.3 days, 24.9±16.8 days and 17.7±7.5 days, respectively. After taking duty cycle into account, the number of accurate position estimates (quality class 1–3) per transmission day (i.e., 24 transmission hours) was not significantly different between tag types (Kruskal Wallis $\chi^2 = 2.21$, $df = 2$, $p = 0.33$). Kiwisat 202 tags averaged (± SD) 10.7±3.0 accurate position estimates per transmission day while SPOT5s averaged 9.2±4.2 and Mk-10As averaged 12.0±2.8 accurate position estimates per transmission day (Table S2).

Reactions to biopsy and satellite tagging attempts

All responses corresponded to 'no response' and 'low response' in [1]. Several animals turned on their sides – they seemed to be looking at the arbalester, but may have been looking at the impact site (as described by [58]). Some animals rolled a number of times when tagged. Both such reactions were scored as 2 (Table 1); where the rolls were combined with an extended dive or flight the reactions were scored as 3. The most frequent reaction to a miss was 0 (no reaction), while the most frequent reaction to a hit was 1 (Figure 4). This was typically a slight acceleration, immediate submergence and/or a shake of the body (cf. [58]) (Table 1). Such responses were often so slight that they were difficult to see, even when reviewing video footage.

In the GLMMs, the variance of the individual random effect was effectively zero, indicating either low individual variability in behavioural response, or that we were unable to detect individual variation with this limited data set. The model with the most support included hit/miss and class (adult male, adult female or juvenile) as predictor variables (Table 2). Hit/miss was the most important predictor variable ($\omega_i = 1$), followed by class ($\omega_i = 0.86$) (Table 2). Biopsy/tag had essentially no support, ranking lower than the null model when included as the only predictor variable. Adult females were most likely to respond, followed by juveniles and lastly males. Although the probability of response was highest when hit, behavioural responses were often present when missed (Figure 5).

Results of the Kruskal-Wallis tests support those of the GLMMs. Overall, there were significant response differences in the various categories (Kruskal-Wallis $\chi^2 = 18.48$, $df = 3$, $p < 0.01$). Reactions to tag and biopsy were not significantly different ($\chi^2 = 0.58$, $df = 1$, $p = 0.45$) while reactions to hit and miss were ($\chi^2 = 13.812$, $df = 1$, $p < 0.01$). Post-hoc multiple comparisons showed significant differences between reactions to tag-hit and tag-miss, biopsy-hit and tag-miss, and biopsy-miss and tag-hit (Table S4).

Table 4. Number of biopsy sampling and satellite tagging attempts on killer whales at Marion Island.

	Attempts	Hits	Hit rate (%)[a]	Misses	Successful hits[b]	Sampling/tagging rate (%)[a]
Biopsy	72	44	61.11	28	31	43.06
Tagging (Kiwisat 202)	23	15	65.22	8	7[c]	30.43
Tagging (SPOT5)	11	9	81.81	2	9[c]	81.81
Tagging (Mk10-A)	3	3	100.00	0	3[c]	100.00

Notes: [a]Following [24]; [b]Hit and tissue sample for biopsy sampling, hit and attach for satellite tagging; [c]Tags attached, but did not necessarily penetrate properly.

Effect of arbalester experience

The most supported model included only *range* as a predictor variable ($\beta = -0.13 \pm 0.06$, $p = 0.038$). Models including *experience* and *biopsy/tag* in addition to *range* had $\Delta AIC_c < 2$, but only *range* was a significant or near-significant predictor in these models (Table 3).

Sighting patterns

Changes in sighting proportion were typically small, and mean changes ranged from -0.02–0.68 percentage points (Figure S1). We found no significant differences when comparing sighting ratios before and after tagging/biopsy attempts; there also was no difference if we considered hits only (Table S1). The most frequently observed individual showed very large, positive changes in sighting proportion, but results remained the same if we repeated the comparison without this individual.

Multievent mark-recapture

Models not accounting for heterogeneity performed poorly. The most parsimonious seasonality model allowed detection of both hidden groups to fluctuate independently with season. Removing seasonality from the one mixture group (thus creating a 'resident' group with constant detection throughout the year) increased the $QAIC_c$ score.

When sampling was modelled as a permanent state change, $QAIC_c$ favoured removal of the sampling variable (Table 5). When sampling was modelled as a temporary state change, the sampling variable explained enough variation in detection probability to remain in the top ranked model, although the difference in $QAIC_c$ was only 0.09, indicating that the effect of sampling on detection was weakly supported (Table 6). In that model, individuals seen and sampled during month t-1 had a higher probability of being detected in month t than individuals that were only seen (and not sampled) during month t-1 (Figure 6).

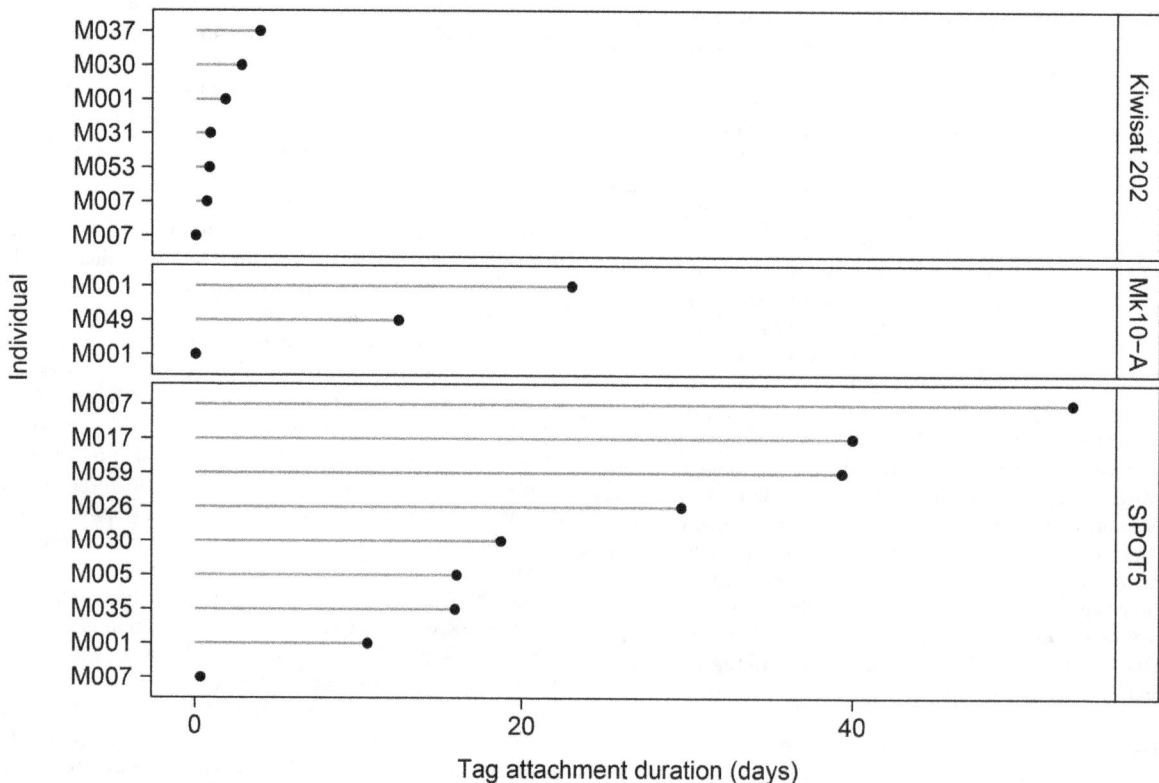

Figure 3. Attachment duration of satellite tags deployed on killer whales at Marion Island.

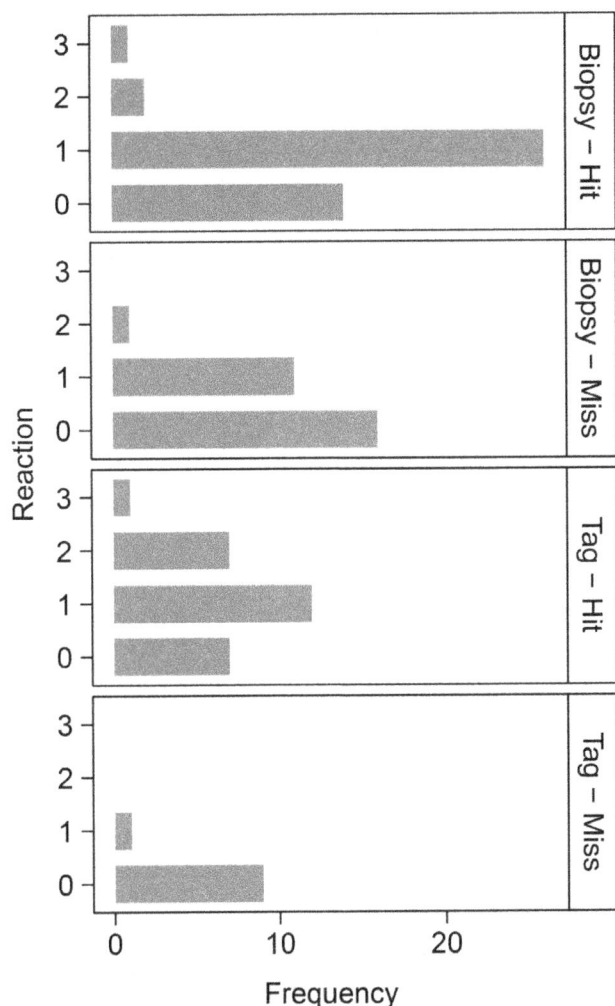

Figure 4. Immediate behavioural reactions to satellite tagging and biopsy sampling. Frequency of different immediate behavioural reactions of killer whales at Marion Island to tagging and biopsy sampling.

Since we corrected for among-individual variation in sighting probability via the mixture model structure, this 'trap-happy' response suggests a possible bias towards sampling (and repeat-sampling) of 'tamer' individuals. Indeed, upon removing the individual that was most often seen and also repeatedly sampled and repeating the analysis, the model including sampling ranked lower than the model without the sampling effect ($\Delta QAIC_c = 1.13$). Finally, the probability of sampling, given detection, was 0.18 (95% confidence interval: 0.14–0.25).

Discussion

Our results suggests that land-based remote biopsy sampling and satellite tagging of killer whales at Marion Island are an effective means of collecting otherwise elusive data and the methods elicit only mild, short-term behavioural responses. We show the potential of multievent trap-dependence models (compared to simpler approaches such as [29–31]) to assess responses to sampling while controlling for intrinsic heterogeneity and other covariates. We found no mid- (1 month) or long-term (<24 months) avoidance of the study site following biopsy or tagging

and conclude that there is no evidence of behavioural changes due to sampling.

Biopsy sampling

Our successful biopsy sampling rate was low compared to biopsy sampling rates of odontocetes in other studies using bows (crossbows and compound bows) (mean $\pm$ SD = 68% $\pm$19 percentage points in [1] compared to our 44%). Biopsy sampling rates of odontocetes with bows are typically lower than for mysticetes or using guns and poles [1], but we further attribute our low biopsy sampling rate to the tether line which worsens the crossbow's already poor performance in wind (of which there is a great deal at Marion Island) and taking less than ideal shot opportunities as necessitated by the shore-based study. Although biopsy sampling opportunities are rare and required many hours of dedicated observations, shore-based work proved viable and we managed to biopsy sample nearly half of all identified whales in our population in the first two years of biopsy sampling. Biopsy sampling rates were lower than tagging rates mainly because tagging was only attempted at much closer ranges (3–9 m, mean = 6 m, compared with 3–20 m, mean = 8 m).

Satellite tagging

Low tagging rates and short attachment durations meant that the Kiwisat 202 tags were not worth deploying (in a cost-benefit sense); this was due largely to poor attachment darts as the tags themselves performed well. The greater size and weight of that configuration probably contributed to their short attachment times – larger tags are subject to greater drag in the water and heavier tags slow the bolt's speed when fired, which may mean that darts do not consistently penetrate to their full depth. This also affected the trajectory of the shot – the heavier tags did not always strike at an appropriate angle, necessitating a single-dart design which further reduced attachment duration. This underlines the importance of using proven techniques and technologies in biopsy and tagging studies. When these are not available, methods and equipment should be developed with the input of those with relevant expertise and experience (e.g., field biologists, engineers, veterinarians) and tested in as realistic a way as possible (e.g., using cetacean carcasses to test tagging and biopsy techniques [59]). When species or populations of special conservation concern are involved, methods and equipment may need to be tested on other species or populations first [12].

Attachment durations were longer but highly variable (like other studies report) for SPOT5 and Mk10-A tags and still short compared to fully implantable tags (e.g., [60,61]). This represents the compromise of a minimally invasive, external tag attachment which can be deployed on smaller species compared to configurations where the tag itself is fully implanted, as used on large whales. Our average SPOT5 and Mk10-A deployment durations were shorter than, but as variable as, other studies using the same tag setup (mean $\pm$ SD = 24$\pm$24 d in [7]; 43$\pm$23 d in [32], 32$\pm$22 d in [8] and 46$\pm$41 d in [34]). At Marion Island killer whales frequently hunt and patrol in dense bull kelp *Durvillaea antarctica* and giant kelp *Macrocystis pyrifera* forests which circle the island inshore, and we suggest that this may shorten attachment durations as tags may become ensnared. We obtained a greater number of accurate position estimates per day than large whale studies using fully implantable tags (e.g., 1.5$\pm$1 in [6], 2$\pm$1.6 in [61]), but we anticipated shorter deployments than those studies and our tags were programmed to transmit more frequently. Killer whales also have shorter dive durations than large whales. The LIMPET setup is thus currently more useful for finer scale movement studies.

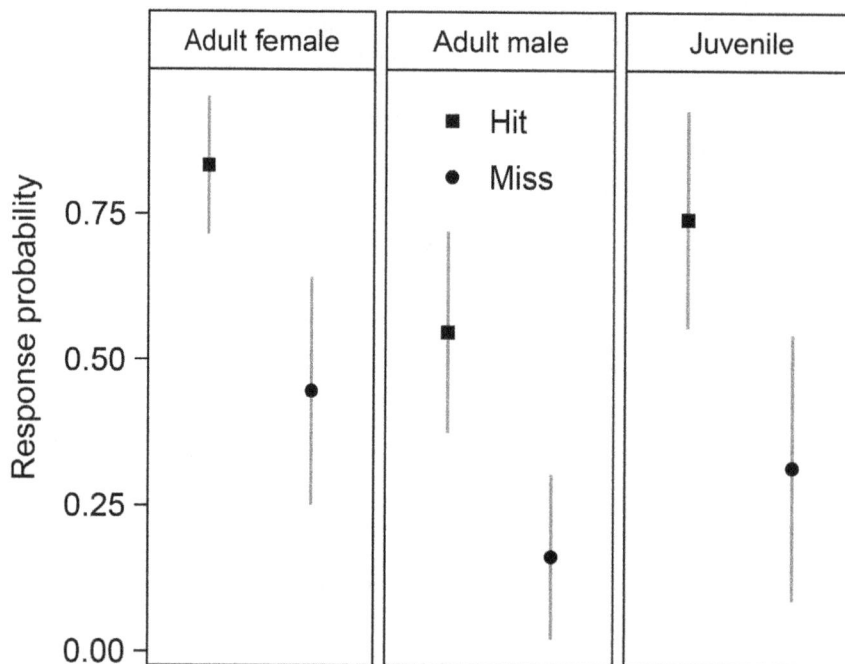

Figure 5. Predicted probability of an immediate behavioural response of killer whales to biopsy and tagging. Response probabilities as predicted by our best generalized linear mixed effects model, which included class (adult male, adult female or juvenile) and method (biopsy or tag); see Table 2.

Reactions

Reactions to tagging were similar to the few responses described in other tagging studies [7,25,26,29,62] and to reactions in other biopsy studies (reviewed by [1]), although there were no 'strong' (*sensu* [1]) reactions in our study. Some authors have attributed responses largely to the research boat rather than the actual tagging or biopsy, but we show that killer whales do respond to shore-based tagging and biopsy (as in [7]).

Although slightly stronger reactions were more frequent in response to tagging, the type of sampling (biopsy sampling or tagging) was not important in determining whether an animal would respond. Similarly, Reeb and Best [63] noted that southern right whales' reactions do not differ when biopsied with deep (11–20.5 cm) darts compared to more superficial darts used in a previous study [24]. This might suggest that, in general, responses to biopsy sampling and tagging are primarily startle, and not pain,

responses. However in our study hit *vs.* miss did influence reactions, indicating that there is an effect of an object hitting the animal's body compared to hitting the water. We cannot say whether hitting the animal's body is simply more startling to the animal or if, and how much, pain plays a role.

Some individual variation in behavioural reactions may be expected, but this was not evident in our study. It is possible that our data were too few to detect consistent individual variation. Sex and age, however, did influence reactions. Adult males were less likely to react than juveniles and adult females. Other studies report that group composition influences reaction but very few studies report sex-differences: Brown et al. [64] reported that female humpback whales responded more often to biopsy sampling, Gauthier and Sears [65] report the same for female fin whales *Balaenoptera physalus*.

Table 5. Selection criteria for multievent capture recapture models of sighting histories of killer whales at Marion Island: long-term (up to 24 months) responses following sampling (tagging or biopsy) attempts.

Model	Np	Deviance	QAIC$_c$[a]	ΔQAIC$_c$	ω_i
DH(2).season + trap + t$_{2008-2011;2011-2013}$	10	1929.22	1122.23	0.00	0.58
DH(2).season + trap + t$_{2008-2011;2011-2013}$+ sampling	11	1927.96	1123.60	1.37	0.29
DH(1).season + trap + t$_{2008-2011;2011-2013}$+ sampling	10	1934.37	1125.17	2.94	0.13
season + trap + t$_{2008-2011;2011-2013}$+ sampling	8	1982.16	1148.30	26.07	0.00
trap + t$_{2008-2011;2011-2013}$+ sampling	7	2042.72	1180.83	58.60	0.00

'Season' refers to the same seasonality affect for all individuals. 'DH(1).season' refers to seasonality applying only to one of two hidden mixture groups (suggesting 'resident' and 'migrant' animals) while 'DH(2).season' refers to seasonality applying to all individuals but independently for two hidden groups (suggesting variation between individuals). 'trap' refers to a trap-dependence effect, 'sampling' refers to a sampling effect and 't$_{2008-2011;2011-2013}$' accounts for two periods with differing field effort.
Notes: [a]$\hat{c}$ = 1.75.

Table 6. Selection criteria for multievent capture recapture models of sighting histories of killer whales at Marion Island: mid-term (1 month) responses following sampling (tagging or biopsy) attempts.

Model	np	Deviance	QAIC$_c$[a]	ΔQAIC$_c$	ω_i
DH(2).season + trap+ $t_{2008-2011;2011-2013}$+ sampling	11	2088.03	1215.01	0.00	0.43
DH(2).season + trap + $t_{2008-2011;2011-2013}$	10	2091.85	1215.10	0.09	0.41
DH(1).season + trap + $t_{2008-2011;2011-2013}$+ sampling	10	2095.21	1217.02	2.01	0.16
season + trap + $t_{2008-2011;2011-2013}$+ sampling	8	2143.99	1240.73	25.72	0.00
trap + $t_{2008-2011;2011-2013}$+ sampling	7	2200.58	1270.98	55.97	0.00

'Season' refers to the same seasonality affect for all individuals. 'DH(1).season' refers to seasonality applying only to one of two hidden mixture groups (suggesting 'resident' and 'migrant' animals) while 'DH(2).season' refers to seasonality applying to all individuals but independently for two hidden groups (suggesting variation between individuals). 'trap' refers to a trap-dependence effect, 'sampling' refers to a sampling effect and '$t_{2008-2011;2011-2013}$' accounts for two periods with differing field effort.
Notes: [a]$\hat{c} = 1.75$.

Effect of arbalester experience

Noren and Mocklin [1] name research team experience as an important factor influencing the success of collecting biopsy samples from cetaceans (although only [58] provides any qualitative support for the statement). We found almost no support for an effect of arbalester experience on sampling success, however such an effect may be obscured by the baseline proficiency of the arbalesters (both had undergone training prior to fieldwork), may only become apparent after even more experience (e.g., hundreds of sampling attempts compared to less than one hundred in this study), or may be stronger in vessel-based studies, where the vessel driver's experience is also relevant (e.g., [58]). Regardless, research team experience remains an important consideration in terms of animal welfare. Consequences of inaccurate shooting may include: hitting non-target animals; hitting target animals at the wrong body location - an important concern for satellite tags which need to be above water to transmit and for biopsy samples where tissue characteristics may vary,

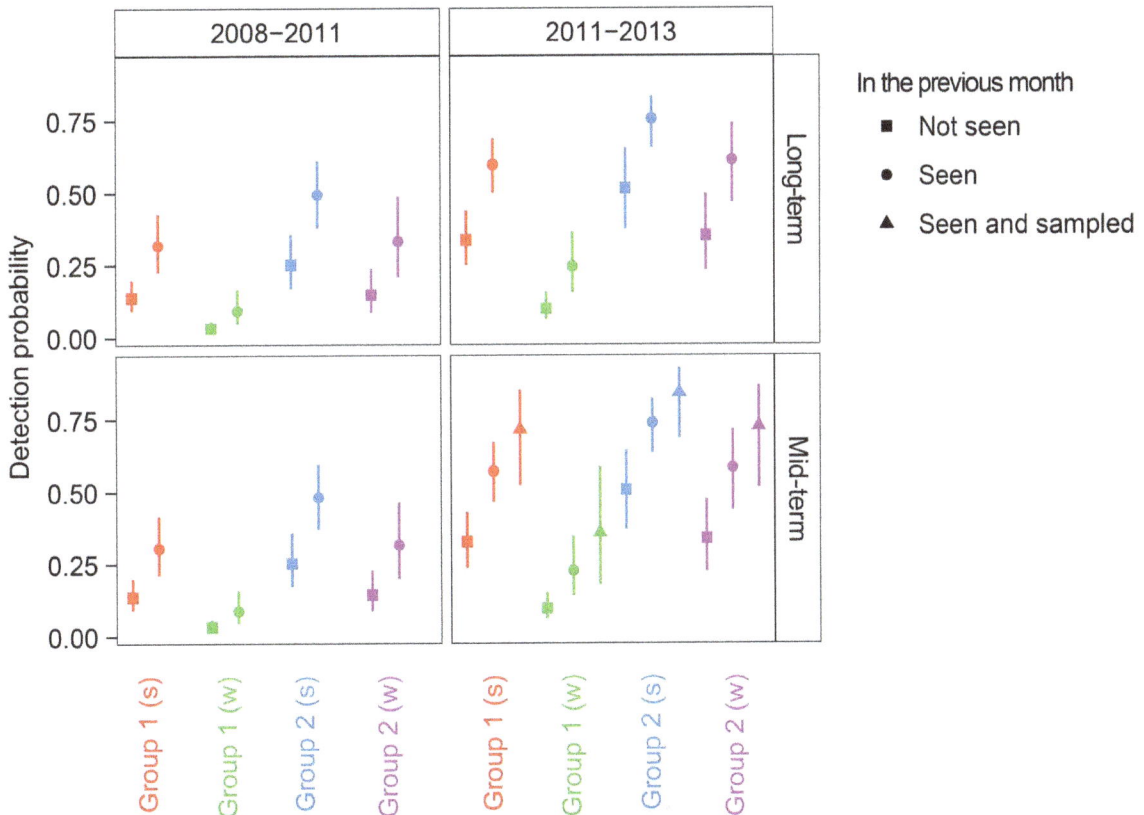

Figure 6. Detection probability of killer whales at Marion Island, given their capture history in the previous month. Detection probability (±95% confidence interval) was estimated using the highest ranked (lowest QAIC$_c$) capture-recapture model in which sampling effect was assumed to be mid-term (1 month). Sampling effect was not in the highest ranked long-term (<24 months) model. The two time periods (2008–2011 and 2011–2013) correspond to different intensities of field effort; we only sampled in 2011–2013. 'Groups' refer to two classes of animals with distinct probabilities of detection (mixture components); s refers to the summer peak in killer whale abundance, w to the winter.

affecting subsequent analyses [3]; and the loss of equipment. Hitting a non-target animal or the wrong place on the body may result in serious injury to the animal.

Sighting rates

Multievent models provided a flexible framework to model the response of individuals to sampling while accounting for demographic processes of the population. The sighting ratio method assumed that 'all animals are equal' with regards to seasonal movement and thus availability for detection; this heterogeneity could confound the results of a simple analysis. In this study the results were not fundamentally different: neither demonstrated a negative response to tagging or biopsy. However, the multievent approach showed the important effect of seasonal occurrence and different residence patterns which influenced sighting probabilities. The weak mid-term (~1 month) positive response to sampling seemed to be caused by a single individual, which underlines the importance of taking individual variation in sighting rates into account. This also highlights potential sampling biases (e.g., sex-biased biopsy sampling [66]) which we could fortunately detect by photographic identification of all sampled individuals. Individuals that centre their home ranges in the study area and have higher sighting rates are more likely to be sampled due to their general availability. Field effort will need to continue in order to generate enough chances to sample animals that occasionally visit the sampling area.

Can sampling lead to mid- or long-term behavioural changes?

Whether or not biopsy sampling and satellite tagging can lead to mid or long-term changes in behaviour depends on several factors. Firstly, an individual must be aware of the sampling attempt. We have shown that individuals do react to sampling attempts (58% of attempts), and are thus often aware of them. However, the absence of a visible behavioural response to a sampling attempt does not necessarily imply that the animal is unaware of the attempt. Several studies have shown physiological responses to human disturbance where there was little or no behavioural response (e.g., [67–69]). This underlines the utility of measuring physiological stress indicators such as glucocorticoid hormones or heart rate, however in many cases such measurement itself will result in stress, confounding the measurements [70,71]. Secondly, the sampling attempt must be perceived negatively by the individual. We assume the immediate behavioural reactions sometimes associated with biopsy sampling - such as defecation, tail slapping, breaching and flight from the area - (see Table 3 in [1]) indicate a negative stimulus, be it fright or pain. Thirdly, in our case where sampling attempts were land-based at two locations, the individual must be able to associate its experience (the sampling attempt) with a spatial location or other cue (seeing the arbalester, for example) and this memory must persist for some length of time. This would seem well within the capabilities of many animals (e.g., [72–74]) and certainly killer whales, which range widely but show strong interannual site fidelity (at Marion Island - [41]) and are cognitively complex [75]. Lastly, given the above, the strength of the negative experience must be sufficient to alter behaviour. Animals may not show a mid-term behavioural response because the motivation to perform an activity (e.g., foraging), or to remain at a high quality site, may exceed the motivation to avoid sampling; individuals may also lack suitable habitat to disperse to in order to avoid sampling. This can be framed as a cost-benefit tradeoff if the disturbance stimulus (in this case sampling) is equated to predation or injury risk [76,77]. This may beg the

question whether killer whales - which do not have significant natural predators - are less sensitive to disturbance stimuli.

Our two sampling locations, <1 km apart, represent a short stretch of the ~50 km stretch of Marion Island coastline patrolled by killer whales [41,78,79]. Breeding colonies of killer whale prey (seals and penguins) at these sites represent a small proportion of the total breeding populations of these species at Marion Island (Table S5). We consider it plausible that an individual killer whale could alter its path by a few hundred meters to avoid the sampling sites, and that this would not represent a considerable energy cost or loss of foraging opportunity. Social bonds may possibly prevent sampling site avoidance, particularly when only some group members have been sampled, but our analyses of the social structure of Marion Island killer whales over 7 years (RRR and PJNdB, in preparation) indicates considerable flexibility in social groups. Half Weight Association Index values – an estimate of the proportion of time two animals spend together – range from 0.21–0.66 (average ± SD = 0.48±0.18) within defined social units, clearly indicating that animals are not constantly associated. Further, 370 (13%) of 2,821 sightings recorded in that study were of single (lone) individuals. This suggests that social bonds between killer whales will not necessarily prevent individuals from avoiding the sampling sites.

The factors we have mentioned which may prevent short term disturbance (sampling) from causing mid-term behavioural changes are intractable in this study, but could stimulate further research in different species or settings. There is debate as to how well behavioural changes signal the sensitivity of animals to disturbance [80]. In cetaceans, documented disturbance is likely largely due to direct or associated noise (e.g., [81] for killer whales). The mid- to long-term sensitivity of cetaceans to satellite tagging and biopsy sampling is unknown, but seems negligible. Best et al. [24] show sensitization to biopsy sampling up to 65 days in female southern right whales with calves, but such cases seem rare [1].

Importantly, we found no significant long-term (<24 months) changes in the sighting probability of tagged or biopsied killer whales. In the only study using a comparable method to ours, Tezanos-Pinto and Baker [23] found no difference in the long-term sighting probabilities between biopsied and non-biopsied bottlenose dolphins *Tursiops truncatus*. Our study supports the idea that cetaceans do not change their long-term behaviour in response to being sampled. However, if such responses are subtle, they may require considerable data and time to detect. We have not tested for physiological responses (e.g., stress) on any temporal scale, nor for an impact on hunting behaviour and demographic performance.

However, one of our stated aims was to 'evaluate whether biopsy sampling and satellite tagging changed the behaviour of individuals, altering mid- (1 month) and long-term (<24 months) sighting patterns.' We wished to evaluate any behavioural changes to our tagging and biopsy sampling protocol, rather than determine the mechanisms affecting such behavioural changes (or lack thereof, as we found). Our results are therefore meaningful independent of any evaluation of intermediate factors, however we recommend longer term monitoring to assess whether satellite tagging and biopsy sampling have any effect on demographic parameters (e.g., [82]).

Conclusions

Remote biopsy sampling and satellite tagging of killer whales from shore is successful at Marion Island and these methods can provide insights into the ecology of this population which is difficult to access at sea. We found that reactions to biopsy sampling and satellite tagging were mild or unnoticeable and we found no significant mid- or long-term changes in the occurrence of killer whales at the study site. However, long-term monitoring of individuals after biopsy sampling and tagging should continue in order to provide continuous assessment of potential impacts on the study animals. Such monitoring should be implemented in other studies where animals are biopsied or tagged, especially considering the increased use of these methods.

Supporting Information

Figure S1 Changes (percentage points) in the sighting proportion of killer whales at Marion Island following various sampling events. a) tag or biopsy – first attempt; b) biopsy – first attempt; c) biopsy – first hit; d) tag – first attempt; e) tag – first hit. Sighting proportion (%) was calculated as the number of sightings of an individual during a given period, divided by the number of sightings of all individuals in the same period. Negative change thus indicates an individual was seen less following a sampling event.

Figure S2 A multinomial tree diagram with arrows denoting the possible transitions between states (solid boxes) from *t* to *t+1*. States occupied are not directly observed, but events (dashed boxes) represent observations following initial capture ('Encounter'). Individuals belong to one of two hidden classes with distinct probabilities of detection; movement between detection groups over time is not allowed. Entry to the population conditions on the first encounter ('Seen') and all individuals are seen once or more prior to sampling ('Initial state' step). Subsequent state transition probabilities are decomposed in three steps as the product of the probabilities of 'Survival', 'Detection' and 'Sampling'. Only individuals that are detected ('Seen') can be sampled. Once sampled, individuals either remain in the sampled state (permanent state change scenario; solid arrows) or may move back to the 'Not Sampled' state at the next occasion (mid-term sampling effect scenario; dashed arrows).

Table S1 Comparisons of sighting proportions before and after tagging and biopsy attempts on killer whales at Marion Island (paired Wilcox rank sum test). The sighting proportion is the number of photographic sightings of an individual in a given period, divided by the number of photographic sightings of all individuals in that period (following [1]). Notes: [a]N is the number of sampling attempts included for each comparison. [b]W is the test statistic. [c]*Tag or biopsy – first attempt* includes only the first attempt (regardless of whether it was a tag or biopsy attempt), hence it is not the sum of *Tag – first attempt* and *Biopsy – first attempt*.

Table S2 Satellite tags deployed on killer whales at Marion Island showing the attachment duration, duty cycle and number of position estimates received. Notes: [a]SA – subadult, A – adult; [b]1– transmit 00:00–24:00 UTC, 2– transmit 00:00–06:00 and 12:00–18:00 UTC, 3– transmit 01:00–22:00 UTC for 30 days, thereafter 01:00–22:00 UTC on every second day, 4– transmit 01:00–22:00 UTC for 25 days, thereafter 01:00–22:00 UTC on every fourth day; [c]Argos position estimate quality class (see text for accuracy); [d]'Accurate' position estimates are quality class 1–3; number of accurate positions estimates per day was corrected for duty cycle (the proportion of time transmitting) and is thus expressed per 'transmission day', i.e., 24 transmission hours.

Table S3 Approximate goodness of fit (GOF) tests for individual capture histories of killer whales at Marion Island. The overdispersion coefficient ($\hat{c}$) for a heterogeneity model including transience and trap-happiness was computed by removing the squared directional test statistics from the time dependant model [1].

Table S4 Multiple comparisons test (kruskalmc in R package pgirmess [1]) results for significant reaction differences to tagging and biopsy attempts of various types.

Table S5 Breeding populations of known killer whale prey at satellite tagging and biopsy sampling locations, and total breeding populations, at Marion Island. Seal numbers refer to pup production and penguin numbers to breeding pairs. Numbers in parentheses are percentage of the total breeding population. Dashes indicate zero animals.

Data S1 Encounter history matrix of killer whales at Marion Island, with temporary state change. Monthly encounter history matrix (May 2008–May 2013) of 48 killer whales. States are indicated as: 0– not seen and not sampled; 1– seen but not sampled; 2– seen and sampled. The sampled state is not permanent (i.e., individuals return to an unsampled state after 1 month).

Data S2 Encounter history matrix of killer whales at Marion Island, with permanent state change. Monthly encounter history matrix (May 2008–May 2013) of 48 killer whales. States are indicated as: 0– not seen and not sampled; 1– seen but not sampled; 2– seen and sampled. The sampled state is permanent (i.e., individuals subsequently remain in the sampled state, if seen).

Data S3 Satellite tagging and biopsy sampling of killer whales at Marion Island. Satellite tagging and biopsy sampling attempts are shown, with associated data. Class: AM – adult male; AF – adult female; J – juvenile. Success: Y – yes (hit and sample for biopsy sampling attempts, hit and attach for satellite tagging attempts); N – no. Reaction: see Table S1 in text. Range – range of the attempt (in meters). Attempt – cumulative attempts by the arbalester.

Acknowledgments

We thank the Marion Island overwinter expedition members of M63-M69 – particularly the 'Sealers' – for providing killer whale photographs and Marthán Bester for his efforts to conduct or support opportunistic killer whale research at Marion Island. Technical advice or assistance was provided by Meredith Thornton, Simon Elwen, Kevin Lay, Johan Bienedell and sons, Jacques Grobelaar, Kimberly Kanapeckas, Fritz Röhr and Rus Hoelzel. The Department of Environmental Affairs supplied logistic support within the South African National Antarctic Programme. Brad Hanson and an anonymous reviewer provided useful comments.

Author Contributions

Conceived and designed the experiments: RRR PJNDB. Performed the experiments: RRR WCO DCT PJNDB. Analyzed the data: RRR WCO GP. Contributed reagents/materials/analysis tools: RRR RDA PJNDB. Contributed to the writing of the manuscript: RRR WCO GP RA PJNDB.

References

1. Noren DP, Mocklin JA (2012) Review of cetacean biopsy techniques: factors contributing to successful sample collection and physiological and behavioral impacts. Mar Mammal Sci 28: 154–199. doi:10.1111/j.1748-7692.2011.00469.x.

2. Hoelzel AR, Dahlheim ME, Stern SJ (1998) Low genetic variation among killer whales (Orcinus orca) in the Eastern North Pacific and genetic differentiation between foraging specialists. J Hered 89: 121–128.

3. Budge SM, Iverson SJ, Koopman HN (2006) Studying trophic ecology in marine ecosystems using fatty acids: a primer on analysis and interpretation. Mar Mammal Sci 22: 759–801.

4. Newsome SD, Clementz MT, Koch PL (2010) Using stable isotope biogeochemistry to study marine mammal ecology. Mar Mammal Sci 26: 509–572. doi:10.1111/j.1748-7692.2009.00354.x.

5. Hunt KE, Moore MJ, Rolland RM, Kellar NM, Hall AJ, et al. (2013) Overcoming the challenges of studying conservation physiology in large whales: a review of available methods. Conserv Physiol 1. doi:10.1093/conphys/cot006.

6. Baumgartner MF, Mate BR (2005) Summer and fall habitat of North Atlantic right whales (Eubalaena glacialis) inferred from satellite telemetry. Can J Fish Aquat Sci 62: 527–543. doi:10.1139/f04-238.

7. Andrews RD, Pitman RL, Ballance LT (2008) Satellite tracking reveals distinct movement patterns for Type B and Type C killer whales in the southern Ross Sea, Antarctica. Polar Biol 31: 1461–1468.

8. Baird RW, Schorr GS, Webster DL, McSweeney DJ, Hanson MB, et al. (2010) Movements and habitat use of satellite-tagged false killer whales around the main Hawaiian Islands. Endanger Species Res 10: 107–121. doi:10.3354/esr00258.

9. Bilgmann K, Möller LM, Harcourt RG, Gales R, Beheregaray LB (2008) Common dolphins subject to fisheries impacts in Southern Australia are genetically differentiated: implications for conservation. Anim Conserv 11: 518–528. doi:10.1111/j.1469-1795.2008.00213.x.

10. Maxwell SM, Hazen EL, Bograd SJ, Halpern BS, Breed GA, et al. (2013) Cumulative human impacts on marine predators. Nat Commun 4: 1–9. doi:10.1038/ncommsS3688.

11. Wilson R, McMahon C (2006) Measuring devices on wild animals: what constitutes acceptable practice? Front Ecol Environ 4: 147–154.

12. Cooke SJ (2008) Biotelemetry and biologging in endangered species research and animal conservation: relevance to regional, national, and IUCN Red List threat assessments. Endanger Species Res 4: 165–185. doi:10.3354/esr00063.

13. Wilson RP, Kreye JM, Lucke K, Urquhart H (2004) Antennae on transmitters on penguins: balancing energy budgets on the high wire. J Exp Biol 207: 2649–2662. doi:10.1242/jeb.01067.

14. Hazekamp AAH, Mayer R, Osinga N (2010) Flow simulation along a seal: the impact of an external device. Eur J Wildl Res 56: 131–140. doi:10.1007/s10344-009-0293-0.

15. Saraux C, Le Bohec C, Durant JM, Viblanc VA, Gauthier-Clerc M, et al. (2011) Reliability of flipper-banded penguins as indicators of climate change. Nature 469: 203–206.

16. Bateson P (1986) When to experiment on animals. New Sci 109: 30–32.

17. McMahon CR, Harcourt RG, Bateson P, Hindell MA (2012) Animal welfare and decision making in wildlife research. Biol Conserv 153: 254–256. doi:10.1016/j.biocon.2012.05.004.

18. Gales NJ, Bowen WD, Johnston DW, Kovacs KM, Littnan CL, et al. (2009) Guidelines for the treatment of marine mammals in field research. Mar Mammal Sci 25: 725–736. doi:10.1111/j.1748-7692.2008.00279.x.

19. Field IC, Harcourt RG, Boehme L, de Bruyn PJN, Charrassin J-B, et al. (2012) Refining instrument attachment on phocid seals. Mar Mammal Sci 28: E325–E332. doi:10.1111/j.1748-7692.2011.00519.x.

20. McMahon CR, Hindell MA, Harcourt RG (2013) Publish or perish: why it's important to publicise how, and if, research activities affect animals. Wildl Res 39: 375–377. doi:10.1071/WR12014.

21. Wells RS, Rhinehart HL, Hansen LJ, Sweeney JC, Townsend FI, et al. (2004) Bottlenose dolphins as marine ecosystem sentinels: developing a health monitoring system. Ecohealth 1: 246–254.

22. Elwen SH, Meyer MA, Best PB, Kotze PGH, Thornton M, et al. (2006) Range and movements of female Heaviside's dolphins (Cephalorhynchus heavisidii), as determined by satellite-linked telemetry. J Mammal 87: 866–877.

23. Tezanos-Pinto G, Baker C (2012) Short-term reactions and long-term responses of bottlenose dolphins (Tursiops truncatus) to remote biopsy sampling. New Zeal J Mar Freshw Res 46: 13–29. doi:10.1080/00288330.2011.583256.

24. Best PB, Reeb D, Rew MB, Palsbøll PJ, Schaeff C (2005) Biopsying southern right whales: their reactions and effects on reproduction. J Wildl Manage 69: 1171–1180.

25. Mate B, Mesecar R, Lagerquist B (2007) The evolution of satellite-monitored radio tags for large whales: one laboratory's experience. Deep Sea Res Part II Top Stud Oceanogr 54: 224–247. doi:10.1016/j.dsr2.2006.11.021.

26. Watkins WA (1981) Reaction of three species of whales Balaenoptera physalus, Megaptera novaeangliae, and Balaenoptera edeni to implanted radio tags. Deep Sea Res Part A Oceanogr Res Pap 28: 589–599. doi:10.1016/0198-0149(81)90119-9.

27. Watkins WA, Tyack P (1991) Reaction of sperm whale (Physeter catodon) to tagging with implanted sonar transponder and radio tags. Mar Mammal Sci 7: 409–413.

28. Goodyear JD (1993) A sonic/radio tag for monitoring dive depths and underwater movements of whales. J Wildl Manage 57: 503–513.

29. Robbins J, Zerbini AN, Gales N, Gulland FMD, Double M, et al. (2013) Satellite tag effectiveness and impacts on large whales: preliminary results of a case study with Gulf of Maine humpback whales. Report SC/65a/SH05 presented to the International Whaling Commission Scientific Committee, Jeju, Korea.

30. Best PB, Mate B (2007) Sighting history and observations of southern right whales following satellite tagging off South Africa. J Cetacean Res Manag 9: 111–114.

31. Mizroch SA, Tillman MF, Jurasz S, Straley JM, Von Ziegesar O, et al. (2011) Long-term survival of humpback whales radio-tagged in Alaska from 1976 through 1978. Mar Mammal Sci 27: 217–229. doi:10.1111/j.1748-7692.2010.00391.x.

32. Schorr GS, Baird RW, Hanson MB, Webster DL, McSweeney DJ, et al. (2009) Movements of satellite-tagged Blainville's beaked whales off the island of Hawai'i. Endanger Species Res 10: 203–213. doi:10.3354/esr00229.

33. Baird RW, Schorr GS, Webster DL, McSweeney DJ, Hanson MB, et al. (2011) Movements of two satellite-tagged pygmy killer whales (Feresa attenuata) off the island of Hawai'i. Mar Mammal Sci 27: E332–E337. doi:10.1111/j.1748-7692.2010.00458.x.

34. Durban JW, Pitman RL (2012) Antarctic killer whales make rapid, round-trip movements to subtropical waters: evidence for physiological maintenance migrations? Biol Lett 8: 274–277. doi:10.1098/rsbl.2011.0875.

35. Reisinger RR, de Bruyn PJN, Tosh CA, Oosthuizen WC, Mufanadzo NT, et al. (2011) Prey and seasonal abundance of killer whales at sub-Antarctic Marion Island. African J Mar Sci 33: 99–105. doi:10.2989/1814232X.2011.572356.

36. Reisinger RR, de Bruyn PJN (2014) Marion Island killer whales: 2006–2013. Mammal Research Institute, University of Pretoria. doi:10.6084/m9.figshare.971317.

37. Williams AJ, Petersen SL, Goren M, Watkins BP (2009) Sightings of killer whales Orcinus orca from longline vessels in South African waters, and consideration of the regional conservation status. African J Mar Sci 31: 81–86. doi:10.2989/AJMS.2009.31.1.7.778.

38. Tixier P, Gasco N, Guinet C (2014) Killer whales of the Crozet Islands: photo-identification catalogue 2014. Villiers en Bois: Centre d'Etudes Biologiques de Chizé - CNRS. doi:10.6084/m9.figshare.1060247.

39. Reisinger RR, de Bruyn PJN, Bester MN (2011) Predatory impact of killer whales on pinniped and penguin populations at the Subantarctic Prince Edward Islands: fact and fiction. J Zool 285: 1–10. doi:10.1111/j.1469-7998.2011.00815.x.

40. Prince Edward Islands Management Plan Working Group (1996) Prince Edward Islands Management Plan. Pretoria: Department of Environmental Affaris and Tourism.

41. Reisinger RR, de Bruyn PJN, Bester MN (2011) Abundance estimates of killer whales at subantarctic Marion Island. Aquat Biol 12: 177–185. doi:10.3354/ab00340.

42. Lambertsen RH (1987) A biopsy system for large whales and its use for cytogenetics. J Mammal 68: 443–445. doi:10.2307/1381495.

43. Collecte Localisation Satellites (2011) Argos User's Manual. Toulouse: Collecte Localisation Satellites.

44. Bates D, Maechler M, Bolker B, Walker S (2013) lme4: Linear mixed-effects models using Eigen and S4. R package version 1.0–4.

45. R Development Core Team (2013) R: A language and environment for statistical computing. Vienna, Austria: R Foundation for Statistical Computing.

46. Burnham KP, Anderson DR (2002) Model selection and multimodel inference: a practical information-theoretic approach. Second edi. New York: Springer.

47. Giraudoux P (2011) pgirmess: Data analysis in ecology. R package version 1.5.1. http://cran.r-project.org/web/packages/pgirmess/index.html.

48. Pradel R (2005) Multievent: an extension of multistate capture recapture models to uncertain states. Biometrics 61: 442–447.

49. Pradel R, Sanz-Aguilar A (2012) Modeling trap-awareness and related phenomena in capture-recapture studies. PLoS One 7: e32666. doi:10.1371/journal.pone.0032666.

50. Pradel R (1993) Flexibility in survival analysis from recapture data: handling trap-dependence. In: Lebreton JD, North PM, editors. Marked individuals in the study of bird populations. Basel: Birkhauser Verlag. 29–37.

51. Lebreton JD, Burnham KP, Clobert J, Anderson DR (1992) Modelling survival and testing biological hypotheses using marked animals: a unified approach with case studies. Ecol Monogr 62: 67–118.

52. Choquet R, Lebreton JD, Gimenez O, Reboulet A-M, Pradel R (2009) U-CARE: Utilities for performing goodness of fit tests and manipulating Capture-REcapture data. Ecography 32: 1071–1074. doi:10.1111/j.1600-0587.2009.05968.x.

53. Péron G, Crochet P, Choquet R, Pradel R, Lebreton JD, et al. (2010) Capture-recapture models with heterogeneity to study survival senescence in the wild. Oikos 119: 524–532. doi:10.1111/j.1600-1706.2009.17882.x.

54. Pledger S, Pollock KH, Norris JL (2003) Open capture-recapture models with heterogeneity: I. Cormack-Jolly-Seber model. Biometrics 59: 786–794.

55. Pradel R, Choquet R, Lima MA, Merritt J, Crespin L (2010) Estimating population growth rate from capture–recapture data in presence of capture heterogeneity. J Agric Biol Environ Stat 15: 248–258. doi:10.1007/s13253-009-0008-8.

56. Pradel R, Hines JE, Lebreton JD, Nichols JD (1997) Capture-recapture survival models taking account of transients. Biometrics 53: 60. doi:10.2307/2533097.

57. Choquet R, Rouan L, Pradel R (2009) Program E-SURGE: a software application for fitting multievent models. In: Thomson DL, Cooch EG, Conroy MJ, editors. Modeling demographic processes in marked populations. New York: Springer. 845–865.

58. Barrett-Lennard LG, Smith TG, Ellis GM (1996) A cetacean biopsy system using lightweight pneumatic darts, and its effect on the behavior of killer whales. Mar Mammal Sci 12: 14–27.

59. Patenaude NJ, White BN (1995) Skin biopsy sampling of beluga whale carcasses: assessment of biopsy darting factors for minimal wounding and effective sample retrieval. Mar Mammal Sci 11: 163–171.

60. Zerbini AN, Andriolo A, Heide-Jørgensen M-P, Pizzorno J, Maia Y, et al. (2006) Satellite-monitored movements of humpback whales *Megaptera novaeangliae* in the Southwest Atlantic Ocean. Mar Ecol Prog Ser 313: 295–304. doi:10.3354/meps313295.

61. Bailey H, Mate B, Palacios D, Irvine L, Bograd S, et al. (2009) Behavioural estimation of blue whale movements in the Northeast Pacific from state-space model analysis of satellite tracks. Endanger Species Res 10: 93–106. doi:10.3354/esr00239.

62. Hauser N, Zerbini AN, Geyer Y, Heide-Jørgensen M-P, Clapham P (2010) Movements of satellite-monitored humpback whales, *Megaptera novaeangliae*, from the Cook Islands. Mar Mammal Sci 26: 679–685. doi:10.1111/j.1748-7692.2009.00363.x.

63. Reeb D, Best PB (2006) A biopsy system for deep-core sampling of the blubber of southern right whales, *Eubalaena australis*. Mar Mammal Sci 22: 206–213. doi:10.1111/j.1748-7692.2006.00015.x.

64. Brown MR, Corkeron PJ, Hale PT, Schultz KW, Bryden MM (1994) Behavioral responses of east Australian humpback whales Megaptera novaeangliae to biopsy sampling. Mar Mammal Sci 10: 391–400. doi:10.1111/j.1748-7692.1994.tb00496.x.

65. Gauthier J, Sears R (1999) Behavioral response of four species of balaenopterid whales to biopsy sampling. Mar Mammal Sci 15: 85–101. doi:10.1111/j.1748-7692.1999.tb00783.x.

66. Kellar NM, Trego ML, Chivers SJ, Archer FI, Minich JJ, et al. (2013) Are there biases in biopsy sampling? Potential drivers of sex ratio in projectile biopsy samples from two small delphinids. Mar Mammal Sci 29: E366–E389. doi:10.1111/mms.12014.

67. Culik B, Adelung D, Woakes AJ (1990) The effect of disturbance on the heart rate and behavior of Adélie penguins (*Pygoscelis adeliae*) during the breeding season. In: Kerry KR, Hempel G, editors. Antarctic Ecosystems: Ecological Change and Conservation. Berlin: Springer-Verlag. 177–182.

68. Wilson R, Culik B, Danfeld R, Adelung D (1991) People in Antarctica– how much do Adélie penguins *Pygoscelis adeliae* care? Polar Biol 11: 363–370. doi:10.1007/BF00239688.

69. Regel J, Pütz K (1997) Effect of human disturbance on body temperature and energy expenditure in penguins. Polar Biol: 246–253.

70. Wikelski M, Cooke SJ (2006) Conservation physiology. Trends Ecol Evol 21: 38–46. doi:10.1016/j.tree.2005.10.018.

71. Tarlow EM, Blumstein DT (2007) Evaluating methods to quantify anthropogenic stressors on wild animals. Appl Anim Behav Sci 102: 429–451. doi:10.1016/j.applanim.2006.05.040.

72. Winter Y, Stich KP (2005) Foraging in a complex naturalistic environment: capacity of spatial working memory in flower bats. J Exp Biol 208: 539–548. doi:10.1242/jeb.01416.

73. Wolf M, Frair J, Merrill E, Turchin P (2009) The attraction of the known: the importance of spatial familiarity in habitat selection in wapiti *Cervus elaphus*. Ecography 32: 401–410. doi:10.1111/j.1600-0587.2008.05626.x.

74. Ban SD, Boesch C, Janmaat KRL (2014) Taï chimpanzees anticipate revisiting high-valued fruit trees from further distances. Anim Cogn. doi:10.1007/s10071-014-0771-y.

75. Marino L, Connor RC, Fordyce RE, Herman LM, Hof PR, et al. (2007) Cetaceans have complex brains for complex cognition. PLoS Biol 5: e139. doi:10.1371/journal.pbio.0050139.

76. Gill J, Norris K, Sutherland W (2001) Why behavioural responses may not reflect the population consequences of human disturbance. Biol Conserv 97: 265–268.

77. Frid A, Dill L (2002) Human-caused disturbance stimuli as a form of predation risk. Conserv Ecol 6: 11.

78. Keith M, Bester MN, Bartlett PA, Baker D (2001) Killer whales (*Orcinus orca*) at Marion Island, Southern Ocean. African Zool 36: 163–175.

79. Pistorius PA, Taylor FE, Louw C, Hanise B, Bester MN, et al. (2002) Distribution, movement, and estimated population size of killer whales at Marion Island, December 2000. South African J Wildl Res 32: 86–92.

80. Beale CM, Monaghan P (2004) Behavioural responses to human disturbance: a matter of choice? Anim Behav 68: 1065–1069. doi:10.1016/j.anbehav.2004.07.002.

81. Williams R, Trites AW, Bain DE (2006) Behavioural responses of killer whales (*Orcinus orca*) to whale-watching boats: opportunistic observations and experimental approaches. J Zool 256: 255–270. doi:10.1017/S0952836902000298.

82. Barbraud C, Weimerskirch H (2011) Assessing the effect of satellite transmitters on the demography of the wandering albatross *Diomedea exulans*. J Ornithol 153: 375–383. doi:10.1007/s10336-011-0752-8.

13

Collapse of a Marine Mammal Species Driven by Human Impacts

Tero Harkonen[1]*, Karin C. Harding[2], Susan Wilson[3], Mirgaliy Baimukanov[4], Lilia Dmitrieva[5], Carl Johan Svensson[2], Simon J. Goodman[5]

1 Swedish Museum of Natural History, Stockholm, Sweden, 2 Department of Marine Ecology, Gothenburg University, Göteborg, Sweden, 3 Tara Seal Research Centre, Killyleagh, County Down, United Kingdom, 4 Institute of Hydrobiology and Ecology, Almaty, Kazakhstan, 5 Institute of Integrative and Comparative Biology, University of Leeds, Leeds, United Kingdom

Abstract

Understanding historical roles of species in ecosystems can be crucial for assessing long term human impacts on environments, providing context for management or restoration objectives, and making conservation evaluations of species status. In most cases limited historical abundance data impedes quantitative investigations, but harvested species may have long-term data accessible from hunting records. Here we make use of annual hunting records for Caspian seals (*Pusa caspica*) dating back to the mid-19th century, and current census data from aerial surveys, to reconstruct historical abundance using a hind-casting model. We estimate the minimum numbers of seals in 1867 to have been 1–1.6 million, but the population declined by at least 90% to around 100,000 individuals by 2005, primarily due to unsustainable hunting throughout the 20th century. This collapse is part of a broader picture of catastrophic ecological change in the Caspian over the 20th Century. Our results combined with fisheries data show that the current biomass of top predators in the Caspian is much reduced compared to historical conditions. The potential for the Caspian and other similar perturbed ecosystems to sustain natural resources of much greater biological and economic value than at present depends on the extent to which a number of anthropogenic impacts can be harnessed.

Editor: Andreas Fahlman, Texas A&M University-Corpus Christi, United States of America

Funding: This study was supported by a grant from the United Kingdom government's Department of Environment, Food and Rural Affairs (DEFRA) Darwin Initiative scheme (grant 162-15-24). KCH was financed by the Centre of Theoretical Biology at the University of Gothenburg, Sweden. The funders had no role in study design, data collection and analysis, decision to publish, or preparation of the manuscript.

Competing Interests: The authors have declared that no competing interests exist.

* E-mail: tero.harkonen@nrm.se

Introduction

High removal levels of keystone species may push ecosystems into new equilibria from which they are unlikely to return to historical states [1]. In marine ecosystems such regime shifts often result from unsustainable harvesting of commercially important fish [1,2,3] or marine mammal species [4]. Determining the past role of such populations can have important implications for reconstructing the historical state of ecosystems in terms of the biomass concentrated at different trophic levels, help with understanding long term human impacts, and provide goals for restoration and management [1,5]. Demographic history is also vital for conservation evaluations since the rate of decline is one of the main criteria used in placing taxa in International Union for the Conservation of Nature (IUCN) threat categories [6]. In contrast to most species, harvested species and populations may have time series of hunting or catch data. In this paper we reconstruct the historical abundance and demography of Caspian seals (*Pusa caspica*) based on exceptionally complete hunting records spanning 140 years from the 1860s to the late 20th century. We chart a catastrophic decline in Caspian seals, primarily driven by over-harvesting, and discuss the implications for the Caspian ecosystem and the current conservation status of the species. Our approach should be applicable for analyses of histories for other key species where some current census, harvesting and life history

data are available, and therefore a tool for assessments of species against IUCN threat criteria and examining historical changes in ecosystem structures.

Caspian seals are endemic to the Caspian Sea, and have been isolated since diverging from the ancestral *Pusa* genus around 1.3 million years ago [7]. They are one of the main large piscivores in the Caspian and large-scale changes in their abundance may therefore impact the structure of the whole ecosystem. The seals range throughout the entire Caspian Sea, which covers an area of 393,000 km^2 [8]. The northern ice fields constitute the critical breeding habitat, where pups are born at the end of January to the beginning of February, and weaning after 4–5 weeks [9]. Ice coverage has gradually diminished over the past three decades [10,11] due to climate warming, and the north-eastern part of the ice-field also overlies one of the world's largest oil fields, which is currently being developed for exploitation. Other issues currently considered as threats to the population include unsustainable levels of hunting and mortality from fisheries by-catch, mass mortalities due to canine distemper virus (CDV), habitat loss and disturbance from industrial development, and possible low prey abundance owing to over-fishing and recent invasion of the Caspian by the comb jellyfish *Mnemiopsis leidyi* [12].

The earliest known evidence for utilisation of Caspian seals by humans dates to around 20,000 years BP in northern Iran [13].

Significant commercial hunting started as early as 1740, and average annual harvests exceeding 100,000 were reported from at least since 1800 [14,15,16]. These numbers indicate that Caspian seals were once abundant, but aerial surveys during the 2005 and 2006 pupping seasons showed a decline to around 21,000 breeding females [17].

In this paper we use the unique and extensive hunting data for Caspian seals to reconstruct the minimum population sizes that could have sustained the recorded hunting pressure over the past 140 years up to the year 2005, when an estimate of pup production was determined from an aerial survey of the breeding population on the ice [17]. We then discuss how changes in abundance of seals may have affected the Caspian ecosystem and vice versa. We consider how the relationship between the seal population and the overall Caspian ecosystem may have altered over this period and consider prospects for recovery of this and other depleted seal populations.

Results

Changes in population size

Using an age-structured projection model (eqns 1–3) and the annually recorded harvest (Fig. 1) of Caspian seals over the period 1867–2005, we estimate the minimum initial female population size in 1867 at 572,800 females, of which 245,830 were breeding (Fig. 2). Given this starting point we estimate about 354,210 females in 1945 of which 193,140 were breeding, and 30,200 in 2005, where the 21,000 breeding females produced the same number of pups, which was approximately the number estimated from the survey in 2005. In simulations employing 20% lower and higher pup survival rates, as a test of the sensitivity to estimates of pup mortality, the estimated population sizes in 1867 were 510,400 and 676,700 females, respectively. Since the sex ratio in Caspian seals is close to parity [9], total initial population size was in the range from 1.0 to 1.6 million seals. Mean population growth was 0.983 for the entire period, using the average juvenile survival rate (Fig. 3), and between 0.982 and 0.986 for low and high pup survival rates, respectively. The population was reduced by about 66% between 1867 and 1964, and by a further 73% between 1965 and 2005 (Fig. 3).

Changes in population structure

The intrinsic population growth for the period 1867–1964 was 1.10. However, due to intense hunting there were great fluctuations in the population structure and the realised population growth rate during that period (Fig. 4). A lower figure for the instrinsic population growth for the period 1965 to 2005 was assumed to allow for reported lower fertility due to OC contamination (Table 1). At the estimated population structure in 2005 the 21,000 pupping females in 2005 would represent 20% of the total population size, which would therefore be about 104 thousand seals. Hunting reduced the simulated mean population growth to 0.971 (Fig. 3). Since hunting after 1965 up until the early 1990s was consistently high and focussed on pups, the age structure in the population during this period was strongly skewed towards adults (Fig. 4). Hunting was reduced in the mid 1990s, resulting in an age structure by 2005 which is close to initial conditions in the 19th century (Fig. 4.).

Generation time

Using the equation 4, the generation time for Caspian seals, measured as the mean age of females giving birth to a cohort, is 18–22 years for adult survival rates ranging between 0.95 and 0.97, respectively.

Discussion

Human impacts on seal populations

Human exploitation of pinnipeds has resulted in the extinction of the Caribbean monk seal (Monachus tropicalis), the Japanese sea lion (Zalophus californianus japonicus) and the extirpation of the Faroese harbour seal (Phoca vitulina) [12].

Grey seals (Halichoerus grypus) were extirpated from the European mainland North Sea coast in the Middle Ages, from the Skagerrak in the 1750s and the Kattegat in the 1930s [18]. A combination of hunting and other human impacts brought northern elephant seals (Mirounga angustirostris) to the brink of extinction [19], and have severely depleted populations of Mediterranean monk seals (Monachus monachus) [20], and Hawaiian monk seals (M. schauinslandi) [21]. Detailed historical hunting records are lacking for many formerly depleated pinnipeds e.g. Antarctic fur seals (Arctocephalus gazella), but such data are available for the Northern fur seal (Callorhinus ursinus), Saimaa ringed seal (Pusa hispida saimensis), Baltic ringed seal (Pusa hispida botnica), Baltic grey seal and the harbour seal (Phoca vitulina vitulina) in the Wadden Sea, Kattegat and Skagerrak [22,23,24,25,26]. Analyses of these hunting records documented collapses in all populations, which were depleted to about 5–10% of pristine abundances before protective measures were taken. The very detailed hunting records for Caspian seals enables a more thorough analysis where we find that numbers of breeding females have decreased from a minimum of 245,800 in 1867 to around 21,000 in 2005, which is a decrease by at least 90%.

Population estimates for the Caspian seal

An earlier reported population size estimate of about one million Caspian seals in the beginning of the 20th century [9], are fairly consistent with our results, which suggest a minimum of 1.2 million seals in 1900 (Fig. 2). However, an estimate of about 360–400,000 for total population size and 46,800 for the size of the reproductive female stock in 1989s [9], which is frequently cited in international compilations, deviates substantially from our calculations (Fig. 2). Data from our study suggest a total population size of about 128,000 and 30,000 for the number of reproductive females for 1989. The estimates for 2005 (104,000 in total and 21,000 reproductive females) therefore indicate a 19% and 30% decline in total numbers of seals and reproductive females respectively, from our estimate for 1989.

In the projections of earlier population sizes we systematically used high parameter values that resulted in under-estimations of population sizes in the past – hence we take a conservative approach and estimate minimum population sizes. We also assumed that the 20th century hunt killed equal numbers of males and females, when in reality the hunting on ice was mainly targeted at females and pups, while the hunting in spring and late autumn was focused on adult animals of both sexes [9]. Consequently, the average rate of decline in numbers of breeding females for the period 1965–2005 of 3.0% ($\lambda = 0.97$) is probably an under-estimate.

Biases in hunting statistics

The annual catch in the Caspian sealing industry in the first half of the 20th century was registered at the seal oil processing plant on the NE coast of the Caspian. This registration was probably reasonably accurate, since the annual harvest fluctuated considerably, rarely reaching the set quota [9], suggesting catches were not over-reported in order to meet targets. The early Russian authors suggest that the inter-annual fluctuations in numbers of seals hunted in the 19th and first part of the 20th Century primarily

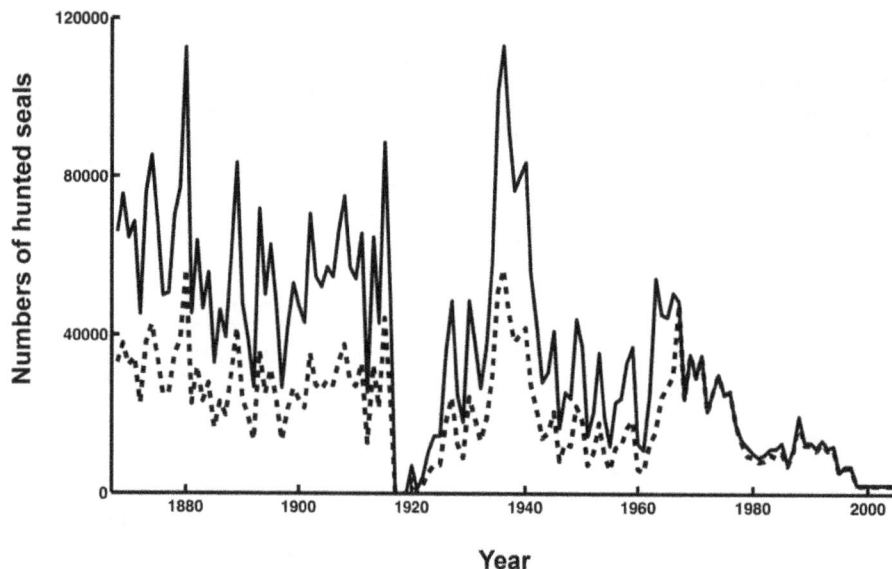

Figure 1. Total registered harvest of Caspian seals (solid line) and the number of pups (dashed line) for the period 1867–2005. Based on published hunting records [14,15,16]. Data for recent years are derived from Russian Federal Fisheries Agency reports [29,30].

reflect variation in hunting effort and access to seals according to winter conditions [15].

Factors affecting recovery of depleted pinniped populations

Many depleted pinniped populations have shared common combinations of factors which have driven their decline. Recovery of populations depends on the extent to which threats persist and on ecological changes following declines [27].

Most formerly over-exploited pinniped populations have recovered when hunting ceased. Examples include the northern elephant seal, most species of fur seals as well as populations of

harbour seals and grey seals. Consequently, protection from hunting has been the single most important factor allowing recovery of formerly depleted seal populations [22,23,24,25,26]. However, in some cases (such as the northern fur seal and the Saimaa ringed seal) recovery has been inhibited by a combination of new threats, such as by-catch and food chain alterations, which were probably less important during the hunting era (e.g. [28]).

The collapse of the Caspian seal population was driven by non-sustainable hunting which caused a rapid decline up to the mid 1990s (Fig. 1). The first steps towards species conservation should logically be a moratorium on hunting. However, although the commercial hunt ceased in 1996 as it was considered economically

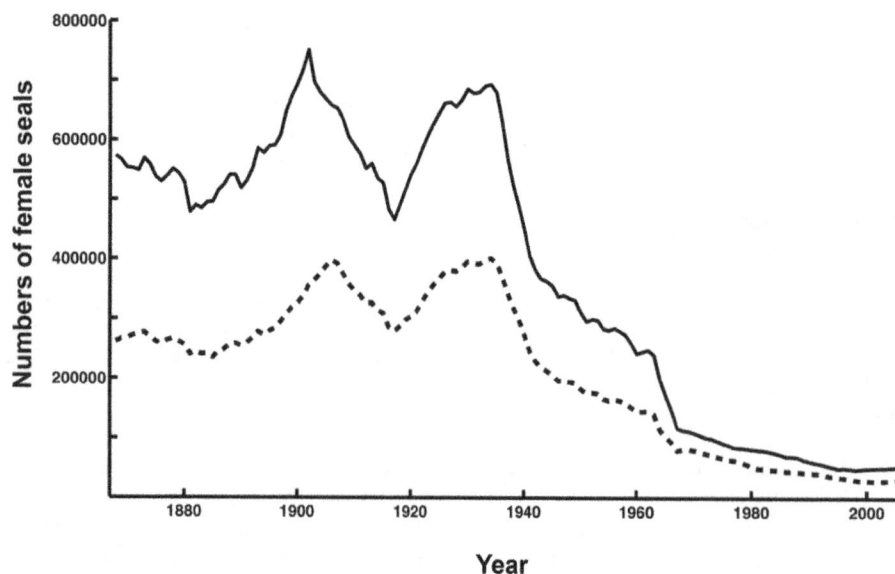

Figure 2. Estimated minimum total female population size (solid line) and the number adult females (dashed line) in the Caspian for the period 1867–2005 as based on historical hunting records (Fig. 1). The hunt during the 1960s led to a rapid decline in population size.

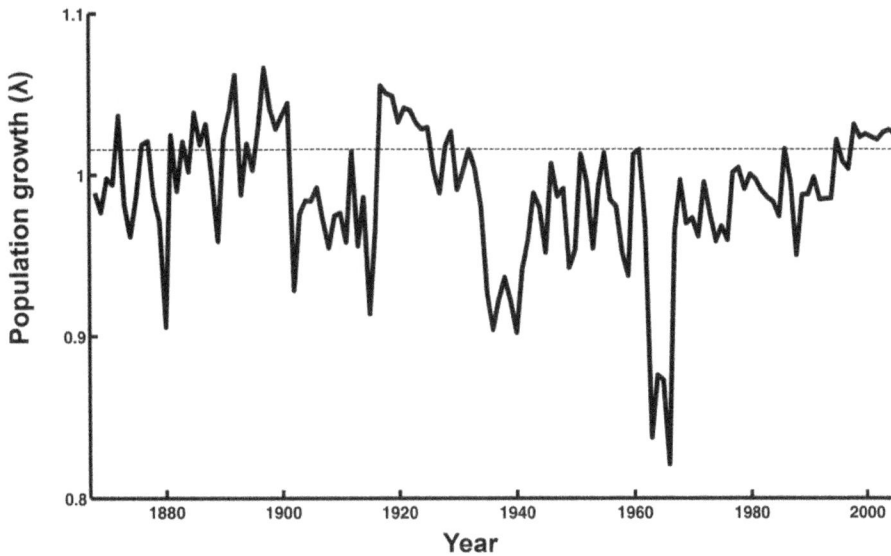

Figure 3. The population growth rate of the Caspian seal population from 1867 to 2006 has fluctuated significantly because of the variable hunting pressure.

unviable, substantial takes for 'scientific purposes' have occurred in the Russian sector in most years since then. Sporadic smaller scale commercial hunting by the Russian Federation resumed in 2003, with takes of around 3–5,000 in some years [29], under annual quotas of 18–20,000 seals (with around 9,000 allocated to Russia) set by The Caspian Bioresources Commission [30].

Static fishing nets, in contrast to active gear [31], pose a serious threat of entanglement to many marine mammals, including juveniles of the critically endangered Saimaa ringed seal and Mediterranean monk seals [20,28], where it is the single most important factor hampering population recovery. By-catch of Baltic ringed and grey seal pups is also substantial [31]. By-catch of Caspian seals, particularly in illegal sturgeon nets is likely to have

been a source of significant mortality for Caspian seals, amounting to several 1,000s of animals per year in recent years [32,33].

Infectious diseases caused by morbilliviruses have resulted in mass mortalities in seals including Antarctic crabeater seals (*Lobodon carcinophaga*) [34] and European harbour seals [35]. Several thousands of seals washed ashore throughout the Caspian in mass mortalities in 1997, 2000 and 2001. Canine distemper virus (CDV) was identified as the primary cause of the 2000 mortality, and the same virus was characterised in 1997 [36,37,38]. CDV is believed to have been recurrent in the seal population during the 1990s [39], although no further mass mortalities have been reported since 2001. Large-scale mortalities of tens of thousands of individuals were also reported from earlier

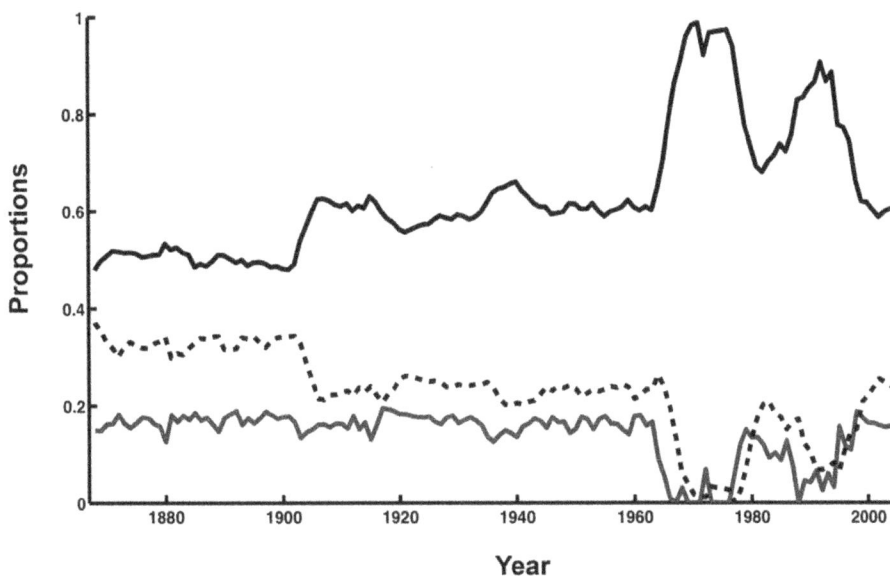

Figure 4. Temporal changes in age structure before pupping of the Caspian seal population. The skewed age structure is mainly due to hunting mortality and, sterility in the 1960s and 1970s. Adults = solid line, sub-adults = dashed line, yearlings = solid grey line.

Table 1. Vital rates for Caspian seals during the periods 1930–1964 and 1965–2005.

Variable	Description	1867–1964	1965–2006
F	Adult fertility	0.94/2	0.7/2
p_p	Pup survival	0.54–0.67–0.80	0.36–0.45–0-54
p_s	Sub-adult survival	0.90	0.90
p_a	Adult survival	0.97	0.97
m	Age at sexual maturity	5	5

Three pup survival rates, i.e. $0.8p_p$, p_p, or $1.2p_p$, are used to attain a realistic span for population size in 1867.

years (1955–56 and 1971 [33], but with no conclusive diagnosis of the cause. The long-term population impacts of such mortalities will depend on the frequency and severity of outbreaks [40,41]. Such catastrophic events reduce the long-term rate of increase of the population and amplify its variance, both leading to dramatically enhanced risks for extinction [40]. It has been suggested that the Caspian seal CDV outbreaks were facilitated or exacerbated by organochlorine contaminant loads [42] but recent re-analysis of these data found no link between OC levels and CDV mortality [43]. Low fertility in Caspian seals (from the 1960s to 1990s) has also been attributed to organochlorine contaminant loads [42], although the levels in 1997–2001 were mostly lower than those found to cause infertility and chronic disease complexes in Baltic ringed and grey seals [44,45].

All pinnipeds require land or ice for breeding, moulting and rest. Human disturbance at land sites result in habitat loss for many species of seals, such as Mediterranean and Hawaiian monk seals [20,11] and protected areas have contributed to the recovery of many populations of seals [25,46].

For Caspian seals the winter ice-field in the north Caspian is essential breeding habitat for which there is no effective substitute. During the past decade industrial shipping in connection with the oil field exploitation has been transiting areas of ice-breeding habitat, resulting in disruption of breeding colonies and habitat destruction [12]. In addition, many terrestrial haul-out sites occupied historically have been abandoned, a process which is probably due to a combination of human disturbance (particularly from human occupation, fishing, and industrial activities), reduction in total population size, and sea level fluctuations.

If the substantial reduction of the Northern Caspian ice fields over the past decades [10,11] continues, Caspian seals are likely to become affected by the loss of breeding habitat. Global warming has already resulted in diminishing ice fields and more variable ice covered periods in the breeding habitats of some ice-breeding seal species [27,47]. Pup production in the southern sub-population of Baltic ringed seals is absent during mild winters and the population growth rate is close to zero [48]. In oceanic habitats, seals may be able to migrate to other ice-covered areas [27], but for land-locked seals migration is not an option.

Unsustainable catches of key species in coastal areas world-wide have led to many marine ecosystem collapses, since depletion of large consumer species can result in cascading effects influencing all trophic levels [1,2,3]. In some ecosystems marine mammals have constituted one or several of these key extracted species. In the Caspian two of the key megafauna predator populations, the seal and the beluga sturgeon (*Huso huso*), have been severely depleted, and other ecosystems may point to potential implications. Similar to the Caspian, hunting reduced Baltic seal

populations by 90–95% during the 20th century. The reduced seal population initially resulted in a ten-fold increase in populations of cod (*Gadus morhua*), but this was followed by commercial over-fishing of cod, which resulted in increased populations of herring (*Clupea harengus*) and sprat (*Sprattus sprattus*) [4]. Current annual catches of commercial fish species in the Baltic (total about one million tons per year) are close to or beyond the maximum sustainable yields [49]. Baltic ringed seals and grey seals, with current populations sizes of 10 and 30 thousand respectively, are unlikely to attain pre-exploitation conditions of 200,000 and 90,000 individuals [22], since this would require half a million tons of fish per year to sustain populations of these sizes.

Reconstruction of historical population sizes of other pinniped populations point to similar changes in productivity and trophic structure in other ecosystems. McClenachan & Cooper [5] estimated the Caribbean monk seal once numbered approximately 300,000 seals. Such a population would consume substantially more fish and invertebrates than are currently produced in present day Caribbean coral reefs, implying a historical productivity matched only by the most remote & pristine Pacific reef systems today.

The Baltic and Caribbean seal cases may have close parallels in the Caspian. Assuming the mean population size of 1.2 million individuals of average mass 60 kg at the beginning of the 20th century, the total biomass of seals would have fallen from around 72,000 tons, to approximately 6,120 tons by 2005. A seal of the size of the Caspian seal requires a daily intake of 3.7–4.5 kg of fish [50], which would suggest the seal population in 1900 would have required between 1.6 and 2.0 million tons of fish per year to sustain it. Food requirements of the 2005 population of around 100 thousand seals would be only about 150 thousand tons of fish per year. In fact, since Russian literature dating from before our hunting dataset suggests intensive commercial hunting of seals began as early as 1740 [14], it is likely the true pre-exploitation population size would have been even higher than the 1.2 million estimated for 1900, with correspondingly higher food requirements at that time.

The influence of recent changes in ecosystem structure [51,52] on the potential for recovery of the Caspian seal population must be substantial. All commercially important fish stocks in the Caspian, including Caspian herring (*Alosa caspica*), salmon (*Salmo salar*) and all sturgeon species have collapsed or are overfished [53]. Anchovy kilka (*Clupeonella engrauliformis*) is by far the most abundant and productive fish species in the Caspian Sea, and it has been harvested commercially since 1925 [54]. Catches peaked in the early 1970s at 420,000 tons, followed by a long-term decline due to overfishing and a collapse to about 75,000 tons in 2005 [55]. Although Caspian seals are opportunistic predators, able to take a wide variety of fish species and not just those targeted by commercial fisheries [56], reduction in availability of high-energy prey and an enforced shift to less energy dense prey items could limit the ability of females to breed successfully. The comb jelly fish *Mnemiopsis leidyi*, accidentally introduced in ship's ballast in the late 1990s, competes with many fish species, including kilka, for zooplankton resources and consumes large quantities of fish larvae, including those of kilka [57]. *Mnemiopsis* may therefore be another factor in the decline of kilka. Primary productivity may also have been affected in the mid-late 20th century due to fluctuations arising from the in-flow of nutrients linked to eutrophification and damming of principle rivers entering the Caspian [58].

There is growing evidence from different ecosystems that large scale ecosystem changes may have profound consequences for the possibilities for large marine vertebrates to recover to former population sizes since pristine conditions of the ecosystems cannot be restored [4,5], and this problem will be especially acute for a species like the Caspian seal which has no possibilities for migration. Reconstruction of potential historical ecosystem states from archive data such as that presented here, could be a valuable tool contributing to a wide variety of impact assessments, setting of management goals, and evaluation of conservation status.

The conservation status of the Caspian seal has been evaluated according to the IUCN criteria A1b: "Population reduction observed, estimated, inferred, or suspected in the past where the causes of the reduction are clearly reversible AND understood AND have ceased, based on … an index of abundance..", where the rate of decline over the past three generations should exceed 70% to fulfil the status "Endangered". The generation time in Caspian seals is approximately 20 years, and three generations (i.e. about 60 years) back from 2005 would therefore suggest 1945 as a reference year for evaluations according to IUCN criteria. The projection in Fig. 2 indicates that the number of reproducing females declined by about 90% between 1945 and 2005. Consequently, the species meets the criteria for the endangered category. The species further fulfils criteria A1d (actual or potential levels of exploitation) and A1e (effects of introduced taxa, hybridization, pathogens, pollutants, competitors or parasites) due to threats of catastrophic reduction in stocks of fish prey, disease epizootics, from continued scientific and commercial hunting, and mortality from fishing by-catch. In addition future climate change may reduce or eliminate altogether the seals' ice-field breeding habitat, while oil industry activity is already disrupting breeding colonies. Industrial and urban development and disturbance around the whole Caspian has led to the abandonment of historical haul-out sites.

Materials and Methods

We estimate the minimum population size of seals that must have been present in the Caspian Sea during the past 140 years to sustain the documented hunt. To perform the modelling we need a) An estimate of current population size, b) Life history data for the Caspian seal, c) Hunting records, and d) A method to project the population in time, requiring e) estimation of the initial population structure in 1867. f) For assesment of the conservation status of Caspian seals according to IUCN criteria we also need to estimate the generation time.

Population size in 2005

Surveys of the entire pupping ice area in the Northern Caspian Sea were carried out in the last two weeks of February 2005 – 2006 [17]. Pup production estimated from the first survey was around 21,000 pups in 2005 and 17,000 pups the following year. Data from the first survey in 2005 were used as input values for estimates of population size.

Life history data

Caspian seal life history data share many characteristics with ringed seals (*Pusa hispida*), since the maximum life-span is about 50 years of age, and they mature relatively late [9,59], with females reported to usually becoming pregnant at 7 years of age, and giving birth at most to a single pup per year [9,59]. However, ages at first pregnancy as late as 7 have only been reported in seal populations that are food limited [60,61]. Since age at sexual maturity also changes with the general nutritive condition or density of the population [60,61], we use an age at first parturition of six years (Smith 1987, Reeves 1998) and therefore of sexual maturity at five years, which will lead to underestimates of the

minimum population sizes that could sustain the recorded hunt (we do this deliberately to ascertain minimum estimates). The maximum rate of increase (which also gives us the smallest population size that can sustain a given hunt) in species with life history features similar to the Caspian seal has been shown to be approximately 10% per year ($\lambda = 1.10$) [62]. Using this scenario, we estimated and inferred age specific survival and fertility rates (Table 1) for Caspian seals between 1867 and 2005 by adjusting these parameters in the projection matrix to give a long-term intrinsic rate of increase at $\lambda \approx 1.10$. Fertility rates have been reported to be lower since 1964 [32] (Table 1). Using this maximum long-term rate of increase result in estimates of minimum population sizes in the past since lower rates of population increase would require greater population sizes to withstand the recorded hunt,

Hunting records

A fleet of sealing vessels provided blubber to a processing plant in Fort Shevchenko in the North-eastern Caspian from the mid 19th century until the 1970s. Annual records of the harvest also included data about the composition of the hunt, where females with pups were targeted up to 1965, after which the hunt was focussed on pups (Fig. 1). Annual records of the harvest also included data about the composition of the hunt. These data are likely to represent fairly accurate hunting records, since the pelts were transported to other factories, where records with comparable figures were kept. Sklabinskij [14] reports that large scale commercial hunting began in 1740, and cites average annual harvests of 160,000 seals prior to 1803 without specifying the period. An average annual catch of 104,651 per year is reported from 1824–1867, with 290,000 in 1844 [14,16]. Until the mid-19th century, seals of all ages and both sexes were taken in summer and autumn on the unpopulated islands of the northern and southern Caspian. This hunt declined by the end of the 19th century, and the hunt turned to targeting females with pups on ice from the last quarter of the 19th century. For 8 years (1933–1940), catches of females and pups were so high (averaging more than 160,000 and increasing to more than 220,000 annually) that this hunting strategy was believed to have been the main cause of the population decline [16]. From 1941 catches were less than 100,000 annually, and after 1965 the hunt was focussed only on pups. For our focal dataset we make use of individual annual statistics available from 1867 onwards (S1) [9,15,29,30].

A population projection model

Using vital rates (survival and fertility) for Caspian seals given in Table 1 and the recorded hunt (Fig. 1), the initial population size in 1867 can be iterated such that the projection matches observed numbers of pups in 2005, which is equivalent to the number of reproducing females in that year. This allows estimates of changes in the minimum population size to be made for the period 1867–2005.

Based on available data (Table 1, [62,63,64]) we assume a stable intrinsic rate of increase at $\lambda \approx 1.10$ at the beginning of the 1860s. We project the Caspian seal population from 1867 onwards to 2005, keeping track of the number of animals in each age class throughout the projection. Since pup survival can vary substantially we investigated effects of low, average, and high levels of pup survival (Table 1) on estimated historical population sizes. However, results in the following refer to the average scenario if not otherwise stated.

In the age-structured population model (Fig. 2), N_i denotes the number of female seals at the end of the ith breeding season, where i is age in years, starting at $i=0$ (new-born female pup) and ranging up to the oldest age class c. We assume even sex ratios for all age classes, but only model the female population. We estimate survival and fertility rates for three age groups; pups, sub-adults and adults. The mean age at sexual maturity is denoted m. Pups correspond to age-class zero, sub-adults to age-classes one to $m-1$, and adults to age-classes m and above.

There are four vital rates in the model: The probability p_p that a new-born pup survives the first year; the probabilities p_s and p_a of sub-adult and adult survival and the probability f for a mature female seal to give birth to a female pup.

The number of new-born pups (N_p) is for each year (t) estimated by:

$$N_{P(t)} = f p_a \sum_{i=m}^{c} \left(N_{i(t-1)} - H_{i(t-1)} \right) \tag{1}$$

where f is the fertility, p_a adult survival, and the sum gives the number of adult female seals that survived the annual hunt $(N\text{-}H)$ in the previous year.

The number of sub-adults in each age class was estimated by including both natural and hunting mortality:

$$N_i(t) = \begin{cases} p_p \left(N_{P(t-1)} - H_{P(t-1)} \right) & i=1 \\ p_s \left(N_{i-1(t-1)} - H_{i-1(t-1)} \right) & 2 \leq i \leq m \end{cases} \tag{2}$$

where the number of one-year-olds (i = 1) is given by the pup survival (p_p) multiplied by the number of pups that survived the hunt the year before. $H_{P(t-1)}$ is the number of pups hunted the year before, and $H_{i-1(t-1)}$ the number of sub-adults of age $i-1$ hunted at $t-1$. In an identical manner the number of adults in each age class was estimated by subtracting the seals killed in the hunt and thereafter multiplying the number of survivors with the natural survival probability for adults.

$$N_{i(t)} = p_a \left(N_{i-1(t-1)} - H_{i-1(t-1)} \right) \quad (m+1) \leq i \leq c \tag{3}$$

where p_a is adult survival.

Initial age structure year 1867

The projection is based on the assumption of a stable age structure, which can be derived from the life history parameters [65]. As the population is projected forward from 1867 to 2005, the annual recorded hunt of adults, sub-adults and pups (Table 1, S1) are removed each year from the population and the resulting age structure is primarily a result of the harvest and the population growth rate. The effect of the assumed initial structure is therefore insignificant for the overall outcome, since the large and biased hunt is the main determinant for the age distribution.

Generation time

Using the parameter values in Table 1 we create two life-tables, one for each of the periods 1867–1964, and 1964–2005. A life table includes the parameters; x, the age, l_x the proportion of a cohort left in age-class x, and m_x the fecundity of each age class. Using the age structure and fecundity from the life tables of Caspian seals the estimated generation time (T), defined as 'the mean age of females giving birth to a cohort', is calculated as:

$$T = \frac{\sum_{x=0}^{x=45} x l_x m_x}{\sum_{x=0}^{x=45} l_x m_x} \tag{4}$$

The denominator is equal to the net reproductive rate. In addition, we vary the transition between the adult age-classes (p_a) between 0.95 and 0.97 to obtain a span of probable generation times.

Conclusions

Our study shows that the collapse in the Caspian seal population was primarily driven by overharvesting. The distribution of the Caspian seal in a completely closed ecosystem, from which individuals cannot disperse to or from adjacent habitats, makes it extremely vulnerable to some or all of the many threats it currently faces, which include mortality caused by hunting and by-catch, reduction in stocks of prey fish and oil industry activities in the ice breeding grounds combined with potential ice-field reductions due to climate change. Until this array of threats can be resolved by the implementation of effective conservation measures, as laid out in the Caspian Seal Conservation Action Plan [66], further rapid declines of this species are likely in the short term.

Acknowledgments

The authors wish to thank the rest of the Caspian International Seal Survey Team (Anders Bignert, Ivar Jüssi, Mart Jüssi, Yesbol Kasimbekov, Vadim Vysotskyi and Mikhail Verevkin) and our colleagues at the Caspian Environment Programme, Hamid Ghaffarzadeh, Igor Mitrofanov, Anders Poulsen, Fidan Kerimova, Elchin Mamedov and Parvin Faschi for their support. We also thank Prof V. Zaitsev of the International Oceanographic Institute, State Technical University, Astrakhan, for drawing our attention to the data contained in Dorofeev and Frejman (1928) [15].

Author Contributions

Conceived and designed the experiments: TH KCH SW SG. Performed the experiments: TH KCH SW SG CJS. Analyzed the data: TH KCH SW SG CJS LD. Contributed reagents/materials/analysis tools: TH KCH SW CJS LD SG MB. Wrote the paper: TH SG SW KCH CJS.

References

1. Jackson JBC, Kirby MX, Berger WH, Bjorndal KA, Botsford LW, et al. (2001) Historical Overfishing and the Recent Collapse of Coastal Ecosystems. Science 293: 629–637. DOI: 10.1126/science.1059199

2. Frank KT, Petrie B, Choi JS, Leggett WC (2005) Trophic cascades in a formerly cod-dominated ecosystem. Science 308: 1621–1623.

3. Myers RA, Baum JK, Shepherd TD, Powers SP, Peterson CH (2007) Cascading effects of the loss of apex predatory sharks from a coastal ocean. Science 315: 1846–1850.

4. Österblom H, Hansson S, Larsson U, Hjerne O, Wulff F, et al. (2007) Human-induced trophic cascades and ecological regime shifts in the Baltic Sea. Ecosystems 10: 877–889.

5. McClenachan L, Cooper AB (2008) Extinction rate, historical population structure and ecological role of the Caribbean monk seal. Proc R Soc B 275: 1351–1358.

6. IUCN (2001) 2001 IUCN Red List Categories and Criteria version 3.1. Available:http://www.iucnredlist.org/technical-documents/categories-and-criteria/2001-categories-criteria. Accessed 2010 Sep 1.

7. Fulton TL, Strobeck C (2010) Multiple fossil calibrations, nuclear loci and mitochondrial genomes provide new insight into biogeography and divergence timing for true seals (Phocidae, Pinnipedia). J Biogeog 37: 814–829.

8. Kosarev AN, Yablonskaya EA (1994) The Caspian Sea. SPB Academic Publishing, The Hague.

9. Krylov VI (1990) Ecology of the Caspian seal. Finn Game Res 47: 32–36.

10. Kouraev AV, Papa F, Byharizin PI, Cazenave A, Cretaux J-F, et al. (2003) Ice cover variability in the Caspian and Aral seas from active and passive microwave satellite data. Pol Res 22: 42–50.

11. Kouraev AV, Papad F, Mognarda NM, Buharizine PI, Cazenaved A et al. (2004) Sea ice cover in the Caspian and Aral Seas from historical and satellite data. J Marine Systems 47: 89–100.

12. Anonymous (2008) Pusa caspica. Available: IUCN Red List of Threatened Species. Version 2010.2. http://www.iucnredlist.org/apps/redlist/details/41669/0/. Accessed 2010 Sep 1.

13. Peasnall B (2002) Iranian Mesolithic. In: The Encyclopaedia of Prehistory, Volume 8 South and Southwest Asia. Peregrine PN, Ember M, editors pp.198–214, ISBN 0-306-46262-1. Published by Human Relations Area Files.

14. Sklabinskij N (1891) Svedenija o tiuleniej promyshlennosti na Kaspijskom more za poslednie 25 let. Rybnoe delo, N6, Astrakhan (in Russian), "Information on seal hunting industry in the Caspian Sea for the last 25 years", Rybnoe delo, N6 (November 1891) (special edition annex to the Astrakhan fisheries data sheet).

15. Dorofeev SV, Frejman SJ (1928) Kaspijskij tjulen' i ego promysel vo l'dah. «Trudy nauchnogo instituta rybnogo hozjajstva», M. – 1928. T. III. Vyp. 3. – S. 5-118 (in Russian), "Caspian seal and its hunting on the ice fields", Publications of Fisheries Research Institute, Moscow, 1928 3: 5–118.

16. Badamshin BI (1961) Zapasy kaspijskogo tjulenja i puti ih racional'nogo ispol'zovanija. Trudy soveshhanij ihtiologicheskoj komissii Akademii nauk SSSR, 12: 170–179 (in Russian), "Caspian seal resources and the ways of their rational usage", Proceedings of Meetings of the Ichthyological Commission of Academy of Science of USSR, Moscow, Russia. 12: 170–179. Translated by M. Slessers, ed. K. Hollingshead, U.S Naval Oceanographic Office, Washington D.C. 20390; 1970. Electronic ed. M. Uhen & M. Kwon, Smithsonian Inst., 2007. http://www.paleoglot.org/files/Badamshin%2061.pdf

17. Härkönen T, Jüssi M, Baimukanov M, Bignert A, Dmitrieva L, et al. (2008) Pup production and breeding distribution of the Caspian seal (Phoca caspica) in relation to human impacts. Ambio 37: 356–361.

18. Härkönen T, Brasseur S, Teilmann J, Vincent C, Dietz R, et al. (2007) Status of grey seals along mainland Europe, from the Baltic to France. NAMMCO Scientific Publications 6: 57–68.

19. Cooper CF, Stewart BS (1983) Demography of northern elephant seals. Science 219: 969–971.

20. Gucu AC, Gucu G, Orek H (2004) Habitat use and preliminary demographic evaluation of the critically endangered Mediterranean monk seal (Monachus monachus) in the Cilician Basin (Eastern Mediterranean). Biol Cons 116: 417–431.

21. Gerrodette T, Gilmartin WG (1990) Demographic consequences of changed pupping and hauling sites of the Hawaiian monk seal. Cons Bio 4: 423–430.

22. Harding KC, Härkönen TJ (1999) Development in the Baltic grey seal (Halichoerus grypus) and ringed seal (Phoca hispida) populations during the 20th century. Ambio 28: 619–627.

23. Lander RH (1980) Summary of northern fur seal data and collection procedures. Vol. 1. Land data from the United States and the Soviet Union. U.S. Dep. Commer., NOAA Technical Memorandum NMFFST/NWC-3: 315 pp.

24. Reijnders PJH (1985) On the extinction of the South Dutch harbour seal population. Biol Conserv 31: 75–84.

25. Heide-Jørgensen M-P, Härkönen T (1988) Rebuilding seal stocks in the Kattegat-Skagerrak. Mar Mamm Sci 4: 231–246.

26. Kokko H, Helle E, Lindström J, Ranta E, Sipilä T, et al. (1999) Backcasting population sizes of ringed and grey seals in the Baltic and Lake Saimaa during the 20th century. Ann Zool Fennici 36: 65–73.

27. Kovacs KM, Aguilar P, Aurioles D, Burkanov V, Campagna C, et al. (2011) Global Threats to Pinnipeds. Mar Mamm Sci 28: 428–436

28. Tonder M, Salmi P (2004) Institutional changes in fisheries governance: the case of the saimaa ringed seal, Phoca hispida saimensis, conservation. Fisheries management and Ecology 11: 283–290.

29. Russian Federal Fisheries Agency reports 2003–2006. Statistics on fishing, extraction of other aquatic biological resources, and products of the fishing industry. Available: http://fish.gov.ru/activities/DocLib/%D0%A1%D1%82%D0%B0%D1%82%D0%B8%D1%81%D1%82%D0%B8%D0%BA%D0%B0%20%D0%B8%20%D0%B0%D0%BD%D0%B0%D0%BB%D0%B8%D1%82%D0%B8%D0%BA%D0%B0.aspx. Accessed 2011 Jan 7.

30. Russian Federal Fisheries Agency Order N825 2010 Approval of the total allowable catch of Aquatic Biological Resources in 2011. Available: http://base.consultant.ru/cons/cgi/online.cgi?req=doc;base=LAW;n=106188;dst=0;ts=1F0DBF9BFAF90EBE5055D8274CF6C780. Accessed 2011 Jan 7.

31. Lunneryd S-G, Königsson S, Sjöberg NB (2004) Bifångst av säl, tumlare och fåglar i det svenska yrkesfisket. Finfo 2004:8. Swedish Board of Fisheries. Sweden

32. Eybatov TM (1997) Caspian seal mortality in Azerbaijan. In: Caspian Environment program Proc. 1st Bionet workshop Bordeaux, H. Dumont H, S. Wilson S, Wazniewicz B. pp 95–100. World Bank.

33. Eybatov T, Asadi H, Erokhin P, Kuiken T, Jepson P, et al. (2002) Caspian seal (Phoca caspica) mortality. Ecotox Final Report, Appendix A2. World Bank. http://www.caspianenvironment.org/ecotoxreport.htm.

34. Laws RM, Taylor RTF (1957) A mass dying of crabeater seals, Lobodon carcinophaga (Gray). Proceedings of the Zoological Society of London 129: 315–324.

35. Härkönen T, Dietz R, Reijnders P, Teilmann J, Harding K et al. (2006) A review of the 1988 and 2002 phocine distemper virus epidemics in European harbour seals. Dis Aquat Org 68: 115–130.

36. Kennedy S, Kuiken T, Jepson PD, Deaville R, Forsyth M, et al. (2000) Mass die-off of Caspian seals caused by canine distemper virus. Emerg Inf Dis 6 637–639.

37. Kuiken T, Kennedy S, Barrett T, Van de Bildt MWG, Borgsteede F, et al. (2006) The 2000 canine distemper epidemic in Caspian seals (Phoca caspica): pathology and analysis of contributory factors. Vet Pathol 43: 321–338.

38. Forsyth MA, Kennedy S, Wilson S, Eybatov T, Barrett T (1998) Canine distemper virus in a Caspian seal. Vet Rec 1998: 143: 662–664 doi:10.1136/vr.143.24.662.

39. Ohashi K, Miyazaki N, Tanabe S, Nakata H, Myura R, et al. (2001) Seroepidemiological survey of distemper virus infection in the Caspian Sea and Lake Baikal. Vet Microbiol 82: 203–210.

40. Harding KC, Härkönen T, Caswell H (2002) The 2002 European seal plague: Epidemiology and population consequences. Ecol Lett 5: 727–732.

41. Harding KC, Hansen J, Goodman S (2005) Acquired immunity and stochasticity in epidemic intervals impede the evolution of host disease resistance. Am Nat 166: 722–730.

42. Kajiwara N, Watanabe N, Wilson S, Eybatov T, Mitrofanov IV, et al. (2008) Persistent organic pollutants (POPs) in Caspian seals of unusual mortality event during 2000 and 2001. Environ Pollut 117: 391–402.

43. Wilson SC, Eybatov TM, Amano M, Asadi H, Jepson PD, et al (in prep) Implications of contamination levels in Caspian seals, Phoca caspica.

44. Helle E (1980) Lowered reproductive capacity in female ringed seals, Pusa hispida, in the Bothnian Bay, northern Baltic, with special reference to uterine occlusions. Ann Zool Fennici 17: 147–158.

45. Bredhult C, Bäcklin B-M, Bignert A, Olovsson M (2008) Study of the relation between the incidence of uterine leiomyomas and the concentrations of PCB and DDT in Baltic gray seals. Reproductive Toxicology 25: 247–255.

46. Reijnders PJH (1978) Recruitment in the harbour seal (Phoca vitulina) population in the Dutch Wadden Sea. Neth J Sea Res 12: 164–179.

47. Meier HEM, Döscher R, Halkka A (2004) Simulated distributions of Baltic Sea-ice in the warming climate and consequences for the winter habitat of the Baltic Ringed Seal. Ambio 33: 249–256.

48. Sundqvist L, Harkonen T, Svensson CJ, Harding KC (in press) Linking climate trends to population dynamics in the Baltic ringed seal - Impacts of historical and future winter temperatures. Ambio.

49. ICES Advice. Available: www.ices.dk/products/icesadvice.asp. Accessed 2012 Sep 15.

50. Härkönen T, Heide-Jørgensen M-P (1991) The harbour seal Phoca vitulina as a predator in the Skagerrak. Ophelia 34:191–207.

51. Dumont H (1995) Ecocide in the Caspian Sea. Nature 377: 673–674.

52. Roohi A, Kideys AE, Sajjadi A, Hashemian A, Pourgholam R, et al. (2010) Changes in biodiversity of phytoplankton, zooplankton, fishes and macrobenthos in the Southern Caspian Sea after the invasion of the ctenophore (Mnemiopsis leidyi). Biol Inv 12: 2343–2361.

53. Strukova E, Guchgeldiyev O (2010) Study of the economics of bio-resources utilization in the Caspian. Report to Caspian Environment Programme, published by World Bank. Available: http://www.caspianenvironment.org/LibRep/Insert/View/MoreInfo.asp?ID=1020. Accessed 2010 Sep 1.

54. Mamedov EV (2006) The biology and abundance of kilka (Clupeonella spp.) along the coast of Azerbaijan, Caspian Sea. Ices J Mar Sci 63: 1665–1673.

55. Daskalov GM, Mamedov EV (2007) Integrated fisheries assessment and possible causes for the collapse of anchovy kilka in the Caspian Sea. Ices J Mar Sci 64: 503–511.

56. Piletskii AM, Krylov VI (1990) Diet of the Caspian seal in the Volga delta. In: Some remarks to biology and ecology of the Caspian seal. Krylov VI editor, pp. 58–78. Published by Russian Federal Research Institute of Fisheries and Oceanography (VNIRO), Moscow, Russia.

57. Ivanov VP, Kamakin AM, Ushivstsev VB, Shiganova T, Zhukova O, et al. (2000) Invasion of the Caspian Sea by the comb jellyfish Mnemiopsis leidyi (Ctenophora). Biol Inv 2: 255–258.

58. Yousefian M, Kideys AE (2003) Biochemical composition of Mnemiopsis leidyi in the southern Caspian Sea. Fish Physiology and Biochemistry 29: 127–131.

59. Popov LA (1979) Caspian seal. In Mammals of the Seas, Vol. II. Pinniped species summaries and report on sirenians. FAO Fisheries series 59: 74–75.

60. Harding KC, Härkönen T (1995) Estimating age at sexual maturity in crabeater seals Lobodon carcinophagus. Can J Fish Aquat Sci 52: 2347–2352.

61. Kjellqwist SA, Haug T, Øritsland T (1995) Trends in age-composition, growth and reproductive parameters of Barents Sea harp seals, Phoca groenlandica. ICES J Mar Sci 52: 197–208.

62. Harding KC, Härkönen T, Helander B, Karlsson O (2007) Status of Baltic grey seals: Population assessment and risk analysis. NAMMCO Scientific Publications 6: 33–56.

63. Reeves RR (1998) Distribution, abundance and biology of ringed seals (Phoca hispida): an overview. NAMMCO Scientific Publications 1: 9–45.

64. Smith TG (1987) The ringed seal, Phoca hispida, of the Canadian Western Arctic. Canadian Bulletin of Fisheries and Aquatic Science 216: 1–81.

65. Caswell H (2001) Matrix population models: Construction, Analysis, and Interpretation. 2nd edn. Sinauer Associates Incorporated, Sunderland, Mass. USA.

66. Caspian Environment Programme (2007) Caspian Seal Conservation Action Plan. Available: http://www.caspianenvironment.org/autoindex/index. php?dir = NewSite/DocCenter/reports/2007/Caspian%20Seal%20Conservation %20Action%20Plan/. Accessed 2011 Mar 10.

14

Red Shift, Blue Shift: Investigating Doppler Shifts, Blubber Thickness, and Migration as Explanations of Seasonal Variation in the Tonality of Antarctic Blue Whale Song

Brian S. Miller[1]*, Russell Leaper[2], Susannah Calderan[1], Jason Gedamke[3]

1 Australian Marine Mammal Centre, Australian Antarctic Division, Kingston, Australia, 2 School of Biological Sciences, University of Aberdeen, Aberdeen, United Kingdom, 3 Ocean Acoustics Program, NOAA Fisheries Office of Science and Technology, National Oceanic and Atmospheric Administration, Silver Spring, Maryland, United States of America

Abstract

The song of Antarctic blue whales (*Balaenoptera musculus intermedia*) comprises repeated, stereotyped, low-frequency calls. Measurements of these calls from recordings spanning many years have revealed a long-term linear decline as well as an intra-annual pattern in tonal frequency. While a number of hypotheses for this long-term decline have been investigated, including changes in population structure, changes in the physical environment, and changes in the behaviour of the whales, there have been relatively few attempts to explain the intra-annual pattern. An additional hypothesis that has not yet been investigated is that differences in the observed frequency from each call are due to the Doppler effect. The assumptions and implications of the Doppler effect on whale song are investigated using 1) vessel-based acoustic recordings of Antarctic blue whales with simultaneous observation of whale movement and 2) long-term acoustic recordings from both the subtropics and Antarctic. Results from vessel-based recordings of Antarctic blue whales indicate that variation in peak-frequency between calls produced by an individual whale was greater than would be expected by the movement of the whale alone. Furthermore, analysis of intra-annual frequency shift at Antarctic recording stations indicates that the Doppler effect is unlikely to fully explain the observations of intra-annual pattern in the frequency of Antarctic blue whale song. However, data do show cyclical changes in frequency in conjunction with season, thus suggesting that there might be a relationship among tonal frequency, body condition, and migration to and from Antarctic feeding grounds.

Editor: Jean-Benoit Charrassin, Musee National d'Histoire Naturelle, France

Funding: Deployment of long-term recorders was supported by the Australian Antarctic Division. This work was supported by the International Whale and Marine Mammal Conservation Initiative of the Australian Government via Southern Ocean Research Partnership of the International Whaling Commission. The funders had no role in study design, data collection and analysis, decision to publish, or preparation of the manuscript.

Competing Interests: The authors have declared that no competing interests exist.

* Email: brian.miller@aad.gov.au

Introduction

Antarctic blue whales (*Balaenoptera musculus intermedia*) produce repeated, stereotyped, low-frequency song comprising three units: an approximately 10 second tonal unit with a frequency of maximum power (henceforth referred to as peak-frequency) around 28–26 Hz and two shorter frequency-modulated downsweeps [1,2]. In addition to this three-part song, it is believed that Antarctic blue whales also produce songs consisting of only the first tonal unit [2]. The calls of the three-part song have been named 'z-calls' because of their characteristic shape when viewed as a spectrogram (Figure 1). Comparison of z-calls recorded in different years has revealed both long-term [3,4] and seasonal [4] patterns in the tonal frequency of these sounds (Figure 2). Gavrilov et al. [4] reported a linear inter-annual decline of the tonal component of these calls of 0.135 Hz/year (R²=0.99), and an intra-annual decline between 0.4–0.5 Hz from March to December (R²>0.8).

McDonald et al. [3] discussed a number of hypotheses for the long-term inter-annual decline, including changes in population structure, ambient noise, physical environment, and whale

behaviour. They concluded that the most likely explanation of the trend was related to increasing population density, and suggested that the tonal decline was an anatomical constraint of the mechanism of sound production that also resulted in a decreased call source level. A key driver of this theory was that the source levels required for whales to keep in acoustic contact with a constant number of conspecifics would not have to be so high if population density were increasing. However, presently there are not enough estimates of the source level of calls (let alone population density) of Antarctic blue whales to test whether source levels have decreased in a manner similar to that predicted by McDonald et al. [3].

Gavrilov et al. [4] proposed that the mechanism behind the intra-annual pattern (Figure 2) might be explained by a gradual decrease in the depth at which songs are produced. They suggested that this decrease in depth could arise from changes in dive behaviour over the length of each season, or that it could be due to other factors such as variations in water temperature or change in blubber mass. However, they considered that such an explanation was not likely to apply to the long-term trend and

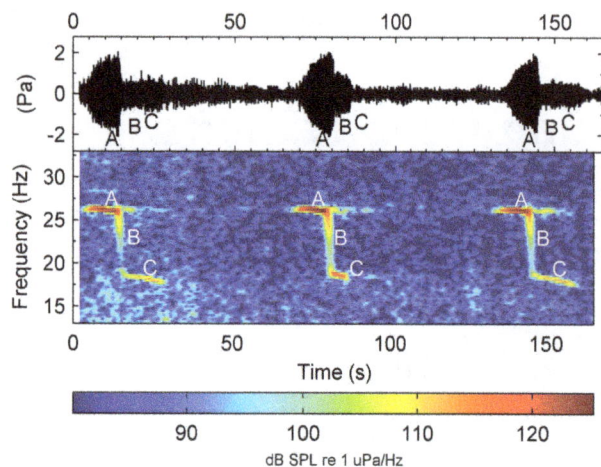

Figure 1. Visualisation of Antarctic blue whale song. Pressure waveform and spectrogram of Antarctic blue whale "z-calls" recorded off Antarctic ice-edge during February 2013. The call is divided into 3 units labelled A, B, and C. Spectrogram was produced using a sample rate of 250 Hz, 1024 point FFT, and 87.5% overlap between time slices. Colors indicate received power spectral density (dB re 1 μPa/Hz).

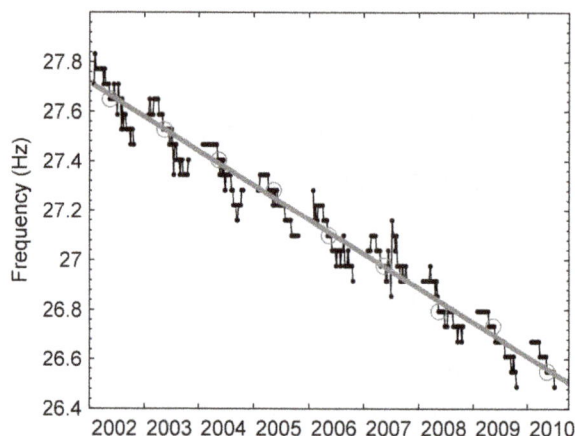

Figure 2. Long-term and intra-annual trends in tonality of Antarctic blue whale song. Long-term trend and intra-annual pattern in tonal frequency of Antarctic blue whale calls. Reprinted with permission from. Gavrilov et al. (2012). Copyright 2012 Journal of the Acoustical Society of America, American Institute of Physics.

suggested that changes in whale vocal behaviour remained the most parsimonious explanation for the long-term inter-annual decline.

Here we investigate the Doppler effect [5] as an additional explanation for some of the intra-annual patterns in observations of tonal frequency. Doppler shift is the change in frequency of a wave that arises from relative motion between the source and the receiver of the wave. The equation for Doppler shift can be written as the ratio, r, of the measured frequency, f_m, to the true (i.e. non-shifted) frequency f_w:

$$r = \frac{f_m}{f_w} = \frac{v+c}{c} \qquad (1)$$

where v is the relative speed between the whale and the receiver, and c is the speed of sound along the path between source and receiver. If observations are made at a fixed receiver, such as the hydrophone array used by Gavrilov et al. [4], then any potential shift in frequency due to the Doppler effect must arise from movement of the sound source, in this case vocalising Antarctic blue whales.

Seasonal movements of Antarctic blue whales are not well described; however it has been proposed that they, like most baleen whale species, migrate between high latitude summer feeding grounds and low-latitude wintering grounds [6]. There is strong evidence that Antarctic blue whales have a circumpolar Antarctic distribution during the austral summer [7]. In contrast, there are few visual observations of Antarctic blue whales during austral winter [6]. However, acoustic detections of z-calls (distinctive to Antarctic blue whales) provide some of the most compelling evidence that these animals do migrate to mid-or low latitudes in austral winter [4,8–10], despite year-round acoustic detections in the Antarctic [1].

The temporal aspect of these acoustic detections suggests a mid or low-latitude winter destination for Antarctic blue whales. Stafford et al. [8] reported that low and mid-latitude detections begin in April, and continue through November in the South Pacific, South Atlantic, and Indian Oceans. Samaran et al. [9]

found year-round acoustic detections of Antarctic blue whale calls at a mid-latitude site in the Indian Ocean (46°S, 53°E), but proportionally more days with detections in austral winter. Gavrilov et al. [4] also reported near-year round acoustic detection of Antarctic blue whales at Cape Leeuwin, a mid-latitude Indian Ocean site (35°S, 114°E; Figure 3) with detections having highest intensities from May to September.

The peak in intensity in May at Cape Leeuwin could potentially represent the point of closest approach for the majority of the migrating whales, or it could arise from a peak in the number of whales calling. Samaran et al. also found that the month with the highest proportion of days with detected Antarctic blue whale calls off Crozet Island, another mid-latitude location was May [9]. This peak in calling in May at two widely separated locations is further evidence that at this time of year (vocalising) whales are either migrating through to mid-latitudes, calling more frequently, or possibly a combination of the two.

One implication of the Doppler effect could be the ability to track migrating whale populations using recordings made from widely spaced hydrophones located along a latitudinal gradient. For example, at mid latitudes there should be an increase in frequency early in the migration season as the animals approach the hydrophone and a drop late in the season as the animals move away. Such recordings, especially when combined with amplitude information (e.g. [4]), acoustic propagation models (e.g. [11]) and/ or acoustic bearings to the sound source [12] could potentially allow for passive acoustic tracking of the migration of populations of vocalizing whales [13].

Here we investigate whether Doppler shift could explain the intra-annual pattern in tonal frequency reported by Gavrilov et al. [4]. We first examine a situation where whale movements were observed and z-calls were recorded simultaneously in order to test whether the Doppler effect on tonal frequencies was measurable for small-scale movements. We then re-examine the intra-annual pattern observed by Gavrilov et al. off Cape Leeuwin [4], and supplement this analysis with year-long recordings from two sites in the Antarctic (Figure 3). Next, we examine whether intra-annual changes in frequency fit with existing knowledge of large-scale migrations of Antarctic blue whales. Additionally, we investigate whether intra-annual variation in tonal frequency is correlated with blubber thickness. Finally, we discuss additional

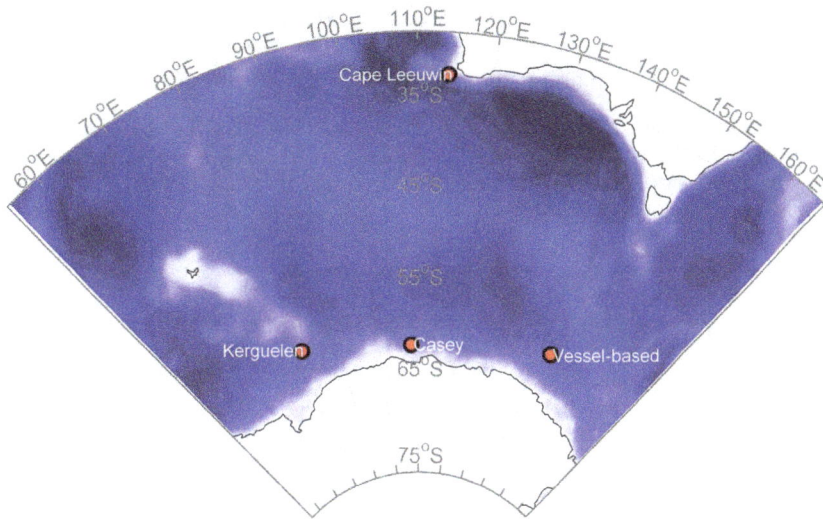

Figure 3. Map of recording sites. Locations of long-term and vessel-based recording stations used in this manuscript for investigation of tonal frequency of the song of Antarctic blue whales. The data from Gavrilov et al. [4] (*i.e.* Figure 2) were observed at Cape Leeuwin.

observations and continued data collection that might further test hypotheses to explain the changes in tonal frequency of blue whale song.

Methods

A. Vessel-based measurements of frequency and whale speed

During the 2013 Antarctic Blue Whale Voyage of the Southern Ocean Research Partnership, acoustic recordings of Antarctic blue whales were collected along with simultaneous visual tracking [14]. Upon approach, the location of surfacing whales was measured using a video-photogrammetric system (described by Leaper and Gordon [15]) to determine their range and bearing relative to the ship. Acoustic recordings were made during approach using Directional Frequency Analysis and Recording (DIFAR) sono-buoys [16].

The acoustic recording chain was calibrated in accordance with procedures outlined by [12,17–19]. Radio signals from the DIFAR 53D sonobuoy (Ultra Electronics Inc. Canada) were received using an omnidirectional VHF antenna (PCTel Inc. MFB1443; 3 dB gain tuned to 144 MHz centre frequency) and pre-amplifier (Minicircuits Inc. ZX60-33LN-S+) mounted on the mast of the ship at a height of 21 m. The preamplifier was connected to a power splitter via LMR400 cable and signals were received with two WiNRaDiO G39WSBe sonobuoy receivers. Received signals were digitised via a sound board (RME Fireface; RME Inc.), and signals were recorded on a personal computer using the software program PAMGuard [20].

Over the course of the voyage there were dozens of high-quality audio recordings and visual tracks of Antarctic blue whales. However there was only one instance (an encounter on 7 February 2013) of simultaneous video and audio recordings where the whale produced z-calls. This data set was used to investigate whether there was a relationship between whale movements and the received tonal frequency of calls (*i.e.* whether our observations were sensitive enough to detect the Doppler effect). We re-arrange Equation 1 in order to obtain the expected linear relationship between measured tonal frequency, f_m, and velocity yielding:

$$f_m = av + b \qquad (2)$$

where $a = f_w/c$ and $b = f_w$. The 'true' (*i.e.* non-Doppler shifted) frequency, f_w, was defined to be the long-term trend described by Gavrilov et al., [4]:

$$f_w(t) = -0.135t + 27.666; \qquad (3)$$
$$\left(R^2 = 0.99, 95\%\text{CI} \pm 0.003\,\text{Hz/year}\right)$$

Here t represents the number of years since the start of the dataset: 12 Mar. 2002. It should be noted that the velocity, v, corresponds only to the component of movement in the direction of the acoustic wavefront such that:

$$v = \|v_w\| \cos \theta \qquad (4)$$

where $\|v_w\|$ is magnitude of the velocity of the whale, and θ is the difference in angle between the direction of motion of the whale and the bearing from the sonobuoy to the whale.

Locations of Antarctic blue whales obtained via photogram-metric video tracking were assumed to correspond to the "true" location of the whale (at the surface) due to the high accuracy and precision of this technique [15]. Average heading and whale speed were then computed between successive photogrammetric loca-tions. All z-calls in this data set were produced when the whale was out of sight underwater, and linear interpolation between successive photogrammetric locations was used to estimate the locations of the whale at the times when z-calls were received.

Sonobuoys were assumed to drift in a constant direction at a constant speed. The direction and speed of drift were estimated by measuring acoustic bearings to the research vessel (*i.e.* a source with a known location) at intervals of 20 s, and solving for the direction and speed that maximised the likelihood of these measurements [21]. A single estimate of constant drift direction and speed was produced for each sonobuoy for the entire duration of the recording.

Acoustic analysis was restricted to the duration over which there were high-quality photogrammetric measurements. Songs originating from the tracked whale were identified and used for further analysis, while songs that were believed to be from other whales were discarded. Several criteria, including the type of call, temporal pattern of calling, and received level, were used in addition to the acoustic bearing to the source of the song (from the DIFAR sensors) to determine whether or not the call should be included for further analysis.

Measurements of peak-frequency were made from audio recordings of z-calls that were selected for analysis. Peak-frequency measurements were made in the frequency domain by computing the power-spectral density (PSD) for acoustic data spanning the duration of the first tonal unit of the z-calls, which we refer to as unit A. Measurements of peak-frequency were restricted to the band between 25 and 27 Hz in order to exclude potential sources of tonal noise (e.g. engine and/or generator noise from vessels).

The frequency resolution (*i.e.* bin-width) of the PSD is equal to the inverse of the duration of the signal. Due to the relatively short duration of the calls compared to the desired frequency resolution, acoustic waveforms were extended with zeros before the start and after the end of the signal to allow for a sufficiently large number of samples in order to more accurately locate the spectral maxima when computing the spectrum via Fast-Fourier Transform (FFT). Before padding each end with zeros, a Hann window was applied to the acoustic waveform in the time domain in order to minimise any spectral distortion that might arise from the impulsive discontinuity that would otherwise occur at the interface between zeros and acoustic signal.

B. Long-term measurements of frequency

1. Intra-annual trends in frequency. In contrast to the vessel-based observations, analysis of the intra-annual pattern in frequency relied solely upon the PSD with no attempt to measure individual whale calls. Thus, our analysis methods were identical to those employed by Gavrilov et al., [4]. Measurements of peak-frequency in the Antarctic blue whale band, f_m, were digitized from Figure 5 in Gavrilov et al., [4]. Again, the long-term trend from Gavrilov et al. [4], was taken to be the 'true' (i.e. non-shifted) frequency, f_w (equation 3). For each weekly observation reported by Gavrilov et al. [4], the frequency ratio, r, of measured frequency to 'true' frequency (i.e. the left side of equation 1) was computed. The frequency ratio (i.e. scaling the peak-frequency by the long-term trend) enabled the comparison of intra-annual trends for data that were recorded in different years.

A similar analysis of peak-frequency was also performed on two data sets recorded off Antarctica: data from Acoustic Recording Packages (ARPS; [22]) off Casey Station from 2004 to 2005, and the Kerguelen Plateau from 2005 to 2007. These data were recorded near the sea floor at approximately 1800 m depth at a sample rate of 500 Hz. Before analysis, these data were filtered and re-sampled to 100 Hz in order to maintain a small memory footprint for computations. PSD was averaged daily and the FFT size was 16384 samples to obtain 0.006 Hz frequency resolution; comparable to that of Gavrilov et al. [4]. Portions of the recordings that contained strong broadband noise sources (e.g. large storms) were excluded from the PSD analysis. Additionally long-term spectral averages were visually inspected for time periods when energy from the 20 Hz calls of fin whales was more intense than that of the tonal component of blue whales, and these time periods were also removed. For each daily PSD, the frequency with maximum energy in the 25–29 Hz band was selected as the peak-frequency. Monthly means and standard deviations of these daily peak-frequencies were computed for each station.

All vessel-based work and long-term acoustic recordings were carried out in strict accordance with the approvals and conditions of the Antarctic Animal Ethics Committee for Australian Antarctic Science projects 2683 and 4102. All data used in this work is publicly available via the Australian Antarctic Data Centre (http://data.aad.gov.au/), and are discoverable through the Catalogue of Australian Antarctic and Sub-antarctic Metadata.

2. Doppler effect. In order to assess whether the Doppler effect was a plausible explanation for the intra-annual trends in peak-frequency, we re-arrange equation 1 in order to obtain the relative velocity, v, of the source i.e. the population of whales emitting z-calls:

$$v = c(r-1) \qquad (5)$$

where positive velocities indicate that the direction of travel is towards the observer and negative velocities indicate the direction of travel is away from the observer. The sound speed, c, was assumed to be 1500 m/s.

3. Changes in whale anatomy. In addition to Doppler shift, we also conducted a preliminary investigation of the relationship between blubber thickness and the frequency ratio, r. Measurements of the blubber thickness of Antarctic blue whales were digitised from the 1929 Discovery Report by Mackintosh and Wheeler [23]. In accord with the original analysis [23], we considered two size-based groups of Antarctic blue whales: those less than 19 m, and those greater than 23 m. For each size class and we applied weighted least-squares linear regression to investigate potential correlation between the monthly measurements of blubber thickness and the monthly variation in frequency ratio from all recording sites. Monthly variation in frequency ratio, m, was computed as the percentage change in peak-frequency from that of the 'true' frequency, m, such that:

$$m = 100\frac{f_m - f_w}{f_w} = 100(r-1) \qquad (6)$$

The variance of m was used as the weights when computing the slope and intercept for the weighted least-squares fit.

Results and Discussion

A. Vessel-based observations

Results. During the recording session on 7 February 2013, the whale passed within a kilometre of a sonobuoy (Figure 4). Maximum received levels of whale calls correlated well with the estimated point of closest approach (c. 660 m). This provided confidence that the calls were produced by the photogrammetrically-tracked whale, and that estimates of direction and speed of drift of the sonobuoy (170 degrees; 0.93 m/s respectively) were also consistent. Song was recorded both as the whale was approaching the sonobuoy, and as the whale moved away from the sonobuoy (Figure 5a).

The average speed of the whale between photogrammetrically-derived positions was approximately 2 m/s throughout the encounter. With respect to the buoy, the velocity of the whale ranged from just above 1 m/s to nearly −2 m/s (with negative sign denoting whale movements away from the sonobuoy; Figure 5b). Whale velocity components along the direction of the acoustic wavefront ranged from 1 to −1 m/s (Figure 5c). Measured peak-frequencies ranged between 26.050 and 26.325 Hz, while frequencies predicted from the Doppler effect

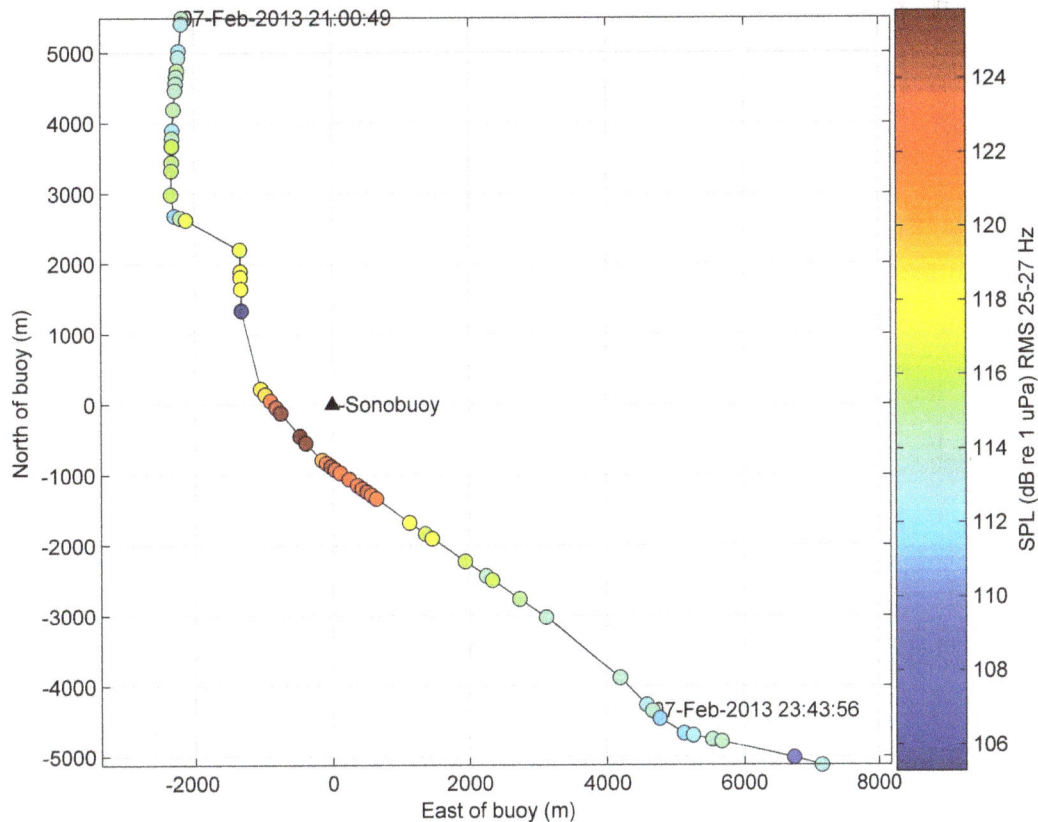

Figure 4. Whale track near a sonobuoy. Whale positions obtained by photogrammetric video tracking (solid black line). All positions are relative to the location of the drifting sonobuoy (black triangle). Filled circles show the estimated location of the whale, relative to the receiver, when z-calls were detected. Color of the circle indicates the received root-mean-square (RMS) sound pressure level (SPL) of call unit A measured in the 25–27 Hz band.

(equation 2) ranged between 26.160–26.220 Hz, assuming the long-term trend reported by Gavrilov et al. [4] (equation 3).

The velocity, v, explained only a very small proportion of the variability in observed peak-frequency in the multiple calls produced by this individual, f_m ($R^2 = 0.07$; p = 0.039; Figure 6). The intercept of the measured peak-frequencies was 26.182 Hz and the standard deviation of the raw data was 0.0814. Applying the Doppler ratio derived from the whale velocity (right-hand side of equation 1), we obtained a base (*i.e.* non-shifted) frequency of 26.181 Hz, and a standard deviation of 0.0784.

Discussion. Simultaneous observation of whale movement and acoustic recordings provided an opportunity to test the degree to which the Doppler effect was responsible for frequency variation in calls recorded from an Antarctic blue whale. The observed relationship between speed and peak-frequency (0.021 Hz m^{-1} s) was significant (p = 0.039) and was also very similar to that predicted by the Doppler effect (0.018 Hz m^{-1} s). Furthermore, by 'correcting' the raw observations of peak-frequency for Doppler effects, the standard deviation of the data was reduced from 0.0814 to 0.0784 Hz demonstrating that we were able to remove the Doppler effect in order to better estimate the 'true' peak-frequency emitted by the whale. However, the variance in measured peak-frequency was greater than would be expected to occur from only Doppler effects due to motion of the whale. This suggests that factors in addition to Doppler shift were responsible for the variation in peak-frequency between independent calls and that these factors dominated the variance.

Change of tonal frequency in blue whale calls may derive from a number of physical factors that are not mutually exclusive. Urick (1983) indicated that both frequency shift and dispersion arise not only from Doppler shift, but also from reverberation of sound as it reflects off the moving sea surface [24]. He further noted that there appeared to be a complex relationship between reverberation, frequency shift, frequency dispersion and wind-speed. Thus whilst the small amount of Doppler shift did undoubtedly occur from the motion of the whale, it appears that it is but one of several factors that contribute to frequency variation between individual calls.

In addition to physical factors in the environment that might have affected the peak-frequency itself, measurement error could also have added to the masking of the contribution of the Doppler effect. Given our careful consideration to use only calls with high-signal-to-noise ratio, the largest source of measurement error is likely to have arisen in estimation of velocities of the whale and sonobuoy. Velocities were estimated by interpolation of surface positions and thus are only an average rather than instantaneous representation of the underwater speed and course of the vocalising whale. Compounding this issue is the fact that the observed swim speeds were all in the same narrow range of approximately 1–2 m/s. Measurement errors in estimating the velocity (of either the whale or sonobuoy) would be expected to increase the deviation of the measured peak-frequency from that predicted by Doppler, but would not necessarily be expected to yield the level of variation observed in the vessel-based measurements. Furthermore, our observed slope of 0.021 Hz m s^{-1} was very similar to that of 0.018 Hz m s^{-1} predicted to arise from

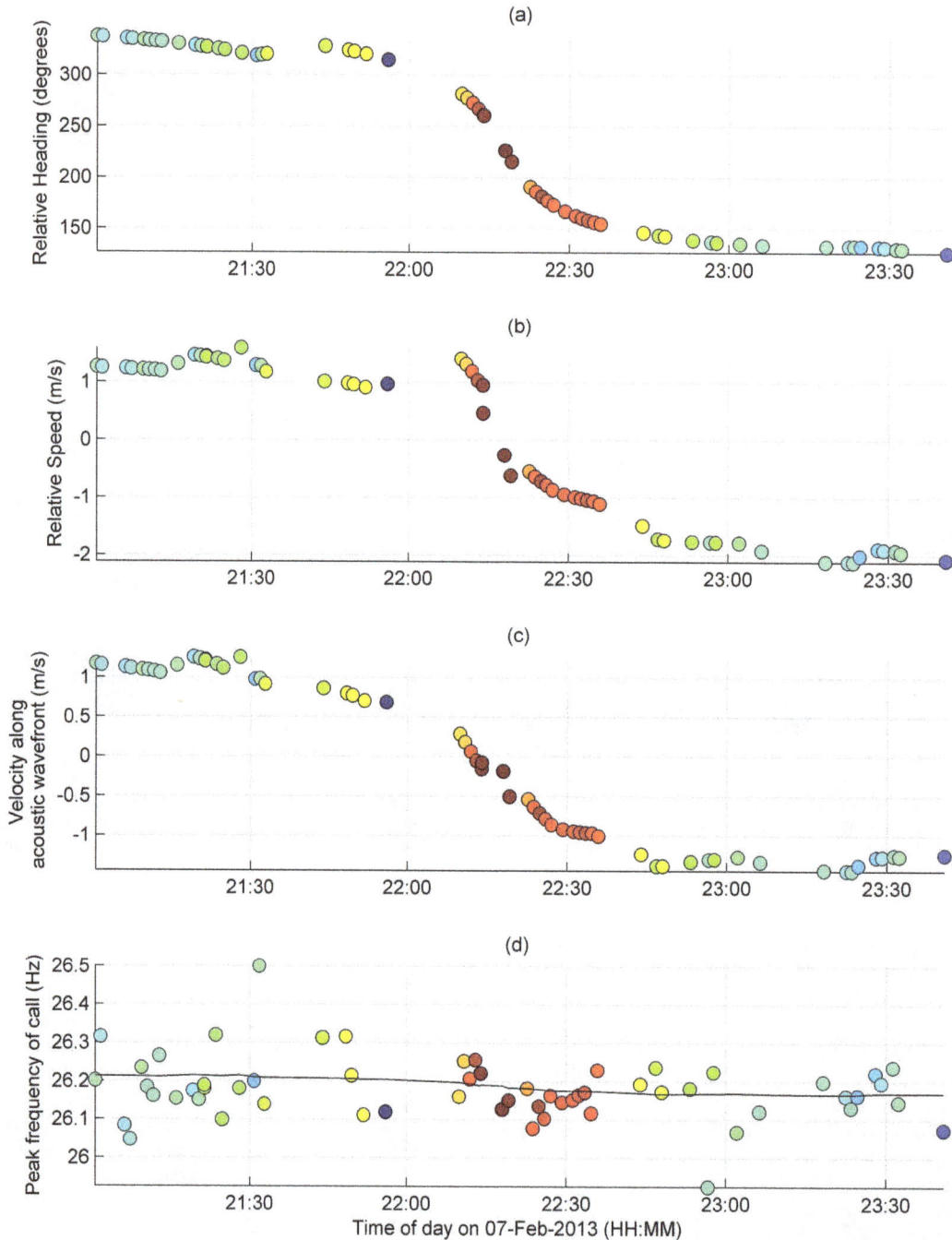

Figure 5. Time series of whale movements. Time series of whale movements shown at the times when z-calls were detected (filled circles). (a) Bearing from sonobuoy to whale. (b) Relative speed between the whale and the buoy. (c) The component of whale velocity in the direction of the acoustic wavefront; (d) Peak-frequency of whale call. The black line in (d) corresponds to the prediction from Equations 2 and 3. Colour of circles corresponds to received level of call as per Figure 4.

Doppler shifts, indicating that measurement errors in both speed and peak-frequency were reasonably small and relatively unbiased.

Lastly, the inherent precision of the whale's sound production was likely a substantial source of variability in peak-frequency. While physical factors and acoustic measurement errors may also contribute to variability, the likelihood that a whale will produce vocalisations which vary in frequency from one call to the next is potentially the largest driver of variation in peak-frequency. While the range of observed peak-frequencies was very small (approx-

imately 0.25 Hz) this range of peak-frequencies is nearly twice as large as the inter-annual decline of 0.135 Hz [4]. Neither the degree to which whales control the pitch of their song (nor the ability of the intended recipient to perceive differences in pitch of said song) have been quantified to date, but further discussion of models of sound production and perception can be found in the following section on whale anatomy and sound production.

Despite these limitations, our results highlight the benefits of combined visual and acoustic observations and demonstrate that

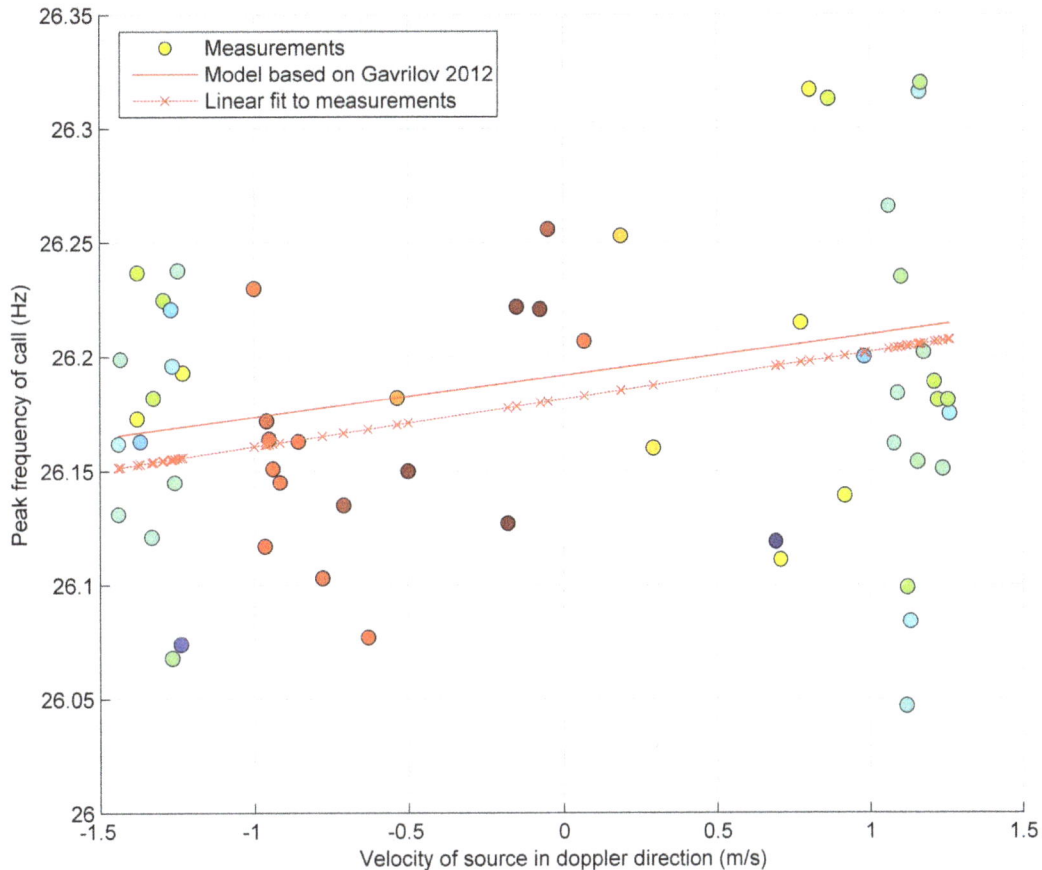

Figure 6. Relationship between observed frequency and movements. Peak-frequency as a function of the velocity of the whale in the direction of the receiver. Filled circles show measured values and colours indicate received level as per Figure 4. Solid line represents the expected frequency shift derived from Equations 2 and 3 (f_w = 26.192; slope = 0.018). Dashed line represents a linear fit to the measurements (f_w = 26.182; slope = 0.021; R^2 = 0.07 (p = 0.039).

we are able to describe the variance in peak-frequency having removed the effect of Doppler shift on the received signals. To our knowledge, the data presented here represent the first successful attempt to measure the Doppler effect in any cetacean vocalisation.

Obtaining more underwater tracks, ideally of higher accuracy and over a wider range of velocities, could help to reduce these confounding effects. Time-depth recorders with yaw-pitch-roll sensors, and acoustic recording capability such as the DTAG or Acousonde could provide one such way to obtain more accurate underwater tracks, and these instruments would also allow comparison of recordings from an instrument moving on the whale with a stationary one. Sonobuoys with integrated GPS receivers and telemetry would also greatly improve the estimation of buoy velocity. Finally, data fusion algorithms could be used to combine position information from video-tracks, DIFAR sonobuoys, acoustic time-depth recording tags, time-differences-of-arrival of sound, and possibly multipath [25,26] in order to obtain more accurate tracks from the existing and future data sets.

B. Long-term observations

1. Results. The peak-frequencies at each of the long-term recording sites (Figure 3) were compared with the long-term trend in frequency (equation 3) in order to obtain a time series of frequency ratios (i.e. left hand side of equation 1) for each site. Computation of the frequency ratios enabled comparison of the

intra-annual trend in frequency among all three sites while accounting for differences caused by the long-term decline in recordings from different years. At all three recording sites the frequency ratios followed the same cyclical pattern over the year, with ratios greater than one more likely to occur from March through June; ratios remaining near one in July and August, and ratios less than one occurring in September and October (Figure 7). The mean annual frequency ratio using measurements from all three sites was 1.0009 with 95% interval between 0.9901 and 1.0077. Mean monthly ratios using data from all sites combined ranged between 0.9971 (October) and 1.0038 (April) (black solid line in Figure 7). Mean monthly ratios and standard deviations showed increased variability compared to the annual mean due to smaller sample size, especially during summer months.

Linear regression revealed correlation between the monthly measurements of blubber thickness and the monthly variation in peak-frequency, but only for male blue whales less than 19 m in length (intercept = −1.55; slope = 4.77; R2 = 0.92; p = 0.004; Figure 8). There was no correlation between blubber thickness and monthly variation in peak-frequency for male blue whales greater than 23 m in length (intercept = −0.23; slope = 0.69; R2 = 0.278; p = 0.594).

2. Intra-annual trend in frequency-ratio. Gavrilov et al., described the intra-annual frequency pattern as declining from March to December and then "resetting" the following March [4].

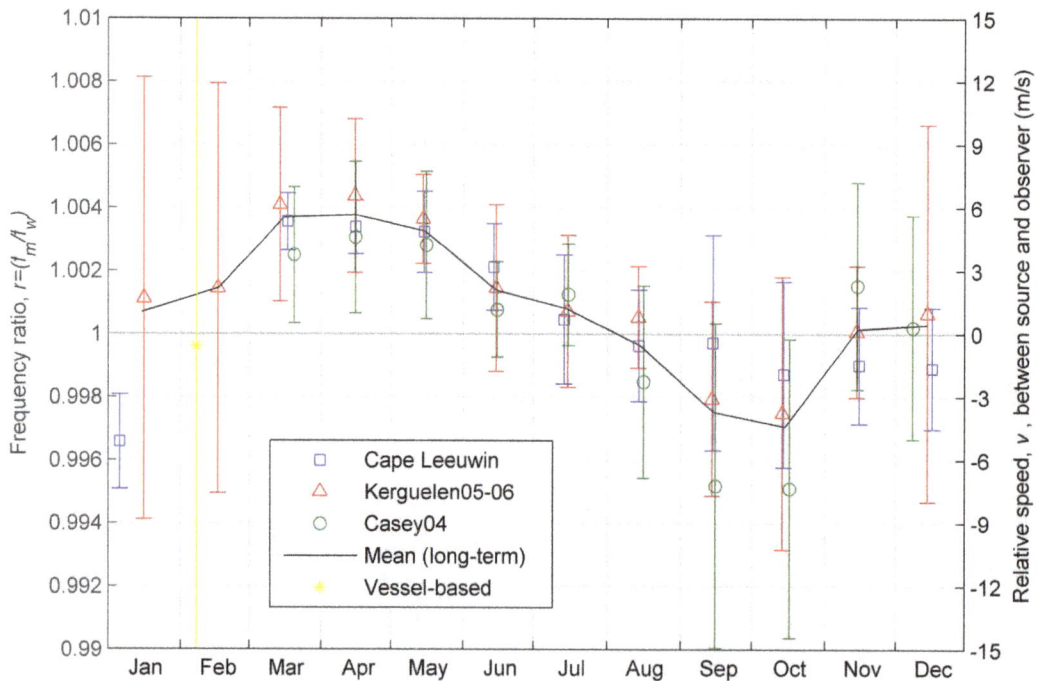

Figure 7. Monthly observations of frequency shift. Markers show the ratio of measured to 'true' frequency of Antarctic blue whale song. Measured frequency and 'true' frequency are calculated from the data from (Gavrilov et al. 2012) and monthly means are pooled from 9 years of acoustic observations (blue dots). The Antarctic recording stations Kerguelen (red triangle), and Casey (green circle) comprise 2 and 1 years of acoustic observations. Error bars show the monthly standard deviation. The black line connects the monthly mean of all observations from all of the long-term recording stations. The yellow star shows the mean of the vessel-based measurements with error bars denoting one standard deviation (note that error bars for the vessel-based observations extend well beyond the range of the vertical axis for this figure).

This sharp "resetting" may have resulted from lack of acoustic observations and measurements at Cape Leeuwin during January and February. By including data from the Kerguelen plateau, we observed a more gradual increase in frequency over January and February that leads to this apparent "reset." This gradual increase in frequency over the austral summer fleshes out the overall intra-

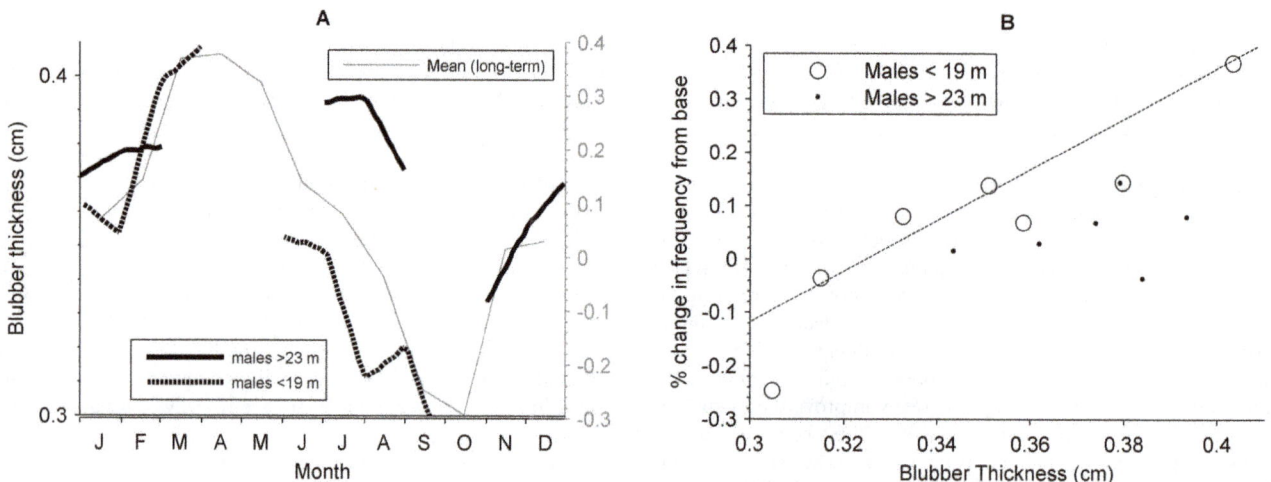

Figure 8. Relationship between blubber and tonal frequency. Seasonal changes around the base frequency measured in this study correlate with seasonal changes in blubber thickness measured by Mackintosh and Wheeler (1929) [23], particularly for males less than 19 m. (**A**) Time series of intra-annual variation in frequency ratio and blubber thickness. Gray line (right vertical axis) represents the monthly change in frequency ratio (equation 6) measured for all recording sites, while black solid and dashed lines are a summary blubber thickness measurements (left vertical axis) digitised from Mackintosh et al., (1929) [23]. (**B**) Relationship between blubber thickness and intra-annual measurements of peak-frequency. Dots represent whales greater than 23 m in length, while open circles represent whales less than 19 m in length again with blubber thickness digitised from Mackintosh et al., (1929). Dashed line shows the least-squares fit at all locations to males less than 19 m weighted by the inverse variance of the monthly frequency ratio (intercept -1.55; slope $= 4.77$; $R^2 = 0.92$; $p = 0.004$). Males greater than 23 m did not have a significant relationship, so no trend line is shown (intercept $= -0.23$; slope $= 0.69$; $R^2 = 0.28$; $p = 0.59$).

annual pattern with a more sinusoidal rather than sawtooth appearance.

3. Doppler effect. The mean swimming speeds estimated from frequency shift were within the range of plausible speeds for blue whales at all three locations [27–30]. However, mean monthly speeds in April and October appear to be too high to be maintained. If they were maintained, then migration would be completed in a matter of days.

Furthermore, the Doppler effect occurs due to the relative speed between the source and the receiver in the direction of the acoustic wavefront, not the absolute speed of the source. This implies that the maximum frequency shifts will occur at the whale's top speed only when the whale's course is directly towards or away from the receiver. If the Doppler effect were responsible for the similar frequency trends in the Antarctic and off Australia, then whales must be simultaneously moving towards both Antarctic and sub-tropical receivers at equal speeds, and this is not consistent with any plausible migration route. Consequently, we believe that it is highly unlikely that the intra-annual pattern in frequency is primarily a result of the Doppler effect during migration.

4. Changes in whale anatomy. After removing the long-term trend, the lowest peak-frequencies produced by whales occurred in October, while the highest occurred in April. The timing of these minima and maxima of peak-frequencies seems to loosely correspond with the arrival to and departure from the Antarctic feeding grounds [6], and thus supports the hypothesis that intra-annual frequency shift may be caused in-part by changes in body condition. Arrival of Antarctic blue whales to the Antarctic feeding grounds is believed to begin in September and increase through December with whales potentially continuing to arrive in the Antarctic into February ([6,31], W.K. de la Mare unpublished data). Peak-frequency of song also increases from October until March in conjunction with arrival (and presumably feeding) in the Antarctic. By April, most of the whales that will migrate are believed to have departed from the Antarctic [6], and peak-frequency decreases during this time as singers are presumably away from their main feeding grounds.

In addition to the co-occurrence of the extrema of peak-frequency with the arrival and departure of whales to the Antarctic, the gradual variation in mean frequency from month-to-month and the increased variability as whales return to the Antarctic also supports a link between intra-annual frequency patterns and whale anatomy (i.e. body condition). Furthermore, linear regression reveals that the cyclical intra-annual pattern in tonal frequency appears to match that of blubber thickness for male blue whales [23], but only those less than 19 m in length (Figure 8). While there is admittedly a temporal disparity between these two data sets (collected nearly a century apart) and presently a lack of understanding of a causal mechanism linking blubber thickness to tonal frequency, this correlation is intriguing and worthy of further investigation.

5. Sound production, tonal frequency, and intensity. While we cannot rule out a purely behavioural reason for the intra-annual change in frequency, throughout the year the mean variation by month rarely exceeds 0.5% of the "base" frequency for that year. At such low frequencies it is unknown if blue whales, like bottlenose dolphins [32], can perceive a difference in frequency of 0.5% despite indications that they have a hypertrophied cochlea indicative of acute low-frequency hearing [33]. However the change in the mean-monthly peak-frequency throughout the year is less than variation between calls observed during an hour of vessel-based measurements of a single whale. If an individual exhibits this much variability between calls in such a short period of time, it seems unlikely that the observed longer term seasonal pattern of such small shifts in peak-frequency is a result of intentional behavioural changes by all vocalising whales.

In further investigations of intra-annual frequency trends of blue whale song, it may be desirable to consider the intensity (i.e. source levels) of calls and the density of blue whales in addition to the number of calls detected. McDonald et al., proposed that calls with lower peak-frequencies would have lower source levels and should occur when population density is high [3]. Catch data indicates that peak-densitiy of blue whales in the Antarctic occurs in December ([33]; W. K. de la Mare unpublished data) or February [6], thus our observations of lowest peak-frequencies in October, rather than December-February suggest that the intra-annual change in frequency may not necessarily be driven by the same factors that McDonald et al. proposed as the reasons for long-term decline [3].

Sound production in blue whales is not well understood, and initial theories [34,35] do not appear to satisfactorily describe the mechanism, observed frequency content, and source levels of blue whale sounds [36]. New models of sound production have recently been proposed for mysticetes [36] and tested for humpback whales [37], but remain untested on blue whales. Adam et al. suggest that their model of sound production for humpbacks not only accounts for both the low tonal frequencies, high-source levels, and long duration, but also the high repetition rate of these calls [37]. However, further data on source-levels, density of whales, and whale behaviour (i.e. the purpose of song) would be required to test the hypotheses of Adam et al. [37] and McDonald et al. [3] for Antarctic blue whales.

While we have detailed a clear seasonal pattern in tonal frequency of Antarctic blue whale calls, it remains to be seen whether these intra-annual patterns, like the long-term decline [3], also occur in other populations of blue whales. Although there are hints that similar intra-annual variation in frequency may occur in southeast Indian ocean pygmy blue whales (Balaenoptera musculus brevicauda) [38], further investigation and quantification of these patterns for other populations of blue whales is required. Comparative studies across different populations may yield further insights into the cause(s) of these seasonal variations.

Conclusions

Variation in the peak-frequency of Antarctic blue whale calls was measured from vessel-based recordings in the Antarctic. This variation was significantly correlated with, but also much greater than, the level that would be predicted by the Doppler effect. This suggests that, at least at low speeds, factors other than the Doppler effect are likely to be the predominant drivers of the seasonal variation in peak-frequency of Antarctic blue whale calls. Furthermore, the fact that the same intra-annual pattern was observed off Cape Leeuwin, Casey Station, and the Kerguelen Plateau makes it unlikely that Doppler shifts coincident with migration are responsible for the intra-annual variation in blue whale peak frequencies. However, this same fact also makes it unlikely that the physical environment (e.g. water temperature, salinity, etc.) is responsible for the pattern, barring extremely long-range acoustic propagation. Thus changes in whale behaviour, or more likely body condition, remain the most parsimonious explanations for the observed intra-annual pattern.

Our results indicate that seasonal patterns in tonal frequency may also yield biological insight into the life-history of Antarctic blue whales complementary to historical [8–10,39,40] and ongoing [41] studies of the spatial variation and seasonality of acoustic detections. Future studies of intra-annual variation in tonal frequency of blue whale song should consider correcting for

Doppler effects, but may only need to do so in situations where whales are moving at high speeds. Further acoustical studies of whale migration should focus on more precise estimates of the number of calling whales, measurements of the intensity (as well as propagation loss and source level of calls) and supplementing acoustical data with anatomical measurements such as length (e.g. [42–44]), girth and body condition (e.g. [45–47]).

Acknowledgments

Thanks to all the scientists and logistics staff at the Australian Antarctic Division and Australian Marine Mammal Centre for their hard work and support both during the 2013 Antarctic Blue Whale Voyage and during the deployment and recovery of moored acoustic recorders. Thanks to John Hildebrand, Sean Wiggins and Scripps Institution of Oceanography for the provision of the long term autonomous Acoustic Recording Packages (ARPs) used in collecting the acoustic data off Casey and the Kerguelen Plateau. Thanks to the excellent and professional crews of the Aurora Australis and the *FV Amaltal Explorer*. Thanks to Christopher Donald from Australian Defence for provision of the expired 53D sonobuoys used in this research. Particular thanks are due to Paul Ensor and Jay Barlow for their encouragement, keen visual observations, and excellent guidance of the ship during close approach and video tracking. Special thanks to Bill de la Mare, Mike Double, Elanor Miller and Victoria Wadley for their feedback and encouragement throughout the development of this manuscript.

Author Contributions

Conceived and designed the experiments: BSM RL SC JG. Performed the experiments: BSM RL SC JG. Analyzed the data: BSM RL SC JG. Contributed reagents/materials/analysis tools: BSM RL SC JG. Wrote the paper: BSM RL SC JG.

References

1. Širović A, Hildebrand JA, Wiggins SM, McDonald MA, Moore SE, et al. (2004) Seasonality of blue and fin whale calls and the influence of sea ice in the Western Antarctic Peninsula. Deep Sea Res Part II Top Stud Oceanogr 51: 2327–2344. doi:10.1016/j.dsr2.2004.08.005.

2. Rankin S, Ljungblad DK, Clark CW, Kato H (2005) Vocalisations of Antarctic blue whales, *Balaenoptera musculus intermedia*, recorded during the 2001/2002 and 2002/2003 IWC/SOWER circumpolar cruises, Area V, Antarctica. J Cetacean Res Manag 7: 13–20.

3. McDonald MA, Hildebrand JA, Mesnick S (2009) Worldwide decline in tonal frequencies of blue whale songs. Endanger Species Res 9: 13–21. doi:10.3354/esr00217.

4. Gavrilov AN, McCauley RD, Gedamke J (2012) Steady inter and intra-annual decrease in the vocalization frequency of Antarctic blue whales. J Acoust Soc Am 131: 4476–4480. doi:10.1121/1.4707425.

5. Ballot B (1845) Akustische Versuche auf der Niederländischen Eisenbahn, nebst gelegentlichen Bemerkungen zur Theorie des Hrn. Prof. Doppler. Ann der Phys und Chemie 11: 321–351.

6. Mackintosh NA (1966) The distribution of southern blue and fin whales. In: Norris KS, editor. Whales, dolphins and porpoises. Berkeley and Los Angeles: University of California Press. pp. 125–145.

7. Branch TA, Stafford KM, Palacios DM, Allison C, Bannister JL, et al. (2007) Past and present distribution, densities and movements of blue whales Balaenoptera musculus in the Southern Hemisphere and northern Indian Ocean. Mamm Rev 37: 116–175. doi:10.1111/j.1365-2907.2007.00106.x.

8. Stafford KM, Bohnenstiehl DR, Tolstoy M, Chapp E, Mellinger DK, et al. (2004) Antarctic-type blue whale calls recorded at low latitudes in the Indian and eastern Pacific Oceans. Deep Sea Res Part I Oceanogr Res Pap 51: 1337–1346. doi:10.1016/j.dsr.2004.05.007.

9. Samaran F, Adam O, Guinet C (2010) Discovery of a mid-latitude sympatric area for two Southern Hemisphere blue whale subspecies. Endanger Species Res 12: 157–165. doi:10.3354/esr00302.

10. Samaran F, Stafford KM, Branch TA, Gedamke J, Royer J-Y, et al. (2013) Seasonal and Geographic Variation of Southern Blue Whale Subspecies in the Indian Ocean. PLoS One 8: e71561. doi:10.1371/journal.pone.0071561.

11. Samaran F, Adam O, Guinet C (2010) Detection range modeling of blue whale calls in Southwestern Indian Ocean. Appl Acoust 71: 1099–1106. doi:10.1016/j.apacoust.2010.05.014.

12. Greene CRJ, McLennan MW, Norman RG, McDonald TL, Jakubczak RS, et al. (2004) Directional frequency and recording (DIFAR) sensors in seafloor recorders to locate calling bowhead whales during their fall migration. J Acoust Soc Am 116: 799–813. doi:10.1121/1.1765191.

13. Sullivan EJ, Holmes JD, Carey WM, Lynch JF (2006) Broadband passive synthetic aperture: Experimental results. J Acoust Soc Am 120: EL49. doi:10.1121/1.2266024.

14. Double MC, Barlow J, Miller BS, Olson P, Andrews-Goff V, et al. (2013) Cruise report of the 2013 Antarctic blue whale voyage of the Southern Ocean Research Partnership. Report SC65a/SH/21 submitted to the Scientific Committee of the International Whaling Commission. Jeju Island, Republic of Korea.

15. Leaper R, Gordon JC (2001) Application of photogrammetric methods for locating and tracking cetacean movements at sea. J Cetacean Res Manag 3: 131–141.

16. Miller BS, Barlow J, Calderan S, Collins K, Leaper R, et al. (2013) Long-range acoustic tracking of Antarctic blue whales. Rep Submitt to Sci Comm Int Whal Comm SC/65a/SH1: 1–17.

17. Maranda BH (2001) Calibration Factors for DIFAR Processing.

18. Miller BS, Collins K, Barlow J, Calderan S, Leaper R, et al. (2014) Blue whale vocalizations recorded around New Zealand: 1964–2013. J Acoust Soc Am 135: 1616–1623. doi:10.1121/1.4863647.

19. Miller BS, Gedamke J, Calderan S, Collins K, Johnson C, et al. (2014) Accuracy and precision of DIFAR localisation systems: Calibrations and comparative measurements from three SORP voyages. Submitt to Sci Comm 65b Int Whal Comm Bled, Slov SC/65b/SH08: 14.

20. Gillespie D, Mellenger DK, Gordon JC, Mclaren D, Mchugh R, et al. (2008) PAMGUARD: Semiautomated, open source software for real-time acoustic detection and localisation of cetaceans. Proceedings of the Institute of Acoustics. Southhampton, UK: Conference on Underwater Noise Measurement, Impact and Mitigation 2008., Vol. 30. pp. 54–62.

21. Miller BS, Wotherspoon S, Calderan S, Leaper R, Collins K, et al. (2014) Estimating drift of DIFAR sonobuoys when localizing blue whales. Submitt to Sci Comm 65b Int Whal Comm Bled, Slov SC/65b/SH09: 8.

22. Wiggins S (2003) Autonomous acoustic recording packages (ARPs) for long-term monitoring of whale sounds. Mar Technol Soc J 37: 13–22.

23. Mackintosh N, Wheeler J, Clowes A (1929) Southern blue and fin whales. Discov Reports 1: 257–540.

24. Urick RJ (1983) Principles of underwater sound. 3rd ed. New York: McGraw-Hill.

25. Nosal E-MM, Frazer LN (2007) Sperm whale three-dimensional track, swim orientation, beam pattern, and click levels observed on bottom-mounted hydrophones. J Acoust Soc Am 122: 1969. doi:10.1121/1.2775423.

26. Valtierra RD, Glynn Holt R, Cholewiak D, Van Parijs SM (2013) Calling depths of baleen whales from single sensor data: Development of an autocorrelation method using multipath localization. J Acoust Soc Am 134: 2571–2581. doi:10.1121/1.4816582.

27. Bailey H, Mate B, Palacios D, Irvine L, Bograd S, et al. (2009) Behavioural estimation of blue whale movements in the Northeast Pacific from state-space model analysis of satellite tracks. Endanger Species Res 10: 93–106. doi:10.3354/esr00239.

28. Oleson EM, Calambokidis J, Burgess WC, McDonald MA, LeDuc CA, et al. (2007) Behavioral context of call production by eastern North Pacific blue whales. Mar Ecol Prog Ser 330: 269–284.

29. Sears R, Perrin WF (2009) Blue whale. pp 112–116. In: Perrin WF, Würsig B, Thewissen JGM, editors. Encyclopedia of Marine Mammals. New York: Academic Press. pp. 120–124.

30. Andrews-Goff V, Olson PA, Gales NJ, Double MC (2013) Satellite telemetry derived summer movements of Antarctic blue whales. Report SC/65a/SH03 submitted to the Scientific Committee of the International Whaling Commision. Jeju Island, Republic of Korea.

31. De la Mare WK (2014) Estimating relative abundance of whales from historical Antarctic whaling records. Can J Fish Aquat Sci 71: 106–119. doi:10.1139/cjfas-2013-0016.

32. Thompson R, Herman L (1975) Underwater frequency discrimination in the bottlenosed dolphin (1–140 kHz) and the human (1–8 kHz). J Acoust Soc Am 57: 943–948.

33. Ketten DR (1997) Structure and function in whale ears. Bioacoustics Int J Anim Sound its Rec 8: 103–135.

34. Thode AM, D'Spain GL, Kuperman WA (2000) Matched-field processing, geoacoustic inversion, and source signature recovery of blue whale vocalizations. J Acoust Soc Am 107: 1286–1300.

35. Aroyan J, McDonald MA, Webb SC, Hildebrand JA, Clark D, et al. (2000) Acoustic models of sound production and propagation. In: Au WWL, Popper AN, Fay RR, editors. Hearing by whales and dolphins. Springer. pp. 409–469.

36. Reidenberg JS, Laitman JT (2007) Discovery of a low frequency sound source in Mysticeti (baleen whales): anatomical establishment of a vocal fold homolog. Anat Rec (Hoboken) 290: 745–759. doi:10.1002/ar.20544.

37. Adam O, Cazau D, Gandilhon N, Fabre B, Laitman JT, et al. (2013) New acoustic model for humpback whale sound production. Appl Acoust 74: 1182–1190. doi:10.1016/j.apacoust.2013.04.007.

38. Gavrilov AN, Mccauley RD, Salgado-Kent C, Tripovich J, Burton CLK (2011) Vocal characteristics of pygmy blue whales and their change over time. J Acoust Soc Am 130: 3651–3660. doi:10.1121/1.3651817.

39. Širović A, Hildebrand JA, Wiggins SM, Thiele D (2009) Blue and fin whale acoustic presence around Antarctica during 2003 and 2004. Mar Mammal Sci 25: 125–136. doi:10.1111/j.1748-7692.2008.00239.x.

40. Gedamke J, Gales N, Hildebrand JA, Wiggins S (2007) Seasonal occurrence of low frequency whale vocalisations across eastern Antarctic and southern Australian waters, February 2004 to February 2007. Report SC/59/SH5 submitted to the Scientific Committee of the International Whaling Commission. Anchorage, Alaska.

41. SORP Acoustic Trends Project (2014). Available: http://www.marinemammals.gov.au/sorp/antarctic-blue-whales-and-fin-whales-acoustic-program. Accessed 2014 Feb 28.

42. Growcott A, Miller BS, Sirguey P, Slooten E, Dawson SM (2011) Measuring body length of male sperm whales from their clicks: the relationship between inter-pulse intervals and photogrammetrically measured lengths. J Acoust Soc Am 130: 568–573. doi:10.1121/1.3578455.

43. Gordon JC (1990) A simple photographic technique for measuring the length of whales from boats at sea. Reports Int Whal Comm 40: 581–587.

44. Jaquet N (2006) A simple photogrammetric technique to measure sperm whales at sea. Mar Mammal Sci 22: 862–879. doi:10.1111/j.1748-7692.2006.00060.x.

45. Moore M, Miller C, Morss M (2001) Ultrasonic measurement of blubber thickness in right whales. J Cetacean Res Manag Special Is: 301–309.

46. Nousek-McGregor AE, Miller CA, Moore MJ, Nowacek DP (2014) Effects of body condition on buoyancy in endangered north atlantic right whales *. Physiol Biochem Zool 87: 160–171. doi:10.1086/671811.

47. Miller C, Reeb D, Best P, Knowlton A, Brown M, et al. (2011) Blubber thickness in right whales Eubalaena glacialis and Eubalaena australis related with reproduction, life history status and prey abundance. Mar Ecol Prog Ser 438: 267–283. doi:10.3354/meps09174.

15

Mercury and Selenium in Stranded Indo-Pacific Humpback Dolphins and Implications for Their Trophic Transfer in Food Chains

Duan Gui[1], Ri-Qing Yu[2], Yong Sun[1], Laiguo Chen[3], Qin Tu[1], Hui Mo[4], Yuping Wu[1]*

1 Guangdong Provincial Key Laboratory of Marine Resources and Coastal Engineering, School of Marine Sciences, Sun Yat-Sen University, Guangzhou, China, **2** Department of Biology, University of Texas at Tyler, Tyler, Texas, United States of America, **3** Urban Environment and Ecology Research Center, South China Institute of Environmental Sciences (SCIES), Ministry of Environmental Protection, Guangzhou, China, **4** South China Botanical Garden, Chinese Academy of Sciences, Guangzhou, China

Abstract

As top predators in the Pearl River Estuary (PRE) of China, Indo-Pacific humpback dolphins (*Sousa chinensis*) are bioindicators for examining regional trends of environmental contaminants in the PRE. We examined samples from stranded *S. chinensis* in the PRE, collected since 2004, to study the distribution and fate of total mercury (THg), methylmercury (MeHg) and selenium (Se) in the major tissues, in individuals at different ages and their prey fishes from the PRE. This study also investigated the potential protective effects of Se against the toxicities of accumulated THg. Dolphin livers contained the highest concentrations of THg (32.34 ± 58.98 μg g^{-1} dw) and Se (15.16 ± 3.66 μg g^{-1} dw), which were significantly different from those found in kidneys and muscles, whereas the highest residue of MeHg (1.02 ± 1.11 μg g^{-1} dw) was found in dolphin muscles. Concentrations of both THg and MeHg in the liver, kidney and muscle of dolphins showed a significantly positive correlation with age. The biomagnification factors (BMFs) of inorganic mercury (Hg$_{inorg}$) in dolphin livers (350$\times$) and MeHg in muscles (18.7$\times$) through the prey fishes were the highest among all three dolphin tissues, whereas the BMFs of Se were much lower in all dolphin tissues. The lower proportion of MeHg in THg and higher Se/THg ratios in tissues were demonstrated. Our studies suggested that *S. chinensis* might have the potential to detoxify Hg via the demethylation of MeHg and the formation of tiemannite (HgSe) in the liver and kidney. The lower threshold of hepatic THg concentrations for the equimolar accumulation of Se and Hg in *S. chinensis* suggests that this species has a greater sensitivity to THg concentrations than is found in striped dolphins and Dall's porpoises.

Editor: Dwayne Elias, Oak Ridge National Laboratory, United States of America

Funding: This research was supported by the National Natural Science Foundation of China (Grant no. 41276147); a Science and Research Project of the Marine Non-profit Industry (Grant no. 201105011-5); the Ocean Park Conservation Foundation, Hong Kong; the National Key Technology R & D Program (Grant no. 2011BAG07B05-3); and the Sousa chinensis Conservation Action Project from the Administrator of Ocean and Fisheries of Guangdong Province, China. The funders had no role in study design, data collection and analysis, decision to publish, or preparation of the manuscript.

Competing Interests: The authors have declared that no competing interests exist.

* Email: exwyp@mail.sysu.edu.cn

Introduction

Mercury (Hg) in its inorganic form is a ubiquitous pollutant that is globally distributed by atmospheric transportation. Less toxic Hg(II) in environments can easily be converted into the highly toxic methylmercury (MeHg), primarily by sulfate- and iron-reducing bacteria and methanogens [1–7] through the putative Hg methylation genes *hgc*A and *hgc*B via the acetyl CoA pathway [5]. MeHg is a strong neurotoxic substance; once it is bioavailable in aquatic ecosystems, it can be bioaccumulated and biomagnified quickly through aquatic food webs, which creates a health threat to humans and aquatic mammals such as dolphins [3].

The estuary of the Pearl River, which is the third longest river in China, is a traditional nursery for fisheries and provides an ideal habitat for Indo-Pacific humpback dolphins (*Sousa chinensis*). The Pearl River Estuary (PRE) region contains a group of cities that include Guangzhou, Shenzhen and Hong Kong, forming one of the largest local and global economic hubs in southern China. Rapid industrial development and urbanization in recent decades have undermined the habitats of local fish and dolphins in the estuary. *S. chinensis* is considered one of the most endangered species (the National Key Species for Protection, Grade 1) in China. The Chinese government has delimitated the PRE as a national nature reserve for *S. chinensis*. Recent studies have indicated a decreasing trend of total mercury (THg) concentration in sediments with distance away from the PRE and toward the South China Sea [8] and showed an accelerated input of THg in sediment cores in recent decades [9]. These results suggest that THg contamination in this region has been strongly correlated with industrial development. Contamination by Hg(II) and MeHg was also observed in surface water in the tributary (e.g., Dong River) of the Pearl River Delta. However, THg and MeHg contamination in Indo-Pacific humpback dolphins and the interaction between THg toxicity and Se accumulation in their bodies have not been systematically studied in this ecosystem [10,11].

In aquatic ecosystems, dolphins are top predators that have long lifespans, which increases their potential to accumulate Hg(II) and MeHg [12]. MeHg is easily accumulated in the livers of cetaceans, likely as a result of the ability of the liver to store and biotransform toxic contaminants [13]. The accumulation of Hg(II) and MeHg might cause detrimental effects on the reproduction system, immune responses, central nervous system and organs such as the liver and kidneys [14,15].

Previous studies have reported that selenium (Se) can mitigate the toxicity of MeHg in the liver of cetaceans by forming a highly insoluble Se–Hg compound after demethylating MeHg [16,17]. The presence of the nontoxic Se–Hg compound was confirmed in the cytoplasm of hepatic cells in *Stenella coeruleoalba*, and an equimolar ratio between Se and THg was reported in liver tissue with a high THg concentration [18,19]. This phenomenon likely explains the ability of cetaceans to tolerate high THg concentrations without directly observable adverse effects and exhibit a low MeHg residue in the brain tissue. Se is a micronutrient that plays an important role in enzymes (e.g., glutathione peroxidase) in the maintenance of normal organ functions. Adverse biological effects occur when there is a deficiency of bioavailable Se, and an Hg-induced Se deficiency can lead to MeHg or THg toxicity [20]. The present investigation examined samples from dolphins stranded in the PRE, which have been collected since 2004, to study the distribution and fate of toxic pollutants (THg, MeHg and Se) in the major organs, in individuals at different ages and in their prey fishes from the PRE. The aims of the study were also to investigate the potential protective effects of Se against the toxicities of accumulated THg, and to provide evidence for the conservation of this endangered species.

Materials and Methods

Ethics Statement

This study on the Indo-Pacific humpback dolphins was approved by the Ministry of Agriculture of Chinese government under permit number 2003-54. The protocol was specifically verified by the Administration of Ocean and Fisheries of Guangdong Province, China under permit number 1999-583. No issue on ethics was concerned in this study.

Samples and chemical analysis

The tissue samples of *S. chinensis* were collected from dead and stranded animals along the PRE of the South China Sea from 2004 to 2012 (**Fig. 1**). Before performing the necropsies, the biological parameters were measured and tooth specimens were acquired to determine the age of specific individuals according to the method described by Jefferson [21] and Myrck et al. [22]. According to the report from Jefferson et al. [23], male dolphins inhabiting the PRE reach sexual maturity at 12–14 years, whereas females generally reach maturity at 9–10 years. Accordingly, the specimen assigned ID numbers from 21 to 28 in the present study were considered adults, and those with assigned numbers 1 to 20 were considered juveniles (**Table S1**). The liver, kidney and muscle tissue samples were placed in clean and acid-washed plastic bags and stored at −20°C immediately after collection. Information based on previous studies [21,24] indicated that the Indo-Pacific humpback dolphins have specific preferences for prey fishes. Based on that, 13 species of fish were collected from the PRE. Whole-body fish samples were smashed for metal analysis. All samples (dolphin and fish) were kept frozen until processing.

Samples were processed and prepared by a method similar to that of previous studies [25], in which portions of the different tissue samples were freeze-dried (Freeze-drying system, Labconco, Kansas City, Missouri, USA) for 48 h at 40–133×10^{-3} mBar and −49°C. The dried samples were then ground with an automatic agate mortar (Retsch, Germany) for 10 min. The concentration of THg in the dried tissue samples was measured without pretreatment or digestion by using the Hydra-C Automated Direct Hg Analyzer (Teledyne Instruments, Leeman Labs, USA). All specimens were analyzed in batches that included a procedural blank and standard reference material DORM-3 (National Research Council of Canada, Canada). The procedural blank and the reference material were treated and measured in the same way as the tissue samples. To analyze the MeHg concentration, tissue samples were digested in a KOH–methanol solution at 65°C for 4 h. The extractants were subjected to aqueous ethylation, separation by gas chromatography and detection by cold vapor atomic fluorescence spectrometry (CVAFS) (TEKRAN Model 2700 with an automated MERX purge and trap system, Brooks Rand Labs, USA) modified from USEPA method 1630 [26].

The digestion and preparation of dried tissue samples for Se analysis followed the methods described by Hung et al. [10]. Approximately 0.2 g of tissue samples was weighed in Teflon digestion tubes and soaked overnight in a mixture of 2 ml of double-distilled deionized water (3-D water) and 5 ml of 70% nitric acid (Merck, Germany). The digestion tubes were then sealed, placed in a microwave oven (Xin Tuo, model XT-9912, China) and subjected to a pressure increase of 65 psi for 15 m. After cooling, 2 ml H_2O_2 was added to the sample solution. The pressure was then increased to 65 psi for 15 m. The resulting digests were cooled and filtered through disposable syringe filter discs (0.45 μm pores, 25 mm diameter, with a mixed cellulose-ester filtering material, Jing Teng, China) equipped with 50-ml plastic syringes. The filtrates were transferred to 25-ml volumetric flasks and diluted with 3-D water. The samples were kept in acid-washed PVC tubes at 4°C prior to trace element analysis. The concentration of Se was measured by an inductively coupled plasma mass spectrometer (ICP-MS) (Agilent 7700, USA) with the procedural blank and reference TORT-2 included with every batch of tissue samples.

Quality assurance/quality control (QA/QC)

The precision and accuracy of the analytical methods were determined and monitored using the certified material TORT-2 (lobster hepatopancreas) and DORM-3 (fish protein) from the National Research Council of Canada. The recovery rates for THg, MeHg and Se were approximately 95.3%, 94.9% and 94.3%, respectively.

Statistical analysis

Statistical analyses were performed using SPSS (Statistical Package for the Social Sciences) software (StatSoft, ver. 22, USA), and the level of statistical significance was defined as $p < 0.05$. Grubbs' test was used to identify outliers, which were removed before further calculations. The Kolmogorov–Smirnoff test was used to assess the normality of the data distribution; if the data were not in normality, all the data were log_{10}-transformed to improve normality prior to analysis to best fit the underlying assumptions of the analysis of variance. The correlation between measured parameters was assessed by coefficient of determination (R^2). Differences in the concentrations of MeHg, THg and Se were analyzed among the groups (adult males, adult females, juvenile males and juvenile females) by using one-way ANOVA followed by Tukey's post-hoc tests. The contaminant concentrations of the

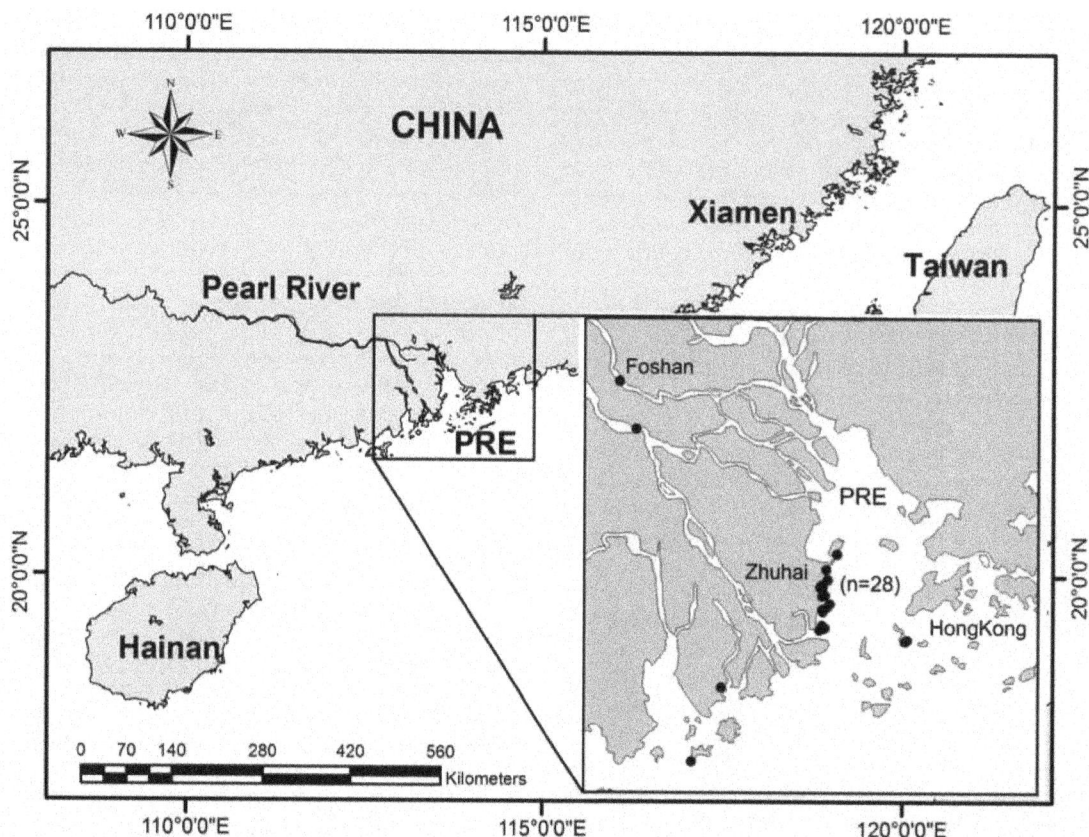

Figure 1. Sampling sites in the Pearl River Estuary (PRE) where the stranded Indo-Pacific humpback dolphins (n = 28) were collected from 2004 to 2012.

stranded dolphins that exhibited a large dispersion were analyzed by non-parametric tests.

Results

The total Hg, Se and MeHg concentrations and their related ratios were determined in the tissue samples from the liver (n = 28), kidney (n = 22) and muscle (n = 15) of Indo-Pacific humpback dolphins (**Table 1**). The mean concentrations of THg, Se and MeHg in 13 fish species from the PRE were summarized in **Table 2**. The average THg concentration in the prey fish species was 0.146 µg g^{-1} dw, with a range from 0.062 to 0.303 µg g^{-1} dw, and the average Se concentration was 2.23 µg g^{-1} dw, with a range from 1.93 to 3.34 µg g^{-1} dw. No significant differences in concentrations were found for the different fish species except the predatory species *Arius sinensis* and *Pampus argenteus*, which showed high THg concentrations. The potential biomagnification factor of inorganic mercury (Hg$_{inorg}$) in the dolphin tissue from the prey fishes was the highest in the liver tissue (350-fold) and the lowest in the muscle tissue (4.78-fold) (**Table 2**). For MeHg, dolphin muscle tissue had the highest biomagnification factor (18.7-fold), followed by liver tissue (14.3-fold) and kidney tissue (9.6-fold). Overall, Se in dolphin tissues showed the lowest potential for biomagnification of the three contaminants. For all dolphin individuals, muscle represents around 30% of total body mass, while other tissues (e. g., liver and kidney) contribute much less to the total body weight (<5%).

The highest mean concentration of THg was found in dolphin liver tissue (32.3±59.0 µg g^{-1} dw), whereas a lower concentration

was found in the kidneys (4.52±5.53 µg g^{-1} dw), with the smallest residue level shown in the muscle tissue (1.45±1.62 µg g^{-1} dw). The difference in THg accumulation among the three tissues was significant ($p<0.05$) (**Table 1**). Conversely, a significant difference in MeHg accumulation among different tissues was not found. Dolphin muscle tissue contained the highest MeHg concentration (1.02±1.11 µg g^{-1} dw), followed by liver (0.79±0.61 µg g^{-1} dw) and kidney (0.53±0.49 µg g^{-1} dw). Se concentrations displayed a similar profile as THg, with the highest mean value found in the liver tissue (15.2±19.4 µg sg^{-1} dw), followed by kidney (7.57±3.47 µg g^{-1} dw) and muscle tissue (1.89±1.69 µg g^{-1} dw).

The log$_{10}$-transformed concentrations of both THg and MeHg in the liver, kidney and muscle tissue of dolphins showed a significantly positive correlation with the log$_{10}$-transformed values of age (**Fig. 2A, B**). However, the slopes between tissue MeHg with age were much lower than those between THg and age for the three tissues. The MeHg/THg ratio in the livers and kidneys significantly decreased with age, while the MeHg/THg ratio in the muscles showed no trend with age (**Fig. 2C**). The mean percentage of MeHg/THg in the liver, kidney and muscle tissue was 18±15, 29±22 and 77±23, respectively (**Table 1**), indicating that the MeHg concentrations in the liver and kidney tissue represented 30% or less of the THg, which was generally lower than the MeHg/THg fraction that appeared in the prey fish (ranging from 18% to 83%, with an average of 42%, **Table 2**). The dolphin muscle tissue showed the highest MeHg/THg ratio. Se accumulation in the liver and kidney tissue significantly increased with age, whereas the concentration of Se in muscle tissue was not obviously affected by age (**Fig. 2D**).

Table 1. The mean concentrations and standard deviations (SD), in $\mu g\ g^{-1}$ dry weight, of total mercury (THg), selenium (Se) and methyl mercury (MeHg); the molar ratio (%) of Se to THg; and the percentage (%) of MeHg/THg in liver, kidney and muscle tissue of Indo-Pacific humpback dolphins from the Pearl River Estuary (PRE), China.

		Liver (n = 28)					Kidney (n = 22)					Muscle (n = 15)				
		THg	Se	MeHg	Se:THg	MeHg:THg	THg	Se	MeHg	Se:THg	MeHg:THg	THg	Se	MeHg	Se:THg	MeHg:THg
JM (n = 13)	Mean	4.24	5.63	0.49	8.51	25	1.27	5.67	0.26	37.7	43.4	0.58	1.52	0.48	10.6	78.7
	SD	5.12	3.54	0.35	5.90	14.5	1.65	2.86	0.21	27.8	22.2	0.40	0.42	0.36	6.67	12.9
JF (n = 7)	Mean	2.27	5.09	0.47	7.14	23.4	1.73	6.95	0.36	15.46	27.5	1.29	2.78	0.43	7.82	66.9
	SD	1.36	1.20	0.15	2.59	6.66	1.14	1.83	0.16	8.6	11.9	1.91	2.67	0.34	2.64	32.3
	p values	n.s.	n.s.	n.s.	n.s.	n.s.	n.s.	n.s.	n.s.	n.s.	n.s.	n.s.	n.s.	n.s.	n.s.	n.s.
AM (n = 3)	Mean	84.3	35.6	1.56	1.07	1.88	8.21	10.3	0.98	3.13	11	2.27	1.61	2.03	0.90	89.4
	SD	8.19	6.93	0.44	0.14	0.64	1.19	2.68	0.35	0.35	3.39	0	0	0	0	0
AF (n = 5)	Mean	116	41.8	1.56	0.99	1.89	13.4	11.1	1.06	2.29	7.42	3.18	1.43	2.72	0.79	85.6
	SD	81.8	24.9	0.46	0.11	0.91	5.16	2.45	0.60	0.47	2.18	1.05	0.24	0.94	0.21	6.47
	p values	n.s.	n.s.	n.s.	n.s.	n.s.	n.s.	n.s.	n.s.	n.s.	n.s.	n.s.	n.s.	n.s.	n.s.	n.s.
M (n = 16)	Mean	3.19	5.63	0.60	9.08	20.7	2.87	6.73	0.43	29.7	36.1	0.82	1.53	0.70	9.01	80.4
	SD	3.76	3.54	0.44	5.76	15.9	3.32	3.42	0.39	28.4	23.7	0.70	0.39	0.64	7.07	12.5
F (n = 12)	Mean	31.6	20.4	0.93	4.92	14.4	6.90	8.79	0.67	9.61	18.6	1.29	2.2	1.29	4.81	73.9
	SD	50.3	24.2	0.62	3.90	11.8	6.77	2.96	0.54	9.17	13.4	1.88	2.13	1.28	4.01	27.4
	p values	n.s.	0.016	n.s.	n.s.	n.s.	n.s.	n.s.	n.s.	n.s.	n.s.	n.s.	n.s.	n.s.	n.s.	n.s.
All (n = 28)	Mean	32.3	15.2	0.79	6.03	18.5	4.52	7.57	0.53	21.5	29.0	1.45	1.89	1.02	6.75	76.9
	SD	59.0	19.4	0.61	5.40	15.3	5.53	3.47	0.49	25.3	22.5	1.62	1.69	1.11	6.26	23.3

JM: juvenile male (<12 years); JF: juvenile female (<9 years); AM: adult male (>12 years); AF: adult female (>9 years).
n.s.: not significant.

Table 2. The mean concentrations and standard deviations (SD), in µg g^{-1} dry weight, of total mercury (THg), inorganic mercury (Hg$_{inorg}$), methylmercury (MeHg), selenium (Se), and the percentage (%) of MeHg/THg in the prey fishes (whole body) for Indo-Pacific humpback dolphins and their average biomagnification factors (BMFs) in the dolphin tissues collected from the Pearl River Estuary (PRE), China.

Species	Family	Sample number	THg		Hg$_{inorg}$		MeHg		Se		MeHg/THg
			Mean	SD	Mean	SD	Mean	SD	Mean	SD	Mean
Johnius belengerii	Sciaenidae	3	0.146	0.075	0.100	0.079	0.046	0.01	2.51	0.41	25
Collichthys lucidus	Sciaenidae	4	0.134	0.106	0.073	0.082	0.060	0.026	2.61	0.47	45
Clupanodon thrissa	Clupeidae	2	0.182	0.129	0.147	0.127	0.036	0.002	1.87	0.58	20
Coilia mystus	Engraulidae	5	0.158	0.152	0.121	0.144	0.037	0.008	1.99	0.43	23
Pampus argenteus	Stromateidae	3	0.221	0.111	0.189	0.119	0.032	0.008	2.22	0.22	15
Harpadon nehereus	Synodontidae	2	0.079	0.024	0.030	0.016	0.048	0.009	2.06	0.63	61
Cynoglossus bilineatus	Cynoglossidae	3	0.077	0.024	0.039	0.015	0.038	0.015	2.21	0.33	50
Sillago sihama	Sillaginidae	3	0.166	0.036	0.029	0.017	0.137	0.044	1.99	0.27	83
Arius sinensis	Ariidae	2	0.303	0.007	0.247	0.006	0.056	0.001	1.93	0.08	19
Selaroides leptolepis	Carangidae	2	0.118	0.065	0.079	0.058	0.039	0.007	1.99	0.40	33
Mugil cephalus	Mugilidae	2	0.062	0.002	0.033	0.005	0.030	0.003	1.94	0.32	48
Odontamblyopus rubicundus	Taenioididae	2	0.161	0.027	0.053	0.002	0.107	0.028	3.34	0.35	67
Odontamblyopus lacepedii	Taenioididae	3	0.077	0.000	0.033	0.009	0.044	0.009	2.39	0.53	57
Mean biomagnification factor	Dolphin liver				350		14.3		6.8		
Mean biomagnification factor	Dolphin kidney				44.3		9.6		3.4		
Mean biomagnification factor	Dolphin muscle				4.78		18.7		0.8		

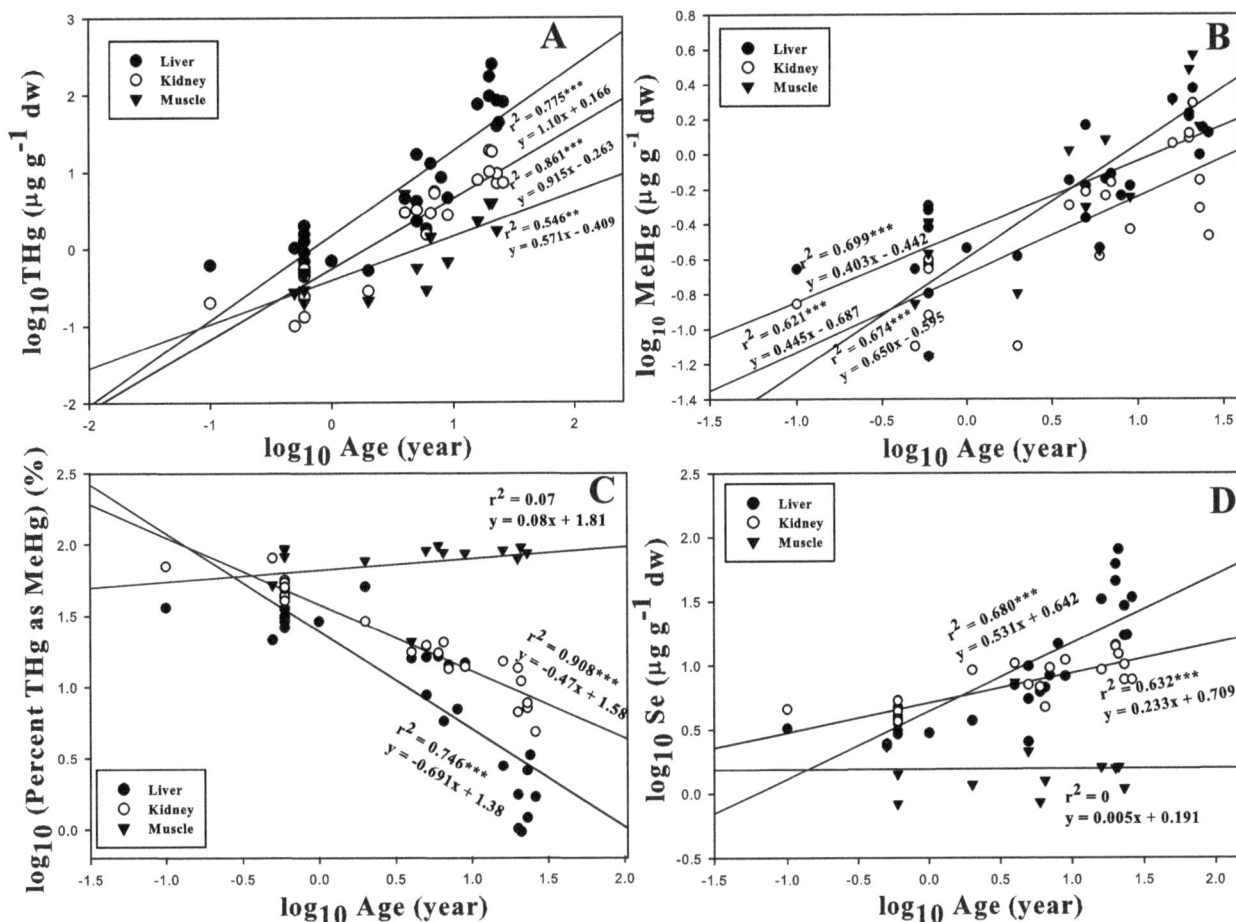

Figure 2. Relationships of dolphin age with the concentrations of THg, MeHg, Se and MeHg/THg ratio in the liver, kidney and muscle tissue, respectively, in the Indo-Pacific humpback dolphins stranded in the Pearl River Estuary (PRE) region.

The mean molar ratio of Se to THg was highest in the kidney tissue (21.5 ± 25.3) and the lowest in the liver (6.04 ± 5.40), with a medium ratio observed in the muscle tissue (6.75 ± 6.26) (**Table 1**). The molar ratios of Se/THg in the livers and kidneys of the juveniles were eight-fold higher than in the adult livers and 14-fold higher than in the adult kidneys, respectively. A significant positive correlation was found between the concentrations of Se and THg in both the livers and kidneys (**Fig. 3**A, B). The $\log_{10}$ Se/THg molar ratios showed strongly negative regressions with the $\log_{10}$ MeHg levels in the livers ($p<0.05$) and kidneys ($p<0.05$) (**Fig. 3**C, D). A significantly positive relationship between THg and MeHg was observed in the livers, with an inflection range of 8.4–16.9 µg g^{-1} dw of THg (**Fig. 4**).

No significant differences were found in the THg and MeHg concentrations in the liver and muscle tissues between males and females. Only the mean Se concentrations in the liver tissue were significantly different between males and females ($p<0.05$) (**Table 1**).

Discussion

Compared with the kidneys and muscles, extremely high concentrations of THg were found in the livers of *S. chinensis* (32.3 ± 59.0 µg g^{-1} dw), although the upper limits of the THg concentrations in the livers were lower than those found in dolphins from the waters of Hong Kong (906 µg g^{-1} dw) [27].

This phenomenon may be related to the detoxification function of marine mammal livers in terms of storage and biotransformation. The tolerance limit of THg in mammalian hepatic tissues appears to be within the range of 100 to 400 µg g^{-1} wet weight (ww) [28]. The THg concentrations shown in **Table 1** are below this range; however, a dolphin analyzed in our laboratory was found with a liver THg concentration of 1374 µg g^{-1} dw. Albeit not as high as the results from Japan (1600 µg g^{-1} dw) [29] and the Mediterranean (13150 µg g^{-1} dw) [30], the THg concentrations detected in the liver samples of *S. chinensis* in the PRE were high enough to cause damage to the internal organs of the contaminated individuals.

The accumulation of MeHg tended to increase with age (**Fig. 2**) but at a lower rate than that of THg. No obvious trend existed between the MeHg/THg ratio (%) and age in the muscle samples, whereas the liver and kidney samples showed a decreasing trend of MeHg with age, which could be ascribed to a slow demethylation process that occurs in livers and kidneys but not in muscles [29,31]. Among the three tissue types, muscle tissue accumulated the highest MeHg concentrations, which is consistent with the hypothesis that lack of demethylation mechanisms occur in muscles [32]. Based on studies of several small mammals [33], the lethal level of MeHg in brain tissue was proposed to be in the range of 12–30 µg g^{-1} ww (equivalent to 60–150 µg g^{-1} dw). However, the highest MeHg concentration in the brain tissue

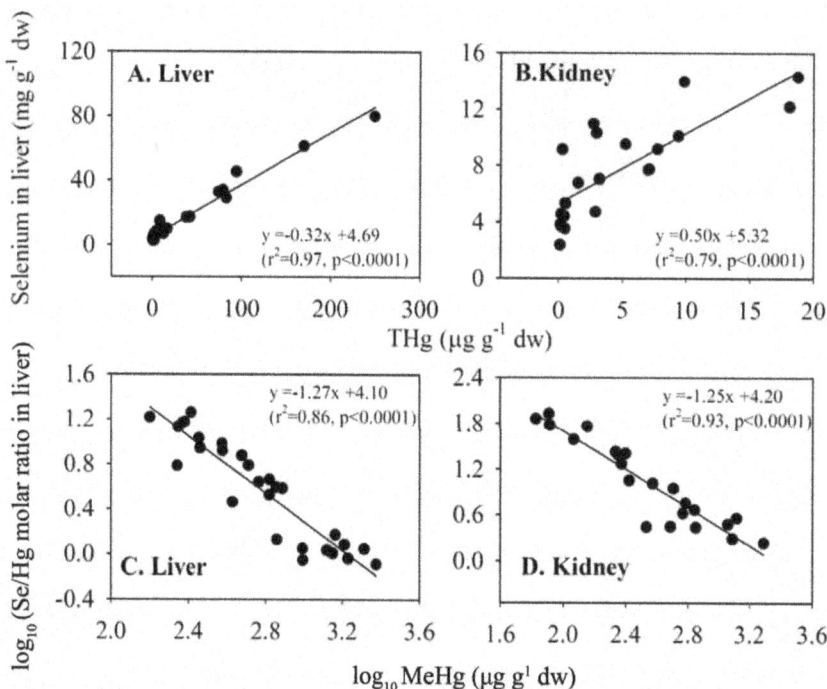

Figure 3. Regression analysis of Se with THg, and of log$_{10}$ (Se/Hg) with log$_{10}$ MeHg in the liver and kidney of Indo-Pacific humpback dolphins (n = 28) stranded in the Pearl River Estuary (PRE) region.

(0.41 µg g^{-1}) from our previous study (unpublished data) was significantly lower than the lethal level.

The relationships between age and THg concentration in livers and kidneys of different species of marine mammal have been extensively examined [12,30,34], and indicate that the hepatic and renal THg concentrations increase with age. This correlation implies a higher capability for the bioaccumulation of toxic elements than for their elimination throughout these animals' life span. In addition, as the size of the prey and the quantities of food tends to increase in proportion with the growth of the dolphins, the trophic transfer of the toxic metals may also progressively increase

[35]. In the present study, the THg concentration increased with age in the livers, kidneys and muscles (**Fig. 2**).

Because of the difficulties in tracking down the prey fish species of dolphins in natural habitats to analyze their tissue contaminant residues, the trophic transfer of THg, MeHg and Se from prey fish to dolphin has scarcely been studied, even though dolphins are considered one of the top predators in the ocean. Through a comparison of the mean concentrations of THg, MeHg and Se from the prey fish to dolphin tissues, our studies outline the first direct evidence of the biomagnification processes of contaminants in this cetacean (**Table 2**). For Hg$_{inorg}$, the average concentrations

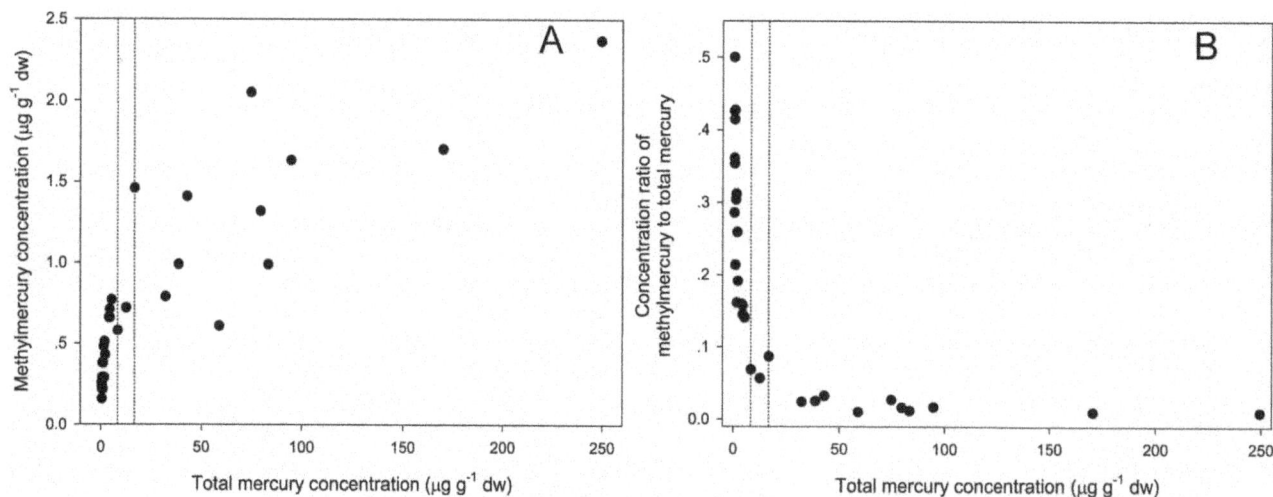

Figure 4. MeHg concentration (µg g^{-1} dw) (A) and concentration ratio of MeHg to THg (B) against THg concentration (µg g^{-1} dw) in the liver samples of the stranded Indo-Pacific humpback dolphins in the Pearl River Estuary region (n = 28).

in the three dolphin tissues were 4.78- to 350-fold greater than the concentrations detected in their prey fish, suggesting that a significant biomagnification process could occur for Hg, especially in the liver. The present study also showed that concentrations of MeHg in the cetacean tissue were 9- to 18-fold higher than the concentrations in the prey fish because of the high assimilation and bioaccumulation capability of MeHg across trophic levels. Conversely, trophic transfer of Se was weak except in dolphin livers. The concentrations of Se in the dolphin muscle tissue were even lower than the concentrations found in the prey fish (BMF< 1), which is likely due to the lack of accumulation or rapid elimination of Se in this tissue. The comparison of THg, Se and MeHg between the dolphin organ tissues and prey fish indicated that THg and MeHg could significantly accumulate in dolphin organs. However, whether THg and MeHg can threaten the health of this cetacean is best determined by the molar ratio of Se/THg, which will be discussed below.

In the marine environment, the dominant portion of Hg in fish and squid is present in the form of MeHg [36]. However, the majority of Hg accumulated in the internal organs of cetaceans appears as inorganic Hg(II), especially in the liver, indicating that a demethylation of MeHg occurs in the liver [34,37]. The formation of the compound Se–Hg, resulting from the combination of a demethylation product (i.e., Hg(II)) and Se, appears to be the last step in the demethylation processes, leading to the fossilization of THg and Se in the form of an inert compound [17,19,38]. In the present study, the relationship between the $\log_{10}$ Se/THg value and $\log_{10}$ MeHg concentration in the livers and kidneys of Indo-Pacific humpback dolphins showed a significant negative correlation (**Fig. 3**). This finding strongly suggests that the Se–Hg compound was formed, causing the fraction of Se/THg in the form of the Se–Hg compound to decrease with respect to the increased MeHg levels [39]. The significant positive relationship between Se and THg and the significant negative correlation between the proportion of MeHg in THg and the THg concentration further confirms that detoxification processes act on MeHg in the livers and kidneys of the dolphins (**Fig. 3** and **Fig. S1**), similar to the results reported for other cetaceans [18,37].

The molar ratio of Se/THg can be considered an indicator of the extent of toxicity caused by Hg contamination because the toxic effects are alleviated after the formation of the Se–Hg compound, which is stable and has no relevant biological impact. Se can be combined with other elements to form various compounds. A molar ratio between Se and THg higher than 1 implies that Se is providing potential protection against Hg toxicity. However, a ratio below 1 suggests limited Se protection against Hg toxicity [40,41]. In the present study, the average molar ratios of Se to THg in the liver, kidney and muscle tissue were 6.03, 21.48 and 6.75, respectively, which are all significantly higher than 1. The high Se/THg ratios in tissues were due to the fact that juveniles comprised most of the analyzed samples and low THg concentrations were found in these samples. However, certain adults had molar ratios below 1 and were in danger of suffering Hg toxicity. According to Caurant et al. [38], young animals were still unable to demethylate MeHg efficiently, leading to higher mean percentages of MeHg/THg concentrations in juveniles than adults. We concluded that a molar ratio of Se/THg of 1 in the organs of adults can be used as an indicator of demethylation processes in the organs. The mean molar ratios of Se/THg in the liver samples of adult dolphins were approximately 1, whereas the mean value in the kidneys was 2.65. However, this pattern does not preclude the possibility of the demethylation process occurring in the kidneys. Se, as the constituent of selenoprotein P and selenoenzymes, also plays an important role in maintaining the normal functions of the kidneys in addition to its detoxification function against Hg.

The relationship between THg and MeHg in dolphin livers may indirectly reveal how demethylation occurs. In this study (**Fig. 4**A), when the THg concentration was lower than 8.4 µg g^{-1} dw, the MeHg concentration increased significantly, indicating that no demethylation occurred when the body burden of THg was low. However, when the THg concentration increased from 8.4 µg g^{-1} dw to 16.9 µg g^{-1} dw, the increase of MeHg in the livers slowed down because the demethylation process was presumably occurring, although we lack the direct chemical and biochemical evidence to confirm this assumption. The concentration ratio of MeHg to THg was $1.86 \pm 0.8\%$ (Mean ± SD, n = 9) when the THg concentration exceeded 16.9 µg g^{-1} dw, whereas the concentration ratio was $27 \pm 11\%$ (Mean ± SD, n = 16) when the THg concentration was below 8.4 µg g^{-1} dw (**Fig. 4**B). Therefore, it is assumed that the demethylation processes could be activated when the THg concentration reaches a threshold range (i.e. 8.4–16.9 µg g^{-1} dw). According to Palmisano et al. [42], mercury is generally stored as MeHg in the livers of dolphins initially, and after a threshold value (100 µg g^{-1} ww) is reached, demethylation takes place with the co-accumulation of Se/THg at a molar ratio of 1:1. In the livers of S. chinensis, Se was significantly related to THg ($R^2 = 0.9740$, $p<0.05$), and the molar ratios of Se/THg in most our samples were far higher than 1 (**Fig. S2**), suggesting that Se was involved in the detoxification process of MeHg.

The co-accumulation of Se/THg with a molar ratio of 1:1 in the livers of cetaceans after a threshold value is reached has been reported by several authors in striped dolphins and Dall's porpoises [42,43]. Together, these results indicated that exceeding the threshold value of THg was a necessarily initial step before triggering the demethylation processes. Compared with the threshold value of striped dolphins (100 µg g^{-1} ww) and Dall's porpoises (20–30 µg g^{-1} dw), a lower THg concentration was found in the present study for S. chinensis. These differences could be attributed to the metabolism and varying foraging habits among the cetacean species, which might also influence the accumulation of Hg and Se, as reflected in the Se/THg molar ratios [12,44]. The lower threshold of hepatic THg concentrations for the equimolar accumulation of Se and Hg in S. chinensis suggests that this species has a greater sensitivity to THg concentrations than is found in striped dolphins and Dall's porpoises.

Conclusions

This study is the first comprehensive investigation on the distribution and fate of THg, MeHg and Se in livers, kidneys and muscles of Indo-Pacific Humpback Dolphins collected for almost 10 years from the PRE of China, and in the prey fishes from the PRE. The results clearly showed that THg and Se were mainly accumulated in dolphin livers while the highest residue of MeHg was found in dolphin muscles. This study contains the first experimental evidence for the potential trophic transfer from the fish to Indo-Pacific Humpback dolphins, showing a high biomagnification factor of Hg$_{inorg}$ in liver (350×) and MeHg in muscle (18.7×). Our studies suggest that the Indo-Pacific humpback dolphins may have the potential to detoxify Hg via the demethylation of MeHg and the formation of tiemannite (HgSe) in the liver and kidney.

Acknowledgments

We thank the Guangdong Pearl River Estuary Chinese White Dolphin National Nature Reserve for the logistic service; Mr. Wenzhi Lin, Mr. Xi Chen and Mr. Yinku Wang for collecting samples.

Author Contributions

Conceived and designed the experiments: YW. Performed the experiments: YS QT HM. Analyzed the data: DG RY YS. Contributed reagents/materials/analysis tools: RY DG. Contributed to the writing of the manuscript: DG YW RY YS. Supervised the research: YW RY LC. Revised the manuscript: DG YR YW.

References

1. Compeau G, Bartha R (1985) Sulfate-reducing bacteria: principal methylators of mercury in anoxic estuarine sediment. Appl Environ Microbiol 50: 498–502.

2. Fleming EJ, Mack EE, Green PG, Nelson DC (2006) Mercury methylation from unexpected sources: molybdate-inhibited freshwater sediments and an iron-reducing bacterium. Appl Environ Microbiol 72: 457–464.

3. Hong YS, Hunter S, Clayton LA, Rifkin E, Bouwer EJ (2012) Assessment of mercury and selenium concentrations in captive bottlenose dolphin's (Tursiops truncatus) diet fish, blood, and tissue. Sci Total Environ 414: 220–226.

4. Gilmour CC, Podar M, Bullock AL, Graham AM, Brown SD, et al. (2013) Mercury Methylation by Novel Microorganisms from New Environments. Environ Sci Technol 47: 11810–11820.

5. Parks JM, Johs A, Podar M, Bridou R, Hurt RA, et al. (2013) The genetic basis for bacterial mercury methylation. Science 339: 1332–1335.

6. Yu RQ, Flanders J, Mack EE, Turner R, Mirza MB, et al. (2012) Contribution of coexisting sulfate and iron reducing bacteria to methylmercury production in freshwater river sediments. Environ Sci Technol 46: 2684–2691.

7. Yu RQ, Reinfelder JR, Hines ME, Barkay T (2013) Mercury methylation by the methanogen Methanospirillum hungatei. Appl Environ Microbiol 79: 6325–6330.

8. Shi JB, Ip C, Zhang G, Jiang G, Li X (2010) Mercury profiles in sediments of the Pearl River Estuary and the surrounding coastal area of South China. Environ Pollut 158: 1974–1979.

9. Yu X, Li H, Pan K, Yan Y, Wang WX (2012) Mercury distribution, speciation and bioavailability in sediments from the Pearl River Estuary, Southern China. Mar Pollut Bull 64: 1699–1704.

10. Hung CL, So MK, Connell DW, Fung CN, Lam MH, et al. (2004) A preliminary risk assessment of trace elements accumulated in fish to the Indo-Pacific Humpback dolphin (Sousa chinensis) in the northwestern waters of Hong Kong. Chemosphere 56: 643–651.

11. Wu Y, Shi J, Zheng GJ, Li P, Liang B, et al. (2013) Evaluation of organochlorine contamination in Indo-Pacific humpback dolphins (Sousa chinensis) from the Pearl River Estuary, China. Sci Total Environ 444: 423–429.

12. Kunito T, Nakamura S, Ikemoto T, Anan Y, Kubota R, et al. (2004) Concentration and subcellular distribution of trace elements in liver of small cetaceans incidentally caught along the Brazilian coast. Mar Pollut Bull 49: 574–587.

13. Thompson DR (1990) Metal levels in marine vertebrates. CRC PRESS, BOCA RATON, FL(USA): 143–182.

14. Endo T, Hisamichi Y, Kimura O, Haraguchi K, Baker CS (2008) Contamination levels of mercury and cadmium in melon-headed whales (Peponocephala electra) from a mass stranding on the Japanese coast. Sci Total Environ 401: 73–80.

15. Dietz R, Riget F, Born E (2000) An assessment of selenium to mercury in Greenland marine animals. Sci Total Environ 245: 15–24.

16. Moreira I, Seixas T, Kehrig H, Fillmann G, Di Beneditto A, et al. (2009) Selenium and mercury (total and organic) in tissues of a coastal small cetacean, Pontoporia blainvillei. J Coast Res 81: 866–870.

17. Lailson-Brito J, Cruz R, Dorneles PR, Andrade L, Azevedo AdF, et al. (2012) Mercury-Selenium Relationships in Liver of Guiana Dolphin: The Possible Role of Kupffer Cells in the Detoxification Process by Tiemannite Formation. PLoS ONE 7: e42162.

18. Nakazawa E, Ikemoto T, Hokura A, Terada Y, Kunito T, et al. (2011) The presence of mercury selenide in various tissues of the striped dolphin: evidence from μ-XRF-XRD and XAFS analyses. Metallomics 3: 719–725.

19. Nigro M (1994) Mercury and selenium localization in macrophages of the striped dolphin, Stenella coeruleoalba. Journal of the Marine Biological Association of the United Kingdom Plymouth 74: 975–978.

20. Khan MAK, Wang F (2009) Mercury-selenium compounds and their toxicological significance: Toward a molecular understanding of the mercury-selenium antagonism. Environ Toxicol Chem 28: 1567–1577.

21. Jefferson TA (2000) Population biology of the Indo-Pacific hump-backed dolphin in Hong Kong waters. Wildlife monographs: 1–65.

22. Myrck AC Jr., Hohn AA, Sloan PA, Kimura M, Stanley DD (1983) Estimating age of spotted and spinner dolphins (Stenella attenuata and Stenella longirostris) from teeth.

23. Jefferson TA, Hung SK, Robertson KM, Archer FI (2012) Life history of the Indo-Pacific humpback dolphin in the Pearl River Estuary, southern China. Mar Mammal Sci 28: 84–104.

24. Barros NB, Jefferson TA, Parsons E (2004) Feeding habits of Indo-Pacific humpback dolphins (Sousa chinensis) stranded in Hong Kong. Aquat Mamm 30: 179–188.

25. Ruelas-Inzunza JR, Horvat M, Pérez-Cortés H, Páez-Osuna F (2003) Methylmercury and total mercury distribution in tissues of gray whales (Eschrichtius robustus) and spinner dolphins (Stenella longirostris) stranded along the lower Gulf of California, Mexico. Cienc Mar 29: 1–8.

26. U.S.EPA Method 1630 Methyl Mercury in Water by Distillation, Aqueous Ethylation, Purge and Trap, and Cold Vapor Atomic Fluorescence Spectrometry. US Environmental Protection Agency Office of Water Office of Science and Technology Engineering and Analysis Division (4303) 401 M Street SW Washington, DC 20460.

27. Parsons E (1998) Trace metal pollution in Hong Kong: Implications for the health of Hong Kong's Indo-Pacific hump-backed dolphins (Sousa chinensis). Science of the Total Environment 214: 175–184.

28. Wagemann R, Muir DCG (1984) Concentrations of heavy metals and organochlorines in marine mammals of northern waters: overview and evaluation. Western Region, Department of Fisheries and Oceans Canada.

29. Honda K, Tatsukawa R, Itano K, Miyazaki N, Fujiyama T (1983) Heavy metal concentrations in muscle, liver and kidney tissue of striped dolphin, Stenella coeruleoalba, and their variations with body length, weight, age and sex. Agric Biol Chem 47: 1219–1228.

30. Leonzio C, Focardi S, Fossi C (1992) Heavy metals and selenium in stranded dolphins of the Northern Tyrrhenian (NW Mediterranean). Sci Total Environ 119: 77–84.

31. Martoja R, Berry JP (1981) Identification of tiemannite as a probable product of demethylation of mercury by selenium in cetaceans. A complement to the scheme of the biological cycle of mercury [detoxification]. Vie Milieu.

32. Storelli M, Ceci E, Marcotrigiano G (1998) Comparison of total mercury, methylmercury, and selenium in muscle tissues and in the liver of Stenella coeruleoalba (Meyen) and Caretta caretta (Linnaeus). Bull Environ Contam Toxicol 61: 541–547.

33. Wren CD (1986) A review of metal accumulation and toxicity in wild mammals: I. Mercury. Environ Res 40: 210–244.

34. Meador J, Ernest D, Hohn A, Tilbury K, Gorzelany J, et al. (1999) Comparison of elements in bottlenose dolphins stranded on the beaches of Texas and Florida in the Gulf of Mexico over a one-year period. Arch Environ Contam Toxicol 36: 87–98.

35. André J, Ribeyre F, Boudou A (1990) Mercury contamination levels and distribution in tissues and organs of delphinids (Stenella attenuata) from the Eastern Tropical Pacific, in relation to biological and ecological factors. Mar Environ Res 30: 43–72.

36. Das K, Debacker V, Bouquegneau JM (2000) Metallothioneins in marine mammals. Cell Mol Biol 46.

37. Holsbeek L, Siebert U, Joiris CR (1998) Heavy metals in dolphins stranded on the French Atlantic coast. Sci Total Environment 217: 241–249.

38. Caurant F, Navarro M, Amiard JC (1996) Mercury in pilot whales: possible limits to the detoxification process. Sci Total Environ 186: 95–104.

39. Cáceres-Saez I, Dellabianca NA, Goodall RNP, Cappozzo HL, Guevara SR (2013) Mercury and Selenium in Subantarctic Commerson's Dolphins (Cephalorhynchus c. commersonii). Biol Trace Elem Res 151: 195–208.

40. Peterson SA, Ralston NV, Peck DV, Sickle JV, Robertson JD, et al. (2009) How might selenium moderate the toxic effects of mercury in stream fish of the western US? Environ Sci Technol 43: 3919–3925.

41. Sørmo EG, Ciesielski TM, Øverjordet IB, Lierhagen S, Eggen GS, et al. (2011) Selenium Moderates Mercury Toxicity in Free-Ranging Freshwater Fish. Environ Sci Technol 45: 6561–6566.

42. Palmisano F, Cardellicchio N, Zambonin P (1995) Speciation of mercury in dolphin liver: a two-stage mechanism for the demethylation accumulation process and role of selenium. Mar Environ Res 40: 109–121.

43. Yang J, Kunito T, Tanabe S, Miyazaki N (2007) Mercury and its relation with selenium in the liver of Dall's porpoises (*Phocoenoides dalli*) off the Sanriku coast of Japan. Environ Pollut 148: 669–673.

44. Seixas TG, Kehrig HdA, Fillmann G, Di Beneditto APM, Souza CM, et al. (2007) Ecological and biological determinants of trace elements accumulation in liver and kidney of *Pontoporia blainvillei*. Sci Total Environ 385: 208–220.

16

The Gut Bacterial Community of Mammals from Marine and Terrestrial Habitats

Tiffanie M. Nelson[1,2]*, **Tracey L. Rogers[2]**, **Mark V. Brown[3]**

1 Evolution and Ecology Research Centre, School of Biological, Earth and Environmental Sciences, University of New South Wales, Kensington, Australia, **2** Australian Institute of Marine Science, Water Quality and Ecosytem Health, Arafura Timor Research Facility, Casuarina, Australia, **3** Evolution and Ecology Research Centre; School of Biotechnology and Biomolecular Sciences, University of New South Wales, Kensington, Australia

Abstract

After birth, mammals acquire a community of bacteria in their gastro-intestinal tract, which harvests energy and provides nutrients for the host. Comparative studies of numerous terrestrial mammal hosts have identified host phylogeny, diet and gut morphology as primary drivers of the gut bacterial community composition. To date, marine mammals have been excluded from these comparative studies, yet they represent distinct examples of evolutionary history, diet and lifestyle traits. To provide an updated understanding of the gut bacterial community of mammals, we compared bacterial 16S rRNA gene sequence data generated from faecal material of 151 marine and terrestrial mammal hosts. This included 42 hosts from a marine habitat. When compared to terrestrial mammals, marine mammals clustered separately and displayed a significantly greater average relative abundance of the phylum *Fusobacteria*. The marine carnivores (Antarctic and Arctic seals) and the marine herbivore (dugong) possessed significantly richer gut bacterial community than terrestrial carnivores and terrestrial herbivores, respectively. This suggests that evolutionary history and dietary items specific to the marine environment may have resulted in a gut bacterial community distinct to that identified in terrestrial mammals. Finally we hypothesize that reduced marine trophic webs, whereby marine carnivores (and herbivores) feed directly on lower trophic levels, may expose this group to high levels of secondary metabolites and influence gut microbial community richness.

Editor: Jack Anthony Gilbert, Argonne National Laboratory, United States of America

Funding: Funding for the field support was provided by the Secretaría de Ciencia y Tecnología and the Dirección Nacional del Antártico (PICTO No. 36054), the Australian Research Council, Winifred Scott Foundation, and the Evolution and Ecology Research Centre, University of New South Wales, Sydney, Australia. The funders had no role in study design, data collection and analysis, decision to publish, or preparation of the manuscript.

Competing Interests: The authors have declared that no competing interests exist.

* E-mail: t.nelson@aims.gov.au

Introduction

Bacteria inhabiting the gastro-intestinal tract of mammals expand their host's metabolic potential by harvesting energy that would otherwise be inaccessible [1–3]. This symbiosis between mammals and bacteria has contributed, in part, to the success of the class *Mammalia*, allowing them to radiate in large numbers to occupy a variety of environmental niches [3,4]. Herbivorous mammals, for instance, were able to survive on plant material after acquiring a gut bacterial community with the capability to digest cellulose in plant cell walls [5].

Mammalian hosts first acquire their gut bacterial community during transport through the birth canal and subsequently through maternal, social and environmental transmissions [6–8]. Genetic factors within the host also shape the gut bacterial community, a result of their long history of co-evolution [9–11]. This is evident in the strong physiological effects which the gut bacterial community can exert on the host mammal, such as modulating the immune response system or affecting brain development [12,13].

In a pioneering study, Ley et al. 2008 [3] compared the faecal bacterial community of a variety of terrestrial mammals and identified that host phylogeny, diet and, to a lesser extent, gut morphology influenced the composition of the gut bacterial community. Since then, studies have further confirmed that the composition of the gut bacterial community follows along evolutionary lineages [11,14]. Recently, diet has been shown to be the primary driver of functional capacity in the gut, resulting in a convergence of microbial communities between phylogenetically un-related hosts [15].

Further insight could be gained by comparing a diverse range of extant mammals with differing life history traits. One group of mammals that have been relatively understudied are marine mammals. Their comparatively recent evolution and differing life history traits and adaptation to a marine habitat [16], suggest they are a necessary addition to the current understanding of drivers shaping the mammalian gut bacterial community.

To understand patterns in gut bacterial communities, the faecal bacterial communities of a broad range of terrestrial and marine mammals were compared. Marine mammals included two species of seals inhabiting the Antarctic [17]; three species of seals inhabiting the Arctic [18] and data from one dugong (a marine herbivore) [19]. Terrestrial mammals included carnivores, omnivores and herbivores from a range of phylogenetic groups (number of individuals, n = 109). The aim of this study was to identify broad scale patterns of gut bacterial communities of mammals from marine and terrestrial habitats.

Materials and Methods

Ethics Statement

Samples collected from southern elephant seals and leopard seals were carried out in strict accordance with the recommendations in the Australian Code of Practice for the Care and Use of Animals for Scientific Purposes. Protocols used in the study were approved by the University of New South Wales Animal Care and Ethics Committee (permit number 08/83B and 03/103B). The southern elephant seal is listed as vulnerable under the Environment Protection and Biodiversity Conservation Act 1999 (EPBC Act) and listed under the Convention on International Trade in Endangered Species of Wild Fauna and Flora (CITES). Permission to export southern elephant seal biological materials was obtained from the Australian Government Department of the Environment, Water, Heritage and the Arts (permit number 2008-AU-534289).

Permission to access regions in Antarctica where seals were located was approved by the Ministry of Foreign Affairs and the Dirección Nacional del Antártico. Southern elephant seals in this study were located in Antarctic Special Protected Area 132 'Peninsula Potter' and additional permissions were obtained through the Dirección Nacional del Antártico under Article 3, Annex II of the Madrid Protocol to the Antarctic Treaty (no permit number). For southern elephant seal males and leopard seals, individuals were anaesthetised using a mixture of tiletamine and zolazepam (Zoletil – Virbac Australia) at a combined dose of 1 mg/kg. Female southern elephant seals were anaesthetised with 3–6 mg/kg ketamine hydrochloride. On all occasions, procedures were performed by qualified personnel and all efforts were made to minimize suffering.

Selection of Studies for Comparison

Studies for comparison were selected on the basis that analysis methods for bacterial community composition sequenced a region of the 16S rRNA gene using highly conserved bacterial or universal primer sets. The specific methods each study used to generate this data are outlined in Table S1.

Individual sequence data were obtained from the National Centre for Biotechnology Information website (http://www.ncbi.nlm.nih.gov/) or directly from locations specified in source articles. The dataset comprising Antarctic seal gut bacterial community by the same authors is deposited to the database Dryad (www.datadryad.org) under the provisional DOI:10.5061/dryad.42f2q [20]. In total the combined dataset contained samples from 151 individual mammal hosts.

Preparation of Datasets for Comparison

Sequence taxonomy was assigned using the Ribosomal Database Project II (RDP) v.10 Classifier tool [21]. Taxonomy of sequences was assigned using RDPs Naïve Bayesian Classifier which classifies sequences to the genus level [21]. As the total number of sequences in each dataset differed, datasets were randomly sub-sampled to a maximum of 100 sequences using the software Daisy-Chopper (www.genomics.ceh.ac.uk/GeneSwytch/). The result was a total of 13,848 sequences from 151 mammalian hosts (Table S2).

Meta-analyses such as this may be prone to study effects, whereby similarity or differences in methodology, rather than ecology, generate the observable patterns. To address concerns over any potential study effects, we selected and analysed data from four host phylogenetic families, the *Phocidae*, *Ursidae*, *Hominidae* and *Canidae*, (Figure S1) as each contained samples from different studies. When analysed in the same manner as the main dataset (see methods) the gut bacterial community of the host cluster more closely based on phylogenetic family of the host (Figure S1A) than by which study the host originated from (Figure S1B, Table S1).

In addition, to examine the effect of data standardisation versus data rarefaction (to enable incorporation of datasets that contained fewer than 100 sequences per host), the complete dataset was rarefied to 24 sequences per host, which was the lowest number present in any study used. When analysed using the same techniques as described below (see methods) this subsampled dataset generated the same statistically significant overall patterns as those observed when using the expanded dataset (see Table S2 and Figure S2).

The assessments of study effects give provide the author's with confidence that the results reported herein describe ecological effects rather than methodological effects.

Statistical Analyses of Data

The combined dataset was standardised prior to transformation. This involved converting abundance counts into relative percentages for each individual host. A Bray-Curtis dissimilarity matrix [22] was generated from square-root transformed data. To facilitate comparisons of the gut bacterial community between hosts, hosts were assigned to *a priori* groups based on habitat

Table 1. Difference of the gut bacterial community of host mammals based on groupings of diet, habitat, phylogeny, and gut morphology using ANOSIM.

Source of variation	Pair-wise comparisons	R	p
Diet		0.39*	<0.01
	Herbivores, omnivores	0.33	<0.01
	Herbivores, carnivores	0.49*	<0.01
	Omnivores, carnivores	0.26	<0.01
Diet and habitat		0.52**	<0.01
	Terrestrial herbivores, terrestrial omnivores	0.33*	<0.01
	Terrestrial herbivores, terrestrial carnivores	0.62**	<0.01
	Terrestrial herbivores, marine carnivores	0.65**	<0.01
	Terrestrial omnivores, terrestrial carnivore	0.32	<0.01
	Terrestrial omnivores, marine carnivores	0.50**	<0.01
	Terrestrial carnivores, marine carnivores	0.69**	<0.01
Phylogenetic order		0.19	<0.01
Phylogenetic family		0.50**	<0.01
Gut morphology		0.11	<0.01
	Hindgut fermenters, simple guts	0.14	<0.01
	Hindgut fermenters, foregut fermenters	0.17	<0.01
	Simple guts, foregut fermenters	0.15	<0.01

ANOSIM of gut bacterial abundance data was used to generate a permutated Global R statistic (R) and permutated p-value (p). Significance level: **R = >0.5, *R = 0.3< R <0.5. Comparison of these results with rarefied data is displayed in Table S3.

(marine or terrestrial), phylogenetic family, dominant dietary source (carnivorous, omnivorous or herbivorous) or gut morphology (simple, hind or foregut fermenters). These factors have been identified previously as potential drivers of gut bacterial community composition [3,15]. Similarities between hosts and host groups were visualised using non-metric multi-dimensional scaling (nMDS) [23]. The result of nMDS ordination is a two-dimensional plot where the position of each sample is determined by its distance from all other points in the analysis. The contribution of classified genera to the observed dissimilarity between groups in the nMDS were calculated using SIMPER (similarity percentages procedure) [24]. SIMPER decomposes average Bray-Curtis dissimilarities between all pairs of hosts into percentage contributions from each classified genus and therefore identifies which genera are characteristic of bacterial community structure [24]. The non-parametric estimator of species richness, Chao1 [25], was calculated using genus abundance data for each host using the online software program EstimateS [26].

Differences in the composition of the gut bacterial community between hosts were tested with the non-parametric permutation procedure ANOSIM (Analysis of Similarity) as it is more robust to heterogeneous dispersion of data [24]. ANOSIM is applied to the dissimilarity matrix and generates a test statistic, R. The magnitude of R is indicative of the difference within groups and between groups and is scaled to lie between −1 and +1, a value of zero represents the null hypothesis (no difference between groups) and value towards one represents the alternative hypothesis (all similarities within groups are less than any similarity between groups) [24]. R-values >0.50 (**) were interpreted as well separated; R >0.30 (*) as overlapping but clearly different; and, R >0.20 as barely separable at all. Results were considered significant where $P\text{-}value = <0.025$. Significant differences in

bacterial richness and phyla abundance between groups were tested using a Student's T-Test in Excel 2010 (Microsoft Pty Ltd). All other statistical tests were performed using the software PRIMER-E v6 [24].

Results

Patterns in the Gut Bacterial Community of Marine and Terrestrial Mammals

Herbivores and carnivores displayed significant differences in the composition of their gut bacterial communities (ANOSIM: $R = 0.49$, $p = <0.01$; Table 1; Figure 1). The gut bacterial community of marine carnivores and terrestrial carnivores was significantly different ($R = 0.69$, $p = <0.01$; Table 1). Con-specific hosts from the same family were more similar than non-con-specific hosts ($R = 0.50$, $p = <0.01$). Across all mammals, the gut morphology of hosts did not contribute to significant differences in the gut bacterial community (Table 1). Previously, the influence of captivity in leopard seals was identified as a strong driver of the gut bacterial community (see Nelson et al. 2012) [17]. However, in this study, six host species were sampled from both captive and wild habitats and compared in an nMDS plot (Figure S3) and it was clear that captive and wild con-specifics clustered closer to one another than they did to unrelated hosts.

Taxonomic Differences in Gut Bacterial Community Composition between Marine and Terrestrial Mammals

Marine carnivores possessed a significantly lower average relative abundance of the phylum *Firmicutes* in their gut bacterial community with 43.2 ± 4.0 compared to $65.6 \pm 2.3\%$ in the gut bacterial community of terrestrial mammals (Student's paired t-test $p = <0.001$; Table 2 and Figure 2) and 68.9% in the gut

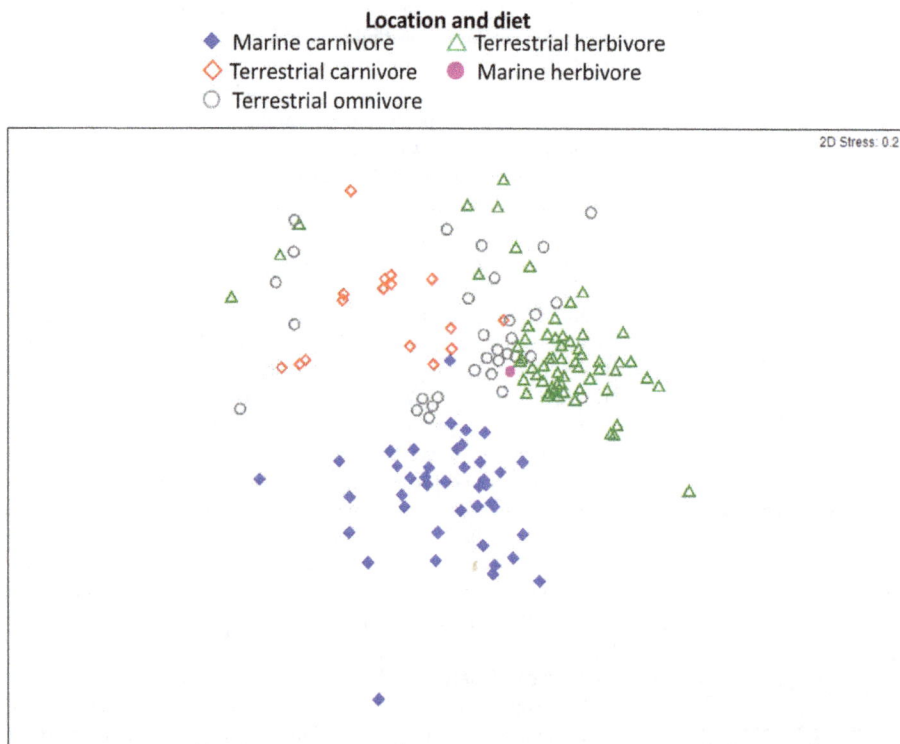

Figure 1. Influence of diet and habitat on the gut bacterial community of mammals. nMDS ordination plot dislaying similarity of the gut bacterial community in the host mammal as grouped by diet and habitat. See Figure S4 for detailed display of this figure.

Table 2. Differences in presence of dominant phyla in the gut bacterial community of mammals based on groupings of habitat and diet using Student's paired t-test.

Comparison	P			
	Firmicutes	Bacteroidetes	Proteobacteria	Fusobacteria
Terrestrial herbivores, terrestrial omnivores	0.473	0.996	0.423	0.001**
Terrestrial herbivores, terrestrial carnivores	0.033	0.001**	0.828	0.002**
Terrestrial herbivores, marine carnivores	<0.001***	0.502	0.018*	<0.001***
Terrestrial omnivores, terrestrial carnivores	0.165	0.001**	0.266	0.735
Terrestrial omnivores, marine carnivores	<0.001***	0.536	0.004**	<0.001***
Terrestrial carnivores, marine carnivores	<0.001***	0.002**	0.151	0.005**
Terrestrial mammals, marine mammals (all diet types)	<0.001***	0.991	0.002**	<0.001***

Student's paired t-test of gut bacterial abundance data to generate a p-value (p). Significance level: $p = \leq 0.001$ (***), $p = 0.01$ (**), $p = 0.025$ (*). The lack of replication in the marine herbivore grouping does not allow for significance testing.

bacterial community of the marine herbivore (Figure 2). The phylum *Proteobacteria* was significantly more abundant in marine carnivores with an average relative abundance of $15.6 \pm 3.0\%$ compared with $5.9 \pm 1.6\%$ in the gut bacterial community of

terrestrial omnivores and herbivores (t-test $p = 0.004$ and 0.018, respectively; Table 2 and Figure 2). The phylum *Bacteroidetes* was similar for each of the dietary groups of terrestrial and marine mammals with an average relative abundance of $19.2 \pm 1.8\%$ with the exception of terrestrial carnivores, which displayed a significantly reduced abundance of $3.7 \pm 2.3\%$ (t-test $p = <0.002$; Table 2 and Figure 2).

The phylum *Fusobacteria* was significantly greater in the gut bacterial community of marine carnivores with an average relative abundance of $22.0 \pm 3.5\%$ compared with $2.1 \pm 0.8\%$ in other dietary groups (t-test $p = <0.001$; Figure 2 and Table 2). Domestic dogs, *Canis lupus familiaris*, were the only non-marine carnivore with a higher than average abundance of the phylum *Fusobacteria* with $32.7 \pm 1.6\%$. The domestic dog clustered closest to the marine carnivores in the nMDS plot (Figure 1 and S4) and this is further highlighted when observing the plot with only members of the order *Carnivora* (Figure 3).

Members of the family *Ursidae* are also seen to cluster closely together regardless of dietary preference, which includes herbivores, omnivores and carnivores (Figures 3 and S4). Overlap in the presence of particular genera between host dietary groups was evident. Marine and terrestrial herbivores, as well as omnivores, displayed overlap in the genera *Anaerotruncas*, *Ruminococcus* and *Roseburia* from the phylum *Firmicutes* (Table S4). These were less abundant in marine or terrestrial carnivores. Some genera, including the genus *Oscillibacter* from the phylum *Firmicutes*, and the genera *Prevotella* and *Bacteroides* from the phylum *Bacteroidetes* were abundant in all groups except for terrestrial carnivores (Table S4). Likewise, some genera, such as *Coprococcus* and *Blautia* from the phylum *Firmicutes* were abundant across all groups with the exception of the marine carnivores. The genus *Lactobacillus* from

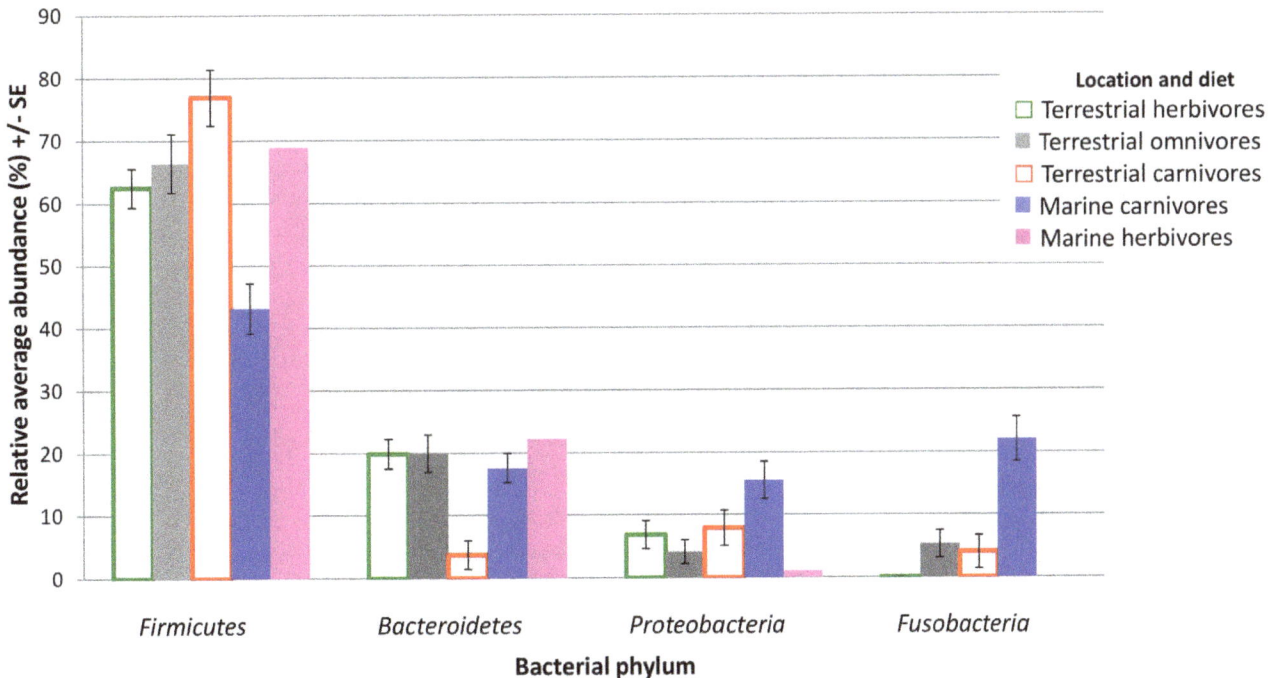

Figure 2. Differences in abundance of dominant phyla in the gut bacterial community of mammals based on groupings of habitat and diet. Average relative abundance of each major phyla in the gut bacterial community of host mammals grouped by habitat and diet. Error bars represent standard errors (SE). The lack of replication in the marine herbivore grouping does not allow for estimation of SE or significance testing. Student's paired t-test were conducted between groups as displayed in Table 2.

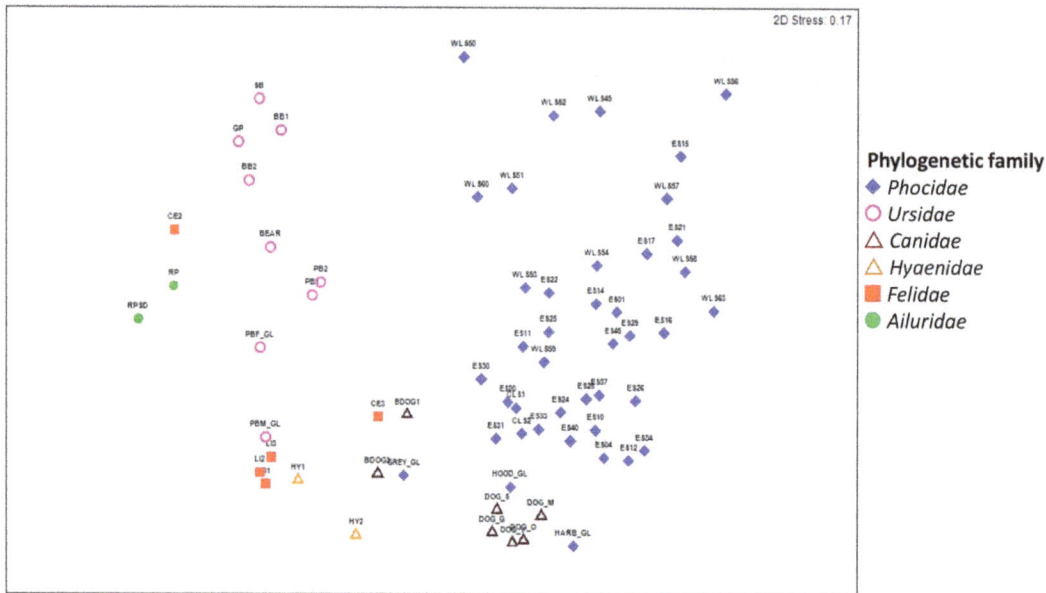

Figure 3. Influence of phylogenetic family on the gut bacterial community of mammals from the order *Carnivora*. nMDS plot displays gut bacterial community of all host mammals in the order *Carnivora* grouped by family. Host species labels are as follows: (RP) red panda; (BB) black bear; (PB) polar bear; (SB) spectacled bear; (GP) giant panda; (BEAR) bear from Norway; (CE) cheetah; (BDOG) bushdog; (DOG) domestic dog; (LI) lion; (HY) spotted hyena; (GREY) grey seal; (HOOD) hooded seal; (HARB) harbour seal; (ES) southern elephant seal; (WLS) leopard seal.

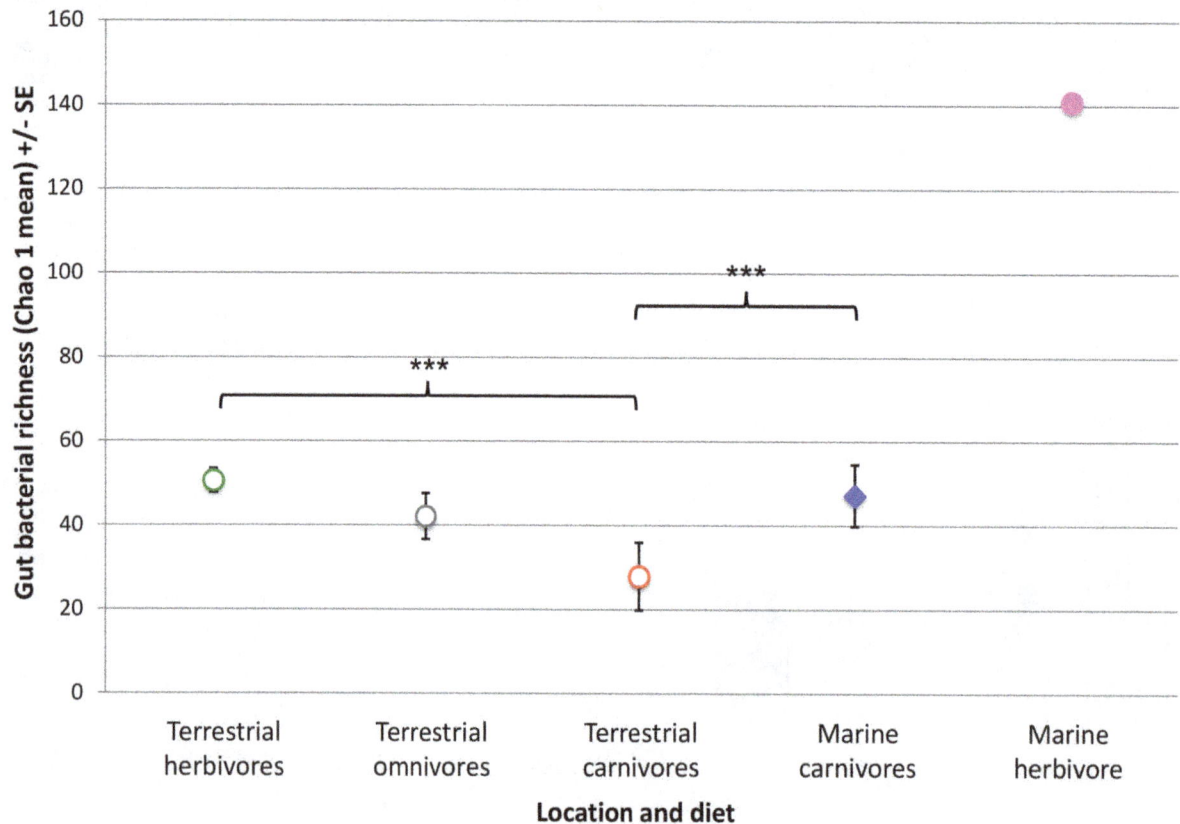

Figure 4. Richness of the gut bacterial community of mammals grouped by habitat and diet. Richness of the gut bacterial community was measured using Chao 1 mean. Error bars represent standard error (SE). Student's paired t-test were conducted between groups with significance level: $p = \leq 0.001$ (***), $p = 0.01$ (**), $p = 0.025$ (*). The lack of replication in the marine herbivore grouping does not allow for estimation of SE or significance testing.

the phylum *Firmicutes* was commonly shared between carnivorous hosts compared with other dietary groups (Table S4).

Richness of Mammal Guts

Herbivores possessed a faecal bacterial community significantly richer than that of carnivores or omnivores (Figure 4). The single marine herbivore displayed a gut bacterial community richer (Chao 1 mean = 141.0) than that of terrestrial herbivores (Chao 1 = 51.0±2.8) or marine carnivores (Chao 1 = 47.0±7.4), although insufficient replication did not allow for significance testing of this pattern (Figure 4). The marine carnivores possessed significantly richer faecal bacterial community than the terrestrial carnivores (Figure 4). Additionally, hindgut fermenters displayed a significantly richer gut bacterial community than hosts with simple gut morphology (t-test: $p = 0.0028$).

Discussion

Comparative Marine and Terrestrial Mammals Gut Bacterial Communities

The composition of the gut bacterial community of the marine mammals available at the time of this study is clearly distinct from that of the available terrestrial mammals. Differences in the gut bacterial community of carnivorous marine mammals appears to be due, in part, to their considerably reduced abundance of *Firmicutes* and increased abundance of *Fusobacteria* compared to terrestrial mammals. Members of the phylum *Fusobacteria* range from facultative anaerobes to obligate anaerobes that ferment carbohydrates or amino acids to produce various organic acids including acetic, formic and butyric acid [27–29]. Species from the phylum *Fusobacteria* occur in a range of habitats, including sediments as well as the oral or intestinal habitats of animals [28,30–32]. Future functional analysis will provide insight as to the specific roles of these phyla in the represented hosts.

The occurrence of greater than average abundance of *Fusobacteria* within the gut of dogs and also within the gut of marine carnivores suggests an interesting trend (see Figure S4). The *Canidae* (dogs) are located with the *Phocidae* in the order *Carnivora* [33] (see Figure S5). Canids and phocids possess shared immune system receptors and diseases, such as *Morbillivirus*, that are capable of passing between dog and seal hosts [34,35]. Although it is beyond the scope of this study to make conclusions about the strength of this pattern of shared *Fusobacteria*, future investigations may help to understand if evolutionary links between seals and dogs can be identified through host-associated microbes.

Composition of the leopard seal gut bacterial community was previously shown to differ significantly in captivity compared with wild hosts, as a result of specific dietary items and local habitat differences [17]. However, this study suggests that the influence of captivity is reduced when comparisons are made at a broader scale (see Figure S3). Host phylogeny and dietary types were stronger indicators of gut bacterial community similarity than was captivity. However, it seems apparent that captive and wild con-specific hosts do display differences in their gut bacterial community at a finer scale. Specific dietary items, antibiotic administration and local exposures result in altered gut bacterial communities in captive individuals [15,36,37] and it is likely these factors that are contributing to the observed differences in the wild and captive hosts compared in this study.

Several challenges are faced when consuming plant material as a primary food source due to the indigestible cell walls [5]. The need to access complex carbohydrates in plants, such as cellulose and starch, is thought to be the driver of a rich gut bacterial community of herbivores [3]. The diversity of gut bacterial communities associated with terrestrial herbivores has been previously identified as significantly richer than those of terrestrial omnivores and carnivores [3]. This finding is also supported by our data. Further, our data indicate that marine carnivores have a richer gut bacterial community than terrestrial carnivores, whilst the one marine herbivore sampled had the richest bacterial community of all terrestrial herbivores. Taken together these results suggest that mammals living in marine habitats may generally possess a richer gut bacterial community than their diet-equivalent terrestrial mammals. In the marine environment, secondary metabolites produced by plants and other primary producers may be considerably higher [5,38,39], which could impact the gut bacterial community of herbivores or carnivores at higher trophic levels. Dugongs primarily consume seagrass and are known to occasionally supplement their diet with macro invertebrates and algae [40,41]. The dugong sampled in this case was in captivity during sampling and had been fed a diet of the eelgrass, *Zostera marina* [19]. Eelgrass has been identified as producing secondary chemical defences, specifically phenolic acids, which have the capacity to cause considerable reduction of bacteria at even low dosages in experimental studies [42,43]. The marine carnivores in this study are also known to feed directly on lower trophic levels, causing them to be exposed to secondary metabolites [44,45]. One of the consequences of these different traits specific to marine based diets is that marine mammals may require a gut bacterial community with a greater diversity of functions enabling the breakdown of excess chemical compounds. Increased sampling of marine herbivores would enable us to unravel these processes.

Supporting Information

Figure S1 Mammals in families from different studies. To display the impacts of possible study effects due to the different techniques used across studies, these nMDS plots display relationships between gut bacterial communities generated using different methods from the phylogenetic families *Phocidae, Canidae, Ursidae* and *Hominidae* labelled by family (A) and by study (B).

Figure S2 Repeat figures from study with dataset rarefied to minimum number sequences per host. Figure 1 (A) and Figure 3 (B) are repeated here to show the similarity in structure when using the minimum number of 24 rarefied sequences per host.

Figure S3 Similarity of the gut bacterial community of host mammals with captive and wild representatives. Non-metric multidimensional scaling ordination plot displays similarity of the gut bacterial community of host mammals with representatives from captive (c) and wild (w) habitats.

Figure S4 Similarity of the gut bacterial community of mammals grouped by diet and habitat. Detailed nMDS ordination plot of Figure 1 with host labels and enlarged region for clarity.

Figure S5 Descriptive phylogeny of families from the order *Carnivora*. Adapted from Agnarsson *et al.* 2010 [33]. Members of the *Phocidae* are marked with red circle.

Table S1 Overview of main methods employed by included studies.

Table S2 Characteristics of mammalian hosts used in the study. Abbreviated table data is as follows: number of sequences used (No. of seq.); gut morphology (Gut morph.); hindgut fermenter (HG); foregut fermenter (FG); simple gut (S); marine (M); terrestrial (T); carnivore (C); herbivore (H); and omnivore (O).

Table S3 Comparison of ANOSIM results between the genera rarefied to the minimum of 24 and those subsampled to 100. Results display those reported in Table 1 and the comparable results when the dataset was rarefied to the minimum number of sequences represented by any one host which was 24. ANOSIM of gut bacterial abundance data was used to generate a permutated Global R statistic (R) and permutated p-value (p). Significance level: **R = >0.5, *R = 0.3< R <0.5.

Table S4 Characteristic genera in the gut bacterial community of mammal hosts grouped by diet and habitat. The foremost ten characteristic genera in the gut bacterial community of hosts identified using SIMPER analysis. Hosts are grouped based on diet and habitat.

Acknowledgments

The authors are grateful for the field assistance and sample collection provided by Instituto Antártico Argentino, the Australian Marine Mammal Research Centre and Taronga Zoo. Three independent reviewers are thanked for their assessment of the study.

Author Contributions

Conceived and designed the experiments: TMN TLR MVB. Performed the experiments: TMN. Analyzed the data: TMN MVB. Contributed reagents/materials/analysis tools: TMN TLR MVB. Wrote the paper: TMN.

References

1. Savage DC (1977) Microbial ecology of the gastrointestinal tract. Annu Rev Microbiol 31: 107–133.
2. Hooper L V, Gordon JI (2001) Commensal host-bacterial relationships in the gut. Science (80-) 292: 1115–1118.
3. Ley RE, Hamady M, Lozupone C, Turnbaugh PJ, Ramey RR, et al. (2008) Evolution of mammals and their gut microbes. Science (80-) 320: 1647–1651.
4. Collinson ME, Hooker JJ, Skelton PW, Moore PD, Ollerton J, et al. (1991) Fossil evidence of interactions between plants and plant-eating mammals. Philos Trans R Soc London Ser B Biol Sci 333: 197–208. Available: http://rstb.royalsocietypublishing.org/content/333/1267/197.abstract.
5. Choat JH, Clements KD (1998) Vertebrate herbivores in marine and terrestrial environments: a nutritional ecology perspective. Annu Rev Ecol Syst 29: 375–403. Available: http://www.jstor.org/stable/221713.
6. Palmer C, Bik EM, DiGiulio DB, Relman DA, Brown PO (2007) Development of the human infant intestinal microbiota. PLoS Biol 5: 1556–1573.
7. Lombardo MP (2008) Access to mutualistic endosymbiotic microbes: an underappreciated benefit of group living. Behav Ecol Sociobiol 62: 479–497.
8. Trosvik P, Stenseth NC, Rudi K (2009) Convergent temporal dynamics of the human infant gut microbiota. ISME J 4: 151–158.
9. Rawls JF, Mahowald MA, Ley RE, Gordon JI (2006) Reciprocal gut microbiota transplants from zebrafish and mice to germ-free recipients reveal host habitat selection. Cell 127: 423–434.
10. Kovacs A, Ben-Jacob N, Tayem H, Halperin E, Iraqi F, et al. (2011) Genotype is a stronger determinant than sex of the mouse gut microbiota. Microb Ecol 61: 423–428. Available: http://dx.doi.org/10.1007/s00248-010-9787-2.
11. Ochman H, Worobey M, Kuo C-H, Ndjango J-BN, Peeters M, et al. (2010) Evolutionary relationships of wild hominids recapitulated by gut microbial communities. PLoS Biol 8: e1000546. Available: http://dx.doi.org/10.1371/journal.pbio.1000546.
12. Heijtz RD, Wang S, Anuar F, Qian Y, Björkholm B, et al. (2011) Normal gut microbiota modulates brain development and behavior. Proc Natl Acad Sci 108: 3047–3052. Available: http://www.pnas.org/content/108/7/3047.abstract.
13. Hooper L V (2009) Do symbiotic bacteria subvert host immunity? Nat Rev Microbiol 7: 367–375.
14. Yildirim S, Yeoman CJ, Sipos M, Torralba M, Wilson BA, et al. (2010) Characterization of the fecal microbiome from non-human wild primates reveals species specific microbial communities. PLoS One 5: e13963. Available: http://dx.doi.org/10.1371/journal.pone.0013963.
15. Muegge BD, Kuczynski J, Knights D, Clemente JC, González A, et al. (2011) Diet drives convergence in gut microbiome functions across mammalian phylogeny and within humans. Science (80-) 332: 970–974. Available: http://www.sciencemag.org/content/332/6032/970.abstract. Accessed 8 August 2013.
16. Davis CS, Delisle I, Stirling I, Siniff DB, Strobeck C (2004) A phylogeny of the extant Phocidae inferred from complete mitochondrial DNA coding regions. Mol Phylogenet Evol 33: 363–377. Available: http://www.sciencedirect.com/science/article/pii/S1055790304001940.
17. Nelson TM, Rogers TL, Carlini AR, Brown M V (2012) Diet and phylogeny shape the gut microbiota of Antarctic seals: a comparison of wild and captive animals. Evol Ecol Res Cent 15: 1132–1145. Available: http://dx.doi.org/10.1111/1462-2920.12022.
18. Glad T, Kristiansen VF, Nielsen KM, Brusetti L, Wright A-DG, et al. (2010) Ecological characterisation of the colonic microbiota in Arctic and sub-Arctic seals. Microb Ecol 60: 320–330.
19. Tsukinowa E, Karita S, Asano S, Wakai Y, Oka Y, et al. (2008) Fecal microbiota of a dugong (Dugong dugong) in captivity at Toba Aquarium. J Gen Appl Microbiol 54: 25–38.
20. Nelson T, Rogers T, Carlini A, Brown M (2012) Data from: Diet and phylogeny shape the gut microbiota of Antarctic seals: a comparison of wild and captive animals. Dryad Digit Repos doi105061/dryad42f2q.
21. Wang Q, Garrity GM, Tiedje JM, Cole JR (2007) Naive bayesian classifier for rapid assignment of rRNA sequences into the new bacterial taxonomy. Appl Environ Microbiol 73: 5261–5267. Available: http://aem.asm.org/cgi/content/abstract/73/16/5261.
22. Bray JR, Curtis JT (1957) An ordination of the upland forest communities of Southern Wisconsin. Ecol Monogr 27: 325–349.
23. Guttman L (1968) A general nonmetric technique for finding the smallest coordinate space for configuration of points. Psychometrika 33: 469–506.
24. Clarke KR, Gorley RN (2006) PRIMER v 6: User Manual/Tutorial.
25. Chao A (1984) Non-parametric estimation of the number of classes in a population. Scand J Stat 11: 783–791.
26. Colwell R (1997) EstimateS - Statistical Estimation of Species Richness and Shared Species from Samples. Available: http://viceroy.eeb.uconn.edu/EstimateSPages/AboutEstimateS.htm.
27. Bennett KW, Eley A (1993) Fusobacteria: new taxonomy and related diseases. J Med Microbiol 39: 246–254. Available: http://jmm.sgmjournals.org/content/39/4/246.abstract.
28. Staley JT, Whitman B W (2005) Phylum XIX. Fusobacteria Garrity and Holt 2001, 140. In: Krieg NR, Ludwig W, Whitman WB, Hedlund BP, Paster BJ, et al., editors. Bergey's Manual of Systematic Bacteriology Volume 4: The Bacteroidetes, Spirochaetes, Tenericutes (Mollicutes), Acidobacteria, Fibrobacteres, Fusobacteria, Dictyoglomi, Gemmatimonadetes, Lentisphaerae, Verrucomicrobia, Chlamydiae, and Planctomycetes. New York: Springer, Vol. 4. pp. 747–774.
29. Potrykus J, Mahaney B, White RL, Bearne SL (2007) Proteomic investigation of glucose metabolism in the butyrate-producing gut anaerobe Fusobacterium varium. Proteomics 7: 1839–1853. Available: http://dx.doi.org/10.1002/pmic.200600464.
30. Kapatral V, Anderson I, Ivanova N, Reznik G, Los T, et al. (2002) Genome sequence and analysis of the oral bacterium Fusobacterium nucleatum strain ATCC 25586. J Bacteriol 184: 2005–2018. Available: http://jb.asm.org/cgi/content/abstract/184/7/2005.
31. Nagaraja TG, Narayanan SK, Stewart GC, Chengappa MM (2005) Fusobacterium necrophorum infections in animals: pathogenesis and pathogenic mechanisms. Anaerobe 11: 239–246. Available: http://www.sciencedirect.com/science/article/pii/S1075996405000090.
32. Suau A, Rochet V, Sghir A, Gramet G, Brewaeys S, et al. (2001) Fusobacterium prausnitzii and related species represent a dominant group within the human fecal flora. Syst Appl Microbiol 24: 139–145. Available: http://www.sciencedirect.com/science/article/pii/S072320200470017X.
33. Agnarsson I, Kuntner M, May-Collado LJ (2010) Dogs, cats, and kin: A molecular species-level phylogeny of Carnivora. Mol Phylogenet Evol 54: 726–745. Available: http://www.sciencedirect.com/science/article/pii/S1055790309004424.
34. Schreiber A, Eulenberger K, Bauer K (1998) Immunogenetic evidence for the phylogenetic sister group relationship of dogs and bears (Mammalia, Carnivora: Canidae and Ursidae). Exp Clin Immunogenet 15: 154–170. Available: http://www.karger.com/DOI/10.1159/000019067.
35. Osterhaus ADME, Vedder EJ (1988) Identification of virus causing recent seal deaths. Nature 335: 20. Available: http://dx.doi.org/10.1038/335020a0.

36. Dethlefsen L, Huse S, Sogin ML, Relman DA (2008) The pervasive effects of an antibiotic on the human gut microbiota, as revealed by deep 16S rRNA sequencing. PLoS Biol 6: 2383–2400.

37. De Filippo C, Cavalieri D, Di Paola M, Ramazzotti M, Poullet JB, et al. (2010) Impact of diet in shaping gut microbiota revealed by a comparative study in children from Europe and rural Africa. Proc Natl Acad Sci USA 107: 14691–14696. Available: http://www.pnas.org/content/107/33/14691.abstract.

38. Engel S, Jensen PR, Fenical W (2002) Chemical ecology of marine microbial defense. J Chem Ecol 28: 1971–1985. Available: http://dx.doi.org/10.1023/A:1020793726898.

39. Hay ME, Fenical W (1988) Marine plant-herbivore interactions: the ecology of chemical defense. Annu Rev Ecol Syst 19: 111–145. Available: http://www.jstor.org/stable/2097150.

40. Marsh H, Channells PW, Heinsohn GE, Morrissey J (1992) Analysis of stomach contents of dugongs from Queensland. J Wildl Res 9: 55–67.

41. Preen A (1995) Diet of dugongs: are they omnivores? J Mammal 76: 163–171.

42. Vergeer LHT, Develi A (1997) Phenolic acids in healthy and infected leaves of Zostera marina and their growth-limiting properties towards Labyrinthula zosterae. Aquat Bot 58: 65–72. Available: http://www.sciencedirect.com/science/article/pii/S0304377096011151.

43. Harrison PG (1982) Control of microbial growth and of amphipod grazing by water-soluble compounds from leaves of Zostera marina. Mar Biol 67: 225–230. Available: http://dx.doi.org/10.1007/BF00401288.

44. McClintock JB, Baker BJ (1997) A Review of the chemical ecology of Antarctic marine invertebrates. Am Zool 37: 329–342. Available: http://icb.oxfordjournals.org/content/37/4/329.abstract.

45. Lippert H, Brinkmeyer R, Mülhaupt T, Iken K (2003) Antimicrobial activity in sub-Arctic marine invertebrates. Polar Biol 26: 591–600. Available: http://dx.doi.org/10.1007/s00300-003-0525-9.

17

Molecular Epidemiology of Seal Parvovirus, 1988–2014

Rogier Bodewes[1]*, **Rebriarina Hapsari**[1], **Ana Rubio García**[2], **Guillermo J. Sánchez Contreras**[2], **Marco W. G. van de Bildt**[1], **Miranda de Graaf**[1], **Thijs Kuiken**[1], **Albert D. M. E. Osterhaus**[1,3,4,5]

1 Department of Viroscience, Erasmus MC, Rotterdam, the Netherlands, 2 Seal Rehabilitation and Research Centre, Pieterburen, the Netherlands, 3 Viroclinics Biosciences BV, Rotterdam, the Netherlands, 4 Research Center for Emerging Infections and Zoonoses, University of Veterinary Medicine, Hannover, Germany, 5 Artemis One Health, Utrecht, the Netherlands

Abstract

A novel parvovirus was discovered recently in the brain of a harbor seal (*Phoca vitulina*) with chronic meningo-encephalitis. Phylogenetic analysis of this virus indicated that it belongs to the genus *Erythroparvovirus*, to which also human parvovirus B19 belongs. In the present study, the prevalence, genetic diversity and clinical relevance of seal parvovirus (SePV) infections was evaluated in both harbor and grey seals (*Halichoerus grypus*) that lived in Northwestern European coastal waters from 1988 to 2014. To this end, serum and tissue samples collected from seals were tested for the presence of seal parvovirus DNA by real-time PCR and the sequences of the partial NS gene and the complete VP2 gene of positive samples were determined. Seal parvovirus DNA was detected in nine (8%) of the spleen tissues tested and in one (0.5%) of the serum samples tested, including samples collected from seals that died in 1988. Sequence analysis of the partial NS and complete VP2 genes of nine SePV revealed multiple sites with nucleotide substitutions but only one amino acid change in the VP2 gene. Estimated nucleotide substitution rates per year were 2.00×10^{-4} for the partial NS gene and 1.15×10^{-4} for the complete VP2 gene. Most samples containing SePV DNA were co-infected with phocine herpesvirus 1 or PDV, so no conclusions could be drawn about the clinical impact of SePV infection alone. The present study is one of the few in which the mutation rates of parvoviruses were evaluated over a period of more than 20 years, especially in a wildlife population, providing additional insights into the genetic diversity of parvoviruses.

Editor: Jianming Qiu, University of Kansas Medical Center, United States of America

Funding: This work was financially supported in part by a grant from the Niedersachsen-Research Network on Neuroinfectiology (N-RENNT) of the Ministry of Science and Culture of Lower Saxony, Germany, and the Virgo consortium. The funding sources had no role in study design, data collection and analysis, decision to publish, or preparation of the manuscript.

Competing Interests: One author of the manuscript has interests to declare: Prof. Dr. ADME Osterhaus is part time chief scientific officer of Viroclinics Biosciences B.V. There are no patents, products in development or marketed products to declare. The other authors have no competing interests to declare.

* Email: r.bodewes@erasmusmc.nl

Introduction

Parvoviruses are small, non-enveloped, viruses with linear, single-stranded DNA genomes of approximately 5 kb. The family *Parvoviridae* comprises two subfamilies: *Parvovirinae* and *Densovirinae*. Members of the *Parvovirinae* infect vertebrates, while members of the *Densovirinae* infect invertebrates [1]. At present, eight genera have been recognized by the International Committee on Taxonomy of Viruses (ICTV) within the subfamily *Parvovirinae*: *Amdoparvovirus, Aveparvovirus, Bocaparvovirus, Copiparvovirus, Dependoparvovirus, Erythroparvovirus, Protoparvovirus* and *Tetraparvovirus* [2,3].

The genus *Erythroparvovirus* currently consists of six species, *Primate erythroparvoviruses 1–4, Rodent erythroparvovirus 1* and *Ungulate erythroparvovirus 1* [2,3]. Human parvovirus B19 or *Primate erythroparvovirus 1* is the best studied member of this genus. It is prevalent worldwide; more than 15% of pre-school children, 50% of younger adults and 85% of the elderly have serologic evidence of infection based on data from Australia, Denmark, England, Japan, and the USA [4]. Infection by human parvovirus B19 can be asymptomatic or symptomatic, with clinical

manifestations ranging from mild erythema infectiosum (fifth disease) in healthy children to chronic arthropathy in adults, transient aplastic crisis in patients with underlying hematologic disease, and hydrops fetalis [4–6]. In addition, involvement of human parvovirus B19 in encephalitis and other neurologic diseases was suggested by serology and viral DNA detection by PCR of the cerebrospinal fluid of patients [7]. Other members of the genus *Erythroparvovirus*; simian parvovirus, rhesus macaque parvovirus, and pig-tailed macaque parvovirus (*Primate erythroparvoviruses 2, 3, 4*), are associated with anemia in the respective non-human primates [8,9]. Chipmunk parvovirus (*Rodent erythroparvovirus 1*) and bovine parvovirus type 3 (*Ungulate erythroparvovirus 1*) are also members of this genus and were identified in apparently healthy animals [10,11].

Recently, a novel parvovirus, tentatively called seal parvovirus (SePV), was identified in the brain and lungs tissues of a young harbor seal (*Phoca vitulina*) with severe chronic neurological signs [12]. Phylogenetic analysis suggested that this virus belonged to the genus *Erythroparvovirus*. Histological examination of the brain of the infected seal confirmed the presence of a chronic non-suppurative meningo-encephalitis, while other causes of brain disease, such as neoplasia, physical trauma, and infections with herpesvirus or morbillivirus, were excluded. *In situ* hybridization showed that the novel SePV DNA was present in the Virchow-Robin spaces and in the brain parenchyma adjacent to the meninges of the seal [12]. This discovery showed that parvoviruses indeed can enter the brain parenchyma.

During a previous study using tissue samples collected from necropsies in 2008–2012, SePV DNA was detected in only two out of 94 additional seals [12]. In the present study, this prevalence study was extended using tissues and sera collected from both harbor and grey seals (*Halichoerus grypus*) from Northwestern European coastal waters from 1988 to 2014. In addition, we investigated the genetic diversity, estimated nucleotide substitution rates and evaluated the clinical relevance of SePV infections.

Materials and Methods

Ethics statement

Samples of seals used in the present study were provided by the Seal Research and Rehabilitation Centre (SRRC), and the SRRC provided permission to the Department of Viroscience, Erasmus Medical Centre to use the samples for the present study. Admission and rehabilitation of wild seals at the SRRC is permitted by the government of the Netherlands (application number FF/75/2012/015). In the present study, only samples were used that were collected from dead seals or blood samples that were collected for routine diagnostics during rehabilitation. These samples were collected from seals that were either found dead in the wild by the SRRC, died at the SRRC despite intensive care or were euthanized at the SRRC due to the presence of severe clinical signs in the absence of any indication of future recovery. In case of euthanasia, this was performed with T-61 (0.3 ml/kg) after sedation as described previously [13]. Tissue samples that were used in the present study have been described previously [12,14–17].

Sample collection

Blood samples were collected from harbor (n = 131) and grey seals (n = 69) of different age groups rehabilitated from 2002–2014 at the SRRC, Pieterburen, the Netherlands and after clotting, samples were centrifuged briefly and serum was stored at −20°C until further processing. Samples of spleen tissue (n = 110) were collected from both harbor seals (n = 99) and grey seals (n = 11) that died during outbreaks of phocine distemper virus (PDV) among seals of Northwestern Europe in 1988 and 2002 [16,18] and in-between years (1988–2002) and stored at −80°C until further processing. An overview of samples used in the present study is listed in **Table S1**. Seals were defined into three different age categories; pups (estimated to be less than two months of age), juveniles (estimated to be between two and twelve months of age) and (sub)adults (estimated to be older than one year of age).

SePV DNA detection

In previous studies of humans infected with parvovirus B19, viral DNA could not be detected in blood samples, but was detected in tissues such as bone marrow, tonsils, synovium, and lymphoid tissues [19,20]. Since replication of SePV was observed *in vitro* in seal bone marrow cells, this might be also applicable for SePV [12]. However, since bone marrow samples were not available, spleen tissue samples were selected. These samples were homogenized in 1 ml Hank's minimal essential medium (HMEM) containing 0.5% lactalbumin, 10% glycerol, 200 U/ml penicillin, 200 µg/ml streptomycin, 100 U/ml polymyxin B sulfate, 250 µg/ml gentamycin, and 50 U/ml nystatin (ICN Pharmaceuticals)

(transport medium) using a Fastprep-24 Tissue Homogenizer (MP Biomedicals) and centrifuged briefly. Total nucleic acids were extracted from serum (50 µl) and homogenized spleen tissue (200 µl) using the High Pure Viral Nucleic Acid kit (Roche) according to the instructions of the manufacturer. Samples were screened for SePV DNA using a real-time PCR targeting the VP1 gene as described previously [12]. Each sample was tested at least in two independent experiments and to avoid false positive samples due to cross-contamination during mass necropsies, only samples that yielded a cycle threshold (Ct) value below 35 were considered positive.

Sequencing of the partial genome of SePV variants

On samples that tested positive by real-time PCR, additional specific PCRs were performed to amplify the partial NS (924 bp) and the complete VP2 gene. The partial NS gene was amplified with forward primer TAGAATGGCTTGTGCGGTGT and reverse primer GTGGGTTTCAATGGCCTACT. The complete VP2 gene was amplified using two partial overlapping primer sets, with forward primer TTGCCGGCCATCTCGTCGTA and reverse primer AGCCTGTCCTTCATCTGACC targeting the 5′end of the gene and forward primer TCTCAGGCT-CAATGGCGTAG and reverse primer GTGAAGCTT-TATTTTTGGGCAC targeting the 3′ end of the gene. PCR products were separated by gel electrophoresis, bands of the appropriate size were extracted using the MinElute Gel Extraction Kit (Qiagen) and cloned using the TOPO TA Cloning Kit for Sequencing (Invitrogen). Multiple clones were sequenced with an ABI Prism 3130xl genetic analyzer (Applied Biosystems). For sequence analysis, a consensus sequence was deduced from at least three colonies for each sample.

Phylogenetic analysis, estimation of substitution rates and selection pressure

Obtained nucleotide sequences of the complete VP2 and partial NS genes of SePV variants were aligned with various other viruses belonging to the genus *Erythroparvovirus* using MAFFT7 [21]. Subsequently, a maximum likelihood phylogenetic tree was built in MEGA6 [22] with 500 bootstrap values using the Jukes Cantor model, which was selected by analysis with jModeltest 2.1.3 [23].

The number of nucleotide substitutions per site per year was estimated with a Bayesian Markov Chain Monte Carlo (MCMC) method implemented in the BEAST 1.7 package [24]. Dates of sequences were used in combination with the HKY substitution model and a log-normal relaxed clock. The analysis was conducted using a time-aware linear Bayesian skyride coalescent tree prior [25]. Three independent MCMC analyses were performed for 10 million states. These analyses were combined and analyzed with Tracer, version 1.5. Uncertainty in parameter estimates was reported as the 95% highest posterior density (HPD). The selective pressures on the NS and VP2 genes were assessed by calculating the ratio between non-synonymous substitutions (dN) and synonymous substitutions (dS) per site using the single likelihood ancestor counting (SLAC) method implemented in HyPhy platform accessed via the DataMonkey webserver (www.datamonkey.org) [26].

Analysis of the clinical relevance of SePV infection

The clinical or pathological reports of seals that tested positive for SePV DNA were retrospectively analyzed. Available information included age group (pups <2 months of age, juveniles >2 months and <1 year of age, (sub)adults >1 year of age), sex, gross pathology results and possible cause of death (only for tissue

Table 1. Overview of characteristics of seals with SePV.

Sample ID	Sex	Age	Sample	Location of stranding[1]	Year	Ct-value SePV	SePV PCR confirmation	Co-infection with PDV[2]	Co-infection with PhHV-1	Pathological diagnoses and/or major clinical signs
PV88023.6	NI[3]	NI	spleen	UK	1988	27.8	+	NI	-	NI
PV88927.20	NI	NI	spleen	NL	1988	15.4	+	NI	-	NI
HG020628.02	F	(subadult)	spleen	NL	2002	33.7	+	-	-	Acute non-infectious cause of dead (possible drowning)
PV020628.13	M	(subadult)	spleen	NL	2002	29.4	+	-	+	Intestinal volvulus
PV020718.12	F	juvenile	spleen	NL	2002	19.9	+	+	+	Extensive, acute, severe, interstitial pneumonia
PV020719.08	F	(subadult)	spleen	NL	2002	34.1	+	+	+	Extensive, acute, severe, interstitial pneumonia
PV020719.09	M	(subadult)	spleen	NL	2002	24.3	-	+	-	Extensive, acute, severe, interstitial pneumonia
PV020919.03	F	(subadult)	spleen	NL, Terschelling	2002	25.0	-	+	-	Euthanized due to presence of severe neurological signs, PDV detected in the brain
PV021004.1	M	(subadult)	spleen	NL, Vlieland	2002	28.0	-	-[4]	-	Emaciation; diffuse, subacute, marked, purulent bronchopneumonia
HG06-130	F	juvenile	serum	NL, South-Holland	2006	28.6	+	-	-	Several wounds on flippers and skin lesions suggestive of pox virus infection
PV12-410[5]	M	juvenile	serum and spleen	NL, Ameland	2012	29.6	+	-	-	Chronic, mild, non-suppurative meningo-encephalitis
PV12216.01[5]	F	juvenile	spleen	NL, Terschelling	2012	21.1	+	-	+	Acute, moderate necrotizing bronchitis and bronchiolitis and acute interstitial pneumonia

[1]:UK: United Kingdom, NL: the Netherlands.
[2]: based on results described previously [14].
[3]: NI: no information available.
[4]: No PDV was detected [14], although observed lesions were highly suggestive of infection with PDV.
[5]: samples from a previous study [12].

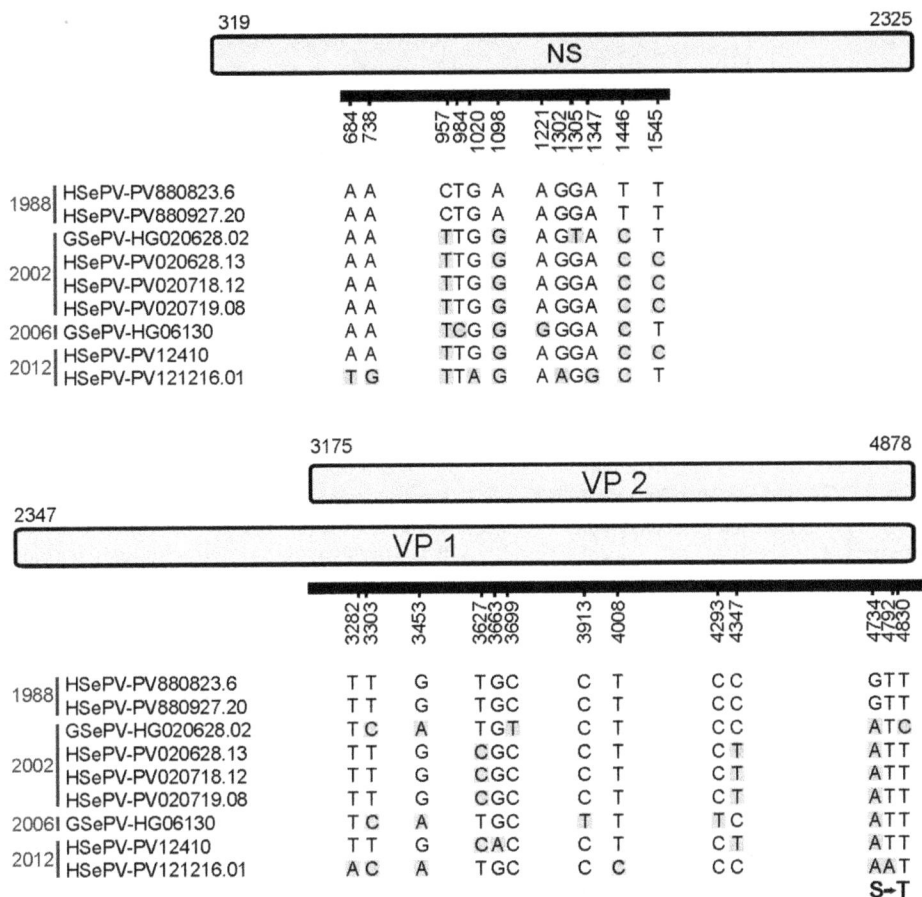

Figure 1. Variation in SePV sequences. Schematic overview of the partial NS gene and the complete VP2 gene that was amplified of 9 SePV variants. The consensus sequence of each virus was compared to the oldest sample available (HSePV-PV880823.6). Locations with nucleotide mutations are indicated in grey. Only in sample HSePV-121216.01, a nucleotide mutation resulted in an amino acid substitution (S540T).

samples), and hematology results (only for the serum samples). Since in a previous study co-infection with herpesvirus was detected in most cases, samples positive for SePV DNA were tested for the presence of co-infection with herpesvirus using a degenerate nested pan-herpesvirus PCR targeting the conserved DNA polymerase gene, as described previously [27]. PCR products with the appropriate band size were sequenced and obtained sequences were analyzed by BLAST search to determine which herpesvirus was detected.

Results

Prevalence of SePV infection

Out of 200 seal serum samples from 2002–2014 analyzed, only one sample had a Ct-value <35 (0.5%), which was the serum of a grey seal from 2006 (rehabilitation number: HG 06-130). In addition, in 9 out of 110 spleen tissue samples, the Ct-value was lower than 35 (8%). Of those nine samples, two were collected from harbor seals that had died during the 1988 and seven during the 2002 PDV outbreaks (six harbor seals and one grey seal). Combination of the data obtained in the present study with data from a previous study [12], resulted in a SePV prevalence in spleen samples of 6.9% for harbor seals (based on in total 174 animals) and 9.1% for grey seals (n = 11) (**Figure S1, Table 1**).

Genetic analysis of SePV

Out of 12 samples in which SePV DNA was detected by real-time PCR (including samples from a previous study [12]), the partial NS (Genbank accession numbers: KM252699-KM252706) and the complete VP2 gene (Genbank accession numbers KM252691-KM252698) could be amplified in only 9 samples, possibly due to fragmentation of the DNA or inhibiting factors present in the other samples. Analysis of the obtained nucleotide sequences showed that SePV from 1988 to 2012 displayed minimal sequence variation. Using the parvovirus detected in the oldest sample (SePV-PV880823.6) as a reference, 12 (1.3%) variant nucleotide positions were present in the partial NS gene (**Figure 1**), of which ten were transitions and two were transversions, and all 12 were synonymous. At three nucleotide positions (positions 957, 1098, and 1446) mutations were fixed in the viruses detected after 1988. All the nucleotide substitutions occurred in the third codon base. Compared to the viruses detected in 1988, the virus with the highest number of nucleotide substitutions was SePV-PV121216.01, which was detected in 2012 (**Figure 1**). Estimated nucleotide substitution rates of the partial NS were 2.00×10^{-4} (HPD, 1.36×10^{-5}–4.04×10^{-4}) nucleotide substitutions per site per year.

In the complete VP2 gene, 13 (0.76%) variant nucleotide positions were present, including 11 transitions and 2 transversions. Among those, 11 occurred in the third codon base and 2 in the first codon base. In the SePV detected in a harbor seal

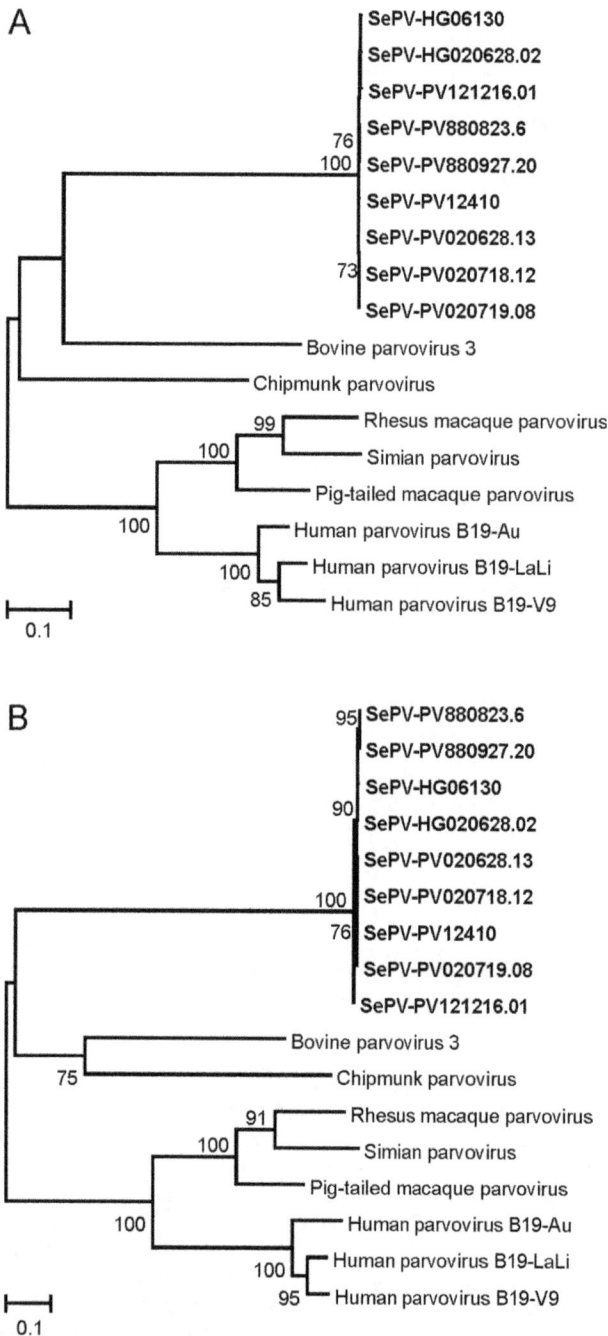

A

B

Figure 2. Maximum likelihood tree of the VP2 gene and partial NS1 gene of SePV. Phylogenetic maximum-likelihood tree with 500 bootstrap replicates of the nucleotide sequence of the VP2 genes (A) and partial NS genes (B) of SePV variants and various viruses of the genus *Erythroparvovirus*. Only bootstrap values >70 are indicated. Genbank accession numbers: SePV-HG06130: KM252691, SePV-HG020628.02: KM252694, SePV-PV121216.01: KM252698, SePV-PV880823.6: KM252692, SePV-PV880927.20: KM252693, SePV12410: KF373759, SePV-PV020628.13: KM252695, SePV-PV020718.12: KM252696, SePV-PV020719.08: KM252697, bovine parvovirus 3: AF406967, chipmunk parvovirus: GQ200736, rhesus macaque parvovirus: AF221122, simian parvovirus: U26342, pig-tailed macaque parvovirus: AF221123, human parvovirus B19-Au: M13178, human parvovirus B19-LaLi: AY044266, human parvovirus B19-V9: AJ249437.

(identification number PV121216.01), a mutation at nucleotide position 4792 was present, which resulted in an amino acid change from serine to threonine (amino acid residue 540). For the VP gene, we found nucleotide substitution rates of 1.15×10^{-4} (HPD, 1.43×10^{-5}–2.22×10^{-4}) nucleotide substitutions per site per year with a dN/dS value of 0.33.

The high similarity of different SePV variants was reflected by phylogenetic and pairwise identity analysis of the NS and VP2 nucleotide sequences. Pairwise identities between SePV NS variants were 99.3% or higher and between SePV VP2 variants 99.5% or higher, while pairwise identities between SePV and other parvoviruses belonging to the genus *Erythroparvovirus* were 51.9% or lower for the NS gene and 50.6% or lower for the VP2 gene (**Figure 2**).

Evaluation of the clinical relevance of SePV infection

Available clinical and pathology data of the seals from which positive samples were obtained, including two positive seals detected in a previous study [12], showed that SePV DNA was detected in young and (sub)adult seals, but not in pups. Both male and female seals were infected. In both seals in which SePV was detected in the serum, the numbers of red blood cells present in the blood sample was determined. Grey seal HG06-130 had a red blood cell (RBC) count of 3.66×10^{12} cells/liter and harbor seal PV12-410 had a mean RBC value of 3.69×10^{12}/liter (ranged from 3.03 to 4.29×10^{12} cells/liter during stay at the SRRC), which were both lower than normal (reference values for a normal RBC count are for both seal species of this age 3.9–5.7×10^{12} cells/liter). Despite a low RBC count, the hemoglobin level of grey seal HG06-130 was 9.2 mmol/liter, which was within the reference values (8.4–14.9 mmol/liter). The mean hemoglobin level of harbor seal 12–410 was somewhat lower than normal (mean value 8.4 mmol/liter, ranged from 6.7 to 9.6 mmol/liter during stay at the SRRC).

Since samples from 1988 and 2002 were collected from seals that had died during PDV outbreaks, the death of at least four seals was related to infection with PDV and the role of SePV could not be evaluated. In one grey seal that died in 2002 (HG020628.02) and in which SePV DNA was detected, the cause of death was non-infectious and no other significant abnormalities were observed upon necropsy of this animal (**Table 1**). One harbor seal that died in 2002 (PV020628.13) had an intestinal volvulus and no significant abnormalities were reported that could be related to a viral infection, although it was positive for phocine herpesvirus-1 (PhHV-1). Besides this seal, PhHV-1 was detected in three additional spleen samples (**Table 1**).

Discussion

SePV was discovered in 2012 in a young harbor seal with severe chronic meningo-encephalitis [12]. In the present study, we demonstrated that SePV was already circulating among both harbor and grey seals of Northwestern Europe since 1988, although SePV was detected only in a relatively small number of samples.

The partial NS and the complete VP2 gene analyzed in the present study corresponded with highly variable regions of the parvovirus B19 genome [28]. However, sequences of SePV variants analyzed in the present study showed only one amino acid substitution, which was detected in the most recent variant in 2012. This amino acid substitution was present in the VP2 protein (S540T), but since both serine and threonine are polar amino acids it is unlikely that this would have led to substantial conformational change of the VP2 protein. Of interest, there was also only minimal sequence variation between SePV detected in grey seals and harbor seals, similar to what was reported for parapox viruses in two species of pinnipeds and herpesviruses among phocids [29,30]. These findings suggest that the metapopulation of harbor and grey seals act as a single reservoir for these viruses.

Estimation of nucleotide substitutions rates of parvoviruses has been the focus of multiple studies [31–34]. However, the number of studies in which evolutionary dynamics of parvoviruses were evaluated over a period of more than 20 years is limited, especially in a wildlife population [32,35,36]. In the present study, we evaluated the annual nucleotide substitutions rates of SePV using samples from 1988 to 2012, which is to our knowledge only the second study performed for viruses of the genus *Erythroparvovirus* over such a long period [35]. However, only a limited number of samples was included in the present study and it could not be excluded that the observed genetic variations were due to polymorphisms of the NS and VP2 genes. In addition, only the partial NS gene was sequenced and observed nucleotide substitution rates might not be representative for the complete NS gene.

Of interest, the relatively low prevalence of this virus in combination with the rapid population growth of both harbor and grey seals from 1988 to 2012 [37] allowed to evaluate genetic diversity of a parvovirus in a population without or with only low herd immunity. Furthermore, harbor seals are generally known to have a limited migration range [38], which minimizes the gene flow of seals that might have differences in susceptibility to this virus and limits the chance of recombination with other parvoviruses from seals of other areas [39]. Although only a relatively low number of different virus variants was studied, observed annual nucleotide substitutions rates were similar to those of other parvoviruses [31–34]. Despite this relatively high mutation rate, only one amino acid substitution was detected in the VP2 gene and none in the NS gene, suggesting that natural selection suppressed amino acid changes for both genes. Of interest, also no nucleotide changes were detected in the consensus SePV sequence that was detected in two serum samples collected from seal 12–410 with an interval of five weeks.

SePV was first discovered in a seal with meningo-encephalitis [12], while in the present and previous study SePV was also detected in at least two seals without neurological signs. Since in most cases SePV DNA was detected in seals that were also infected with PDV and/or PhHV-1, the spectrum of clinical signs associated with SePV infection remained unclear. Of interest, results of the present study and a previous study [12] suggested a higher prevalence of SePV during the PDV epizootic in 2002. This might be related to immunosuppression in the seal population following infection with PDV, which predisposes for other viral infections [40]. Anemia or associated signs of low red blood cell counts is one of the manifestations associated with *Erythroparvovirus* infection in humans and non-human primates

[6,8,9]. In the present and previous study [12], in total two positive samples were identified for which hematology parameters were available. Both seals had relatively low red blood cells counts. However, about 6% of the seals admitted to SRRC have low red blood cells counts upon admission and there are various causes of anemia or low red blood cell counts in seals, such as malnutrition, hemorrhage, chronic disease, and intoxication [41]. Additional studies need to be performed to elucidate the exact impact of SePV infection in both harbor and grey seals.

In summary, the present study showed that SePV has been circulating among both harbor and grey seals from Dutch coastal waters at least from 1988 onwards. Analysis of sequence data suggested that SePV circulating in the Dutch coastal waters showed mutation rates similar to other parvoviruses while this resulted in only one amino acid change. Although the number of virus variants that was analyzed was limited, these results indicate that SePV circulating among seals of the Dutch coastal waters undergoes gradual alteration similar to human parvovirus B19 [35] but with strong negative selection for amino acid changes. These results provide additional insights into the genetic diversity of parvoviruses.

Supporting Information

Figure S1 Prevalence of SePV. Spleen tissues of harbor and grey seals were tested for the presence of SePV DNA. Indicated is the percentage of samples of spleens of seals in which SePV DNA was detected by real-time PCR for all samples of both harbor and grey seals (A), or for specific years for which samples were available from grey seals (B) or harbor seals (C). Numbers above the x-axis represent the number of samples that was tested.

Acknowledgments

The authors wish to thank all people who helped with collecting samples during the PDV outbreaks in 1988 and 2002 for their commitment. In addition, the authors wish to thank the Natural Environment Research Council's Sea Mammal Research Unit (United Kingdom) for providing seal sample PV880823.6.

Author Contributions

Conceived and designed the experiments: RB RH MdG ADMEO. Performed the experiments: RB RH ARG GJSC MWGvdB MdG. Analyzed the data: RB RH MdG. Contributed reagents/materials/analysis tools: ARG GJSC MWGvdB TK. Wrote the paper: RB RH MdG TK ADMEO.

References

1. Cotmore SF, Tattersall P (2007) Parvoviral Host Range and Cell Entry Mechanisms. In: Karl Maramorosch AJS, Frederick AM, editors. Advances in Virus Research: Academic Press. pp. 183–232.

2. International Committee on Taxonomy of Viruses (2013) Virus Taxonomy: 2013 Release. Available: http://www.ictvonline.org/virusTaxonomy.asp.

3. Cotmore SF, Agbandje-McKenna M, Chiorini JA, Mukha DV, Pintel DJ, et al. (2014) The family Parvoviridae. Arch Virol 159: 1239–1247.

4. Broliden K, Tolfvenstam T, Norbeck O (2006) Clinical aspects of parvovirus B19 infection. J Intern Med 260: 285–304.

5. Bonvicini F, Puccetti C, Salfi NC, Guerra B, Gallinella G, et al. (2011) Gestational and fetal outcomes in B19 maternal infection: a problem of diagnosis. J Clin Microbiol 49: 3514–3518.

6. Heegaard ED, Brown KE (2002) Human parvovirus B19. Clinical microbiology reviews 15: 485–505.

7. Douvoyianni M, Litman N, Goldman DL (2009) Neurologic Manifestations Associated with Parvovirus B19 Infection. Clin Infect Dis 48: 1713–1723.

8. Green SW, Malkovska I, O'Sullivan MG, Brown KE (2000) Rhesus and pig-tailed macaque parvoviruses: identification of two new members of the erythrovirus genus in monkeys. Virology 269: 105–112.

9. O'Sullivan MG, Anderson DC, Fikes JD, Bain FT, Carlson CS, et al. (1994) Identification of a novel simian parvovirus in cynomolgus monkeys with severe anemia. A paradigm of human B19 parvovirus infection. J Clin Invest 93: 1571–1576.

10. Yoo BC, Lee DH, Park SM, Park JW, Kim CY, et al. (1999) A novel parvovirus isolated from Manchurian chipmunks. Virology 253: 250–258.

11. Allander T, Emerson SU, Engle RE, Purcell RH, Bukh J (2001) A virus discovery method incorporating DNase treatment and its application to the identification of two bovine parvovirus species. Proc Natl Acad Sci U S A 98: 11609–11614.

12. Bodewes R, Rubio Garcia A, Wiersma LC, Getu S, Beukers M, et al. (2013) Novel B19-like parvovirus in the brain of a harbor seal. PLoS ONE 8: e79259.

13. Greer LL, Whaley J, Rowles TK (2001) Euthanasia. In: CRC Handbook of Marine Mammal Medicine (Dierauf, L and Gulland, FMD eds), 2nd edition. CRC Press. Chapter 32, pp. 729–737.

14. Rijks JM, Read FL, van de Bildt MW, van Bolhuis HG, Martina BE, et al. (2008) Quantitative analysis of the 2002 phocine distemper epidemic in the Netherlands. Vet Pathol 45: 516–530.

15. Rijks JM, Van de Bildt MW, Jensen T, Philippa JD, Osterhaus AD, et al. (2005) Phocine distemper outbreak, The Netherlands, 2002. Emerg Infect Dis 11: 1945–1948.

16. Osterhaus AD, Vedder EJ (1988) Identification of virus causing recent seal deaths. Nature 335: 20.

17. Osterhaus AD, Groen J, Spijkers HE, Broeders HW, UytdeHaag FG, et al. (1990) Mass mortality in seals caused by a newly discovered morbillivirus. Vet Microbiol 23: 343–350.

18. Jensen T, van de Bildt M, Dietz HH, Andersen TH, Hammer AS, et al. (2002) Another phocine distemper outbreak in Europe. Science 297: 209.

19. Kerr JR (2000) Pathogenesis of human parvovirus B19 in rheumatic disease. Ann Rheum Dis 59: 672–683.

20. Manning A, Willey SJ, Bell JE, Simmonds P (2007) Comparison of tissue distribution, persistence, and molecular epidemiology of parvovirus B19 and novel human parvoviruses PARV4 and human bocavirus. J Infect Dis 195: 1345–1352.

21. Katoh K, Standley DM (2013) MAFFT multiple sequence alignment software version 7: improvements in performance and usability. Mol Biol Evol 30: 772–780.

22. Tamura K, Stecher G, Peterson D, Filipski A, Kumar S (2013) MEGA6: Molecular Evolutionary Genetics Analysis version 6.0. Mol Biol Evol 30: 2725–2729.

23. Posada D (2008) jModelTest: phylogenetic model averaging. Mol Biol Evol 25: 1253–1256.

24. Drummond AJ, Suchard MA, Xie D, Rambaut A (2012) Bayesian phylogenetics with BEAUti and the BEAST 1.7. Mol Biol Evol 29: 1969–1973.

25. Minin VN, Bloomquist EW, Suchard MA (2008) Smooth skyride through a rough skyline: Bayesian coalescent-based inference of population dynamics. Mol Biol Evol 25: 1459–1471.

26. Kosakovsky Pond SL, Frost SD (2005) Not so different after all: a comparison of methods for detecting amino acid sites under selection. Mol Biol Evol 22: 1208–1222.

27. VanDevanter DR, Warrener P, Bennett L, Schultz ER, Coulter S, et al. (1996) Detection and analysis of diverse herpesviral species by consensus primer PCR. J Clin Microbiol 34: 1666–1671.

28. Norja P, Eis-Hubinger AM, Soderlund-Venermo M, Hedman K, Simmonds P (2008) Rapid sequence change and geographical spread of human parvovirus B19: comparison of B19 virus evolution in acute and persistent infections. J Virol 82: 6427–6433.

29. Martina BE, Jensen TH, van de Bildt MW, Harder TC, Osterhaus AD (2002) Variations in the severity of phocid herpesvirus type 1 infections with age in grey seals and harbour seals. The Veterinary Record 150: 572–575.

30. Nollens HH, Gulland FMD, Jacobson ER, Hernandez JA, Klein PA, et al. (2006) Parapoxviruses of seals and sea lions make up a distinct subclade within the genus Parapoxvirus. Virology 349: 316–324.

31. Shackelton LA, Holmes EC (2006) Phylogenetic evidence for the rapid evolution of human B19 erythrovirus. J Virol 80: 3666–3669.

32. Shackelton LA, Parrish CR, Truyen U, Holmes EC (2005) High rate of viral evolution associated with the emergence of carnivore parvovirus. Proc Natl Acad Sci U S A 102: 379–384.

33. Parrish CR, Aquadro CF, Strassheim ML, Evermann JF, Sgro JY, et al. (1991) Rapid antigenic-type replacement and DNA sequence evolution of canine parvovirus. J Virol 65: 6544–6552.

34. Zehender G, De Maddalena C, Canuti M, Zappa A, Amendola A, et al. (2010) Rapid molecular evolution of human bocavirus revealed by Bayesian coalescent inference. Infect Genet Evol 10: 215–220.

35. Suzuki M, Yoto Y, Ishikawa A, Tsutsumi H (2009) Analysis of nucleotide sequences of human parvovirus B19 genome reveals two different modes of evolution, a gradual alteration and a sudden replacement: a retrospective study in Sapporo, Japan, from 1980 to 2008. J Virol 83: 10975–10980.

36. Allison AB, Kohler DJ, Fox KA, Brown JD, Gerhold RW, et al. (2013) Frequent cross-species transmission of parvoviruses among diverse carnivore hosts. J Virol 87: 2342–2347.

37. Trilateral Seal Expert Group (TSEG) (2013) Aerial surveys of Harbour Seals in the Wadden Sea in 2013. Available: www.waddensea-secretariat.org/sites/default/files/downloads/TMAP_downloads/Seals/aerial_surveys_of_harbour_seals_in_the_wadden_sea_in_2013.pdf. Site accessed July 30, 2014.

38. Dietz R, Teilmann J, Andersen SM, Riget F, Olsen MT (2013) Movements and site fidelity of harbour seals (Phoca vitulina) in Kattegat, Denmark, with implications for the epidemiology of the phocine distemper virus. Ices J Mar Sci 70: 186–195.

39. Goodman SJ (1998) Patterns of extensive genetic differentiation and variation among European harbor seals (Phoca vitulina vitulina) revealed using microsatellite DNA polymorphisms. Mol Biol Evol 15: 104–118.

40. Rijks JM (2008) Phocine distemper revisited: Multidisciplinary analysis of the 2002 phocine distemper virus epidemic in the Netherlands [PhD. thesis]: Erasmus University Rotterdam.

41. Bossart GD, Reidarson TH, Dierauf LA, Duffield DA (2001) Clinical pathology. In: CRC Handbook of Marine Mammal Medicine (Dierauf, L and Gulland, FMD eds) 2nd edition. CRC Press. Chapter 19, pp. 394.

18

Australian Fur Seals (*Arctocephalus pusillus doriferus*) Use Raptorial Biting and Suction Feeding When Targeting Prey in Different Foraging Scenarios

David P. Hocking[1,2,3]*, **Marcia Salverson**[3], **Erich M. G. Fitzgerald**[2], **Alistair R. Evans**[1,2]

1 School of Biological Sciences, Monash University, Melbourne, Victoria, Australia, **2** Geosciences, Museum Victoria, Melbourne, Victoria, Australia, **3** Wild Sea Precinct, Zoos Victoria, Melbourne, Victoria, Australia

Abstract

Foraging behaviours used by two female Australian fur seals (*Arctocephalus pusillus doriferus*) were documented during controlled feeding trials. During these trials the seals were presented with prey either free-floating in open water or concealed within a mobile ball or a static box feeding device. When targeting free-floating prey both subjects primarily used raptorial biting in combination with suction, which was used to draw prey to within range of the teeth. When targeting prey concealed within either the mobile or static feeding device, the seals were able to use suction to draw out prey items that could not be reached by biting. Suction was followed by lateral water expulsion, where water drawn into the mouth along with the prey item was purged via the sides of the mouth. Vibrissae were used to explore the surface of the feeding devices, especially when locating the openings in which the prey items had been hidden. The mobile ball device was also manipulated by pushing it with the muzzle to knock out concealed prey, which was not possible when using the static feeding device. To knock prey out of this static device one seal used targeted bubble blowing, where a focused stream of bubbles was blown out of the nose into the openings in the device. Once captured in the jaws, prey items were manipulated and re-oriented using further mouth movements or chews so that they could be swallowed head first. While most items were swallowed whole underwater, some were instead taken to the surface and held in the teeth, while being vigorously shaken to break them into smaller pieces before swallowing. The behavioural flexibility displayed by Australian fur seals likely assists in capturing and consuming the extremely wide range of prey types that are targeted in the wild, during both benthic and epipelagic foraging.

Editor: Brian Lee Beatty, New York Institute of Technology College of Osteopathic Medicine, United States of America

Funding: This study was supported by the Holsworth Wildlife Research Endowment and Monash University. ARE acknowledges the support of the Australian Research Council. The funders had no role in study design, data collection and analysis, decision to publish, or preparation of the manuscript.

* Email: email@david-hocking.com

Introduction

Raptorial biting or pierce feeding was traditionally thought to be the primary prey capture tactic used by most pinnipeds when hunting underwater [1,2]. This hunting tactic involves actively pursuing prey, before striking out with the head or accelerating the whole body to seize prey in the teeth and jaws by biting or snapping [3]. This has been suggested to be the original feeding mode in pinnipeds, as it requires few changes from the original terrestrial bauplan [3]. This is supported by observed tooth condition in some of the earliest fossil pinnipeds, including *Enaliarctos* and *Pteronarctos*, which were found to have distinct wear facets on their teeth, indicating their use in piercing and cutting food [1]. However, study of the feeding mode in extant phocid seals has shown a number of species to instead use suction feeding for prey capture underwater, including the crabeater seal (*Lobodon carcinophaga*) [4], bearded seal (*Erignathus barbatus*) [5], leopard seal (*Hydrurga leptonyx*) [6] and harbour seal (*Phoca vitulina*) [7]. Among extant pinnipeds, the otariid seals (fur seals

and sea lions) are often considered to perform raptorial feeding rather than suction feeding [7]; however, no detailed description has been published documenting their foraging behaviours from first-hand observations.

In this present study we therefore sought to test the hypothesis that otariid seals primarily perform simple raptorial biting when capturing prey underwater. To do this we performed controlled feeding trials with captive Australian fur seals (*Arctocephalus pusillus doriferus*). When hunting at sea this species consumes a wide range of prey, including both schooling epipelagic prey and benthic fish and cephalopods [8–11]. In one study, prey DNA collected from scat samples represented 54 species of bony fish, four cartilaginous fish and four species of cephalopod [12]. Australian fur seals are unusual amongst fur seals in that they primarily perform benthic foraging near the seafloor over the continental shelf, rather than epipelagic foraging in open water, which is more typical of other fur seal species [13–15]. This has been identified though use of time-depth recorders that show them to perform U-shaped dives, where the seal descends directly to the seafloor before tracking along the bottom in search of prey [16]. Arnould and Hindell [16] found that 78% of foraging dives performed by Australian fur seals hunting in central Bass Strait, between Tasmania and mainland Australia, were U-shaped dives close to the seafloor, while the remaining foraging dives were V-shaped dives associated with epipelagic foraging. Given the different conditions faced when hunting near the seafloor versus in open water, it is therefore possible that Australian fur seals may vary their prey capture behaviours when hunting in different settings. Hence, it was also the aim of this study to explore the range of foraging behaviours displayed by captive fur seals when encountering prey that has been presented in different ways.

Methods

Study Animals

Detailed observations were made during feeding trials carried out in the main seal pool in the 'Wild Sea' precinct of Melbourne Zoo (Elliott Ave, Parkville 3052, VIC, Australia). Two adult female Australian fur seals (Bay and Tarwin) were observed in this study (Table 1). Both were brought into captivity after being rescued from the wild in poor condition as juveniles and being deemed unsuitable for release. They lived on display as part of the permanent collection at Melbourne Zoo, where they had also been trained using positive reinforcement to take part in educational displays for the zoo's visiting public. All observations and protocols carried out as part of this work were done under the approval of the Zoos Victoria Animal Care and Ethics Committee (ZV12007; ZV12012).

Foraging Mode and Feeding Cycle

To make direct observations of the foraging tactics used by captive fur seals, we presented each seal individually with dead prey items dropped down a 15 cm diameter PVC pipe positioned in front of the pool's underwater viewing window. This ensured that the seals would encounter the prey items free-floating in the water column approximately 1 meter underwater. Multiple prey items were dropped down the pipe at once to encourage the seals to capture and consume the prey items consecutively underwater during the dive. If the prey was found to float in the pipe or was sinking too slowly, a bucket of pool water was poured down the pipe to flush it out. Two species of fish were used in this trial: yellowtail scad (*Trachurus novaezelandiae*, mean fork length ± s.e.m: 152.9±2.15) and pilchard (*Sardinops* sp., mean fork length ± s.e.m: 144.5±2.71). Unfortunately it was not possible to use live prey during these trials.

We used a high-definition video camera (Sony NX70), filming at 50 frames/second, to record the feeding events through the pool's underwater viewing window. The camera's frame rate was used to measure the duration of some important components of the feeding cycle by counting the number of frames between text markers placed into the video footage using Adobe Premiere Pro CC (Adobe Systems Inc., San Jose, California). We measured the time duration from the first frame where the mouth is seen to open until the frame where maximum gape is achieved (duration of jaw opening). Maximum gape was measured as the distance between the tip of the upper and lower lips; this should be considered an estimate as variation in the orientation of the animal made it very difficult to make precise measurements. Gape was measured using ImageJ 1.45 s (National Institutes of Health, USA). We then measured the duration from maximum gape until the frame where the jaws were fully closed to grip the prey item between the seal's teeth (duration of jaw closing). After the initial capture, prey was often manipulated during further jaw movements until it was

Table 1. Experimental subjects.

Subject	ARKS #	Species	Sex	Est. Date of birth	Body Length (cm)	Girth (cm)	Mass (kg)
Bay	A70598	Australian fur seal *Arctocephalus pusillus doriferus*	F	November 2006	136	88	48
Tarwin	980419	Australian fur seal *Arctocephalus pusillus doriferus*	F	November 1997	145	98	58

transported fully into the mouth and swallowed. The number of mouth movements or chews was tallied and the duration was measured from the last frame of the first jaw closure until the last frame of the final jaw closure that marked the end of the feeding event (duration of prey manipulation). The total duration of the feeding event was measured from the beginning of jaw opening until the end of prey manipulation (event duration). Due to variation in the orientation of the animal it was only possible to make these measurements in a subset of feeding events filmed. Only 11 out of 49 events filmed for Tarwin, and 20 out of 45 for Bay, were included in this analysis.

Variation in Feeding Behaviours

To explore the range of foraging behaviours used in different scenarios, we presented each seal with three different methods of prey presentation. In each method we presented the seal with six dead fish (three whiting *Sillago* sp. and three pilchards *Sardinops* sp.), drawn from their regular daily diet. Again, feeding trials were performed individually for each seal to prevent competition between animals. In the first method we threw the six prey items loosely around the surface of the pool so that the seal encountered them free-floating in open water (scatter feed). For the second presentation method we placed the six fish into a hollow plastic ball with small, round openings in its side. The ball was attached to an elastic bungee cord that allowed it to be manipulated and pushed around in the water by the seals (mobile ball device).

In the third method, prey were concealed within a static feeding device that could not be manipulated to knock out the hidden prey (static box device). The device was made from a plastic storage box to which a plastic front-plate had been attached into which we had set recessed PVC tubes. These tubes were arranged in four columns that alternated between 5 cm and 10 cm diameter tubes. The top and bottom tubes were connected on the inside of the device, while the middle tube opened into the internal cavity of the box. To prevent fish falling into the connecting tube or the inside of the box, fiberglass fly-wire mesh was glued to the inside ends of the recessed tubes. The box was attached to a solid wooden frame that was temporarily tied to the pool fence with Velcro straps. The box could be raised or lowered on wooden beams that slid over the frame, before being locked into a static position either above or below the water. When in use, the six fish were placed into the recessed tubes and a separate plastic board was hooked over the front of the device. The box was then lowered into the water and locked in position, before the plastic board was pulled away. This board prevented fish falling out of the device with the flow of bubbles as it was lowered into the water.

To conduct these feeding trials one seal was given access to the main pool and left for 10 minutes to acclimate before the experimental session began. After acclimation, one of the seal's keepers entered the exhibit to present it with its prey using one of these three methods. We then filmed the seal's behaviour at 25 frames/second from above and below the water for a 20-minute observation period to document all of the behaviours associated with capturing and handling their food. After the first seal completed its experimental session the seals were swapped so that the second seal could participate. Each seal only participated in one experimental session per day and all sessions were carried out between 0730–1030 h before their first training session. We carried out five replicates of each experimental treatment for each seal, so in total the two Australian fur seals were presented with 30 prey items for each of the three prey presentation methods.

The video footage for each experimental session was viewed in Adobe Premiere Pro CC so that the behaviours used to capture and handle prey could be compared among treatments. Where

visible, the initial prey capture tactic for each prey item was recorded as either biting or suction. The frequency of these was compared among the three prey presentation methods for each seal using a Pearson's chi-squared test using the standardized residuals to assess significance based on the critical value of ± 1.96, which corresponds to an alpha (α) of 0.05. Statistical analyses were conducted using R statistical and graphical environment (R version 3.0.2, R development core team, 2013) [17].

Results

Foraging Mode and Feeding Cycle

When capturing free-floating prey in open water both subjects primarily used raptorial biting to secure their prey (Fig. 1; Video S1). This involved opening and snapping their jaws shut over the prey item so that it was caught between the teeth (Fig. 2). Most prey items were captured from the side so that they were held between the postcanine teeth at the end of the first bite (Fig. 1c). In many events suction was used in combination with biting to draw prey to within range of the teeth as part of the initial capture (Video S1). This was identified through movement of the prey item towards the mouth before the jaws snapped shut. Following the use of suction, lateral water expulsion was often visible, with water that had been drawn into the mouth during suction being purged via the sides of the mouth. However, in some events the prey item showed no movement towards the mouth and no water expulsion was visible after the initial bite, indicating that little or no suction was used in these events. If only a small amount of water was drawn into the mouth it is possible that this could have been swallowed along with the prey rather than being expelled, or expelled without being visible. In a small number of events extremely rapid suction appeared to play a major role in drawing prey almost fully into the mouth before the jaws closed (Fig. 3; Video S1). This occurred when the seal approached the prey item head on rather than from the side, possibly because the item could be more easily drawn directly into the oral cavity when in this orientation, rather than being first caught laterally between the postcanine teeth.

Once captured between the teeth, further mouth movements or chews were used to re-orient the prey item before it was drawn fully into the oral cavity (Video S1). Suction was likely used to transport the prey item from where it was held by the teeth into the oral cavity, as indicated by the performance of lateral-water expulsion following transport of the prey item. Unfortunately it was not possible to identify when swallowing occurred within the feeding cycle. Kinematic variables are summarized in Table 2.

Variation in Feeding Behaviours

Both subjects varied their foraging behaviours for the three methods of prey presentation. When targeting prey during the scatter feed treatments, they swam rapidly as they approached and captured each of the six prey items near the surface soon after they landed in the water. Both seals used raptorial biting as the primary prey capture tactic in 100% of feeding events where the initial capture tactic was clearly visible. However, although raptorial biting was considered the primarily capture tactic used, it is likely that some suction was also used in combination with biting to draw prey within range of the teeth, as observed in the first part of this study. Most prey items were captured in the jaws by biting near the head of the fish so that the postcanine teeth pierced the prey item. If first captured near the tail or from the side, the prey item was manipulated underwater using mouth movements or chews to re-orient it before it was swallowed head first.

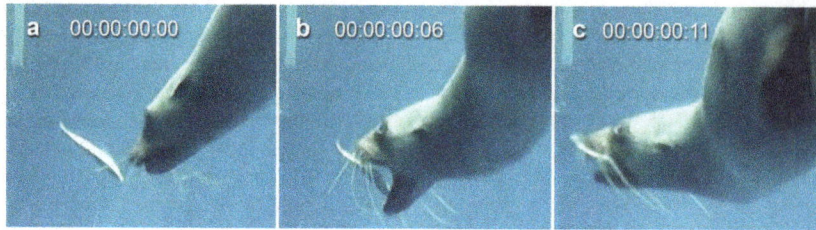

Figure 1. Raptorial biting used to capture a prey item floating in open water filmed at 50 frames/second. a) Jaws are closed as the seal approaches the prey item, b) Jaws open to maximum gape and overtake the prey item, c) jaws snap shut over prey item capturing it between the postcanine teeth. Prey capture event performed by Bay (ARKS# A70598), a female Australian fur seal (*Arctocephalus pusillus doriferus*). Time displayed as hours:minutes:seconds:frames. For the footage see Video S1.

We found both seals to manipulate the mobile ball device by pushing it with their muzzles in an effort to knock out concealed prey items (Fig. 4). Bay in particular would carefully manipulate the ball with her muzzle, while looking through the hole in its side to see where the prey items were located (Fig. 4a). When a fish floated close to the opening, she moved her mouth over the hole and rapidly pushed the ball forward to knock the fish into her mouth (Video S1). If a fish started to fall from the ball it was gripped by biting with the anterior teeth, before being pulled from the device and consumed. If the prey item remained concealed within the ball it was extracted using suction. This involved the seal positioning its mouth over the opening in the ball (possibly using the vibrissae to locate the opening) and pushing it, while also generating suction to draw out the fish. Use of suction was indicated by lateral water expulsion, where water drawn into the mouth was expelled via the sides of the mouth after the initial capture (Fig. 4c). When using this device Bay used biting as the initial capture tactic in 64.3% of feeding events, while suction feeding was used in 35.7%. Tarwin was similar with 68.8% of prey initially captured by biting, while 31.3% was captured by suction.

Both seals also used suction when capturing prey from the static box device. They explored each tube in turn with their eyes and vibrissae until they located a hidden prey item. If a prey item floated partially out of the device it was captured by biting, using the anterior teeth, before being pulled out of the device and consumed. If fully concealed, the seal placed its mouth over the targeted recessed tube and sucked the prey item out (Fig. 5; Video S1). Water that had been drawn into the mouth along with prey item was purged via lateral water expulsion as the jaws closed over the fish (Fig. 5d). For both subjects, suction feeding was used as the main prey capture tactic when drawing prey from the static box device. Bay used suction feeding in 67.9% of feeding events while Tarwin used it in 79.3% of feeding events. Remaining prey

captures were performed using biting after the prey item had floated partly out of the device and to within range of the teeth.

We found that prey presentation method had a significant effect on whether prey was initially captured by raptorial biting or suction for both subjects (Bay: χ^2 (2) = 21.4, p<0.01, Tarwin: χ^2 (2) = 27.07, p<0.01). Significantly more prey items were captured by biting and fewer by suction, during the scatter feed where no suction feeding events were observed (standardized residuals = Bay 3.92, Tarwin 4.18; Fig. 6). In contrast, the opposite was true when capturing prey from the static box device, where both seals used significantly more suction (and significantly less biting) to draw out the concealed prey items (standardized residuals = Bay 3.94, Tarwin 4.9; Fig. 6). While raptorial biting was used more frequently than suction when using the mobile ball device, this difference was not found to be statistically significant for either seal (standardized residuals = Bay 0.48, Tarwin 1.44; Fig. 6).

In addition to suction and biting, both subjects were also observed to blow bubbles out of their noses towards or into the mobile ball and static box devices. When using the mobile ball device, bubbles were blown either below the ball or into the hole in its side. As they flowed around and through the device they may have helped to knock out concealed prey items. When using the static box device Bay seldom used bubbles and blew them in the direction of the device without targeting a specific prey item or recessed tube. Tarwin in contrast appeared to use bubble blowing as a targeted foraging tactic where she pressed her nose against the tube and blew a focused stream of bubbles directly into it at the hidden prey item (Video S1). This behaviour was used in all of Tarwin's static box sessions; however, it was generally only used after most of the prey items had already been captured by suction and so only one prey item was successfully captured using this tactic.

It is interesting to note that regardless of the enrichment treatment or prey capture tactic used, the prey items caught in this

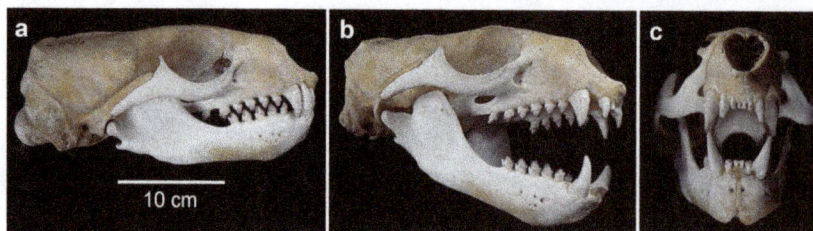

Figure 2. Australian fur seal skull and dentition (*Arctocephalus pusillus doriferus***; NMV C5717).** a) jaws in occlusion showing simple interlocking postcanines, b) jaws opened to approximate gape used during raptorial biting, c) circular opening at front of the mouth formed by the canines and incisors when gape is small.

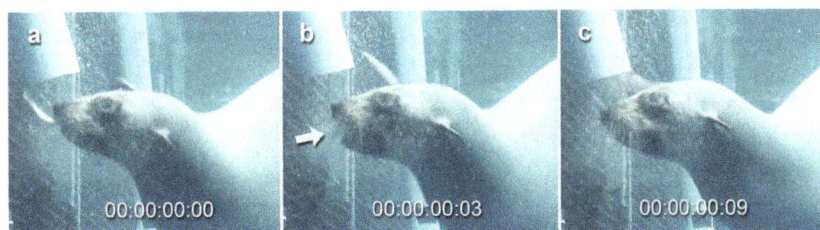

Figure 3. Suction used to draw a prey item into the oral cavity filmed at 50 frames/second. a) prey item held loosely between the lips, b) prey item sucked into the oral cavity as the jaws opened to maximum gape, the arrow indicates the direction of prey movement towards the mouth, c) mouth closes and water is expelled via the lateral sides of the mouth, although this is difficult to see in cloudy water. Prey capture event performed by Bay (ARKS# A70598), a female Australian fur seal (*Arctocephalus pusillus doriferus*). Time displayed as hours:minutes:seconds:frames. For the footage see Video S1.

study were almost all swallowed head first. It was possible to identify whether the prey were swallowed head or tail first in 88 of the 155 fish fed out and of these, 85 were eaten head first, while only three were eaten tail first. Some of the remaining prey items (that could not be distinguished as having been either swallowed head or tail first) were processed by shaking at the surface before being swallowed. This involved the seal holding the prey item in its postcanine teeth before vigorously shaking its head from side to side until the prey item ripped in two (Video S1). When it broke, half the prey item was thrown across the pool while the other half was retained in the mouth and swallowed. The seal then collected and consumed the remaining pieces from around the pool.

Discussion

These results show that while fur seals do indeed use raptorial biting as their primary prey capture mode when targeting prey in open water, in other scenarios suction feeding is more frequently used where it allows them to capture prey that cannot be caught by biting alone. Even when performing classic raptorial biting, suction was often still used in combination with biting to draw prey to within range of the teeth. This may be very important during wild feeding when pursuing evasive prey, as suction generation would counteract any water flow away from the seal's mouth as a product of the compressive bow wave generated by the seal's movement through the water [18].

As we would expect, the phases of the feeding cycle when performing raptorial biting were found to differ from those identified in the feeding cycle of primarily suction feeding marine mammals [5,7,19–21]. No preparatory phase was identified where the jaws were opened to a partial gape (10–30% of maximum gap) prior to the jaw opening fully. This type of preparatory phase has

been identified in a number of species of suction feeding odontocete cetaceans [19–21] as well as in bearded seals [5]. In this aspect Australian fur seals are similar to leopard seals and harbour seals, which have also been found to lack a distinct preparatory phase prior to the main jaw opening [6,7].

Jaw opening was similar to that observed in other marine mammal species, but rather than the prey item being drawn directly into the oral cavity during the subsequent gular or hyolingual depression phase, where suction is generated by retraction of the tongue and hyoid apparatus, the jaws were instead snapped shut on the prey item so that it was caught between the teeth. Once captured, the prey item was manipulated with further mouth movements or chews, before it was drawn fully into the oral cavity for swallowing. The use of suction for transporting the prey item from where it was held by the teeth into the oral cavity was very similar to that described for captive leopard seals [6].

When targeting prey in open water the fur seals consistently aimed for the fish's head. This presumably allows the seal to more easily swallow the prey head first, minimizing risk to the seal of being pierced by the fish's spines as it passes down the seal's throat. When pursuing a fleeing prey item in the wild, the fish's head would presumably also be the body part furthest away from the seal's mouth. Therefore it is possible that by initially aiming for the head, seals are more likely to capture the prey item, even if they hit it further down the body. Targeting the head might also function to kill or disable prey more quickly, allowing the seal to subsequently consume it using as little energy as possible.

At small gapes, the canines and incisors of an Australian fur seal form a circular mouth opening (Fig. 2c). It is possible that this formation assists with the generation of focused suction at the front of the mouth when the gape is small (Fig. 3b). Formation of a

Table 2. Summary of kinematic variables.

Kinematic Variable	Bay	N	Tarwin	N
Jaw opening (sec)	0.095±0.006	20	0.093±0.006	11
Jaw closing (sec)	0.068±0.005	20	0.082±0.011	11
Prey manipulation (sec)	1.125±0.116	20	1.980±0.172	11
Event duration (sec)	1.288±0.116	20	2.155±0.169	11
Maximum gape (cm)	6.753±0.341	20	5.600±0.342	11
Number of Jaw movements	4.800±0.485	20	6.818±0.352	11

Values are means ± s.e.m.

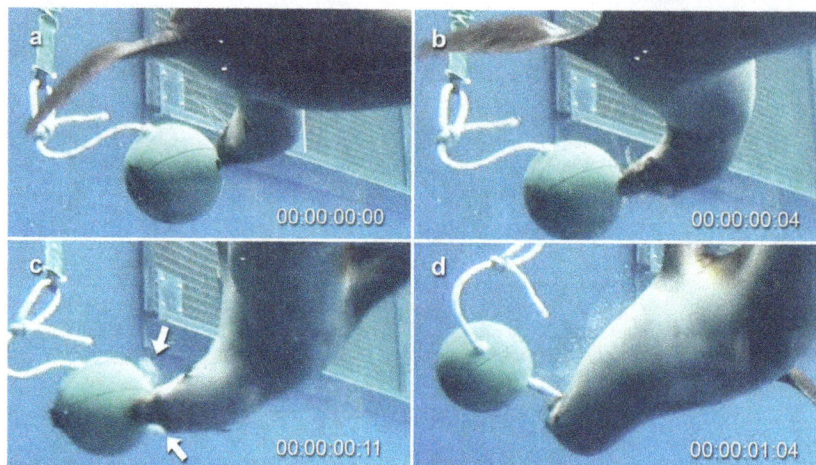

Figure 4. Object manipulation and suction used to draw prey from the mobile ball device filmed at 25 frames/second. a) ball was carefully manipulated using the muzzle, while looking through the hole in its side for concealed prey, b) when a prey item is seen near the opening the seal moved its mouth over the hole before pushing the ball forward, while also generating suction to draw out the prey item, c) water expulsion visible as it is expelled via the sides of the mouth following suction (arrows indicate cloud of turbid water being expelled), d) once protruding from the hole the prey item was gripped using the anterior teeth before being pulled out and consumed. Prey capture event performed by Bay (ARKS# A70598), a female Australian fur seal (*Arctocephalus pusillus doriferus*). Time displayed as hours:minutes:seconds:frames. For the footage see Video S1.

circular mouth opening has been found to be important in other suction feeding pinnipeds, including bearded seals [5], harbour seals [7] and walruses (*Odobenus rosmarus*) [22,23]. Leopard seals were found to seal off the lateral sides of their mouth using their cheeks when performing suction feeding [6]. The fur seals in this study likely used their cheeks in a similar way, especially when using suction alone or when using it in combination with raptorial biting (Video S1).

The preference shown by Australian fur seals for using suction feeding when drawing prey from the static box device is similar to observations in other pinniped species. In captive feeding trials, bearded seals were found to only use suction when capturing prey, even when that prey protruded from the surface of a static feeding apparatus so that it was within range of the teeth [5]. When using a feeding apparatus, harbour seals were found to use suction feeding in 84% of feeding events, with remaining prey items captured by biting, before being drawn out and consumed in a similar manner to what we observed in captive Australian fur seals

[7]. In contrast, when capturing live, free-swimming prey harbour seals were found to use raptorial biting to capture large prey, while small prey was captured by suction [24]. When we observed our subjects to use rapid suction alone to capture prey in open water, it was to capture the smaller pilchards, suggesting that Australian fur seals may also show a preference towards suction feeding when capturing smaller prey. In these events the prey items were captured from directly in front of the mouth. This is similar to what was observed in captive leopard seals by Hocking et al. [6] where prey was sucked into the mouth from within approximately 5 cm of the tip of the muzzle. In contrast, crabeater seals were observed to suck pilchards out of a channel (section of pipe cut in half) from up to 50 cm [4].

In both subjects the eyes remained open throughout the prey capture event as seen in Figs 1-5. This contrasts with observations in phocids, where bearded seals [5], harbour seals [7] and leopard seals [6] were found to close their eyes during the initial capture. Maintaining visual contact throughout the prey capture event may

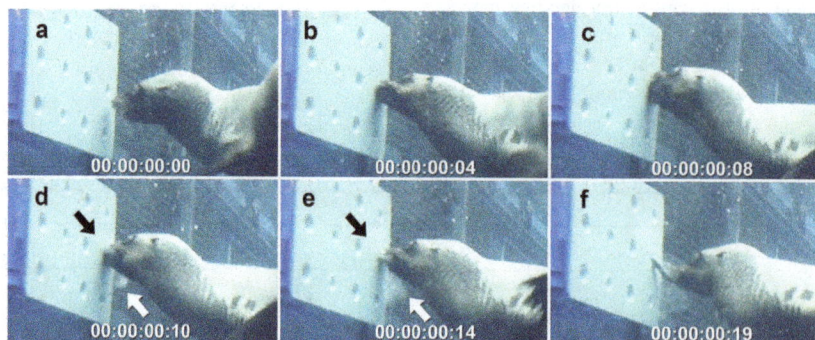

Figure 5. Suction used to draw prey out of the static box device filmed at 25 frames/second. a) prey item was found by carefully looking into each recessed tube, b) mouth was positioned over the opening using eyes and whiskers, c) suction generated by widening the gape and retracting tongue and hyoid, d) jaws closed to grip prey with anterior teeth, while performing lateral water expulsion where seawater drawn into mouth during suction was expelled via sides of the mouth (arrows indicate cloud of turbid water being expelled), e-f) pull prey out of device while holding it with the anterior teeth before performing further manipulation and swallowing. Prey capture event performed by Tarwin (ARKS# 980419), a female Australian fur seal (*Arctocephalus pusillus doriferus*). Time displayed as hours:minutes:seconds:frames. For the footage see Video S1.

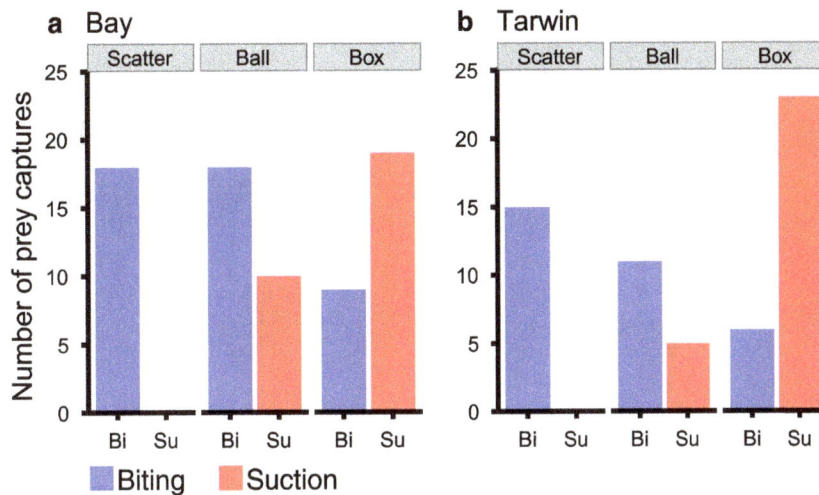

Figure 6. Biting versus suction as initial prey capture tactic when encountering prey in different ways. Number events where biting (Bi) or suction (Su) was the initial prey capture tactic when prey was presented as part of a scatter feed or when using the mobile ball or static box feeding device. Data were recorded for each prey item except where the capture tactic was unclear from the video playback.

assist when targeting fast or evasive prey during daytime epipelagic foraging. When targeting benthic prey in association with the seafloor we might expect Australian fur seals to close their eyes to protect them against substrates that might be abrasive to the eyes [5]. There is also less light in deep water and Australian fur seals are known to forage both during daylight hours and at night [16]. Marshall et al. [5] suggested that bearded seals might close their eyes to increase the tactile sensory modality, by closing down their visual sense. Both subjects were observed to use their vibrissae as well as their vision to locate their prey; however, it is possible that maintaining vision during prey capture is more important in Australian fur seals than it is in these other pinnipeds. Alternatively, there may simply be less risk to the eyes during wild foraging in Australian fur seals, lessening the need to close their eyes during the initial capture, although this seems less likely given their benthic foraging habits.

Following use of suction, lateral water expulsion was observed. This likely allows the seals to swallow prey without ingesting large quantities of seawater. It also functions to re-set the feeding apparatus by emptying it of seawater in preparation for the next suck. In these trials water expulsion was clearly observed as a cloud of bubbles and turbid water being ejected from the lateral sides of the mouth. This was very similar to that described in captive leopard seals [6] and harbour seals [7]. It is possible that some water expulsion could occur without suction, as it is likely that a small volume of water would be captured in the mouth even during raptorial feeding; however, strong lateral water expulsion of a larger volume was not observed in the absence of suction. The presence of clear lateral water expulsion seems to be a good indicator of whether suction was used during the prey capture. When water expulsion is forcefully directed out of the front of the mouth, rather than laterally via the cheeks and postcanine tooth row, it is known as hydraulic jetting and can be considered "the opposite behaviour to suction" [5]. This behaviour has been recorded in a number of pinniped species including bearded seals [5] and harbour seals [7] where it has been used to knock concealed prey out of feeding apparatuses. In these studies it was detected as a positive spike in ambient water pressure. It is possible that the Australian fur seals used in this study may also have used hydraulic jetting when knocking prey from concealment within the

static box device. Lateral water expulsion was clearly visible in alternation with suction as water was expelled via the sides of the mouth in preparation for the next suction event. However, it is possible that some water was also directed anteriorly into the recessed tube, where it may have assisted in dislodging the prey item before it was drawn out with the next suck. Unfortunately, we were not able to make direct measurements of the ambient water pressure within the recessed tubes as part of this trial. Hydraulic jetting is used by walruses and bearded seals to help excavate benthic infauna from sediment during wild foraging [5,22,23]. Harbour seals have also been observed in the wild digging in soft sediment with their flippers and muzzle when hunting sand lance (Ammodytes dubius) [25]. Given the benthic foraging habits of the Australian fur seal, it is possible that they too use suction feeding and hydraulic jetting when targeting cryptic prey.

Another behaviour that may be useful when targeting cryptic prey is the use of focused bubble blowing to knock or scare prey out of hiding. This type of bubble blowing has previously only been observed in wild Weddell seals (Leptonychotes weddelli), which having been found to blow bubbles into crevices in the ice to flush out fish so that they could be captured in open water [26]. Given that only one of our two subjects displayed targeted bubble blowing as a foraging tactic, it is possible that this individual learned this behaviour in captivity. But given the benthic foraging habits of this species, it is still possible that Australian fur seals hunting near the seafloor use this type of behaviour.

It is unclear why some of the prey items were taken to the surface to be processed by shaking prior to being swallowed in pieces. All of the items used in this study were small fish of a similar size, so that we could determine if prey presentation method, rather than the properties of the prey item, altered their foraging behaviours. In the wild, Australian fur seals have been observed to process prey that is too large to swallow whole using shake processing. Given that all of the prey items presented in this study were of a size that could easily be swallowed whole, it seems likely that the prey items were processed by shaking simply as a form of play behaviour. We are further exploring the use of prey processing when handling different types and sizes of prey as part of future research.

The results of these captive trials confirm the hypothesis that Australian fur seals primarily use raptorial biting as their default feeding mechanism when capturing free-floating prey; however, rather than simply involving snapping at prey with the jaws, this study showed that raptorial feeding is instead a complex foraging behaviour that involves the combined use of biting and suction to efficiently capture prey. Given that these seals are also able to use strong suction alone to draw prey into the oral cavity, we must conclude that rather than being a less efficient prey capture strategy [3], raptorial biting must be adaptive for otariid seals that have evolved to favour this feeding mode. The seals in this study also displayed great flexibility in their foraging behaviours, with the ability to employ a range of other tactics, including suction feeding, bubble blowing and possibly hydraulic jetting, when encountering prey under different conditions. While raptorial biting was the default tactic used when hunting free-floating prey, focused suction feeding was the most common tactic used when uncovering or extracting hidden prey. Given the similarity in morphology between the different species of otariid seals, it is likely that many others perform an equally broad range of behaviours when hunting at sea. For the Australian fur seal, this combination of behaviours is likely important to their success as top predators, allowing them to successfully exploit a huge diversity of prey species in environments ranging from the seafloor to the water's surface.

Acknowledgments

We thank our two anonymous reviewers for critically reading early versions of this manuscript, Zoos Victoria and their seal keepers for providing access to the animals in their care and for their assistance in carrying out this research, Marcus Salton, Monique Ladds and Travis Park for feedback on early drafts of this manuscript, Museum Victoria and David Paul for the photographs in figure 2.

Author Contributions

Conceived and designed the experiments: DH AE. Performed the experiments: DH MS AE. Analyzed the data: DH AE. Contributed reagents/materials/analysis tools: DH MS AE. Wrote the paper: DH MS EMGF AE.

References

1. Adam PJ, Berta A (2002) Evolution of prey capture strategies and diet in the Pinnipedimorpha (Mammalia, Carnivora). Oryctos 4: 83–107.
2. Jones KE, Ruff CB, Goswami A (2013) Morphology and Biomechanics of the Pinniped Jaw: Mandibular Evolution Without Mastication. The Anatomical Record 296: 1049–1063.
3. Werth A (2000) Feeding in marine mammals. In: Schwenk K, editor. Feeding: form function and evolution in tetrapod vertebrates. San Diego: Academic Press. pp. 487–526.
4. Klages NTW, Cockroft VG (1990) Feeding behaviour of a captive crabeater seal. Polar Biology 10: 403–404.
5. Marshall CD, Kovacs KM, Lydersen C (2008) Feeding kinematics, suction and hydraulic jetting capabilities in bearded seals (Erignathus barbatus). The Journal of Experimental Biology 211: 699–708.
6. Hocking DP, Evans AR, Fitzgerald EMG (2013) Leopard seals (Hydrurga leptonyx) use suction and filter feeding when hunting small prey underwater. Polar Biology 36: 211–222.
7. Marshall CD, Wieskotten S, Hanke W, Hanke FD, Marsh A, et al. (2014) Feeding kinematics, suction, and hydraulic jetting performance of Harbor seals (Phoca vitulina). PLoS ONE 9(1): e86710. doi:10.1371/journal.pone.0086710.
8. Gales R, Pemberton D (1994) Diet of the Australian fur seal in Tasmania. Australian Journal of Marine and Freshwater Research 45: 653–664.
9. Gales R, Pemberton D, Lu CC, Clarke MR (1993) Cephalopod diet of the Australian fur seal: Variation due to location, season and sample type. Australian Journal of Marine and Freshwater Research 44: 657–671.
10. Hume F, Hindell MA, Pemberton D, Gales R (2004) Spatial and temporal variation in the diet of a high trophic level predator, the Australian fur seal (Arctocephalus pusillus doriferus). Marine Biology 144: 407–415.
11. Kirkwood R, Hume F, Hindell MA (2008) Sea temperature variations mediate annual changes in the diet of Australian fur seals in Bass Strait. Marine Ecology Progress Series 369: 297–309.
12. Deagle BE, Kirkwood R, Jarman SN (2009) Analysis of Australian fur seal diet by pyrosequencing prey DNA in faeces. Molecular Ecology 18: 2022–2038.
13. Arnould JPY, Costa DP (2006) Sea lions in drag, fur seals incognito: insights from the otariid deviants. In: Trites AW, Atkinson SK, DeMaster DP, Fritz LW, Gelatt TS, et al., editors. Sea Lions of the World. Fairbanks: Alaska Sea Grant College Program. pp. 309–323.
14. Arnould JPY, Kirkwood R (2008) Habitat selection by female Australian fur seals (Arctocephalus pusillus doriferus). Aquatic Conservation: Marine and Freshwater Ecosystems 17: S53–S67.
15. Kirkwood R, Goldsworthy S (2013) Fur seals and sea lions. Collingwood: CSIRO Publishing.
16. Arnould JPY, Hindell MA (2001) Dive behaviour, foraging locations, and maternal-attendance patterns of Australian fur seals (Arctocephalus pusillus doriferus). Canadian Journal of Zoology 79: 35–48.
17. R Core Team (2013) R: A language and environment for statistical computing. Vienna, Austria: R Foundation for Statistical Computing.
18. Werth A (2006) Odontocete suction feeding: experimental analysis of water flow and head shape. Journal of Morphology 267: 1415–1428.
19. Kane EA, Marshall CD (2009) Comparative feeding kinematics and performance of odontocetes: belugas, Pacific white-sided dolphins and long-finned pilot whales. The Journal of Experimental Biology 212: 3939–3950.
20. Bloodworth B, Marshall CD (2005) Feeding kinematics of Kogia and Tursiops (Odontoceti: Cetacea): characterization of suction and ram feeding. The Journal of Experimental Biology 208: 3721–3730.
21. Werth A (2000) A kinematic study of suction feeding and associated behaviour in the long-finned pilot whale, Globicephala melas (Traill). Marine Mammal Science 16: 299–314.
22. Kastelein RA, Mosterd P (1989) The excavation technique for molluscs of Pacific walrusses (Odobenus rosmarus divergens) under controlled conditions. Aquatic Mammals 15: 3–5.
23. Kastelein RA, Gerrits NM, Dubbeldam JL (1991) The anatomy of the walrus head (Odobenus rosmarus): part 2. Description of the muscles and of their role in feeding and haul-out behavior. Aquatic Mammals 17: 156–180.
24. Ydesen KS, Wisniewska DM, Hansen JD, Beedholm K, Johnson M, et al. (2014) What a jerk: prey engulfment revealed by high-rate, super-cranial accelerometry on a harbour seal (Phoca vitulina). The Journal of Experimental Biology 217: 2239–2243
25. Bowen WD, Tully D, Bones DJ, Bulheier BM, Marshall GJ (2002) Prey dependent foraging tactics and prey profitability in a marine mammal. Marine Ecology Progress Series 244: 235–245.
26. Davis RW, Fuiman LA, Williams TM, Collier SO, Hagey WP, et al. (1999) Hunting behavior of a marine mammal beneath the Antarctic fast ice. Science 283: 993–996.

19

Patterns of Spatial and Temporal Distribution of Humpback Whales at the Southern Limit of the Southeast Pacific Breeding Area

Chiara Guidino[1], Miguel A. Llapapasca[2], Sebastian Silva[3], Belen Alcorta[3], Aldo S. Pacheco[4]*

1 University of Technology of Sydney, Sydney, Australia, **2** Oficina de Investigaciones en Depredadores Superiores, Instituto del Mar del Perú, Callao, Peru, **3** Pacifico Adventures-Manejo Integral del Ambiente Marino S.A.C., Los Organos, Peru, **4** Instituto de Ciencias Naturales Alexander von Humboldt, Universidad de Antofagasta, Antofagasta, Chile

Abstract

Understanding the patterns of spatial and temporal distribution in threshold habitats of highly migratory and endangered species is important for understanding their habitat requirements and recovery trends. Herein, we present new data about the distribution of humpback whales (*Megaptera novaeangliae*) in neritic waters off the northern coast of Peru: an area that constitutes a transitional path from cold, upwelling waters to warm equatorial waters where the breeding habitat is located. Data was collected during four consecutive austral winter/spring seasons from 2010 to 2013, using whale-watching boats as platforms for research. A total of 1048 whales distributed between 487 groups were sighted. The spatial distribution of humpbacks resembled the characteristic segregation of whale groups according to their size/age class and social context in breeding habitats; mother and calf pairs were present in very shallow waters close to the coast, while dyads, trios or more whales were widely distributed from shallow to moderate depths over the continental shelf break. Sea surface temperatures (range: 18.2–25.9°C) in coastal waters were slightly colder than those closer to the oceanic realm, likely due to the influence of cold upwelled waters from the Humboldt Current system. Our results provide new evidence of the southward extension of the breeding region of humpback whales in the Southeast Pacific. Integrating this information with the knowledge from the rest of the breeding region and foraging grounds would enhance our current understanding of population dynamics and recovery trends of this species.

Editor: Syuhei Ban, University of Shiga Prefecture, Japan

Funding: The authors received no specific funding for this work. The tourism company Pacifico Adventures the collection of the data for this research during whale-watching trips. Co-authors Sebastian Silva and Belen Alcorta are affiliated with Pacifico Adventures. Pacifico Adventures provided support in the form of salaries for authors SS and BA, but did not have any additional role in the study design, data collection and analysis, decision to publish, or preparation of the manuscript. The specific roles of these authors are articulated in the author contributions section.

Competing Interests: The authors have the following interests: Co-authors Sebastian Silva and Belen Alcorta are affiliated with Pacifico Adventures-Manejo Integral del Ambiente Marino S.A.C. There are no patents, products in development or marketed products to declare.

* Email: aldo.pacheco@uantof.cl

Introduction

Several taxa of marine megafauna (e.g., sea turtles, albatrosses, cetaceans) undertake long distance migrations between functionally different habitats types, usually from breeding areas to foraging grounds and vice versa. Oceanographic structures such as thermal fronts are key components driving baleen whales movements and foraging patterns as prey aggregates within and surrounding the fronts [1,2]. The functional role of thermal fronts in breeding areas is less understood since baleen whales feed little during their breeding season [3]. It is thought that humpback whales (*Megaptera novaeangliae*) may use thermal fronts as an environmental cue indicating the proximity of warm neritic waters during their breeding migration from oceanic waters [4,5]. However, little is known about the distributional patterns of these whales in such habitats.

Among humpback whale populations, the individuals inhabiting the Southeast Pacific region perform the longest migration (ca., 8000 km), moving from Antarctic and Magellanic feeding grounds to the breeding region in neritic waters from the coast of Ecuador up to Costa Rica, crossing the Equator during the austral winter/spring [6–8]. The travelled distance is approximately 3000 to 4000 km longer than that estimated for other migrating humpback whale populations worldwide [9]. This extended migration is explained because humpbacks whales seem to search for warmer habitats to avoid the influence of the cold upwelling waters of the Humboldt Current ecosystem extending from central-southern Chile (~40°S) to northern Peru (~4°S) during their breeding migration [10,11]. The average sea surface temperature along neritic waters of the Humboldt Current system during the austral winter and spring months varies between 14–18°C [10], which is colder than the estimated thermal range i.e., 21.1–28.3°C. This is characteristic of breeding and calving areas for humpback whales worldwide [11].

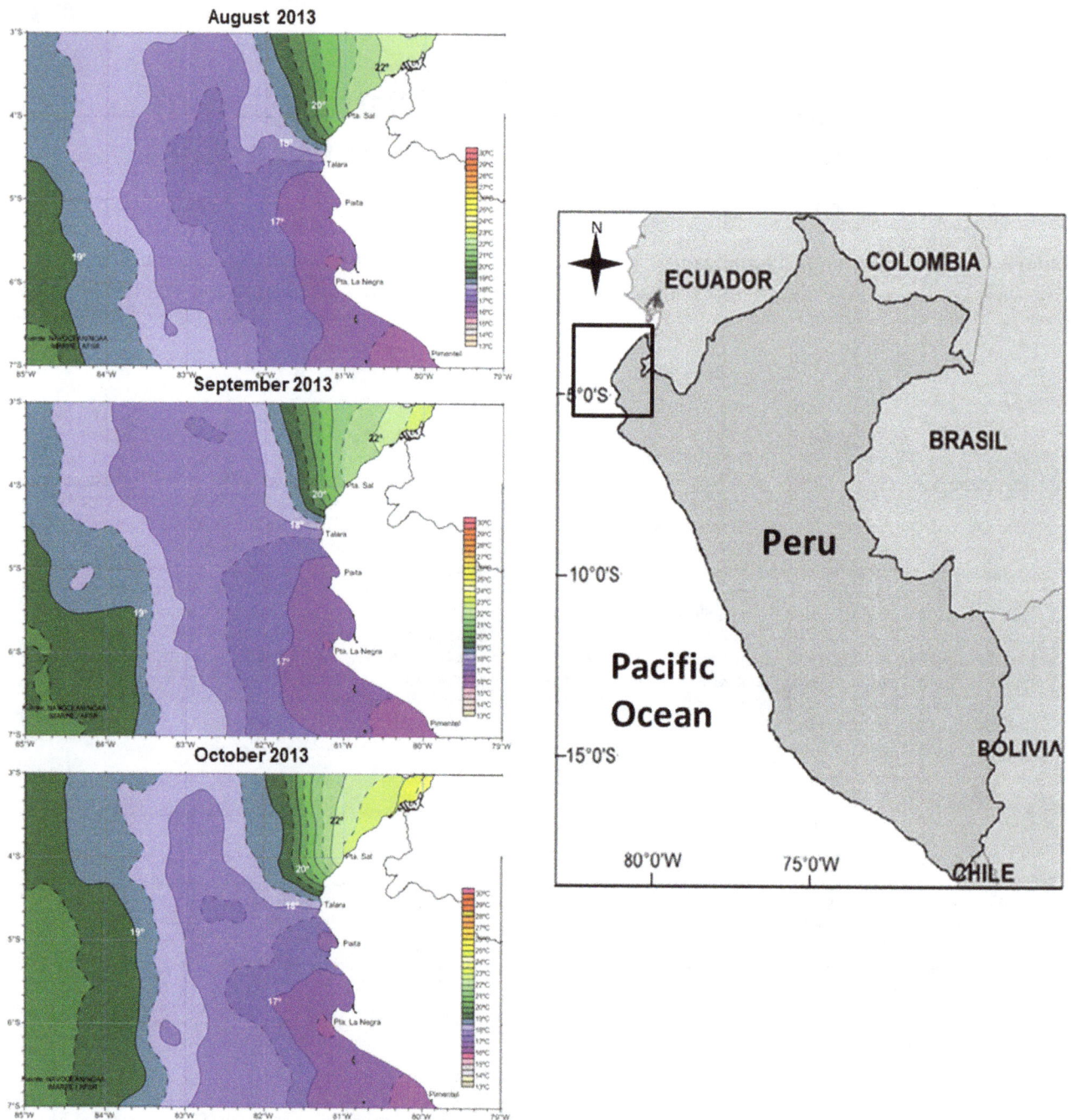

Figure 1. The map at the right shows the location of the northern Peru region. Maps at the left show the distribution of sea surface isotherms (monthly means) during the three months surveys in 2013. The thermal front formed by the convergence of warm equatorial waters coming from the equator and waters of the Humboldt Current system flowing northward is observed. Isotherm images are freely available in http://satelite.imarpe.gob.pe/uprsig/sst_prov.html.

Humpback whales are thought to breed in calm, warm waters because their blubber layer at birth is thin, allowing energy to be invested in growth and development [12], which ultimately enhances calf development and survival. In breeding/calving regions, mother and calf pairs usually prefer to inhabit calm, shallow waters because this may prevent disturbance by competitive males [13,14] and potential predation from killer whales [15,16]. This habitat preference explains the conspicuous distributional pattern observed in breeding/calving regions in which

groups of whales segregate according to their size/age composition and social context. Although all individuals are present in neritic waters, mother and calf groups typically occupy shallower waters closer to shore, while groups composed of adult (e.g., competitive groups) and sub-adult individuals are widely distributed between shallow, moderate depths and the continental shelf break. This distributional pattern has been recorded in wintering areas of the Southeast Pacific population off the coast of Ecuador [5], [17] and Colombia [18], resembling the patterns observed in breeding

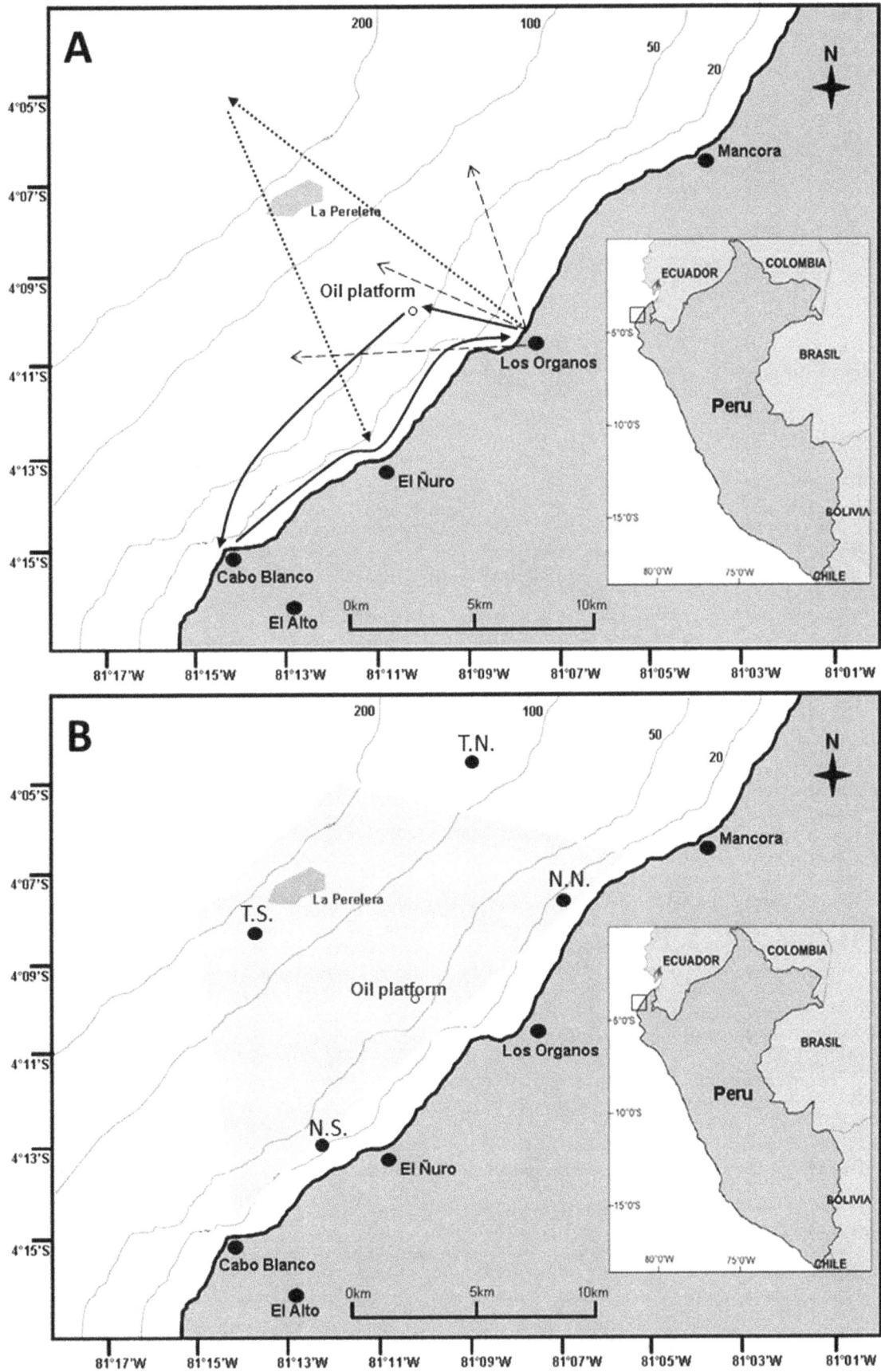

Figure 2. Routes followed during surveys. (A) Solid and dotted arrows are routes during 2010. Dashed arrows are representative of the boats displacement during 2011–2013. (B) Four zones where satellite sea surface temperature values were recorded; neritic north (N.N.), neritic south (N.S.), transitional north (T.N.), transitional south (T.S.). The grey trapeze represents the survey area covering ca. 168 km². Maps were redrawn from the oficial navigation chart of Peru (Carta Náutica del Perú, Dirección de Hidrografía y Navegación, Marina de Guerra del Perú).

regions elsewhere e.g., Costa Rica [19,20], Hawaii [21,22], Brazil [23], Madagascar [24] and eastern Australia [25]. However, the segregated pattern may not be evident in all wintering areas as it may depend on site-specific characteristics related to a suite of abiotic factors and the extent of human disturbance. For example, Cartwright et al. [26] found a non-segregated pattern in mother-calf and adult groups' distribution as the whales were present in similar depth ranges close to a boat harbor at the Au'au Channel, Hawaii.

The northern coast of Peru (between 3°–6°S) may constitute a stepping stone during the seasonal migration of humpback whales into the equatorial breeding/calving region, because in this area cold upwelling waters that flow northward of the Humboldt Current ecosystem converge with warm waters coming southwards from the Equator, forming a thermal front (see Fig. 1) that may serve as an indicator for moving from oceanic waters to the neritic breeding/calving region [4,5], when migrating from high latitudes towards the Equator. However, the available information does not allow clear inferences about the functionality of this region i.e., whether the area serves as a transitional passage during migration or if the region is indeed a functional area for breeding and calving. The assessment of this aspect is important because humpback whales may breed during migration and not only in a defined breeding ground [22]. Peruvian whaling data of the 20th century reports landings of humpback whale adults captured between 80 and 200 miles off the coast in the northern region (~3°30'S to 8°S) [27], without providing catches or sightings in neritic waters. Humpback whale research since the cessation of whaling has been poor in this region and conducted only opportunistically with very limited spatial and temporal effort, precluding a thorough assessment of habitat functionality [28]. Recently, surveys conducted during individual seasons (from July to November) have reported the presence of mother-calf pairs in shallow waters, very close to the shore along the coast of Los Organos (~4°S) [29] and Sechura bay (~5.6°S) [30] (see Fig. 1). In addition, no consistent pattern of directional movement within this region [29] and the conspicuous surface activity throughout the winter/spring season has been reported [31]. Collectively, this information provides new insights about the southern extension of the breeding and calving region in the Southeast Pacific.

Herein, we present new information describing the distribution of humpback whales in neritic waters off northern Peru, encompassing four consecutive breeding seasons (2010–2013). The objective was to assess the functionality of this area for breeding purposes under the prediction that the area is indeed a southward extension of the breeding and calving region. Thus, a segregated distributional pattern according to size/age group will be evident.

Materials and Methods

Ethics statement

This study only used noninvasive observational data. No tissues from live or dead whales were collected thus no specific permits were required for the described fieldwork as dictated by Ley Forestal y de Fauna Silvestre del Perú N°29763 (Forestry and Wildlife Law of Peru) and its Reglamento de la Ley Forestal y de Fauna Silvestre through a Decreto Supremo DS No 014-2001-AG

(Supreme Decree for regulation). During the course of this study no whales were injured by any means of human interaction (e.g., vessel collision). During surveys, whales were carefully approached following a precautionary set of navigation rules (see details in the boat survey section).

Study area

We conducted surveys in the coastal area between Los Organos (4°10'38.23"S, 81°8.27'4.83"W) and Cabo Blanco (4°15'1.36"S, 81°13'50.17"W) in northern Peru, during August, September and October from 2010 to 2013. This area is located within a transitional zone between the Tropical Eastern Pacific and the Temperate Southeastern Pacific ecoregions [32]. The transitional zone is made by the convergence between the cold, nutrient-rich Humboldt Current which flows northward and the warm, less productive Equatorial Countercurrent which flows to the East and the South. The coastline in this area is straight without the presence of main inlets or embayments. Surveys were conducted so that neritic waters up to the transition to the oceanic realm over the continental shelf break were covered.

Boat surveys

Sightings of humpback whales were conducted from whale-watching platforms of research. During 2010, daily surveys were conducted using a boat of 6.7 m length and 2.4 m width with twin outboard engines (85HP each). Navigation from Los Organos started at 7:30 h and took one of two main routes. The first route consisted of sailing to an oil platform as a navigational reference point, heading south to Cabo Blanco, finally returning to Los Organos navigating parallel to the coastline (Fig. 2A). The second route headed to La Perelera bank and further offshore to the north-westernmost point at 14 km, and then returning inbound to El Ñuro and finally back to Los Organos (Fig. 2A). Navigation usually finished at 11.00. In 2011, 2012 and 2013 two more boats (7.9 m length and 2.3 m width with a 150 HP engine and 8.8 m length and 3 m width with twin 200 HP engines, respectively) were added to the sightseeing effort. During these years humpback whales were located by one or two persons sighting whales from the top of a 30 m rocky cliff with the aid of binoculars and directing boat skippers towards the position of the groups via radio communication. The start and end times of the navigation were the same as in 2010. We followed a precautionary set of whale-watching rules during the surveys [33] in order to minimize the potential effect of the whale-watching boats on whales' behavior. Once a whale or group was located, they were approached by skippers that maintained a distance of 30 to 100 m while traveling in the same direction at the same speed as the whale(s). If a whale surfaced close to the boat (i.e. within 30 m), the engine was put in neutral gear until the animal moved away from the boat. Observation time ranged from 10 to 40 min. Observation time for mother-calf pairs were less than 25 min. During surveys, information about the number of whales, relative age/sex class and geographic position (GPS with WGS 84 system) at the closest position of the skipper closest to whales was taken. The depths of the sighted groups were derived from the recorded GPS positions, plotted in a bathymetric map of the area using the application ArcMap in ArcGIS version 10. A group was defined as the total number of animals within ca. 100 m radius, moving in the same

Table 1. Summary of the sighting effort and total number of groups per year registered during the study period.

	N°- Trips	Hrs. Observation	N°-Whales (Total)	Single	Dyad	Three or more	Mother-calf	Mother-calf/escort	Mother-calf/>escort
2010	71	65.9	241	29	23	5	33	19	4
2011	49	51.3	163	19	23	6	20	8	3
2012	76	84.1	273	15	43	18	30	10	3
2013	83	120.9	387	55	33	24	44	18	2
Total	279	322.2	1064	118	122	53	127	55	12

direction and usually exhibiting similar displacement/breathing pattern. Occasionally, groups of whales were sighted at close range, but they were not included as part of the group unless they showed obvious interaction with the first sighted group. Even though we used fixed routes during 2010, our data set may be biased because during 2011, 2012 and 2013 surveys were directed by observers on land. At the beginning of the trip, the observers guided the boats to the first visible whale group. After the first sighting boats navigated the area randomly searching for more humpback whales. Thus, the survey effort was never concentrated in a particular location or time. When two or three groups of whales were visible to the land observer, boats were directed to each group individually. In this way, all visible groups were recorded by the boat crew.

Group composition

The size and composition of the group was determined *in situ* using the following criteria; single, dyad, trio or groups of more than three individuals which generally consisted of relatively large animals (likely adult and/or subadult or a combination of both) following a fairly synchronized breathing and swimming pattern or engaging in competitive behavior (i.e., intense surface activity involving repetitive breaching or whales charging each other, *sensu* [31]). Groups involving the presence of calves include mother-calf pairs, consisting of a fully grown female and its calf, which is an individual with light grey body coloration, measuring at half (or less) the length of a large individual always swimming together. A second type of group arises when the mother-calf pairs are accompanied by an escort: is an escort an individual of equivalent or smaller size than the mother. The final group type is simply a mother-calf pair accompanied by more than one escort.

Sea surface temperatures

To characterize the thermal variability in the study area, satellite sea-surface temperature data (MODIS-Aqua satellite) was obtained from IMARPE (Instituto del Mar del Peru/Peruvian Sea Institute), which is provided by the National Oceanic and Atmospheric Administration (NOAA) and the NAVOCEANO agency of the U.S. Navy. The information consisted of average daily sea surface temperature taken from August to October during 2010, 2011, 2012 and 2013 at two sampling sites located in the northern part of the study area over a neritic and a transitional location (i.e. from neritic to oceanic waters), and two additional sites located in the southern area in similar waters (Fig. 2B).

Statistical analysis

Temporal patterns of variation of the group types were studied using simple linear regression with time as a predictor and the abundance of groups (expressed as the number of sighted groups per number of trips during the respective survey month) as the dependent variable. This variable was used in order to reduce the variation due to the differences in sampling effort, i.e., different number of trips in each month. To assess the existence of a segregate pattern in the spatial distribution of humpback whale group types, we followed the approach of Cartwright et al. [26]. We used Neu test [34–35] to evaluate the degree of preference for a given bathymetric range (i.e., 0–20, 20–50, 50–100, and 100–200 m depth) considering such ranges as proxies of habitat use. First, the area between two isobaths (in km^2) in each bathymetric range was estimated and Chi-square test was used to assess the preference of each group type for a given area. Then, Bonferroni ranks were used to estimate whether bathymetric ranges were disproportionally used by humpback whale groups (see [26]). These ranks are confidence intervals of the "Z" Bonferroni's

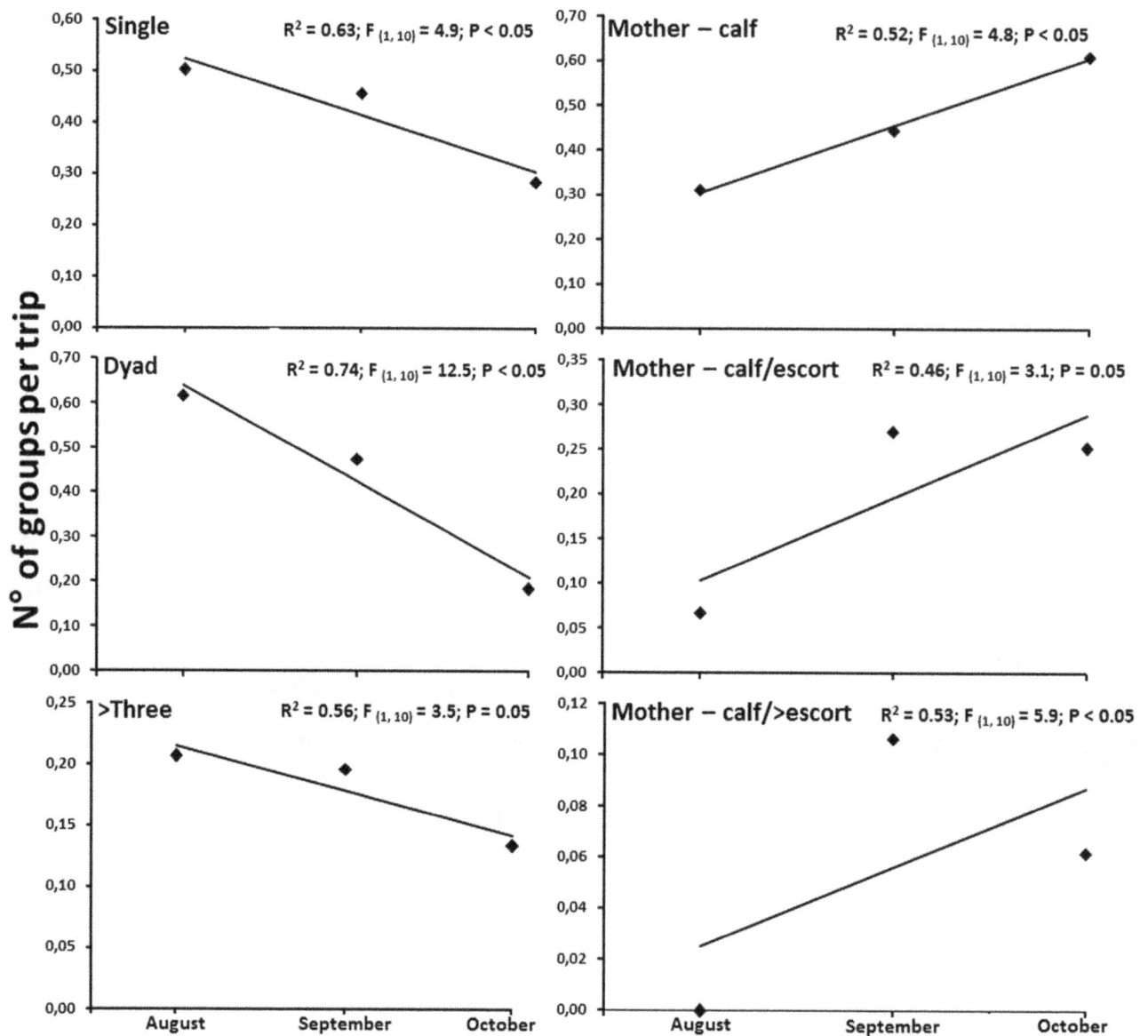

Figure 3. Humpback whales group types. Monthly means (all years) of the number of groups per trip per each sampling month.

formula which includes the proportion of observed groups and the values of "Z" distribution. The proportions of the expected groups are estimated as function of the area for each bathymetric range. Then, habitat use was designated in a function of the relationship between the expected and observed proportion of the groups and the 95% confidence interval according to the following criteria: (a) avoided; if of the observed proportion of groups in each bathymetric rage was entirely below the expected proportion of groups. (b) preferred, if the observed proportion of groups in each bathymetric rage was entirely above the expected proportion of groups or (c) neutral if the observed groups were contained in the expected proportion of groups within the confidence intervals. Neu's standardized selection indices were also calculated since these provide comparable indices of habitat use. Higher values were assumed to be stronger indicator of preference for a given habitat. Neu's analyses were conducted using pooled data (all months and years) of sighting positions since such analysis need many observations to produce meaningful statistics. Mother-calf

pairs with more than one escort group were excluded from this analysis due to the low number of observations. All statistical analyses were performed using the software Statistica 6.0 and PAST version 2.17.

Results

A total of 279 trips were conducted during the four winter-spring seasons between 2010 and 2013. A total of 1064 whales distributed between 487 groups were recorded in a total of 322.3 hours of observation (see Table S1 for details of groups sighting positions). The most abundant group type was mother-calf pairs (n = 127) followed by dyads (n = 122) and singles (n = 118). Of the total number of groups composed by three or more whales (n = 53), 28.5% (n = 15) were competitive groups. It should be noted that there was some variation in the observational effort between years for example, during 2012 and 2013 the number of

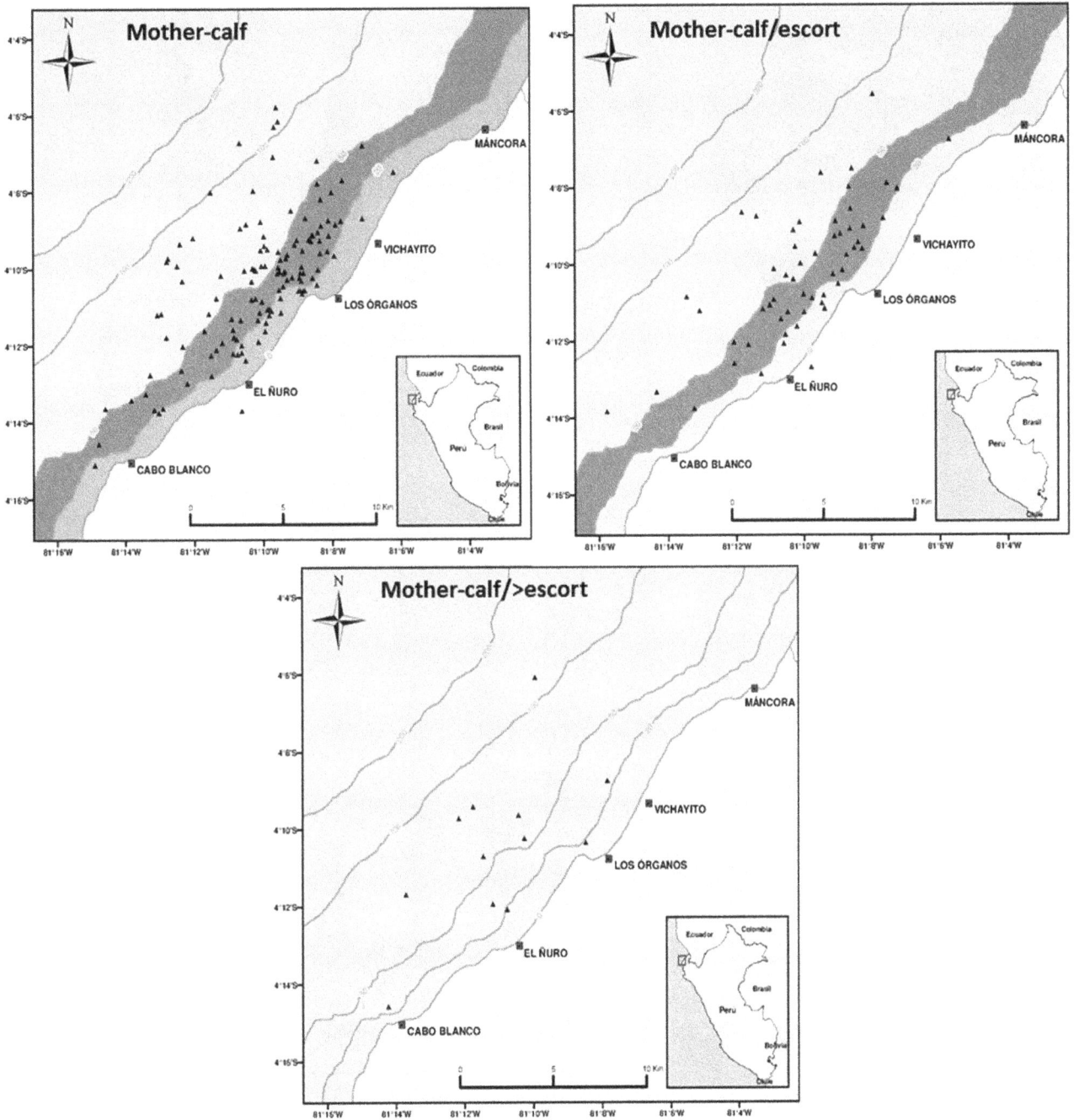

Figure 4. Maps showing GPS-positioned sightings of humpback whales groups; mother-calf pairs, mother-calf and escort and mother-calf and more than one escort during the study period. Most preferred (dark grey) and preferred (light grey) bathymetric ranges are marked after Neu's habitat index.

trips were twofold compared to the breeding season in 2010 and 2011 (Table 1).

Temporal variation of whale group types

Two temporal patterns were clear; singles, dyads and groups made by three or more individuals were more abundant during the first month of the season but tended to significantly decrease towards the end of the study period (Fig. 3). An opposite temporal pattern was evident for mother-calf pairs, mother-calf and escort and mother-calf and more escorts groups, which showed a trend to

increase in abundance during the second half of the season (Fig. 3).

Spatial distribution and habitat preference

Of the sightings, ninety eight percent were distributed in the neritic zone *i.e.*, from the shore to the 200 m isobath, but whales density was specific to depth ranges depending on group composition (Table 2). According to the Neu test, mother-calf pairs, and mother-calf and escort groups showed a strong preference for the 20–50 m depth range (Table 3, Fig. 4),

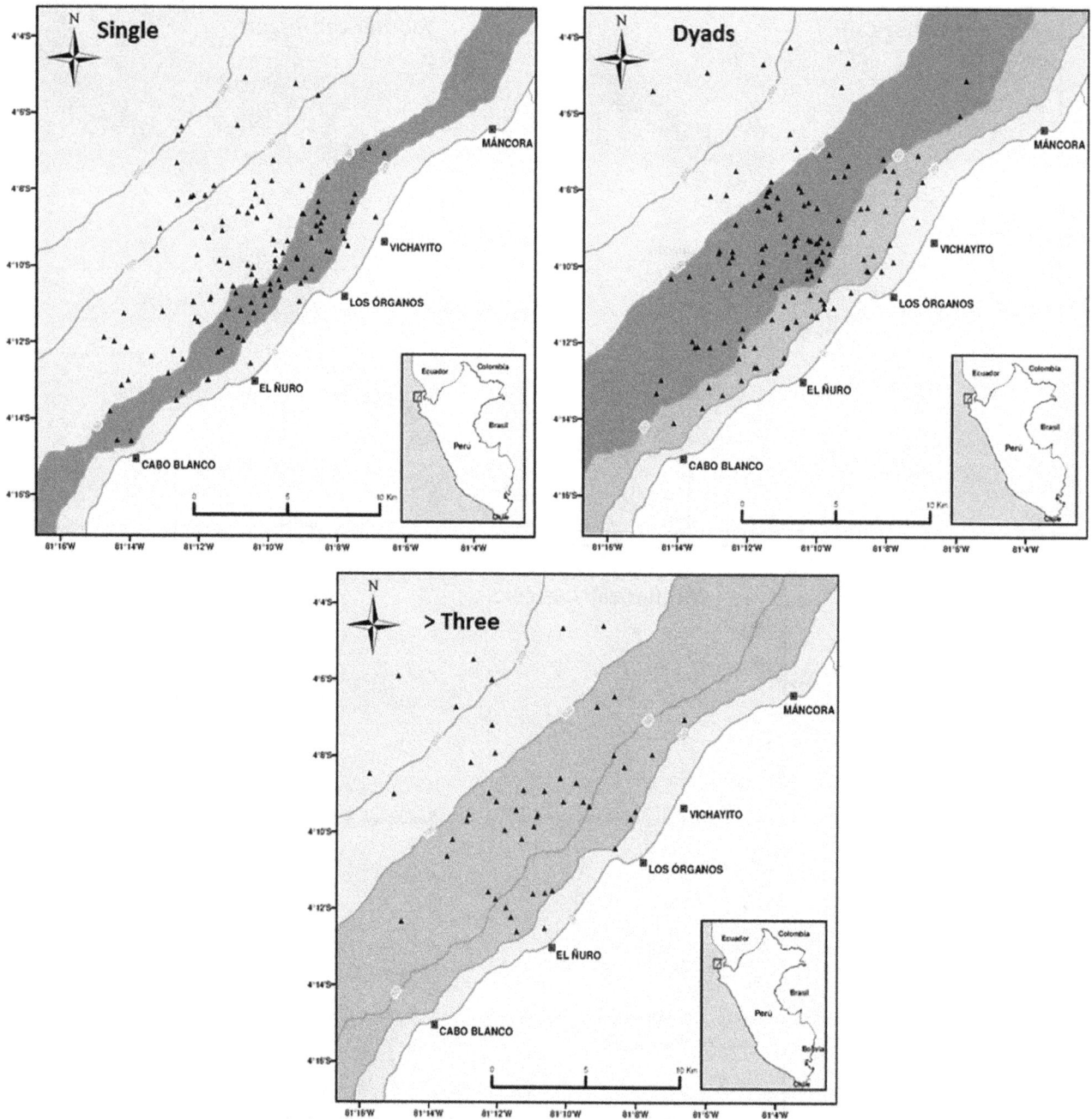

Figure 5. Maps showing GPS-positioned sightings of single, dyads, three or more individuals of humpback whales group during the study period. Most preferred (dark grey) and preferred (light grey) bathymetric ranges are marked after Neu's habitat index.

although mother-calf groups also occurred in the 0–20 m range. No test was conducted for mother-calf and more than one escort groups but the sighting positions suggest a wide distribution over the survey area (Fig. 4). Singles preferred the 20–50 m range while dyads showed a preference for the 20–50 and 50–100 m depth ranges although a strong preference for the latter range was detected (Fig. 5). Groups composed by three or more individuals showed equal preference for 20–50 and 50–100 m depth ranges (Fig. 5). Overall, a segregated spatial pattern was evident; group of whales with calves, particularly mother-calf pairs preferred shallow areas while groups with no calves were more widely distributed throughout the habitat range.

Sea surface temperature variability

The mean value of sea surface temperature for all study seasons was 22.09°C, with the 2012 season being the coldest and the 2013 the warmest (Table 4). Temperature values were warmer at northern locations compared to the southern part of the study area (Table 4). The south-neritic area was the coldest of all with a mean value of 21.3°C with 18.25°C as minimum value. The warmest area was the transitional neritic-oceanic zone where the mean was 23°C. The neritic-north and transitional neritic-oceanic areas showed fairly similar sea surface temperature values (Table 4). It is worth noting that the ca. 7°C of difference between maximum and minimum sea surface temperature values may account for the

Table 2. Percentage and total number of groups in each depth range registered during the study period.

Depth (m)	Single		Dyad		Three or more		Mother-calf		Mother-calf/escort		Mother-calf/>escort	
	%	n	%	n	%	n	%	n	%	n	%	n
0–20	8	10	8	10	4	2	20	25	16	9	17	2
20–50	41	48	23	28	23	12	50	64	49	27	25	3
50–100	42	49	59	72	51	27	28	35	35	19	50	6
100–200	9	11	19	12	23	12	2	3	-	-	8	1

strong thermal variability produced by the convergence of the cold upwelling system and warm equatorial waters in the study region.

Discussion

Our results provide new evidence that the northern coast of Peru may not only represent a migratory route from oceanic waters to the neritic realm during their seasonal breeding migration as suggested in the early literature [4]. Humpback whales in our study area depicted the segregate pattern of spatial distribution characteristic of this species in their breeding/calving grounds i.e., groups made up by dyads, trios and more than three individuals were present from moderated depths towards the continental shelf while mother and calf pairs were present in very shallow waters. The record of 194 groups with calves, particularly the presence of mother – calf and escort groups which are conspicuous in wintering regions [36], the presence of large competitive groups adds further evidence of the breeding and calving function of the northern coast of Peru.

The results of this study complement previous information pointing to the breeding functionality of this area such as the absence of a consistent direction in the displacement of the whales suggesting constant movement within the area (e.g., males actively moving searching for receptive females) [29]. The high abundance of single, dyad and mother and calf groups during August, September and October during four consecutive seasons corroborate the functionality of this area as a breeding and calving habitat. Our data shows that groups without calves progressively decrease towards the second half of the season and instead groups with calves increase as the seasons ends. These suggest that although mating activity occurs during the whole season, breeding and calving seems to be more important during the second half of the season [20]. This type of temporal change has been observed in breeding areas elsewhere e.g., Abrolhos archipelago, Brasil, [37]. Groups of humpback whales, including many calves, usually leave the breeding area at the end of the season. Very young calves and new mothers stay in calm waters until the calves are strong enough to undertake the migration south. Unfortunately, our data does not allow us precise information regarding the type of sex/ age composition of the individuals at the beginning of the season. We speculate that these may be principally adult females leaving the area after mating, while males may stay and change their reproductive strategy to searching for post-partum mating chances in mother and calf pairs [38–39]. This is supported by the increase of groups involving escorts towards the end of the season. Further studies using molecular techniques are still necessary to understand the sequence of arrival and departures of humpback whales in this region.

Spatially, the 98% of the sighted humpbacks were in the neritic zone (<200 m depth) which is similar to the sighting percentage reported in breeding areas elsewhere e.g., [19], [40–43]. The existence of a segregated pattern of distribution is a characteristic of breeding and calving regions and such pattern occurs regardless of the width, slope or the distance of the continental shelf break to the coast. For example, the segregated patterns have been observed in locations with a wide and shallow continental shelf such as the breeding region off Puerto Lopez (Ecuador) [43]. While at Salinas (also Ecuador) [5] and Antogil Bay (Madagascar) [23] humpback whales depicted the segregated pattern even though in these locations the continental shelf is narrow with a steep slope. Behavior and social organization finally determine habitat preference of this species: the need for calm, warm waters for calving away from competitive groups is often reported as the main explanation for the coastal distribution of mother and calf

Table 3. Summary of Neu's test statistics for humpback whales groups' habitat preference.

Habitat	Depth ranges	Area (km²)	Expected groups	Observed groups	Observed proportions	95% C.I.	Proportion of total study area	Neu's Index	Inference
Mother-calf	0–20	43.46	18.3	25	0.197	(0.155–0.238)	0.144	0.282	Preferred
	20–50	56.23	23.7	64	0.504	(0.452–0.556)	0.186	0.558	Preferred*
	50–100	121.39	51.1	35	0.276	(0.229–0.322)	0.402	0.141	Avoided
	100–200	80.69	34	3	0.024	(0.008–0.040)	0.267	0.018	Avoided
Mother-calf/escort	0–20	43.46	7.9	9	0.164	(0.125–0.202)	0.144	0.245	Neutral
	20–50	56.23	10.2	27	0.491	(0.439–0.543)	0.186	0.569	Preferred*
	50–100	121.39	22.1	19	0.345	(0.296–0.395)	0.402	0.185	Avoided
	100–200	80.69	14.7	0	0	0	0.267	0	Avoided*
Single	0–20	43.46	16.8	10	0.086	(0.057–0.116)	0.144	0.144	Avoided
	20–50	56.23	21.8	47	0.405	(0.354–0.457)	0.186	0.525	Preferred*
	50–100	121.39	47.1	49	0.422	(0.371–0.474)	0.402	0.253	Neutral
	100–200	80.69	31.3	10	0.086	(0.057–0.116)	0.267	0.078	Avoided
Dyad	0–20	43.46	18.7	10	0.0787	(0.051–0.107)	0.144	0.150	Avoided
	20–50	56.23	24.2	32	0.2520	(0.207–0.297)	0.186	0.370	Preferred
	50–100	121.39	52.3	76	0.5984	(0.547–0.650)	0.402	0.407	Preferred*
	100–200	80.69	34.8	9	0.0709	(0.044–0.098)	0.267	0.073	Avoided
>Three	0–20	43.46	43.4	2	0.0426	(0.021–0.064)	0.144	0.082	Avoided
	20–50	56.23	56.2	12	0.2553	(0.210–0.301)	0.186	0.381	Preferred
	50–100	121.39	20.9	26	0.5532	(0.501–0.605)	0.402	0.382	Preferred
	100–200	80.69	13.9	7	0.1489	(0.112–0.186)	0.267	0.155	Avoided

"*" denotes higher index values thus regarded as strong preference.

Table 4. Summary of sea surface temperature values recorded during the study period.

	Mean °C	Max °C	Min °C	N.N. °C	N.S. °C	T.N. °C	T.S. °C
2010	22.3±1.4	25.9	19.1	22.7±1.3	21.8±1.5	22.8±1.2	22.1±1.4
2011	21.7±1.03	25.4	18.2	21.9±1	21.1±1.1	22.2±0.9	21.6±1.05
2012	21.3±0.9	23.8	19.2	21.7±0.8	20.1±0.9	22±0.8	21.4±0.9
2013	22.9±1.2	24.6	20	23±1.07	22.3±1.1	23.7±1.02	22.6±1.1

Neritic north (N.N.), neritic south (N.S.), transitional north (T.N.) and transitional south (T.S.).

groups [12–13]. However, geomorphology and anthropogenic impact in the breeding region also play a role in humpback whales local distribution. Cartwright et al. [26] studying the distribution of humpbacks whales at the Au'au channel (an important breeding area of the Hawaiian islands) found a spatial overlap between mother-calf pairs and adults groups, both distributed over 40–60 m depth. The narrowness of the channel together with disturbance by recreational boats in shallow waters seemed to be the cause of this non-segregated spatial pattern [26]. Our results suggest that there is a spatial overlap among mother-calf pairs, mother-calf and escort and singles since these groups shared a strong preference for the 20–50 m depth range. In our study region (~7°S and further north), the continental shelf break is located at 5–9 km from the coast [44], thus the shelf is rather narrow and steep which may explain the certain degree of spatial overlap among humpback groups before moving into deeper waters. For this study, the influence of the artisanal fishery boats in this coastal region, particularly when gathering at the ports, has not been considered. As suggested before [26], such a concentration of vessels may prevent whales from coming closer to the shore, especially mother and calf groups.

To reveal the segregated spatial pattern of distribution of humpback whales groups, it was necessary that surveys covered coastal areas as well as waters over the continental shelf break. Such spatial cover was achieved during our trips although not systematically during all sampling seasons. For example, during the 2010 season the first navigation route was fixed and always reached waters over the 200 m isobaths. However, during the 2011–2013 seasons' navigation to deeper waters were less frequent. As we previously stated, such differences in survey effort may bias results since sighting rates of a given group type may differ between random navigation and fixed transect surveys. Despite of this methodological shortcoming, our data revealed significant temporal and spatial patterns of distribution. However, we suggest caution when comparing these results with studies using transect surveys.

The variability of sea surface temperature in our study region revealed that humpback whales experience rather colder conditions; the mean was 21.5°C during all the seasons which is three degrees lower than the average estimate for breeding regions worldwide [11]. Indeed, we registered minimum values of ~18.5°C in the southern-neritic zone. Migratory humpback whales in the Southern Hemisphere oceans breed at ~14°S, (although recent surveys demonstrate the existence of breeding areas near the equator in the eastern tropical Atlantic e.g. Gulf of Guinea for some South Atlantic stock [45]) while the southeast Pacific stock extends its migration close to lower latitudes and even crossing the equator [6], [9]. Early literature suggested that humpback whales would avoid the influence of the cold neritic conditions imposed by the Humboldt upwelling ecosystem and humpback whales would approach neritic waters when encountering warm conditions further north at the equatorial realm [4]. However, our results suggest that the northern coast of Peru may constitute a threshold between upwelling conditions and equatorial waters and such transitional habitat may constitute an extension of the breeding and calving region. Sea surface temperature values of this region ranged between the lowest values reported for other breeding regions e.g. 19–20°C at Bonin and Ryukyu Islands, Japan [11]. However, much research is still needed to reveal the influence of the thermal gradients on humpback whales distribution and habitat requirements, particularly for mother-calf pairs and newborn calves. For example, there are two existing observations of humpback whale parturition [46,47] and only the observation in Brazilian waters reports a sea surface temperature of 24°C at the moment of birth [46]. Although very preliminary, this may suggest that births occur in warmer areas inside the breeding region and that zones with slightly colder temperatures may function principally as habitats for calf development. However, the fact that we recorded calves in relatively colder temperatures indicates that they are capable of surviving these conditions and may not need to be born in warmer waters. Detailed studies describing the morphology and behavior of the calves e.g. [13] along the breeding region are necessary for a better understanding of the dynamics of this delicate life stage during the life of a humpback whale. It is important to mention that the population of humpback whales is increasing [48] and likely their presence in neritic waters off Peru may constitute an indicate of breeding range expansion. It is possible that breeding in more northern areas occurred in the past because these areas were the optimal habitats. Currently, these areas may be saturated and whales are looking for other breeding areas further south. The presence of several calves reported here supports this notion.

While we provide new information about the distribution of humpbacks at the southern limit of the breeding and calving region, there is also renewed evidence of a northward recovery of the former feeding grounds in the Patagonia region [48,49]. Collectively, these apparent extensions of both breeding and feeding regions suggest that this species is likely recovering and returning to its former distributional ranges before the industrial whaling period. This conclusion is supported with recent estimations of population size, suggesting steady increments throughout time during the last *ca.* 20 years [50]. However, it should be noted that these numbers may be underestimated since calculations were based on data obtained in a fraction of the breeding area (i.e., off Ecuador). Enhancing multinational research efforts along the Southeast Pacific would improve our ecological knowledge and population status of this species which also would improve management and conservation efforts for this species.

Outlook

In this study, we have highlighted the neritic presence of humpback whales during their breeding migration off the coast of

northern Peru. However, this coastal distribution makes the species prone to entanglement and mortality with gillnet fisheries [51]. It is mandatory to regulate the use of such fishing gears during the seasonal presence of this species in neritic habitats. In ecological terms, little is known about the displacement patterns of this species within the breeding region. Satellite tracking and photo-identification research may reveal the timing and specific habitat requirements during the breeding season and these are topics that should be addressed in the near future. Although, we acknowledge the use of whale-watching boats as platforms for investigation opportunities for, humpback whale research in this region should be complemented with other types of sampling such as focal-group followed and transect surveys aimed at testing specific hypotheses on ecological and behavioral aspects of this species.

Acknowledgments

We deeply thank the support of the Pacifico Adventures crew and volunteers particularly, F. Sanchez-Salazar, N. Balducci, A. Petit and E. Larrañaga. C. Paulino and J. Ledesma (IMARPE) who kindly provided us the sea surface temperature data. Comments by R. Cartwright, A. Zerbini, K. Van Waerebeek and an anonymous reviewer helped us to improve an early version of the manuscript. N. Burns and K. Rasmussen kindly revised the English of this manuscript. This is study was supported by the participation of hundreds of enthusiastic whale-watchers.

Author Contributions

Conceived and designed the experiments: CG MAL ASP. Performed the experiments: MAL CG. Analyzed the data: MAL CG ASP. Contributed reagents/materials/analysis tools: SS BA. Wrote the paper: ASP MAL CG.

References

1. Tynan CT, Ainley DG, Barth JA, Cowles TJ, Pierce SD, et al. (2005) Cetacean distributions relative to ocean processes in the northern California Current System. Deep Sea Res II 52: 145–167.

2. Doniol-Valcroze T, Berteaux D, Larouche P, Sears R (2007) Influence of thermal fronts on habitat selection by four rorqual species in the Gulf of St. Lawrence. Mar Ecol Progr Ser 335: 207–216.

3. Heithaus MR, Dill LM (2009) Feeding strategies and tactics. In: Perrin WF, Würsing B, Thewissen JGM, editors. Encyclopedia of Marine Mammals. Academic Press, San Diego. pp. 414–423

4. Clarke R (1962) Whale observation and whale marking off the coast of Chile in 1958 and from Ecuador towards and beyond the Galápagos Islands in 1959. Norsk Hvalfangst-tidende 51: 265–287.

5. Félix F, Haase B (2005) Distribution of humpback whales along the coast of Ecuador and management implications. J Cetacean Res Manag 7: 21–31.

6. Stone GS, Flórez L, Katona S (1990) Whale migration record. Nature 346: 705

7. Flórez-González L, Capella AJ, Haase B, Bravo GA, Felix F, et al. (1998) Changes in winter destinations and the northernmost record of Southeastern Pacific humpback whales. Mar Mammal Sci 14: 189–196.

8. Acevedo J, Rasmussen K, Félix F, Castro C, Llano E, et al. (2007) Migratory destinations of the humpback whales from Magellan Strait feeding ground, Chile. Mar Mammal Sci 23: 453–463.

9. Capella JJ, Gibbons J, Flórez-González L, Llano M, Valladares C, et al. (2008) Migratory round-trip of individually identified humpback whales at the Strait of Magellan: clues on transit times and phylopatry to destinations. Rev Chil Hist Nat 81: 547–560.

10. Swartzman G, Bertrand A, Gutiérrez M, Bertrand S, Vasquez L (2008) The relationship of anchovy and sardine to water masses in the Peruvian Humboldt Current System from 1983 to 2005. Prog Oceanogr 79: 228–237.

11. Rasmussen K, Palacios DM, Calambokidis J, Saborio MT, Dalla Rosa L, et al. (2007) Southern Hemisphere humpback whales wintering off Central America:

insights from water temperature into the longest mammalian migration. Biol Lett 3: 302–305.

12. Clapham P (2001) Why do baleen whales migrate? A response to Corkeron and Connor. Mar Mammal Sci 17: 432–436.

13. Cartwright R, Sullivan M (2009) Behavioral ontogeny in humpback whale (*Megaptera novaeangliae*) calves during their residence in Hawaiian waters. Mar Mammal Sci 25: 659–680.

14. Craig AS, Herman LM, Pack AA, Waterman JO (2014) Habitat segregation by female humpback whales in Hawaiian waters: avoidance of males? Behaviour 151: 613–631.

15. Corkeron PJ, Connor RC (1999) Why do baleen whales migrate? Mar Mammal Sci 15: 1228–1245.

16. Flórez-González L, Capella JJ, Rosebaum HC (1994) Attack of killer whales (*Orcinus orca*) on humpback whales (*Megaptera novaeangliae*) on a South American Pacific breeding ground. Mar Mammal Sci 10: 218–222.

17. Félix F, Botero-Acosta N (2011) Distribution and behaviour of humpback whale mother-calf pairs during the breeding season off Ecuador. Mar Ecol Prog Ser 426: 277–287.

18. Flórez-González L (1991) Humpback whales *Megaptera novaeangliae* in the Gorgona Island, Colombian Pacific breeding waters: population and pod characteristics. Mem Queensland Mus 30: 291–295.

19. Oviedo L, Solís M (2008) Underwater topography determines critical breeding habitat for humpback whales near Osa Peninsula, Costa Rica: implications for Marine Protected Areas. Rev Biol Trop 56: 591–602.

20. Rasmussen K, Calambokidis J, Steiger GH (2012) Distribution and migratory destinations of humpback whales off the Pacific coast of Central America during the boreal winters of 1996–2003. Mar Mammal Sci 28: E267–E279.

21. Craig AS, Herman LM (2000) Habitat preferences of female humpback whales *Megaptera novaeangliae* in the Hawaiian Islands are associated with reproductive status. Mar Ecol Prog Ser 193: 209–216.

22. Johnston DW, Chapla ME, Williams LE, Mattila DK (2007) Indentification of humpback whale *Megaptera novaeangliae* wintering habitat in the Northerwestern Hawaiian Island using spatial habitat modeling. Endang Species Res 3: 249–257.

23. Zerbini AN, Andriolo A, Da Rocha JM, Simoes-Lopes PC, Siciliano S, et al. (2004) Winter distribution and abundance of humpback whales (*Megaptera novaeangliae*) off Northeastern Brazil. J Cetacean Res Manag 6: 101–107.

24. Ersts PJ, Rosenbaum HC (2003) Habitat preference reflects social organization of humpback whales (*Megaptera novaeangliae*) on a wintering ground. J Zool 260: 337–345.

25. Smith JN, Grantham HS, Gales N, Double MC, Noad MJ, et al. (2012) Identification of humpback whale breeding and calving habitat in the Great Barrier Reef. Mar Ecol Prog Ser 447: 259–272.

26. Cartwright R, Gillespie B, LaBonte K, Mangold T, Venema A, et al. (2012) Between a rock and a hard place: habitat selection in female-calf humpback whale (*Megaptera novaeangliae*) pairs on the Hawaiian breeding grounds. PLoS ONE 7(5): e38004. doi:10.1371/journal.pone.0038004

27. Ramírez P (1988) La ballena jorobada *Megaptera novaeangliae* en la costa norte del Perú: Períodos 1961–1965 y 1975–1985. Boletin de Lima 56: 91–96.

28. Van Waerebeek K, Alfaro-Shigueto J, Arias-Schreiber M (1996) Humpback whales off Peru: new records and a rationale for renewed research. International Whaling Commission Scientific Committee document SC/48/SH1, Aberdeen, UK, June 1996. Available: http://iwcoffice.org.

29. Pacheco AS, Silva S, Alcorta B (2009) Winter distribution and group composition of humpback whales (*Megaptera novaeangliae*) off northern Peru. Lat Am J Aquat Mammal 7: 33–38.

30. Santillan L (2011) Records of humpback whales (*Megaptera novaeangliae*) in Sechura Bay, Peru, in spring 2009–2010. J Mar Anim Ecol 1: 29–35.

31. Pacheco AS, Silva S, Alcorta B, Balducci N, Guidino C, et al. (2013) Aerial behavior of humpback whales *Megaptera novaeangliae* at the southern limit of the southeast Pacific breeding area. Rev Biol Mar Oceanogr 48: 185–191.

32. Spalding M, Fox H, Allen G, Davidson N, Ferdaña Z, et al. (2007) Marine ecoregions of the world: a bioregionalization of coastal and shelf areas. Bioscience 57: 573–583.

33. Pacheco AS, Silva S, Alcorta B (2011) Is it possible to go whale watching off the coast of Peru? A case study of humpback whales. Lat Am J Aquat Res 39: 189–196.

34. Neu CW, Byers CR, Peek JM (1974) A technique for analysis of utilization availability data. J Wildl Manage 38: 541–545.

35. McClean SA, Rumble MA, King RM, Baker WL (1998) Evaluation of resource selection methods with different definitions of availability. J Wildl Manage 62: 793–801.

36. Cartwright R, Sullivan M (2009) Associations with multiple male groups increase the energy expenditure of humpback whale (*Megaptera novaeangliae*) female and calf pairs on the breeding grounds. Behaviour 146: 1573–1600.

37. Morete ME, Bisi TL, Rosso S (2007) Temporal pattern of humpback whale (*Megaptera novaeangliae*) group structure around Abrolhos Arquipelago breeding region, Bahia, Brazil. J Mar Biol Assoc UK 87: 87–92.

38. Dawbin WH (1997) Temporal segregation of humpback whales during migration in southern hemisphere waters. Mem Queensland Mus 42: 105–138.

39. Craig AS, Herman LM, Gabriele CM, Pack AA (2003) Migratory timing of humpback whales (*Megaptera novaeangliae*) in the Central North Pacific varies with age, sex and reproductive status. Behaviour 140: 981–1001.

40. Urbán J, Aguayo A (1987) Spatial and seasonal distribution of the humpback whale, *Megaptera novaeangliae*, in the Mexican Pacific. Mar Mammal Sci 3: 333–44.

41. Urbán J (1999) La ballena jorobada, *Megaptera novaeangliae*, en la Península de Baja California, México. Universidad Autónoma de Baja California Sur. Área Interdisciplinaria de Ciencias del Mar. Informe final SNIB-CONABIO proyecto No. H035. México D. F.

42. Scheidat M, Castro C, Denkinger J, González J, Adelung D (2000) A breeding area for humpback whales (*Megaptera novaeangliae*) of Ecuador. J Cetacean Res Manag 2: 165–171.

43. Félix F, Haase B (2001) The humpback whale off the coast of Ecuador, population parameters and behaviour. Rev Biol Mar Oceanogr 36: 61–74.

44. Moron O (2000) Características del ambiente marino frente a la costa peruana. Bol Inst del Mar del Perú 19(1–2): 179–202.

45. Van Waerebeek K, Djiba A, Krakstad J-O, Samba Ould Bilal A (2013) New evidence for a South Atlantic stock of humpback whales wintering on the Northwest African continental shelf. Afr Zool 48: 177–186.

46. Ferreira MEC, Maia-Nogueira R, Hubner de Jesus A (2011) Surface observation of a birth of a humpback whale (*Megaptera novaeangliae*) on the northeast coast of Brazil. Lat Am J Aquat Mammal 9: 160–163.

47. Farias MA, DeWeerdt J, Pace F, Mayer FX (2013) Observation of a humpback whale (*Megaptera novaeangliae*) birth in the coastal waters of Sainte Marie Island, Madagascar. Aquat Mammal 39: 296–305.

48. Gibbons J, Capella JJ, Valladares C (2003) Rediscovery of a humpback whales (*Megaptera novaeangliae*) feeding ground in the Straits of Magellan, Chile. J Cetacean Res Manag 5: 203–208.

49. Hucke-Gaete R, Haro D, Torres-Florez JP, Montecinos Y, Viddi F, et al. (2013) A historical feeding ground for humpback whales in the eastern South Pacific revisited: the case of northern Patagonia, Chile. Aquatic Conserv Mar Freshw Ecosyst 23: 858–867.

50. Félix F, Castro C, Laake J, Haase B, Scheidat M (2011) Abundance and survival estimates of the Southeastern Pacific humpback whale stock from 1991–2006 photo-identification surveys in Ecuador. J Cetacean Res Manag 3(Special Issue): 301–307.

51. García-Godos I, Van Waerebeek K, Alfaro-Shigueto J, Mangel J (2013) Entanglements of large cetaceans in Peru: few records but high risk. Pac Sci 67: 523–532.

Permissions

The contributors of this book come from diverse backgrounds, making this book a truly international effort. This book will bring forth new frontiers with its revolutionizing research information and detailed analysis of the nascent developments around the world.

We would like to thank all the contributing authors for lending their expertise to make the book truly unique. They have played a crucial role in the development of this book. Without their invaluable contributions this book wouldn't have been possible. They have made vital efforts to compile up to date information on the varied aspects of this subject to make this book a valuable addition to the collection of many professionals and students.

This book was conceptualized with the vision of imparting up-to-date information and advanced data in this field. To ensure the same, a matchless editorial board was set up. Every individual on the board went through rigorous rounds of assessment to prove their worth. After which they invested a large part of their time researching and compiling the most relevant data for our readers.

The editorial board has been involved in producing this book since its inception. They have spent rigorous hours researching and exploring the diverse topics which have resulted in the successful publishing of this book. They have passed on their knowledge of decades through this book. To expedite this challenging task, the publisher supported the team at every step. A small team of assistant editors was also appointed to further simplify the editing procedure and attain best results for the readers.

Apart from the editorial board, the designing team has also invested a significant amount of their time in understanding the subject and creating the most relevant covers. They scrutinized every image to scout for the most suitable representation of the subject and create an appropriate cover for the book.

The publishing team has been an ardent support to the editorial, designing and production team. Their endless efforts to recruit the best for this project, has resulted in the accomplishment of this book. They are a veteran in the field of academics and their pool of knowledge is as vast as their experience in printing. Their expertise and guidance has proved useful at every step. Their uncompromising quality standards have made this book an exceptional effort. Their encouragement from time to time has been an inspiration for everyone.

The publisher and the editorial board hope that this book will prove to be a valuable piece of knowledge for researchers, students, practitioners and scholars across the globe.

List of Contributors

Zheng Gong
Department of Mechanical Engineering, Massachusetts Institute of Technology, Cambridge, Massachusetts, United States of America

Duong Tran, Fan Wu, Alexander Zorn and Purnima Ratilal
Department of Electrical and Computer Engineering, Northeastern University, Boston, Massachusetts, United States of America

Ankita D. Jain, Dong Hoon Yi and Nicholas C. Makris
Department of Mechanical Engineering, Massachusetts Institute of Technology, Cambridge, Massachusetts, United States of America

Rocio I. Ruiz-Cooley and Matthew D. McCarthy
Ocean Sciences Department, University of California Santa Cruz, Santa Cruz, California, United States of America

Paul L. Koch
Earth and Planetary Sciences Department, University of California Santa Cruz, Santa Cruz, California, United States of America

Paul C. Fiedler
Southwest Fisheries Science Center, National Marine Fisheries Service, National Oceanic and Atmospheric Administration, La Jolla, California, United States of America

Patricia Brosseau-Liard
Department of Psychology, Concordia University, Montreal, Quebec, Canada

Tracy Cassels and Susan Birch
Department of Psychology, University of British Columbia, Vancouver, British Columbia, Canada

Bjørghild B. Seliussen
Dept. of Population Genetics, Institute of Marine Research, Bergen, Norway

María Quintela
Dept. of Population Genetics, Institute of Marine Research, Bergen, Norway

BIOCOST Research Group, Dept. of Animal Biology, Plant Biology and Ecology, University of A Coruña, A Coruña, Spain

Hans J. Skaug
Dept. of Population Genetics, Institute of Marine Research, Bergen, Norway
Department of Mathematics, University of Bergen, Bergen, Norway

Nils Øen
Dept. of Marine Mammals, Institute of Marine Research, Bergen, Norway

Tore Haug
Dept. of Marine Mammals, Institute of Marine Research, Tromsø, Norway

Christophe Pampoulie
Marine Research Institute of Iceland, Reykjavik, Iceland

Naohisa Kanda and Luis A. Pastene
Institute of Cetacean Research, Tokyo, Japan

Kevin A. Glover
Dept. of Population Genetics, Institute of Marine Research, Bergen, Norway
Department of Informatics, Faculty of Mathematics and Natural Sciences, University of Bergen, Bergen, Norway

Jianjun Cheng, Mingwei Leng, Longjie Li, Hanhai Zhou and Xiaoyun Chen
School of Information Science and Engineering, Lanzhou University, Lanzhou, Gansu Province, China

Emily P. Lane
Department of Research and Scientific Services, National Zoological Gardens of South Africa, Pretoria, South Africa

Morné de Wet and Peter Thompson
Epidemiology Section, Department of Production Animal Studies, Faculty of Veterinary Science, University of Pretoria, Pretoria, South Africa

Ursula Siebert
Institute for Terrestrial and Aquatic Wildlife Research, University of Veterinary Medicine, Hannover, Foundation, Germany

Peter Wohlsein
Department of Pathology, University of Veterinary Medicine, Hannover, Foundation, Germany

Stephanie Plön
South African Institute for Aquatic Biodiversity, c/o Port Elizabeth Museum/Bayworld, Port Elizabeth, South Africa
Coastal and Marine Research Unit, Nelson Mandela Metropolitan University, Port Elizabeth, South Africa

Kexiong Wang and Ding Wang
The Key Laboratory of Aquatic Biodiversity and Conservation of the Chinese Academy of Sciences, Institute of Hydrobiology, Chinese Academy of Sciences, Wuhan, P. R. China

Zhitao Wang
The Key Laboratory of Aquatic Biodiversity and Conservation of the Chinese Academy of Sciences, Institute of Hydrobiology, Chinese Academy of Sciences, Wuhan, P. R. China
University of Chinese Academy of Sciences, Beijing, P. R. China

Yuping Wu
School of Marine Sciences, Sun Yat-sen University, Guangzhou, P. R. China

Guoqin Duan and Hanjiang Cao
Hongkong-Zhuhai-Macao Bridge Authority, Guangzhou, P. R. China

Jianchang Liu
Transport Planning and Research Institute, Ministry of Transport, Beijing, P. R. China

Zachary A. Schakner
Department of Ecology and Evolutionary Biology, University of California Los Angeles, Los Angeles, CA, United States of America

Chris Lunsford
Alaska Fisheries Science Center, National Marine Fisheries Service, NOAA, Auke Bay Laboratories, Juneau, AK, United States of America

Janice Straley
University of Alaska Southeast, Sitka, AK, United States of America

Tomoharu Eguchi and Sarah L. Mesnick
Southwest Fisheries Science Center, National Marine Fisheries Service, NOAA, La Jolla, CA, United States of America

Joë l M. Durant
Centre for Ecological and Evolutionary Synthesis (CEES), Department of Biosciences, University of Oslo, Oslo, Norway

Mette Skern-Mauritzen
Institute of Marine Research, Bergen, Norway

Yuri V. Krasnov
Murmansk Marine Biological Institute, Murmansk, Russian Federation

Natalia G. Nikolaeva
White Sea Biological Station, Department of Biology, Lomonosov Moscow State University, Moscow, Russian Federation

Ulf Lindstrøm
Institute of Marine Research, Tromsø, Norway

Andrey Dolgov
Knipovich Polar Research Institute of Marine Fisheries and Oceanography (PINRO), Murmansk, Russian Federation

Richard J. Griffeth
Centro de Investigación Príncipe Felipe, Tissue and Neuronal Regeneration Lab, Valencia, Spain

Daniel García-Párraga and Jose Luis Crespo-Picazo
Oceanogrà fic (grupo Parques Reunidos), Valencia, Spain

Mario Soriano-Navarro
Centro de Investigación Príncipe Felipe, Electron Microscopy Unit, Valencia, Spain

Alicia Martinez-Romero
Centro de Investigación Príncipe Felipe, Cytomics Unit, Valencia, Spain

Maravillas Mellado-López and Victoria Moreno-Manzano
Centro de Investigación Príncipe Felipe, Tissue and Neuronal Regeneration Lab, Valencia, Spain

FactorStem, Ltd. Valencia, Spain

Jennifer L. Miksis-Olds and Laura E. Madden
Applied Research Laboratory, The Pennsylvania State University, State College, Pennsylvania, United States of America

Ryan R. Reisinger, W. Chris Oosthuizen, Dawn Cory Toussaint and P. J. Nico de Bruyn
Mammal Research Institute, Department of Zoology and Entomology, University of Pretoria, Pretoria, South Africa

Guillaume Péron
Centre for Statistics in Ecology, Environment and Conservation, Department of Statistical Sciences, University of Cape Town, Cape Town, South Africa

Russel D. Andrews
School of Fisheries and Ocean Sciences, University of Alaska Fairbanks, Fairbanks, Alaska, United States of America
Alaska SeaLife Center, Seward, Alaska, United States of America

Catherine A. Bliss, Christopher M. Danforth and Peter Sheridan Dodds
Department of Mathematics and Statistics, Vermont Complex Systems Center, The Computational Story Lab, and the Vermont Advanced Computing Core, University of Vermont, Burlington, Vermont, United States of America

Brian S. Miller and Susannah Calderan
Australian Marine Mammal Centre, Australian Antarctic Division, Kingston, Australia

Russell Leaper
School of Biological Sciences, University of Aberdeen, Aberdeen, United Kingdom

Jason Gedamke
Ocean Acoustics Program, NOAA Fisheries Office of Science and Technology, National Oceanic and Atmospheric Administration, Silver Spring, Maryland, United States of America

Duan Gui, Yong Sun, Qin Tu and Yuping Wu
Guangdong Provincial Key Laboratory of Marine Resources and Coastal Engineering, School of Marine Sciences, Sun Yat-Sen University, Guangzhou, China

Ri-Qing Yu
Department of Biology, University of Texas at Tyler, Tyler, Texas, United States of America

Laiguo Chen
Urban Environment and Ecology Research Center, South China Institute of Environmental Sciences (SCIES), Ministry of Environmental Protection, Guangzhou, China

Hui Mo
South China Botanical Garden, Chinese Academy of Sciences, Guangzhou, China

Xianchao Tang
School of Computer Science and Technology, Tianjin University, Tianjin, China

Tao Xu and Xia Feng
School of Computer Science and Technology, Civil Aviation University of China, Tianjin, China
Information Technology Research Base of Civil Aviation Administration of China, Tianjin, China

Guoqing Yang
School of Computer Science and Technology, Tianjin University, Tianjin, China
Information Technology Research Base of Civil Aviation Administration of China, Tianjin, China

Rogier Bodewes, Rebriarina Hapsari, Marco W. G. van de Bildt, Miranda de Graaf and Thijs Kuiken
Department of Viroscience, Erasmus MC, Rotterdam, the Netherlands

Ana Rubio García and Guillermo J. Sánchez Contreras
Seal Rehabilitation and Research Centre, Pieterburen, the Netherlands

Albert D. M. E. Osterhaus
Department of Viroscience, Erasmus MC, Rotterdam, the Netherlands
Viroclinics Biosciences BV, Rotterdam, the Netherlands
Research Center for Emerging Infections and Zoonoses, University of Veterinary Medicine, Hannover, Germany
Artemis One Health, Utrecht, the Netherlands

David P. Hocking
School of Biological Sciences, Monash University, Melbourne, Victoria, Australia

Geosciences, Museum Victoria, Melbourne, Victoria, Australia
Wild Sea Precinct, Zoos Victoria, Melbourne, Victoria, Australia

Erich M. G. Fitzgerald
Geosciences, Museum Victoria, Melbourne, Victoria, Australia

Alistair R. Evans
School of Biological Sciences, Monash University, Melbourne, Victoria, Australia
Geosciences, Museum Victoria, Melbourne, Victoria, Australia

Marcia Salverson
Wild Sea Precinct, Zoos Victoria, Melbourne, Victoria, Australia

Chiara Guidino
University of Technology of Sydney, Sydney, Australia

Miguel A. Llapapasca
Oficina de Investigaciones en Depredadores Superiores, Instituto del Mar del Perú , Callao, Peru

Sebastian Silva and Belen Alcorta
Pacifico Adventures-Manejo Integral del Ambiente Marino S.A.C., Los Organos, Peru

Aldo S. Pacheco
Instituto de Ciencias Naturales Alexander von Humboldt, Universidad de Antofagasta, Antofagasta, Chile

Index